A Manual of

Third Edition

A Manual of Mammalogy

WITH KEYS TO FAMILIES OF THE WORLD

Robert E. Martin
McMurry University

Ronald H. Pine
*Illinois Mathematics and Science Academy,
and Field Museum*

Anthony F. DeBlase
Formerly of Field Museum

Boston   Burr Ridge, IL   Dubuque, IA   Madison, WI   New York   San Fransisco   St. Louis
Bangkok   Bogotá   Caracas   Lisbon   London   Madrid
Mexico City   Milan   New Delhi   Seoul   Singapore   Sydney   Taipei   Toronto

McGraw-Hill Higher Education
A Division of The McGraw-Hill Companies

A MANUAL OF MAMMALOGY: WITH KEYS TO FAMILIES OF THE WORLD, THIRD EDITION

Published by McGraw-Hill, an imprint of The McGraw-Hill Companies, Inc., 1221 Avenue of the Americas, New York, NY 10020. Copyright © 2001, 1981, 1974 by The McGraw-Hill Companies, Inc. All rights reserved. No part of this publication may be reproduced or distributed in any form or by any means, or stored in a database or retrieval system, without the prior written consent of The McGraw-Hill Companies, Inc., including, but not limited to, in any network or other electronic storage or transmission, or broadcast for distance learning.

Some ancillaries, including electronic and print components, may not be available to customers outside the United States.

 This book is printed on recycled, acid-free paper containing 10% postconsumer waste.

7 8 9 0 QPD/QPD 0 9 8 7

ISBN 978-0-697-00643-1
MHID 0-697-00643-3

Vice president and editor-in-chief: *Kevin T. Kane*
Publisher: *Michael D. Lange*
Senior sponsoring editor: *Margaret J. Kemp*
Developmental editor: *Donna Nemmers*
Marketing managers: *Michelle Watnick /Heather K. Wagner*
Project manager: *Mary E. Powers*
Senior production supervisor: *Sandra Hahn*
Coordinator of freelance design: *Rick D. Noel*
Cover designer: *Rick D. Noel*
Image: *©Corbis—Volume 86, Animals in Action*
Compositor: *Precision Graphics*
Typeface: *10/12 Times Roman*
Printer: *Quebecor Printing Book Group/Dubuque, IA*

The credits section for this book begins on page 321 and is considered an extension of the copyright page.

Library of Congress Cataloging-in-Publication Data

Martin, Robert Eugene, 1944–
    A manual of mammalogy : with keys to families of the world / Robert E. Martin,
Ronald H. Pine, Anthony F. DeBlase. — 3rd ed.
        p.   cm.
    Rev. ed. of : Manual of mammalogy / Anthony F. DeBlase, Robert E. Martin. 2nd. © 1981.
    Includes bibliographical references (p.) and index.
    ISBN 0–697–00643–3
    1. Mammals. 2. Mammals—Identification. I. Pine, Ronald H. II. DeBlase, Athony F.
III. DeBlase, Anthony F. Manual of Mammalogy. IV. Title.

QL703 .M375   2001
599—dc21   00–039414
                                                                                        CIP

www.mhhe.com

*George Azro Moore (1899–1998), teacher, mentor, and friend;
and for Patty, with whom I share my life.*
—*Bob*

For my mother, Virgie Ella Pine, for all her support over the years; and for my father, Herbert Raymond Pine, who, along with his siblings, introduced me to the skeptical frame of mind. Also for Aunt Marjorie and Uncle Joe.
—*Ron*

Alyce Myers DeBlase (1939–1976), who worked as hard on the first edition as Bob or I did.
—*Tony*

Brief Contents

Preface xiii

1. Introduction *1*
2. The Skull *6*
3. Teeth *13*
4. The Integument *21*
5. Horns and Antlers *30*
6. The Postcranial Skeleton *36*
7. Locomotor Adaptations *41*
8. Keys and Keying *49*
9. The Orders of Living Mammals *53*
10. The Monotremes *61*
    Order Monotremata *61*
11. The Marsupials *64*
    Order Didelphimorphia *66*
    Order Paucituberculata *67*
    Order Microbiotheria *68*
    Order Dasyuromorphia *68*
    Order Peramelemorphia *70*
    Order Diprotodontia *71*
    Order Notoryctemorphia *77*
12. The Insectivores *78*
    Order Insectivora *78*
13. The Colugos *83*
    Order Dermoptera *83*
14. The Bats *85*
    Order Chiroptera *85*
15. The Tupaiids *94*
    Order Scandentia *94*
16. The Primates *96*
    Order Primates *96*
17. The Xenarthrans: Anteaters, Armadillos, and Sloths *103*
    Order Xenarthra *103*
18. The Pangolins *107*
    Order Pholidota *107*
19. The Carnivores *109*
    Order Carnivora *109*
20. The Whales, Dolphins, and Porpoises *121*
    Order Cetacea *121*
    Suborder Mysticeti *123*
    Suborder Odontoceti *126*
21. The Macroscelideans *133*
    Order Macroscelidea *133*
22. The Rabbits, Hares, and Pikas *135*
    Order Lagomorpha *135*
23. The Rodents *138*
    Order Rodentia *138*
24. The Aardvark *162*
    Order Tubulidentata *162*
25. The Subungulates *164*
    Order Proboscidea *164*
    Order Hyracoidea *165*
    Order Sirenia *166*
26. The Perissodactyls *169*
    Order Perissodactyla *169*
27. The Artiodactyls *172*
    Order Artiodactyla *172*
28. Sign and Habitat Analysis *180*
29. Recording Data *189*
30. Collecting *200*
31. Specimen Preparation and Preservation *211*
32. Collecting Ectoparasites of Mammals *230*
33. Age Determination *244*
34. Diet Analysis *253*
35. Analysis of Spatial Distribution *261*
36. Estimation of Abundance and Density *270*
37. Literature Research *278*

Glossary 285
Literature Cited 307
Credits 321
Index 323

Contents

Preface *xiii*

1. Introduction *1*
   Characters Defining Mammalia 1
   Systematics and Nomenclature 2
   Distribution 4
   The Structure and Use of This Manual 5
   Care of Specimens 5

2. The Skull *6*
   The Mandible 8
   Variation 9
   Determination of Maturity 9
   Measurements 9
   Key to Labeling of Figure 2.1 11

3. Teeth *13*
   Tooth Anatomy and Replacement 13
   The Kinds of Teeth 14
   Dental Formulas 18

4. The Integument *21*
   Scales 21
   Hair Anatomy 22
   Classification of Hair 23
   Hair Replacement 24
   Color 25
   Integumentary Glands 27
   Mammary Glands 28

5. Horns and Antlers *30*
   True Horns 30
   Pronghorns 30
   Antlers 30
   Giraffe "Horns" 33
   Rhinoceros "Horns" 34
   Function 34

6. The Postcranial Skeleton *36*
   Postcranial Axial Skeleton 36
   The Appendicular Skeleton 37
   Claws, Nails, and Hoofs 39

7. Locomotor Adaptations *41*
   Terrestrial Locomotion 41
   Fossorial Adaptations 43
   Aquatic Adaptations 45
   Arboreal Adaptations 46
   Aerial Adaptations 48

8. Keys and Keying *49*
   Selection of Key Characters 49
   Key Construction 51
   The Keys in This Manual 52

9. The Orders of Living Mammals *53*
   Key to the Orders of Living Mammals 55
   Comments and Suggestions on Identification 58

10. The Monotremes *61*
    Order Monotremata 61
    Distinguishing Characters 61
    Living Families of Monotremata 62
    Key to Living Families of Monotremata 62
    Comments and Suggestions on Identification 62

11. The Marsupials *64*
    Order Didelphimorphia 66
    Distinguishing Characters 66
    Living Family of Didelphimorphia 66
    Comments and Suggestions on Identification 66
    Order Paucituberculata 67
    Distinguishing Characters 67
    Living Family of Paucituberculata 67
    Comments and Suggestions on Identification 68
    Order Microbiotheria 68
    Distinguishing Characters 68
    Living Family of Microbiotheria 68
    Comments and Suggestions on Identification 68
    Order Dasyuromorphia 68
    Distinguishing Characters 68
    Living Families of Dasyuromorphia 68
    Key to Living Families of Dasyuromorphia 69
    Comments and Suggestions on Identification 70
    Order Peramelemorphia 70
    Distinguishing Characters 70
    Living Families of Peramelemorphia 71
    Key to Living Families of Peramelemorphia 71
    Comments and Suggestions on Identification 71
    Order Diprotodontia 71
    Distinguishing Characters 71

Living Families of Diprotodontia 71
Key to Living Families of Diprotodontia 72
Comments and Suggestions on Identification 76
Order Notoryctemorphia 77
Distinguishing Characters 77
Living Family of Notoryctemorphia 77
Comments and Suggestions on Identification 77

12 The Insectivores 78
Order Insectivora 78
Distinguishing Characters 78
Living Families of Insectivora 78
Key to Living Families of Insectivora 79
Comments and Suggestions on Identification 82

13 The Colugos 83
Order Dermoptera 83
Distinguishing Characters 83
Living Family of Dermoptera 84
Comments and Suggestions on Identification 84

14 The Bats 85
Order Chiroptera 85
Distinguishing Characters 85
Living Families of Chiroptera 86
Key to Living Families of Chiroptera 87
Comments and Suggestions on Identification 93

15 The Tupaiids 94
Order Scandentia 94
Distinguishing Characters 94
Living Family of Scandentia 94
Comments and Suggestions on Identification 94

16 The Primates 96
Order Primates 96
Distinguishing Characters 97
Living Families of Primates 97
Key to Living Families of Primates 98
Comments and Suggestions on Identification 102

17 The Xenarthrans: Anteaters, Armadillos, and Sloths 103
Order Xenarthra 103
Living Families of Xenarthra 105
Key to Living Families of Xenarthra 105
Comments and Suggestions on Identification 106

18 The Pangolins 107
Order Pholidota 107
Distinguishing Characters 108
Living Family of Pholidota 108
Comments and Suggestions on Identification 108

19 The Carnivores 109
Order Carnivora 109
Distinguishing Characters 110
Living Families of Carnivora 111
Key to Living Families of Carnivora 112
Comments and Suggestions on Identification 120

20 The Whales, Dolphins, and Porpoises 121
Order Cetacea 121
Suborder Mysticeti 123
Distinguishing Characters 123
Living Families of Mysticeti 124
Key to Living Families of Mysticeti 124
Comments and Suggestions on Identification 126
Suborder Odontoceti 126
Distinguishing Characters 127
Living Families of Odontoceti 127
Key to Living Families of Odontoceti 128
Comments and Suggestions on Identification 132

21 The Macroscelideans 133
Order Macroscelidea 133
Distinguishing Characters 133
Living Family of Macroscelidea 133
Comments and Suggestions on Identification 134

22 The Rabbits, Hares, and Pikas 135
Order Lagomorpha 135
Distinguishing Characters 136
Living Families of Lagomorpha 136
Key to Living Families of Lagomorpha 136
Comments and Suggestions on Identification 137

23 The Rodents 138
Order Rodentia 138
Distinguishing Characters 139
Important Taxonomic Characters 139
Living Families of Rodentia 139
Key to Living Families of the World 142
Key to Living Families of North America 159
Comments and Suggestions on Identification 161

24 The Aardvark 162
Order Tubulidentata 162
Distinguishing Characters 163
Living Family of Tubulidentata 163
Comments and Suggestions on Identification 163

25 The Subungulates 164
Order Proboscidea 164
Distinguishing Characters 165
Living Family of Proboscidea 165
Comments and Suggestions on Identification 165
Order Hyracoidea 165
Distinguishing Characters 166
Living Family of Hyracoidea 166
Comments and Suggestions on Identification 166
Order Sirenia 166
Distinguishing Characters 167
Living Families of Sirenia 167
Key to Living Families of Sirenia 167
Comments and Suggestions on Identification 168

26 The Perissodactyls 169
Order Perissodactyla 169
Distinguishing Characters 169
Living Families of Perissodactyla 171
Key to Living Families of Perissodactyla 171
Comments and Suggestions on Identification 171

27 The Artiodactyls 172
Order Artiodactyla 172
Distinguishing Characters 172
Living Families of Artiodactyla 174
Key to Living Families of Artiodactyla 174
Comments and Suggestions on Identification 179

28 Sign and Habitat Analysis 180
Identifying Mammal Sign 180
Habitat Analysis 184

29 Recording Data 189
Equipment 189
Field Notes 190
Locality 191
Date 194

Measurements, Weight, and Sex 194
Reproductive Condition 195
Parasites 195
Portions Preserved (Type of Preservation) 195
Methods of Collection 195
Specimen Labels 196
Study Skins 196
Skulls and Skeletons to Be Cleaned 197
Skins to Be Tanned 197
Fluid Preservation 197
Special Data Forms 198
Recording Measurements 198
Recording Behavior 198
Computers 198
Documentation Standards for Museum Specimens 198

30 Collecting 200

Health and Safety 200
Collecting, Conservation, and the Law 201
Conservation 202
Locating a Collecting Area 202
Records 202
Methods of Trapping Mammals 203
Methods of Trap Placement 206
Timing of Trapping Activity 206
Calculating Trapping Success 207
Baits and Scents 207
Minimizing Damage to Trapped Mammals 207
Adjusting Trapping Methods for Regional and Climatic Differences 207
Other Methods of Capturing or Collecting Mammals 207
Marking Mammals 209
Care of Collected Mammals 210

31 Specimen Preparation and Preservation 211

Supplies and Equipment 211
Prior to Preparation 212
The Standard Study of Skin 215
Special Specimen Preparation Techniques 221
Preparation of Skins to be Tanned 223
Cleaning Skeletal Material 225

32 Collecting Ectoparasites of Mammals 230

Segregation of Hosts 230
Kinds of Ectoparasites 230
Key to Arthropod Ectoparasites of Mammals 232
Supplies and Equipment Needed 236
Searching for and Removing Ectoparasites from the Host 237
Identification of Ectoparasites 241

33 Age Determination 244

Types of Aging Criteria 244
Use of Statistics and Known-Age Samples 245
Growth of Skull, Skeleton, and Body 245
Relative Growth and Morphology of Teeth 247
Growth of Eye Lenses 248
Growth Lines 248
Age Determination in Live Mammals 250
Limitations of Age-Determination Methods 251

34 Diet Analysis 253

Collection of Samples 254
Identification of Food Items 255
Analysis of Dietary Samples 258
Determination of Resource Levels 259
Calculating Indices of Preference 259

35 Analysis of Spatial Distribution 261

Spatial Organization 261
Home Range 261
Territory 266
Habitat Utilization and Preference 267
Movements 267
Methods for Studying Movements 268

36 Estimation of Abundance and Density 270

Basic Population Concepts 270
Types of Population Estimates 270
Use of Computers for Estimation of Animal Abundance 270
Sampling Configurations for Estimating Density 271
Estimating Size of Sampling Area 271
Accuracy of Estimates 272
Recording Data 272
Estimation by Census or Direct Counts 272
Relative Estimates of Density 273
Density Estimates Based on Mark and Recapture 273
Estimates Based on Multiple Marking Occasions (Deterministic) 276
The Jolly-Seber Stochastic Method 276
Removal Trapping and Catch per Effort Methods 276

37 Literature Research 278

The Literature of Mammalogy 278
Bibliographies and Abstracts 279
Making a Search 280
Computerized Literature Databases 281
Computerized Literature Search Strategies 282

Glossary 285
Literature Cited 307
Credits 321
Index 323

Preface

This manual, including laboratory exercises, is suitable for use in an upper-level undergraduate or graduate course in mammalogy. In this book, we include sufficient background material and illustrations so that anatomical information can be understood even without a prior course in comparative anatomy. Having had such a course is recommended, however. The keys in this manual cover *all* of the living families of mammals of the world and have been extensively revised and updated to reflect the latest taxonomic treatments.

When the first edition of this book was published in 1974, and the second edition published in 1980, the intent was to provide students and other readers a way of identifying members of all living families of mammals of the world. Because mammals occur on every major land mass and in every major body of water in the world, we still believe that students of mammalogy should develop a familiarity with mammals of the world rather than only with those of a particular county, state, or nation. Therefore, with this third edition, we maintain the worldwide focus of this manual. We have also rewritten much of the book, revised the keys, updated the taxonomy and nomenclature, added new material, and updated the literature citations. We have, of necessity, deleted some subject matter and some chapters that were in the second edition, to provide space for topics most appropriate for field and laboratory studies of mammals.

Because most institutions do not have examples of many of the world's mammals, we select North American representatives as examples for particular characteristics whenever possible. However, when North American examples will not suffice, we include information on mammals from other parts of the world. We still believe that it is better for students to examine illustrations of mammals in this book, or in one of the several well-illustrated books now available, than to ignore a particular group or characteristic simply because specimens are not readily available. Similarly, in the use of the keys, we believe that the students' use of keys covering all families will increase their awareness of the world's mammal fauna even though they may be keying out North American forms only.

Organization of and Changes in the Third Edition

The 37 chapters in this edition are grouped into three sections: characteristics of mammals (chapters 1 through 8), living orders of mammals (chapters 9 through 27), and techniques for studying mammals in the field and in the laboratory (chapters 28 through 37). A comprehensive **Glossary** and **Index** are provided, along with an extensive **Literature Cited** section. In this edition, the sections "Taxonomic Remarks" and "Fossil History" have been deleted to save space; these topics are included in many other books. Specific changes to the third edition include: **Chapter 1, "Introduction":** Information on reproductive anatomy and on systematics and nomenclature, which was previously included in separate chapters, is added to this chapter. **Chapter 2, "The Skull":** Terminology is updated and revised. **Chapter 3, "Teeth":** Section on tooth replacement is revised. Discussion of teeth of extinct species of mammals is deleted due to its availability in many other sources. **Chapter 4, "The Integument":** All text is revised, and five figures have been added. **Chapter 5, "Horns and Antlers":** This chapter is revised, and three new figures are added. **Chapter 6, "The Postcranial Skeleton":** A new section is added on the postcranial axial skeleton, along with a new figure. To conserve space, we also include information on claws, nails, and hooves that was previously found in a separate chapter. **Chapter 9, "The Orders of Living Mammals":** Twenty-six orders of living mammals are included in this revised edition, as in the treatment by Wilson and Reeder (1993) and by Feldhamer et al. (1999). The marsupials, Metatheria, are grouped into seven orders. In the Eutheria, two groups previously included in the order Insectivora are now treated separately as the orders Scandentia and Macroscelidea. Also, the ordinal name Edentata is dropped in favor of the name Xenarthra, for the sloths, armadillos, and anteaters. The toothed whales and allies, and the baleen whales are grouped as suborders of the order Cetacea, rather than as separate orders. The "Key to Living Orders" and the "Comments and Suggestions on Identification" have been completely rewritten to reflect all of these taxonomic changes. **Chapter 11, "The Marsupials":** The chapter on marsupials is extensively rewritten to reflect that these mammals are now included in seven orders. **Chapter 12, "The Insectivores":** This chapter now includes only the families Solenodontidae, Tenrecidae, Chrysochloridae, Erinaceidae, Soricidae, and Talpidae. The "Key to Living Families" and portions of the text are rewritten to reflect that only these mammals are now treated as members of this order. **Chapter 15, "The Tupaiids":** The tupaiids were formerly treated in the chapter on Insectivora. Consistent with current taxonomy, the tupaiids, order Scandentia, are now discussed in a separate chapter. **Chapter 16, "The Primates":** This text and key are extensively revised. **Chapter 19, "The Carnivores":** Extensive revisions have been made to the text and key. **Chapter 20, "The Whales, Dolphins, and Porpoises":** These mammals are included in a single order, Cetacea, with baleen whales discussed under the suborder Mysticeti, and with toothed whales, dolphins, and porpoises treated in the suborder Odontoceti. Four figures are added. **Chapter 21, "The Macroscelideans":** These animals were formerly included in the chapter on Insectivora. **Chapter 23, "The Rodents":** Extensive revision made to the text and to the keys, with five figures added. For the Muridae, many subfamilies are keyed out. As in previous editions, there is a "Key to Living Families of the World" and a "Key to Living Families of North America." **Chapter 25, "The Subungulates":** In this edition, the subungulates are considered to include only the orders Proboscidea, Hyracoidea, and Sirenia. Two illustrations are also added. **Chapter 27, "The Artiodactyls":** Extensive revisions are made to the text and the key, including the addition of one figure. **Chapter 28, "Sign and Habitat Analysis":** The section on identifying mammal sign is updated, and the illustration on beaver sign is revised. A new section on quantitative methods for analysis of selected habitat features is added, along with three figures. **Chapter 29, "Recording Data":** This chapter is extensively revised, with obsolete methods deleted. New material is added on working with map coordinates and finding your position by use of global positioning (GPS) receivers. Information is included on the latest documentation standards for museum specimens. **Chapter 30, "Collecting":** Extensive revisions include the addition of specific guidelines and references on how to work safely with mammals in both the field and in the laboratory, and how to obtain the latest information on laws and regulations governing the capture, care, and handling of mammals. **Chapter 31, "Specimen Preparation and Preservation":** New material is added on how to work with tissues and other special preparations. A new figure illustrates how to determine the sex of mammals by examination of external genitalia. **Chapter 32, "Collecting Ectoparasites of Mammals":** The text and the "Key to Arthropod Ectoparasites of Mammals" were extensively revised and updated by Eric H. Smith. Many figures are added to provide a convenient single reference for identification of these arthropods. **Chapter 33, "Age Determination":** Extensive revisions include four new figures, plus new sections on how to determine the ages of live mammals. **Chapter 34, "Diet Analysis":** New methods for determining the diets of mammals are provided, as well as four added figures. **Chapter 35, "Analysis of Spatial Distribution":** New material is presented on the use of computer programs to get estimates of home-range size and characteristics, along with many other changes. **Chapter 36, "Estimation of Abundance and Density":** This chapter includes an abbreviated subset of material that was previously in a separate chapter on populations. Because population biology is normally treated in a separate course or as a major component in a course in ecology, we shortened this section in the current edition. Still included is information on estimating relative abundance and density, including references to computer programs for estimating density. **Chapter 37, "Literature Research":** The new name for this chapter indicates that there are many ways to find literature pertaining to mammals. The list of journals is updated, and new sections are added on computerized literature databases and on computerized literature search strategies, including how to perform Boolean searches. One new figure is added. **Glossary:** Obsolete terms are deleted, and many new terms are added. Definitions of most terms are rewritten to improve clarity and understanding.

Acknowledgments

Over 20 years have passed since the manuscript for the second edition was prepared. In the interim, users have inquired as to when the third edition would appear, and we, as authors, could not provide an answer that was satisfactory to us or to the loyal adopters of the book. In 1998, with the strong encouragement of our Sponsoring Editor, Marge Kemp, and a renewed sense of purpose, we set about to jump-start the revision. This current edition is the culmination of those efforts.

For this edition, Ronald H. Pine joined the team and helped with all aspects of the revision, provided detailed comments on the second edition, and revised many of the chapters dealing with orders and families of mammals. Tony DeBlase, despite a change in authorship sequence, nonetheless did much of the work on this edition, and his vision for the book is still apparent.

In response to a questionnaire about mammal textbooks that you received from McGraw-Hill, many of you made comments and suggestions on how to improve the second edition of the book. For these comments, we thank Rick A. Adams, David M. Armstrong, Kathleen B. Blair, William Caire, David Chesemore, Bruce E. Coblentz, Jack A. Cranford, David Ekkens, Carl H. Ernst, G. Lawrence Forman, John D. Harder, Graham C. Hickman, Carl W. Hoagstrom, C. H. Hocutt, Kay E. Holekamp, Robert W. Howe, Jerome A. Jackson, Gary Kwiecinski, Tom Lee, James A. MacMahon, V. Rick McDaniel, Peter L. Meserve, Chris Norment, Jon C. Pigage, Edward Pivorun, Roger A. Powell, Robert K. Rose, Christopher Sanford, Michelle Pellissier Scott, Steven A. Smith, Don Spalinger, Glenn R. Stewart, Michael Stokes, Michael D. Stuart, Gerald E. Svendsen, Gene R. Trapp, Renn Tumlison, Dallas Wilhelm, and Kenneth T. Wilkins.

Laurie Wilkins made substantial revisions to the section on cleaning skeletal material in Chapter 31, "Specimen Preparation and Preservation." Eric Smith completely revised and updated the text, keys, and figures in Chapter 32, "Collecting Ectoparasites of Mammals." At Texas Tech University, Richard Monk and Janie Milner provided information on how to take tissue samples from animals, and Raegan D. King provided information on a method to store and access specimen data.

Michael Gilliland and Terry C. Maxwell prepared several new line drawings. Kenneth T. Matocha revised one of the figures in Chapter 29, "Recording Data." Randall Zavodny reviewed all of the illustrations in the second edition of the book and made suggestions on how to improve the illustrations for the present edition. Linda White, presently with the Texas Natural Resource Conservation Commission, and Beverly Morey provided editorial comments and typed much of the text for this edition.

At McGraw-Hill, we were prodded to action by the rhetoric of our Sponsoring Editor, Marge Kemp, gently led with encouragement by our Developmental Editor, Donna Nemmers, and kept on task by Project Manager, Mary Powers. These and many other individuals at McGraw-Hill helped to bring this book to completion.

We owe our loyal users a debt of gratitude for sticking with us when the book was out of date and in need of a major revision. Your loyalty has been humbling to us, and we hope that the third edition will meet your needs.

First Edition Reviewers

Our special thanks to the individuals, listed below, who provided inspiration and professional contributions for the development of the first edition. Sydney Anderson, Robert J. Baker, Marge Bell, Dale L. Berry, Craig C. Black, Janet Blefeld, Alberto Cadena, Andrew Chien, Milton R. Curd, Walter W. Dalquest, Alyce M. DeBlase, Luis de la Torre, André Dixon, Patricia Gaddis, Barbara Garner, Hugh H. Genoways, Bryan P. Glass, Edward Gray, David L. Harrison, Theodore A. Heist, J. E. Hill, Philip Hershkovitz, Stephen R. Humphrey, Robert Ingersol, John Jahoda, Lee A. Jones, Karl F. Koopman, Thomas H. Kunz, Karl Liem, L. Patricia Martin, Pegge Luken, Kenneth G. Matocha, WIlliams Mohs, George A. Moore, Joseph Curtis Moore, John A. Morrison, Guy G. Musser, Hans N. Neuhauser, Robert L. Packard, Steven Rissman, Richard Roesner, George Rogers, Stanley Rouk, J. Mark Rowland, C. David Simpson, Terry A. Vaughan, James P. Webb, John Whitesell, Robert W. Wiley, Daniel R. Womochel, and Donna Womochel.

Second Edition Reviewers

The persons listed below provided valuable help and suggestions during the development of the second edition. We wish to thank again the following: Sharon Adams, Elmer C. Birney, Keith Carson, Mary Ann Cramer, Sara Derr, Jerran T. Flinders, Patricia Freeman, George Fulk, Robert J. Izor, Laurel E. Keller, Karl F. Koopman, Cliff A. Lemen, Larry C. Marshall, L. Patricia Martin, Chris Maser, Peter L. Meserve, Dale Osborn, Pamela Parker, Ronald H. Pine, Linda Porter, Eric H. Smith, James D. Smith, Mike Smolen, Sandra L. Walchuk, and Laurie Wilkins.

At the request of McGraw-Hill, a number of people provided detailed comments on the second edition of the manual and made suggestions on how to revise the third edition. We greatly appreciated these detailed comments, although we were not able to accommodate all of the suggestions. Our thanks to these reviewers of the second edition: Richard Buchholz, Northeast Louisiana University; A. Christopher Carmichael, Michigan State University; Elissa Miller Derrickson, Loyola College; Margaret Haag, University of Alberta; Kay E. Holekamp, Michigan State University; Joshua Laerm, University of Georgia; Peter L. Meserve, Northern Illinois University; Dorothy B. Mooren, University of Wisconsin at Milwaukee; and Bruce A. Wunder, Colorado State University.

CHAPTER 1

INTRODUCTION

Mammals are animals that possess a hollow dorsal nerve tube throughout life, and that, in early stages of embryonic development, possess gill pouches and a notochord. These three characters place mammals in the phylum **Chordata.** The brain and spinal cord of mammals are enclosed in, and protected by, a bony skeleton. Thus mammals are members of the subphylum **Vertebrata.** Embryos of the class **Mammalia,** like those of the classes Reptilia and Aves, are surrounded by protective amniotic and allantoic membranes. As in crocodilians and birds, the mammalian heart has four distinct chambers, and as in birds, there is no mixing of oxygen-rich and oxygen-poor blood. **Homeothermy** is generally well developed.

CHARACTERS DEFINING MAMMALIA

In defining the class Mammalia, it is necessary to use characters that may be applicable to fossil remains, as well as to living animals. Thus, the official defining characters are skeletal. The lower jaw of mammals is unique among vertebrates in being composed of only a single pair of bones, the **dentaries,** which articulate directly with the **cranium.** In the other vertebrates, the dentary is only one of several bones in the lower jaw, and it usually does not articulate directly with the cranium. The presence of an articulation between the dentary and **squamosal** is the characteristic used to define Mammalia. The **articular** and **quadrate** bones, elements of the lower jaw and cranium, respectively, in other vertebrates, are modified in mammals to form two of the three middle ear bones, or **ossicles.** The articular and quadrate of the other vertebrate classes become the **malleus** and **incus,** respectively, in mammals. The **columella,** the only middle ear bone of other tetrapod vertebrates, becomes the **stapes** in mammals. Thus, mammals are commonly defined by the presence of a single bone in the lower jaw and by the presence of three ossicles in the middle ear (Fig. 1.1).

In addition to these skeletal characters, living mammals have several characteristics of the soft anatomy that are unique to mammals and common to all living mammalian species. Some of the more conspicuous ones follow.

All mammals possess **hair** at some stage of their life cycle. Most have hair throughout life, but several aquatic forms have only a few stiff bristles. In some cetaceans, even these bristles are lost in the adult (see Chapter 4).

All female mammals posses **mammary glands** that produce milk to nourish the young. In male eutherians, these glands are present but rudimentary. In monotremes, the milk flows from pores in the skin, but all other mammals have nipples and the young suckle.

The thoracic and abdominal cavities of mammals are separated by a **muscular diaphragm.** Some other vertebrates have a membranous septum between these cavities, but only in mammals is this structure fully muscular.

Only the left aortic arch is present in adult mammals (Fig. 1.2). The right aortic arch is lost during early embryonic development. Both arches persist in reptiles, whereas only the right aortic arch persists in birds.

Mammals have **enucleate erythrocytes.** No nuclei are observable in mature red blood cells. Other vertebrates have nuclei in mature erythrocytes.

The **neopallium,** or roof of the forebrain, is proportionately larger in mammals than in other vertebrates, and the **corpora quadrigemina,** an elaboration of the midbrain, is found only in mammals.

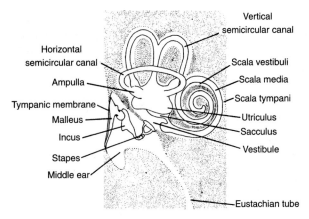

Figure 1.1 The middle and inner ear of a mammal.
(Chiasson 1969)

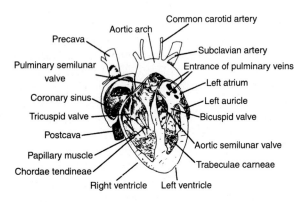

Figure 1.2 Anterior or ventral view of the mammalian heart with the chambers exposed. Note that the aorta exits from the heart and arches to the animal's left.
(Chiasson 1969)

1-A Compare the lower jaw of a dog or coyote (genus *Canis*) with those of a bird, reptile, amphibian, and bony fish. How many bones are present in each jaw? With which bone(s) of the cranium does each mandible articulate?

1-B In demonstration material, locate the middle ear bones in a mammalian skull and compare with the middle ear structure of a bird, reptile, and amphibian.

1-C On a demonstration dissection of a cat or other mammal, examine the structure of the heart and locate the left aortic arch and the muscular diaphragm. Compare these with the heart and diaphragm in demonstration dissections of a frog and a bird.

1-D Compare the erythrocytes in a prepared microscope slide of mammalian blood with those in slides of the blood of a bird, reptile, and/or amphibian.

1-E Compare the structure of the mammalian brain with that of other vertebrates.

SYSTEMATICS AND NOMENCLATURE

The field of science concerned with discovering and expressing the relationships between mammals and other organisms is termed **systematics,** and a scientist in this field is termed a **systematist.** A major subdiscipline of systematics is **taxonomy,** defined by Mayr (1969) as ". . . the theory and practice of classifying organisms," which is practiced by a **taxonomist.** Classification is the process by which living and extinct organisms are arranged into various named groups. These groups of organisms, termed **taxa** (singular, *taxon*) are then ranked in a hierarchy of specific categories or levels of classification. Each taxon is assigned a name, to be universally used by scientists, according to rules governing **nomenclature.** Identification is the assigning of particular specimens to already established and named taxa.

Hierarchies of Classification

Animals and plants are arranged into taxa. These taxa, in turn, are ranked according to a hierarchy of **categories.** The categories generally used for animals are shown in Figure 1.3. The categories printed in all uppercase letters are the major categories used in the classification of all animals. In this ranking, the least inclusive categories and taxa are at

Categories	Examples of Taxa
KINGDOM	Animalia
PHYLUM	Chordata
Subphylum	Vertebrata
CLASS	Mammalia
Subclass	Theria
Infraclass	Eutheria
ORDER	Rodentia
Suborder	Sciuromorpha
Superfamily	Sciur**oidea**
FAMILY	Sciur**idae**
Subfamily	Sciur**inae**
Tribe	Marmot**ini**
GENUS	*Spermophilus*
SPECIES	*Spermophilus spilosoma*
Subspecies	*Spermophilus spilosoma pallescens*

Figure 1.3 Categories of classification generally used for mammals, and taxa to which the spotted ground squirrel, *Spermophilius spilosoma,* belongs as they are compared to these categories. Standardized endings of the names of taxa in the superfamily, family, subfamily, and tribe categories are shown in boldface type. Names of taxa in categories at the generic level and below are italicized. Categories always used in the classification of animals are printed here in uppercase.
(R. E. Martin)

the bottom. Thus a species may be composed of several subspecies, and a genus may be composed of one or more species, etc. If a taxon at any level contains only a single taxon at the next lower level, the taxon is termed monotypic. Thus, the genus *Antilocapra* is **monotypic** because it contains only one living species, *Antilocapra americana,* and the order Dermoptera is monotypic because it includes only one family, Cynocephalidae. A taxon containing more than one taxon at the next lower level is **polytypic.**

Species and Subspecies

To most systematists, the **species** is the basic unit of taxonomy. The most widely accepted current definition of species is that of Mayr (1969), "Species are groups of interbreeding natural populations that are reproductively isolated from other such groups." Thus, species are naturally occurring groups of organisms that a systematist attempts to discover, define, and delimit but does not create. If two sorts of animals are **sympatric** (i.e., have overlapping geographic ranges and do not interbreed), they are thereby demonstrated to be species. Similarly, if they are **parapatric** (i.e., have contiguous but not overlapping ranges) and do not interbreed along this line of contact, they are considered distinct species. However, if morphologically very similar populations are fully **allopatric** (i.e., have nonoverlapping and noncontiguous geographic ranges), potential reproductive isolation is usually not easily demonstrated. If it seems likely that the allopatric populations would interbreed if they were sympatric or parapatric, they are considered to be **conspecific** (members of a single species), but if it seems likely that they would not interbreed they are usually considered to be distinct species. **Sibling species** are sympatric species that are reproductively isolated but morphologically indistinguishable (or very difficult to distinguish).

A subspecies is a relatively uniform and genetically distinct portion of a species, representing a separately or recently evolved lineage with its own evolutionary tendencies, definite geographic range, and a narrow zone (if any) of intergradation (i.e., zone of interbreeding, usually inferred by presence of local, linear steepening in character gradient) with adjacent subspecies (Lidicker 1962). A subspecies may be a unit of evolution, but **speciation** (i.e., splitting off of a new species) occurs only when a group of interfertile individuals become geographically and then reproductively isolated.

A higher taxonomic group contains one or more related species that differ sufficiently from other such groups and share a common lineage. Just what constitutes a sufficient difference, especially at the generic level, is somewhat arbitrary. Ideally, a **genus** is a group containing a single species or several species that differ from species in other genera by marked discontinuities (e.g., different morphological features, behavior, or other features). A **family** is a group of closely related genera that share a common recent exclusive origin and generally exhibit marked differences from genera in other families. An **order** is an assemblage of one or more related families, and a **class** contains one or more related orders. In the class Mammalia, there are 26 living orders.

Zoological Nomenclature

Zoological nomenclature is the system of scientific names applied to taxa of animals, living and extinct. Ideally, any system of nomenclature should promote names that are *unique,* only one name for only one given taxon; *universal,* written in a single language accepted by all zoologists; and *stable,* free of unnecessary or arbitrary name changes. The 10th edition (1758) of *Systema Naturae,* written by the Swedish botanist Carolus Linnaeus (the Latinized version of his name, Carl von Linné), is the starting point for zoological nomenclature. Linnaeus consistently used, in the 10th and later editions, a system of headings consisting of abbreviated names made up of two Latin or Latinized words, the binomen, for species. The Linnaean system, termed binomial nomenclature, has been adopted as the standard for the formation of scientific names of species.

The name of a species is a **binomen** consisting of a capitalized generic name (the first word of the binomen) and an uncapitalized specific name (the second word of the binomen). The species name, as opposed to the specific name, is always a binomen. For example, the species name for humans is *Homo sapiens,* not just the specific (or name of the species) *sapiens.* The name of a subspecies is a trinomen consisting of the generic, specific, and subspecific names. The subspecific name, like the specific name, is never capitalized (e.g., *Lynx rufus baileyi*).

A subgeneric name, if used, is a single word placed in parentheses between the generic and specific names (e.g., *Microtus (Pitymys) pinetorum*). The subgeneric name is capitalized but is not considered a part of the binomen or trinomen. The names of taxa at higher levels (e.g., families, orders) are also single words. Generic, subgeneric, specific, and subspecific names are always printed in italics (or underlined to indicate italics), and the formation and emendation of these names must conform to the rules of the *International Code of Zoological Nomenclature* and Latin grammar.

According to the Code, the names of families and subfamilies are formed by the addition of the suffixes *-idae* and *-inae,* respectively, to the stem word of the type-genus (i.e., a genus that is designated as the type of a family or subfamily). The Code further recommends that the terminations *-oidea* and *-ini* be added to the stems of type-genera to form the names of superfamilies and tribes, respectively. Thus, the family and subfamilies for the type-genus *Sciurus* would be Sciuridae and Sciurinae. The names for the corresponding superfamily and tribe would be Sciuroidea and Sciurini. The names of taxa higher than the genus are capitalized but not written in italics. A writer may thus refer in lowercase letters to the sciurid rodents or to the sciurids, but must capitalize the family Sciuridae.

DISTRIBUTION

Mammals were originally four-footed terrestrial animals, and most living mammals retain this basic plan. But over the millennia of their development, mammals have diversified to fill a great variety of niches—they are now found underground, in marine and fresh waters, and in the air, as well as on the surface of the earth. They exist on all continents, including Antarctica (some seals) and in all oceans. Ignoring humans and the **commensals** that follow them, mammals are absent only from a few remote oceanic islands.

Because of this widespread distribution, it is convenient to refer to ranges of particular mammal groups in terms of **faunal regions** (Fig. 1.4). The faunal regions are based upon broad similarities in animal life. The boundaries of these regions are generally formed by barriers, such as the Himalaya Mountains and the Sahara Desert, which have restricted the distribution of mammals.

The **Neotropical Region** includes all of the South American continent, the islands of the Caribbean, and extends north to central Mexico. The **Nearctic Region** includes the remaining portion of the North American continent. The **Palearctic Region** includes all of Europe, Africa to the southern edge of the Sahara, and Asia north of the south slope of the Himalayas. Because of the great similarities between the Nearctic and Palearctic faunas, these regions are frequently grouped as a single region, the **Holarctic.** The **Ethiopian Region** includes Africa south of the Sahara and most of the Arabian Peninsula. Madagascar, here considered a portion of the Ethiopian, is sometimes considered a distinct region. The **Oriental Region** extends south and east of the Himalayas to **Wallace's Line,** which passes through Indonesia. The **Australian Region** includes the Australian continent, New Guinea, and the Indonesian islands south and east of Wallace's Line. New Zealand and other islands of Oceania and the Antarctic continent are not placed in named faunal regions.

1-F What are the major barriers separating each of the faunal regions from neighboring regions?

1-G List at least three kinds of mammals that may be considered characteristic of each region.

1-H In which faunal region, or regions, is each of the following situated?

Bali	Iceland	Japan	Tunisia
Bolivia	India	Malaysia	Uganda
China	Indonesia	Mexico	Yemen
Greenland	Jamaica	Monaco	

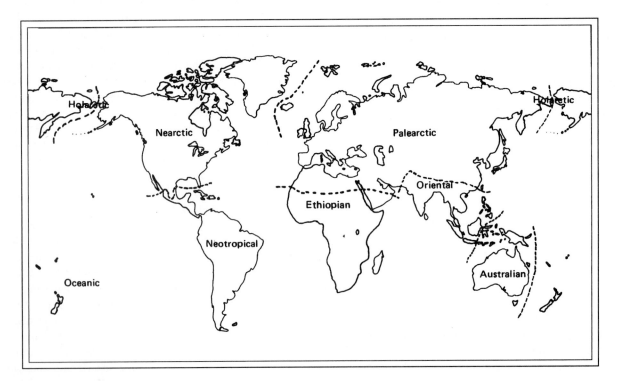

Figure 1.4 The major Faunal Regions of the world.
(A. F. DeBlase)

The Structure and Use of This Manual

As was pointed out earlier, mammals occur on every major land mass and in every major body of water in the world. We believe with today's ease of global travel and with the rapid communications available through the Internet and other electronic media that institutions, instructors, and students must cease to be provincial in their focus only on locally occurring mammals and must consider the world's fauna as a whole. Therefore, we have designed this manual to be worldwide in scope. Recognizing that most institutions will not have examples of many of the world's mammals, we have attempted to select North American representatives as examples for particular characteristics whenever possible. However, when North American examples will not suffice, we have written instructions using specimens from other parts of the world. We feel it is better for the student to examine illustrations of mammals in this manual or in one of the many well-illustrated books now available than to ignore a taxon or characteristic simply because a specimen is not readily available.

Similarly we realize that few North American institutions may have non-North American mammals to use with the keys. However, we believe that the students' use of keys covering all families will increase their awareness of the world's mammalian fauna even though they work only with North American forms.

The chapters in this manual fall into three distinct groupings. Chapters 1 through 8 primarily cover (1) the basic anatomy of mammals, needed to better understand the working of the group and (2) the terminology needed to identify individual groups of mammals. Chapters 9 through 27 are keys to aid in the identification of living orders and families of mammals. Chapters 28 through 37 are guides to the laboratory and field procedures that are most important in the study of mammals.

Care of Specimens

Throughout this manual you will be called upon to closely examine and handle numerous mammal specimens. Whenever possible, we have recommended the use of species most readily available and most easily replaced. However, whether the specimen is expendable or rare, it should always be handled and treated with care.

Skulls and skeletal elements should be handled gently and never picked up by slender processes or other portions that are likely to break. When possible, place skulls on a pad. Never drop them on a hard surface such as a tabletop. Be particularly careful when handling small or delicate skulls. If a tooth falls out of its socket or a bone breaks or becomes disarticulated, report it to the instructor at once so that pieces are not lost.

Study skins require even more careful handling than most skulls. Never pick up a study skin by the feet or tail. A dried skin can be very brittle, and projections such as the feet, ears, or tail can be easily broken off. Most study skins are made to be placed belly down. Do not leave them resting on their backs or sides and do not stack them on top of one another.

Specimens preserved in liquid must be kept moist and not allowed to dry out. If these are being studied for an extended period, moisten them occasionally with the preservative or water. Pay special attention to thin structures such as ears and bat wings.

Take particular care not to detach labels from any specimen. As will be emphasized in Chapter 29, the catalog numbers and data on these labels are exceedingly important to the scientific value of the specimen. If a label should become loose, notify the instructor at once.

Whenever live mammals are studied in the laboratory, they should be treated humanely. Wild mammals need particular attention with respect to housing and diet. Refer to Chapter 30 for further precautions on the handling of wild mammals in field and laboratory settings.

CHAPTER 2
THE SKULL

The mammalian skull is a complex structure. It houses and protects the brain and the receptors for five major senses: smell, taste, vision, hearing, and equilibrium. The braincase has adapted to the changes in the size and proportions of the brain. Specializations in the senses of hearing, smell, and sight have frequently resulted in corresponding changes in the skull, as have various adaptations for gathering food and preparing it for digestion. These, along with the skull's ability to resist decomposition and to fossilize, make it one of the most important anatomical units used in mammalian classification. A knowledge of its anatomy is essential for the identification of mammals. The keys in this volume have been constructed primarily on the basis of skull and tooth characters. Teeth will be discussed in Chapter 3.

The skull is composed of two easily disarticulated elements: an upper **cranium** (with braincase and rostral regions, see below) and the **mandible** or lower jaw. In addition, the tongue is partially supported by the **hyoid apparatus,** a component of the visceral skeleton. Only those bones that are visible externally on a cleaned skull are discussed in detail. For convenience, two major regions of the mammalian cranium may be recognized: the **braincase** and the **rostrum.** The braincase is a "box" of bone protecting the brain. Attached to it or associated with it are the **auditory bullae** (not present in all mammals), which house the middle and inner ears; the **occipital condyles,** which articulate with the first vertebra; and numerous processes and ridges that serve as points of attachment for muscles. Several **foramina** and **canals** penetrate the bones and allow for the passage of nerves and blood vessels. The rostrum is composed of the group of bones projecting anteriorly from a vertical plane drawn through the skull at the anterior edges of the orbits. It includes the upper jaws and the bones that surround the nasal passages and divide these passages from the oral cavity.

2-A On a wolf, coyote, or dog skull (genus *Canis*) locate each of the bones or structures listed below in boldface type. Label each of these on the various views of the coyote skull in Figure 2.1A–C. All terms are listed in the glossary. A key to the numbers on Figure 2.1 is located at the end of this chapter. *Use it only to verify your identifications.*

The dorsal part of the skull is composed mostly of a series of paired bones that meet along the midline. The long slender **nasal bones** roof the **nasal passages.** Posterior to these are the paired **frontals.** Each of these extends down the side of the skull to form the inner wall of the **orbit** or eye socket. The **postorbital process** of the frontal is a projection that marks the posterior margin of the orbit. Posterior to the frontals are the paired **parietals.** A small, unpaired **interparietal** is located between the posterior edges of the parietals; in *Canis,* this is fused posteriorly with the supraoccipital (see below). Low **temporal ridges** arise on the frontals near the postorbital processes and continue posteriorly until they converge to form the sagittal crest. These ridges (including the "crest") increase the area available for attachment of jaw muscles. The posterior portion of the skull is formed by a fused bone, the **occipital.** The **foramen magnum,** through which the spinal cord passes, is located near the center of the occipital and is flanked by two knobs, the

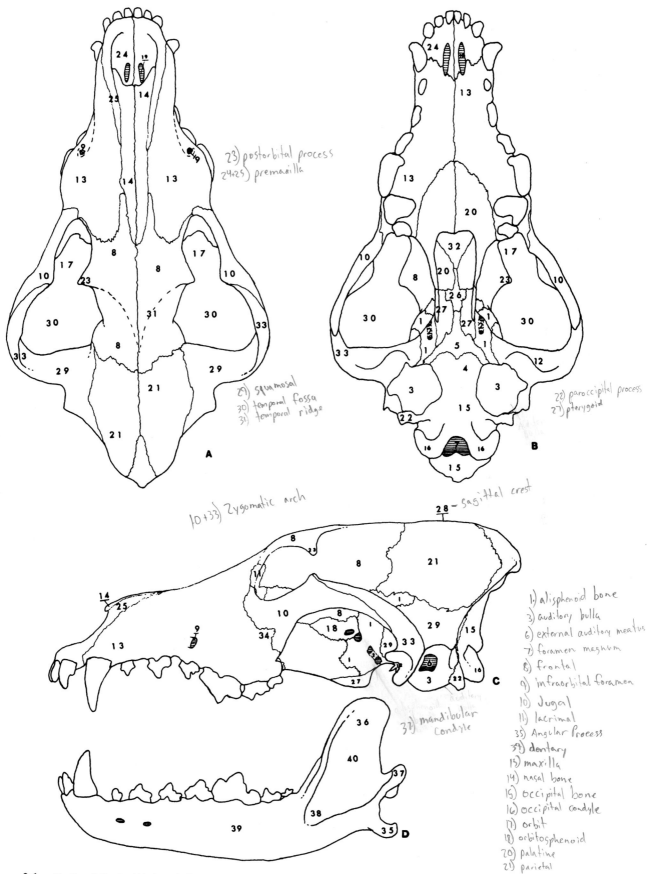

Figure 2.1 Skulls of *Canis*. (A) dorsal view; (B) ventral view; and (C) lateral view of cranium; (D) lateral view of mandible.
(A. F. DeBlase)

occipital condyles, which articulate with the atlas, the first of the neck vertebrae. In young mammals, four bones fuse to form the single occipital bone of the adult. The names for these are used to designate regions of the occipital. Around the foramen magnum, these are the ventral **basioccipital,** the dorsal **supraoccipital,** and the lateral **exoccipitals.** The **occipital crests** extend laterally from the sagittal crest. Branches of the exoccipitals, the **paroccipital processes,** extend ventrally in close association with the auditory bullae. The entire posterior region of the skull is termed the **occiput.**

The tooth-bearing bones of the upper jaws are the paired premaxillae and maxillae. The **premaxillae,** which meet at the anterior end of the skull, have two major branches. The **palatal branches** of the premaxillae meet along the midline of the skull and form the anterior portion of the **hard palate;** the **nasal branches** of the premaxillae project dorsally and posteriorly to form the sides of the **anterior nares** (also called **external nares**). Posterior to the premaxillae, the **maxillae** form the major portions of the sides of the rostrum. A large foramen in each maxilla is the anterior opening of the **infraorbital canal.** Each canal terminates in the orbit and serves for passage of blood vessels and nerves. In some mammals, this opening is not elongated into a canal and is termed the **infraorbital aperture** or **infraorbital foramen.**

The palatal branches of the premaxillae and maxillae together with the paired **palatine bones** form the **hard palate** that separates the **buccal cavity** (mouth) from the nasal passages at this level. A pair of openings at the suture between the premaxillae and maxillae are the **anterior palatal foramina** (also termed the **incisive foramina**). Posterior and dorsal to the palatine bones are the proximal openings of the nasal passages, the **internal nares.** The **vomer** is an unpaired bone forming a septum between the two nasal passages. The highly convoluted bones within these passages are the **turbinals.** Posterior to the internal nares and the palatine bones are the paired **pterygoids.** Between the paired pterygoids and posterior to the vomer is the unpaired **presphenoid.** This complex bone passes beneath the pterygoid, palatine, and maxillary bones to reappear dorsally in the wall of each orbit where it is termed the **orbitosphenoid** and is perforated by the **optic foramen.** The medial **basisphenoid** lies between the basioccipital and the ventral visible portion of the presphenoid.

The conspicuous bony arches forming the ventral and lateral borders of the orbits and temporal fossae are the **zygomatic arches.** Three bones contribute to each zygomatic arch. Anteriorly, the jugal bone articulates with the **zygomatic process of the maxilla.** Posteriorly, the jugal articulates with the **zygomatic process of the squamosal** bone. A short process on the dorsal edge of the zygomatic arch marks the posterior edge of the orbit. In some mammals (but not in *Canis*), this process is continuous with the postorbital process of the frontal, forming a **postorbital bar.** The postorbital bar separates the orbit or eye socket from the temporal fossa, through which some of the muscles of the lower jaw pass. On the ventral side of the base of each zygomatic process of the squamosal, the **mandibular fossa** provides an articulation surface for the lower jaw.

Between the jugal and frontal bones, at the anterior root of each zygomatic arch, is the small **lacrimal bone.** The foramen in this bone is for passage of the tear, or lacrimal, duct. Anterior to the squamosal, and posterior to the frontal and orbitosphenoid, is the **alisphenoid** bone. Ventrally on this bone, near its suture with the basisphenoid, is a small arch of bone surrounding the **alisphenoid canal.**

The bulbous structures between the mandibular fossae and the occipital condyles are the **auditory bullae.** The opening in the side of each bulla is the **external auditory meatus** across which the tympanic membrane, or eardrum, is stretched. In *Canis*, the **tympanic bone** is the only bone visible on the external surface of the bulla, but in some mammals the **entotympanic bone** is also visible externally. Within each bulla is the **middle ear** chamber containing the three ossicles, the **incus, malleus,** and **stapes.** The **otic capsule,** which houses the structure of the inner ear, is covered by the tympanic in *Canis*, but it is visible in primitive mammals that have incomplete auditory bullae. A portion of the **periotic,** one of the bones forming each otic capsule, is frequently exposed between the squamosal and occipital bones. The distal exposed portion of the periotic forms a distinct **mastoid process** in many mammals, but this is not a conspicuous structure in *Canis*. In some mammals (including cats and higher primates), the tympanic and squamosal bones fuse to form a single structure termed the **temporal bone.**

THE MANDIBLE

Compared to the cranium, the mandible is a very simple structure. It is composed of left and right **dentary** bones. The anterior surface of contact between the paired dentaries is the **mandibular symphysis.** This suture is attached fairly firmly in *Canis,* in most other Carnivora, and in many other mammals, and completely fused in primates. But in rodents, most artiodactyls, and many other forms, the two dentaries become easily disarticulated. The horizontal portion of each dentary, the portion that normally bears teeth, is termed the **body,** and the vertically projecting portion is the **ramus.** The **mandibular condyle** is the portion of the mandible that articulates with the mandibular fossa of the cranium. Dorsal to the condyle, the **coronoid process** extends up to fit into the temporal fossa and provides a surface for muscle attachment. Ventral to the condyle, the **angular process** protrudes posteriorly. The shallow depression near the bases of these processes is the **masseteric fossa.** In some mammals (but not in *Canis*), this depression is very deep and occasionally completely penetrates the mandible, forming a **masseteric canal.**

2-B Locate on a *Canis* mandible each of the structures listed above in boldface type. Label these on Figure 2.1D. Check your identifications with the key at the end of this chapter.

Variation

The skulls of species of *Canis* may be considered to represent a "typical" mammal skull. From this "typical" structure are many deviations. The postorbital bar, mastoid process, and other structures conspicuous in some mammals, but absent in *Canis,* have already been mentioned.

The relative lengths of braincase and rostrum vary considerably. Mammals such as certain whales and anteaters have relatively short braincases and long rostra, whereas other species, such as humans, *Homo sapiens,* have large braincases and virtually no rostra.

The orbits may be directed anywhere from laterally, as in the pronghorn, to anteriorly, as in humans. They may be low on the head, as in raccoons, or high on the skull, as in woodchucks.

Nasal bones may be absent, short and broad, or long and narrow. Palatal, nasal, or both branches of the premaxillae may be enlarged, reduced, or lost.

Zygomatic arches may be incomplete, weak, or amazingly robust. Auditory bullae may be complete, incomplete, inflated, or compressed.

Many other such variations can and do exist but are far too numerous to list.

2-C To get an idea of the range of variation that exists in mammalian skulls, make as many of the following comparisons as possible.
  a. Compare the degree of separation of the orbit and temporal fossa in a shrew, human or monkey, raccoon, cat, and horse.
  b. Compare the bone structure of the temporal region in *Canis,* a cat, and *Homo.*
  c. Compare the relative lengths and sizes of the rostrum and braincase in an opossum, shrew, human, coyote, cat, horse, and elephant.
  d. Compare the position of the orbits in a human, raccoon, otter, woodchuck, and deer.
  e. Compare the size and proportion of the nasal bones in an opossum, human, porpoise, elephant, horse, tapir, and moose.
  f. Compare the zygomatic arches of an opossum, shrew, human, *Canis,* rat, North American porcupine, porpoise, and horse.
  g. Compare the structure of the auditory bullae in a hedgehog, human, *Canis,* kangaroo rat, bear, porpoise, and deer.
  h. Compare the placement of the foramen magnum in an opossum, monkey, and deer.

Determination of Maturity

There are several methods of determining the absolute or relative age of an individual. These are discussed in detail in Chapter 33. Because most identification keys, including the ones in this manual, are only for adult mammals, it is necessary for you to be able to distinguish between immature and adult animals. Two cranial characteristics are especially helpful in identifying immature specimens, but neither of these always works. An individual in which it is evident that certain teeth are not yet fully erupted is usually an immature specimen. The degree of fusion of cranial sutures is generally also an indication of age. An immature specimen will have poorly fused sutures, and a very old adult can have sutures that are almost indiscernible. If a skull has a fully erupted dentition and fully fused cranial sutures, it should be possible to identify it using the keys in this manual. The keys may or may not correctly identify a specimen that does not meet these criteria.

Measurements

Several more or less standardized measurements are used in gaining information about mammalian skulls. Because skulls are complex structures that can vary in many ways, different sets of measurements are used for different groups of mammals. The 10 most frequently taken measurements for a *Canis* skull would not be the same as the 10 for a porpoise or a rodent.

Skull measurements are taken in a straight line between two points (or lines or combinations thereof) and are recorded in millimeters. **Calipers** are customarily used for taking these measurements. Dial calipers are the easiest and most efficient type to use. Although various brands and models differ in design, in most models the centimeters are read directly from the bar, and millimeters and tenths of millimeters are read directly from the dial mounted on the movable slide. Vernier calipers are equally accurate but are slightly more difficult to read. Again, models vary in precise design, but in most models centimeters and millimeters are read directly from the bar, and tenths of millimeters are determined by the best match between gradations on the bar and one of the lines on the sliding scale.

When using calipers, take care not to damage specimens. Calipers should be closed to fit snugly against the bone but be careful not to crush, scratch, or puncture the bone.

The following measurements are some of those most frequently taken. An asterisk (*) indicates those that are taken on most species.

Measurements of the Entire Skull

All measurements of length are taken along the midline of the skull.

Basal length. From the anterior edge of the premaxillae to the anteriormost point on the lower border of the foramen magnum (Fig. 2.2, A–B).

Basilar length. From the posterior margin of the alveolus of either of the median upper incisors to the anteriormost point on the lower border of the foramen magnum (Fig. 2.2, C–B).

*****Condylobasal length.** From the anterior edge of the premaxillae to the posteriormost projections of the occipital condyles (Fig. 2.2, A–D).

Condylocanine length. From the anterior edges of the alveoli of the upper canines to the posterior edges of the occipital condyles. (Usually taken instead of condylobasal length in forms in which the premaxillae are frequently lost.) (Fig. 2.2, E–D)

*****Greatest length of skull.** From the most anterior part of the rostrum to the most posterior point of the skull (Fig. 2.3, L–M).

*****Breadth of braincase.** Greatest width across the braincase posterior to the zygomatic arches (Fig. 2.3, A).

*****Least interorbital breadth.** Least distance between the orbits (Fig. 2.3, B).

Mastoid breadth. Greatest distance across mastoid bones, on a line perpendicular to the long axis of the skull (Fig. 2.3, D).

*****Postorbital constriction.** Least distance across the top of the skull posterior to the postorbital process (Fig. 2.3, C).

Rostral breadth. Least breadth of rostrum between designated points on opposite sides of the skull.

*****Zygomatic breadth.** Greatest distance between the outer margins of the zygomatic arches (Fig. 2.3, E–F).

Measurements of Palate and Upper Dentition

Alveolar length and width. Greatest length or width of the alveolus of any specified tooth.

Diastema length. When diastema present, from posterior margin of alveolus of last incisor present to anterior margin of alveolus of first cheek tooth present.

Incisive foramen length. Greatest length of anterior palatal foramen (Fig. 2.2, H–I).

*****Maxillary tooth row.** Length from anterior edge of alveolus of first tooth present in a maxilla to

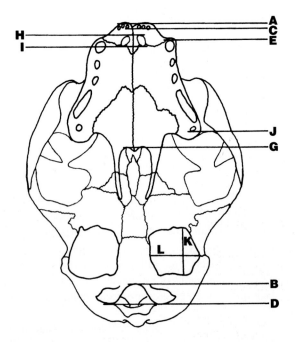

Figure 2.2 *Felis* skull showing points and lines for taking measurements of the ventral side.
(A. F. DeBlase)

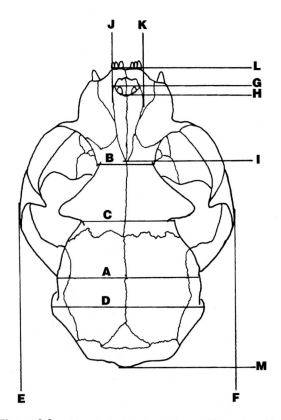

Figure 2.3 *Felis* skull showing points and lines for taking measurements of the dorsal side.
(A. F. DeBlase)

posterior edge of alveolus of last tooth in maxilla (Fig. 2.2, E–J).

*Palatal length. From anterior edge of premaxillae to anteriormost point on posterior edge of palate (Fig. 2.2, A–G).

Palatilar length. From posterior edges of alveoli of first incisors to anteriormost point on posterior edge of palate (Fig. 2.2, C–G).

Palatal width. Usually width of palate between alveoli of some specified pair of teeth. Occasionally includes alveoli or bony outer edge of palate or teeth.

Measurements of Other Portions of the Skull

Nasal length. Length of nasals. From anteriormost point of nasal bones to posteriormost point taken along midline (usually) of nasal bones (Fig. 2.3, G–I).

Nasal width. Greatest width across both nasals (Fig. 2.3, J–K).

Nasal suture length. Greatest length of suture between paired nasal bones (Fig. 2.3, H–I).

Postpalatal length. From anteriormost point on posterior edge of palate to anteriormost point on lower edge of the foramen magnum (Fig. 2.2, G–B).

Tympanic bullae length and width. Greatest length and width of bulla (Fig. 2.2, K and L).

Measurements of Mandible and Lower Dentition

Mandibular diastema. Same as for maxillary diastema (Fig. 2.4, A–B).

*Mandible length. Greatest length of the mandible, usually excluding teeth (Fig. 2.4, D–E).

*Mandibular tooth row. Length from anterior edge of alveolus of canine (if present) or first cheek tooth to posterior edge of alveolus of last tooth. The incisors are not usually included in this measurement (Fig. 2.4, B–C).

2-D Take each of the measurements listed above on the following mammals (not all can be made on all skulls [e.g., length of diastema cannot be taken on an animal without a diastema]). Record measurements to nearest tenth of a millimeter.
    Canis or other carnivore
    Rat or other rodent
    Human or other primate

Compare your figures with those of others in the class.

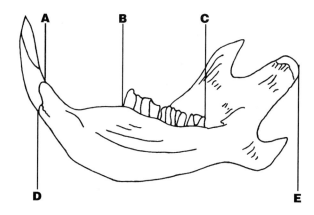

Figure 2.4 Rodent dentary showing points for taking the most commonly used measurements.
(A. F. DeBlase)

How closely do yours agree with those of the others who measured the same specimen? (An indication of accuracy.)

2-E Remeasure one specimen. How does your second set of measurements compare with your first? (An indication of precision.)

KEY TO LABELING OF FIGURE 2.1

1. alisphenoid bone
2. alisphenoid canal
3. auditory bulla (tympanic bone)
4. basioccipital
5. basisphenoid
6. external auditory meatus
7. foramen magnum
8. frontal
9. infraorbital foramen
10. jugal
11. lacrimal
12. mandibular fossa
13. maxilla
14. nasal bone
15. occipital bone
16. occipital condyle
17. orbit
18. orbitosphenoid
19. palatal (=incisive) foramen
20. palatine
21. parietal
22. paroccipital process
23. postorbital process (of the frontal)
24. premaxilla, palatal branch
25. premaxilla, nasal branch
26. presphenoid

27. pterygoid
28. sagittal crest
29. squamosal
30. temporal fossa
31. temporal ridge
32. vomer
33. zygomatic process of squamosal
34. zygomatic process of maxilla
35. angular process
36. coronoid process
37. mandibular condyle
38. masseteric fossa
39. body
40. ramus

CHAPTER 3

TEETH

Although mammalian teeth are similar in basic components, they exhibit great diversity in number, size, and shape. The radiation of mammals into virtually every macrohabitat has resulted in evolutionary adaptations in tooth morphology to cope with varied diets. Teeth are readily fossilized, and many extinct mammals are known only from teeth. Thus, teeth are valuable tools in classifying, identifying, and studying mammals.

TOOTH ANATOMY AND REPLACEMENT

The **crown** is the portion of a tooth exposed above the gumline; the **root** is the portion fitting into the **alveolus** or socket in the jaw. Teeth with a particularly high crown are termed **hypsodont,** and those with a particularly low crown are **brachyodont.** Points and bumps on the crown of the tooth are generally termed **cusps** (see the Premolars and Molars section for nomenclature of crown elements). Teeth may be unicuspid, bicuspid, tricuspid, etc. The side of a tooth closest to the tongue is termed the **lingual** side, and the side closest to the cheek is the **labial** or **buccal** side. The surface of a tooth that meets with a tooth in the opposing jaw is termed the **occlusal** surface.

The major portion of each tooth is made up of a bonelike material called **dentine** (Fig. 3.1). The crown has a thin layer of hard, usually white, **enamel** covering the dentine, and the root is covered by a layer of bonelike **cementum.** The central, living portion of a growing tooth, the **pulp,** is supplied with blood vessels and nerves through one or more openings in the base. In most species, when the tooth has reached a certain size, this opening constricts, the blood supply is much reduced, and growth ceases. Such teeth are termed **rooted.** In some groups, the opening does not constrict and growth of the tooth continues throughout the life of the mammal. Such evergrowing teeth are termed **rootless.**

Most mammals are **diphyodont,** having only two sets of teeth. The **deciduous** or **milk teeth** present in immature mammals are usually replaced by a set of **permanent teeth** that are retained for life. Toothed cetaceans, Odontoceti, and a few other mammals are **monophyodont,** having only one set of teeth. Marsupials and some other mammals have only some of their milk teeth replaced, and others remain as a part of the adult dentition.

Some mammals, such as elephants, manatees, and kangaroos, have a slightly different system of tooth replacement. These mammals feed primarily upon harsh vegetation, and this diet makes for considerable wear on their teeth. In elephants, the alveoli of the cheek teeth converge into a groove, and tooth replacement occurs only at the posterior end of the tooth-row. As the anterior tooth is worn away, a new tooth develops from the rear, and the entire row moves forward (Fig. 3.2). A total of six cheek teeth are available to each quadrant, but only one or parts of two teeth are functional at any one time. Manatees have a similar system of tooth replacement with a potential number of 20 teeth per jaw, but only six to eight function at one time. In kangaroos, tooth replacement is primarily from the rear of the jaw, but the anterior two deciduous cheek teeth are replaced from below by a single tooth.

3-A   Examine the internal structure of a sectioned tooth and compare with Figure 3.1.

3-B   Examine a coyote or dog skull (genus *Canis*) and note the placement of teeth in the alveoli.

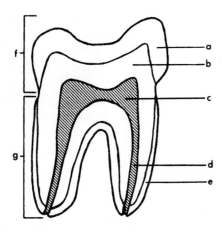

Figure 3.1 Diagrammatic cross section of a mammalian tooth. a, enamel; b, dentine; c, pulp; d, root canal; e, cementum; f, crown; g, root.
(L.P. Martin)

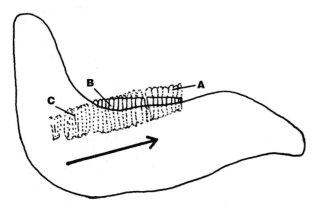

Figure 3.2 Diagram of tooth replacement in an elephant jaw. A, portion of tooth worn away; B, portion exposed; C, portion still embedded in the jaw. Arrow indicates direction of tooth replacement.
(A.F. DeBlase)

Examine a similar specimen from which the teeth have been removed. How many roots does each tooth have? How many cusps?

3-C   Examine skulls of mammals that are in the process of shedding their deciduous teeth. How does replacement occur?

3-D   Examine an elephant jaw. How many cheek teeth are present? Can you notice any difference in wear between the first and last tooth in each jaw (excluding tusks)?

THE KINDS OF TEETH

An individual mammal usually has two or more morphologically different kinds of teeth, a condition termed **heterodont**. This contrasts with the **homodont** dentition of other vertebrates, in which all teeth in an individual resemble each other in shape. In mammals, four basic kinds of teeth are recognized: **incisors, canines, premolars,** and **molars.**

Incisors

Incisors are the teeth rooted in the premaxillary bone and the corresponding teeth in the lower jaw. Placental mammals never have more than three incisors in each jaw quadrant, but marsupials may have up to five in each half of the upper jaw and up to four in each half of the lower jaw. These are usually unicuspid teeth with a single root, but in some groups of mammals, accessory cusps, additional roots, or both may be present.

Incisors are generally chisel-shaped teeth that function primarily for nipping (e.g., a human biting an apple or a horse cropping grass). In cattle, deer, and their relatives, this nipping action has been modified by the loss of the upper incisors. Instead of nipping the vegetation between upper and lower incisors, these animals use their highly mobile lips and prehensile tongue to draw vegetation across the lower teeth, which cut it off in much the same way that a tape dispenser cuts tape. In rodents (see Fig. 23.2), lagomorphs (see Fig. 22.3), and certain other specialized forms, the number of incisors has been reduced, but the first incisors are stout chisel-edged teeth used in gnawing. These incisors are rootless and grow continually as they are worn away at the tips. In vampire bats, the first pair of incisors has a long, sharp edge (see Fig. 14.6B). These teeth are used to shave away a layer of skin to expose blood vessels. The blood that flows to the surface is then ingested. Elephants have incisors that are enlarged to form tusks (see Fig. 25.1). These are rootless and evergrowing and may be used for digging and removing bark from trees. Shrews have incisors that project anteriorly (see Fig. 12.1) and act as forceps in catching and holding insects and other prey.

3-E   Examine the incisors of a shrew, vampire bat, monkey, rodent, horse, and cow, sheep, or deer. What can you deduce about the diet or feeding habits of each of these mammals?

3-F   Examine the pectinate (comblike) lower incisors of a colugo (Dermoptera: Cynocephalidae). Compare these incisors with those of the ringtail lemur (Primates: Lemuridae). In what way are the incisors similar? How do they differ? What is their function?

Canines

Canines are the most anterior teeth rooted in the maxillae and the corresponding teeth of the lower jaw. They never number more than one per quadrant. Canines are usually long, conspicuous, unicuspid teeth with a single root.

However, some mammals may have canines with accessory cusps, additional roots, or both.

Canines are usually used to capture, hold, and kill prey. In herbivorous species, they are frequently reduced or absent. In some groups, such as the hogs and some deer, they are very long and sharp and used for fighting. Pig "tusks" are rootless and in some species arranged so they do not fully occlude. This minimizes wear and allows at least the upper tusks to grow very long. Walruses have been said to use their elongated canines to scrape the mollusks that they feed upon from the ocean floor, but evidence indicates that these conspicuous teeth are not used in this way (Miller 1975; Ray 1973).

Frequently canines and/or other teeth are absent, leaving a wide space between the anterior teeth and the cheek teeth. Any such wide gap between teeth is termed a **diastema.**

Note! In some species, the most conspicuous unicuspid tooth in the anterior part of the jaw is not the canine. Occasionally the last incisor is large and **caniniform,** and the canine is absent or small and resembles a premolar. Conversely, the first premolar is occasionally caniniform, and the canine is small and **incisiform.** In the upper jaw, these teeth are easily identified by locating the suture between the premaxilla and maxilla.

3-G   Examine the canines of the following pairs of mammals. Can you suggest functions (if any) for the specializations?
   a. Peccary (*Tayassu*) and warthog (*Phacochoerus*)
   b. *Canis* and *Felis*
   c. Human and baboon (*Papio*)

Premolars and Molars

Premolars are situated just posterior to the canines and generally some of all of them in each species differ from molars in having deciduous predecessors in the milk dentition. In all placentals (with the exception of tapirs) in which there are four premolars, the first premolar never has a deciduous precursor. In certain other mammals, in which there are fewer than four premolars, the first premolar in the sequence is also not replaced (Slaughter, et al. 1974). Molars are situated posterior to the premolars and never have deciduous predecessors. Authorities disagree as to whether molars are permanent teeth for which there are no corresponding milk teeth, or whether they are milk teeth that erupt late and are not replaced. Premolars are usually smaller than molars and have fewer cusps. However, without embryological investigation or a knowledge of the milk dentition of the species being studied, it is frequently impossible to distinguish between premolars and molars in an adult mammal. Therefore, these two tooth types frequently are referred to together as **cheek teeth, postcanine teeth,** or **molariform teeth.**

Placentals are regarded as having a "late primitive" maximum of four premolars and three molars. Marsupials have only a single tooth in each quadrant of the milk dentition. This milk tooth corresponds to the third premolar, above and below, in the adult dentition. Marsupials are regarded as having a "late primitive" maximum of three premolars and four molars. Teeth are absent in adult monotremes.

Because cheek teeth do the major job of masticating food, they are the teeth that exhibit the greatest diversity correlated with diet. Cheek teeth occur that are adapted for such a variety of foods as mollusks, meat, soft vegetation, tough grasses, hard-bodied insects, worms, and krill. The structure of the cheek teeth is one of the most important criteria in mammalian classification.

A standardized terminology for dental crown elements that is acceptable to all paleontologists and mammalogists is not presently available. The greatest obstacles to the development of a generally accepted terminology are questions of homology of cusps between early and later groups of mammals. Our terminology is derived, in part, from information presented by Patterson (1956), Van Valen (1966), Szalay (1969), and Hershkovitz (1971).

Simple Tribosphenic Cheek Teeth

The earliest known tribosphenic cheek teeth were present in early marsupials and placentals of the Cretaceous. A simple **tribosphenic** upper molar (Fig. 3.3A) has a trigon, whereas the lower molar (Fig. 3.3B) has both a trigonid and a talonid. The triangle-shaped trigon of an upper tribosphenic upper molar has three main cusps with the protocone at the apex along the labial edge of the crown (Fig. 3.3A). The other cusps are an anterior **paracone** and a posterior **metacone** (Fig. 3.3A). The **stylar shelf,** a broad ledge situated labial to the paracone and metacone, has several cusps, including the most anterior, the **parastyle,** which provides a convenient reference point to orient a tooth for study.

The lower tribosphenic molar (Fig. 3.3B) consists of a high-cusped **trigonid** and lower-cusped **talonid,** the latter of which helps to square the outline of the tooth. Three of the cusps of the talonid enclose a depression known as the **talonid basin** that receives the protocone of the trigon during occlusion.

Modified Tribosphenic Cheek Teeth

The simple tribosphenic cheek tooth has been modified in various lineages of mammals. For convenience, some authors (Butler 1941; Hershkovitz 1971; Turnbull 1971) divide the simple and derived tribosphenic molars into three main groups: zalambdodont, dilambdodont, and euthemorphic. Although these modifications apply to molars and some premolars, particularly the most posterior premolars in a series, the discussion that follows is based on molars. A **zalambdodont** upper molar (Fig. 3.4) is characterized by a V-shaped ectoloph. An **ectoloph** is a series of **cristae,** or crests, connecting the paracone

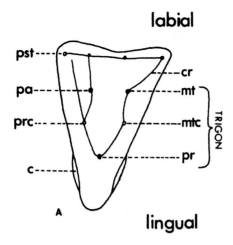

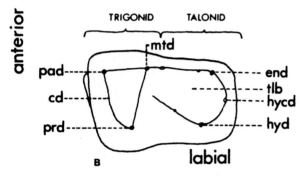

Figure 3.3 Occlusal views, somewhat diagrammatic, of simple tribosphenic left upper (A) and left lower (B) molars. *Upper crown elements:* c, cingulum; cr, crista; mt, metacone; mtc, metaconule; pa, paracone; pr, protocone; prc, paraconule; pst, parastyle. *Lower crown elements:* cd, cristid; end, entoconid; hyd, hypoconid; hycd, hypoconulid; mtd, metaconid; pad, paraconid; prd, protoconid; tlb, talonid basin. Major cusps in solid black. Based on information in Van Valen (1966) and Szalay (1969).
(Modified from Van Valen 1966)

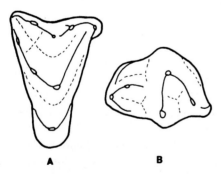

Figure 3.4 Occlusal view of zalambdodont right upper (A) and left lower (B) molars of the otter shrew, *Potamogale*.
(Modified from Butler 1941)

(and sometimes the metacone) with cusps on the stylar shelf. Typically the zalambdodont molar lacks a protocone, and the paracone (sometimes combined with the metacone) is located at the lingual apex of the crown. This type of molar is found in many Insectivora and in

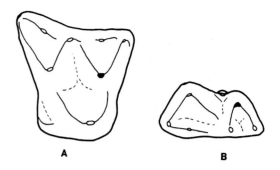

Figure 3.5 Occlusal view of dilambdodont right upper (A) and left lower (B) molars of a tree shrew, *Tupaia*.
(Modified from Butler 1941)

the marsupial "mole," *Notoryctes*. A **dilambdodont** upper molar (Fig. 3.5) has a W-shaped occlusal surface with the protocone near the lingual apex of the trigon. The W-shaped pattern is formed by an ectoloph connecting the metacone and paracone with cusps on the stylar shelf. The molars of the opossums (Didelphidae) and tree shrews (Tupaiidae) are examples of the dilambdodont type. A **euthemorphic** upper molar usually has a square or quadrate crown. The square outline results from the addition of a main cusp, the **hypocone,** to the posterior lingual side of the crown. In certain molars, the hypocone area is identified as the **talon.**

A euthemorphic upper molar with four main cusps is termed **quadritubercular.** Upper and lower molars may become fully **quadrate,** or square, by loss or reduction of some cusps (e.g., in the lower dentition the paraconid is generally lost). Most living mammals have basically euthemorphic molars, although the teeth may be modified in several ways (see next section).

3-H   Examine the upper and lower tribosphenic molars of an opossum (Didelphidae). Identify the trigon, trigonid, talonid, talonid basin, and the major cusps. Observe the shearing and crushing actions that occur as the upper and lower jaws are brought into occlusion.

3-I   Examine a zalambdodont upper molar of Chrysochloridae, Solenodontidae, or Tenrecidae. Locate the ectoloph and position of the paracone (or fused paracone-metacone).

3-J   Examine a dilambdodont upper molar of Didelphidae, Talpidae, or Tupaiidae. Locate the ectoloph, paracone, metacone, and protocone.

3-K   Examine upper and lower euthemorphic molars of a pig, human, various rodents, and a horse. In which species do molars show well-defined cusps? In which are the molars quadritubercular? In which are the molars quadrate?

Specializations of Cheek Teeth

The bunodont tooth is found in many mammals that are basically omnivorous. The **bunodont** tooth is euthemorphic, quadrate, frequently brachyodont, and has four major rounded cusps (Fig. 3.6). It is considered to have developed from a tribosphenic tooth by the bulging out of the side between the protocone and metacone and the development of a new cusp, the hypocone, in this area (hypoconid in the lower teeth). Other smaller cusps may develop between the larger ones. For example, a small **paraconule (=protoconule)** may develop between the protocone and paracone, and a small **metaconule** may develop between the metacone and hypocone. In the lower cheek teeth, a **hypoconulid** is situated on the posterior margin of the talonid between the hypoconid and entoconid. The crowns of bunodont teeth oppose each other directly, and the paraconid is lost. Humans and hogs are examples of mammals with bunodont teeth used for an omnivorous diet.

In mammals that tend toward an herbivorous diet, the cheek teeth are frequently hypsodont. The abrasive action of plant material quickly erodes teeth, so the higher the crown, the longer the tooth will last. Some herbivorous mammals (particularly grazers) have cheek teeth that are rootless and continue to grow throughout life as they are worn away at the top. Many herbivorous mammals have **lophodont** teeth (Fig. 3.7) in which cusps fuse to form elongated ridges termed **lophs.** These ridges create elongated abrasive surfaces for the grinding of plant materials. A **selenodont** tooth (Fig. 3.8) functions in much the same manner but in it each ridge is formed by the elongation of a single cusp. The ridges of selenodont teeth are always crescent-shaped and longitudinally oriented (Fig. 3.8), whereas those of lophodont teeth are variable in shape and may be transversely oriented (see Fig. 3.7). In these teeth, the hard ridges of enamel wear away more slowly than the surrounding tissues and provide a grinding surface similar to that of a millstone. Some mammals such as the horses, Equidae, have complex **selenolophodont** teeth that combine aspects of both lophodont and selenodont teeth.

The cheek teeth of rodents show numerous modifications from the basic quadritubercular plan. These may

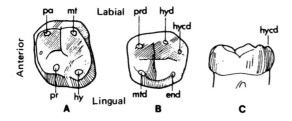

Figure 3.6 Bunodont left upper (A) and right lower (B and C) molars of human, *Homo sapiens,* a modified euthemorphic molar. (A) and (B) are occlusal views: (C) is a labial view. *Upper:* hy, hypocone; mt, metacone; pa, paracone; pr, protocone. *Lower:* end, entoconid; hyd, hypoconid; hycd, hypoconulid; mtd, metaconid; prd, protoconid.
(Modified from Osborn 1907)

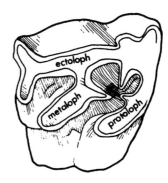

Figure 3.7 Lophodont molar tooth of a rhino (order Perissodactyla), occlusal view. Note the fusion of cusps into transverse and longitudinal lophs.
(Modified from Osborn 1907)

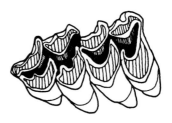

Figure 3.8 Selenodont upper molars of *Capreolus capreolus,* the roe deer. Note the crescent-shaped patterns on the occlusal surface of each molar.
(Modified from Gromova 1962)

include simplification of the occlusal pattern, fusion of cusps, or infoldings along the margins of the teeth. Many of these modifications are described and illustrated in Chapter 23. Taxonomic studies of certain rodents, particularly members of the family Muridae, require a detailed knowledge of cusp and crown morphology, but this knowledge is not required to use the keys in this manual. Further details on the crown elements that may be found in the cheek teeth of murid rodents, where this information is most important, can be found in publications by Hershkovitz (1962) and Reig (1977).

Cheek teeth modified for a carnivorous diet generally are reduced secondarily to two major cusps. The upper and lower teeth, working together, provide a scissors action for shearing flesh (see Fig. 19.2). *Note!* The term *carnassial* has two meanings: **carnassial** or **secodont dentition** is the general type of dentition found in mammals whose principal diet is flesh. The **carnassial pair** or **carnassial teeth,** found only in the order Carnivora, are the two teeth on each side that do most of the shearing. In living carnivores, these teeth are the fourth upper premolar and the first lower molar in the adult dentition and the third upper and fourth lower premolars in the milk dentition.

Many bats have modified tribosphenic teeth (dilambdodont or quadritubercular) in which the three cusps elongate into sharp crescent-shaped cristas (see Fig.14.19), sometimes termed **commissures.** These cristas are useful in cutting and crushing the hard chitinous exoskeletons of insects. Similar specializations are present in the teeth of many Insectivora.

Many fish-eating mammals such as sea lions and porpoises have cheek teeth reduced to a series of sharp unicuspids for holding their slippery food (see Figs. 19.8 and 20.26). The sea otter, *Enhydra lutris,* that feeds primarily on mollusks and echinoderms, and the walrus, *Odobenus rosmarus,* that feeds on mollusks, both have flat brachyodont cheek teeth that crush their food (see Figs. 19.4 and 19.5). A highly specialized cheek tooth is found in the Antarctic crab-eating seal, *Lobodon carcinophagus.* This species feeds upon krill, small planktonic shrimp-like crustaceans, in the cold Antarctic waters. Each cheek tooth of *L. carcinophagus* has three to five long, curving cusps in a straight line reminiscent of the teeth of members of the †Triconodonta. These teeth collectively form a sieve (Fig. 3.9) for straining krill from the ocean.

Many diverse groups of mammals have adapted to diets in which teeth serve little or no major function. In many of these—including bats (e.g., *Leptonycteris nivalis*) and marsupials (e.g., *Tarsipes spenseral*) that feed upon pollen and/or nectar, sloths that feed upon soft buds (see Fig. 17.6), and armadillos (see Fig. 17.5) and aardwolves (see Fig. 19.12) that feed upon soft-bodied insects—the entire dentition is degenerate, and frequently the teeth are reduced to a series of simple flat-topped or unicuspid pegs. The echidnas (see Fig. 10.4), anteaters (see Fig. 17.4), and pangolins (see Fig. 18.2), all of which feed on large numbers of small insects, and the platypus (see Fig. 10.3), which feeds on aquatic invertebrates, tadpoles, and small fish (Nowak 1999), are **edentulate** (i.e., lack teeth entirely). The baleen whales are also edentulate and instead use **baleen** plates to filter krill from the ocean water (see Figs. 20.3 and 20.4).

3-L   Examine a *Canis* skull and identify the incisors, canines, premolars, and molars. For what function is each type of tooth modified?

3-M   Examine the molars of a primate, hog, or both. Locate and identify the four major cusps on each molar. Which, if any, smaller cusps are present? Compare the occlusion of these teeth with those of an opossum.

3-N   Examine the dentition of as many of the following mammals as possible. Identify the kinds of teeth in each species and ascertain the probable diet associated with each dentition.

Mammal	Order	Genus	Probable Diet Based on Dentition
Shrew	Insectivora		
Vampire bat	Chiroptera	*Desmodus*	
Nectar-feeding bat	Chiroptera		
Vespertilionid bat	Chiroptera		
Anteater	Xenarthra		
Armadillo	Xenarthra		
Sloth	Xenarthra		
Rabbit	Lagomorpha		
Rat	Rodentia		
Deer mouse	Rodentia		
Vole	Rodentia		
Sea otter	Carnivora	*Enhydra*	
Mink	Carnivora		
Cat	Carnivora		
Crab-eating seal	Carnivora	*Lobodon* (see Fig 3.9)	
Sea lion	Carnivora		
Walrus	Carnivora	*Odobenus*	
Porpoise	Cetacea		
Aardvark	Tubulidentata	*Orycteropus*	
Elephant	Proboscidea		
Horse	Perissodactyla	*Equus*	
Deer	Artiodactyla		
Cow	Artiodactyla	*Bos*	

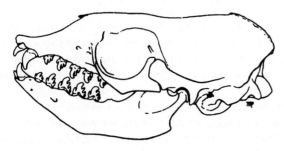

Figure 3.9 Skull of a crab-eating seal, *Lobodon carcinophagus,* showing teeth that resemble those of the †triconodonts.
(Hatt 1946)

DENTAL FORMULAS

The **dental formula** is a shorthand method used by mammalogists to indicate the numbers of each kind of tooth in a particular mammal. The complete dental formula for *Canis,* showing the number of each kind of tooth in each half of each jaw, is written:

incisors $\frac{3\text{-}3}{3\text{-}3}$ canines $\frac{1\text{-}1}{1\text{-}1}$ premolars $\frac{4\text{-}4}{4\text{-}4}$ molars $\frac{2\text{-}2}{3\text{-}3}$ = $\frac{20}{22}$ = 42

The numbers above the line represent the teeth in the upper jaw, and those below the line represent those in the lower jaw. Because the two halves of each jaw normally have identical numbers of teeth, the dental formula is usually written to show only one side. The total number of teeth is found by multiplying by 2. Thus, the above formula may be abbreviated as:

$$I\frac{3}{3} C\frac{1}{1} P\frac{4}{4} M\frac{2}{3} = 42$$

or, because the four kinds of teeth are always recorded in the order shown above, the formula can be further abbreviated by deleting the initials for the tooth types:

$$\frac{3}{3} \frac{1}{1} \frac{4}{4} \frac{2}{3} = 42$$

If a particular type of tooth is not represented in a species, a zero is used. Thus, the dental formula for a Norway rat, *Rattus norvegicus,* is:

$$I\frac{1}{1} C\frac{0}{0} P\frac{0}{0} M\frac{3}{3} = 16$$

$$\text{or } \frac{1}{1} \frac{0}{0} \frac{0}{0} \frac{3}{3} = 16$$

Primitive Dental Formulas

Placental mammals have a maximum of three incisors and one canine per quadrant, and usually have no more than four premolars and three molars per quadrant. Marsupials have a maximum of five upper and four lower incisors on each side and one canine per quadrant. They usually have no more than three premolars and four molars per quadrant. These tooth numbers are considered to represent the ancestral condition. The following dental formulas represent the primitive tooth numbers for marsupials and placentals:

Marsupial   I 5/4 C 1/1 P 3/3 M 4/4 = 50

Placental   I 3/3 C 1/1 P 4/4 M 3/3 = 44

Although reduction in tooth number from the primitive formula is common, an increase in this number is rare. Among marsupials, only the banded anteater, *Myrmecobius fasciatus* (see Fig. 11.11), has more than 50 teeth. Among placental mammals, only the giant armadillo, *Priodontes maximus,* the African bat-eared fox, *Otocyon megalotis,* and most of the toothed whales (e.g., see Figs. 20.24 and 20.30) have more than 44 teeth. *Priodontes* has up to 100 unicuspid, peglike teeth. *Otocyon* has additional molars to make a total of 46 teeth (occasionally 48 to 50). Some toothed whales have up to 260 unicuspid, essentially homodont teeth. In the manatees, Trichechidae, up to 80 teeth develop during the life of the animal, but usually only 24 (occasionally up to 32) are visible at any one time (see Fig. 25.7).

3-O   Examine the teeth in a hog (*Sus,* a placental) skull and write the dental formula. How do you know which cheek teeth are premolars and which are molars?

3-P   Examine the teeth of an opossum (*Didelphis,* a marsupial) skull and write the dental formula. How do you know which cheek teeth are premolars and which are molars?

3-Q   Examine the teeth in a cat (*Felis* or *Lynx*) skull and write the dental formula.

How do you know which cheek teeth are premolars and which are molars?

3-R   Examine the dentition of a porpoise. How many incisors are present? What is the total number of teeth? Can you identify the canines? Why or why not?

Grouped Dental Formulas

Because premolars and molars are frequently impossible to distinguish in the skull of an adult, these two kinds of teeth are sometimes grouped in writing a dental formula. Such a grouped formula for the common harbor seal, *Phoca vitulina,* is:

I 3/2 C 1/1 P + M 5/5 = 34

A similarly grouped formula for a typical nine-banded armadillo, *Dasypus novemcinctus,* is:

I 0/0 C 0/0 P + M 7/7 = 28

However, because the nine-banded armadillo is a placental mammal with a maximum potential of four premolars and three molars, and because seven postcanine teeth are present, it is possible to write a standard dental formula:

I 0/0 C 0/0 P 4/4 M 3/3 = 28

In shrews and some other groups, the posterior incisors, the canines, and the anterior premolars may all be simple, single-cusped teeth that are difficult to distinguish from one another. These are collectively termed the "unicuspids," and a particular tooth may be referred to as, for instance, the third upper unicuspid. Some authors write a standard dental formula for these animals, but Choate (1975) recommended using a formula that identifies the known incisors and premolars and lumps the remaining incisors, canines, and premolars as unicuspids. His formula for *Cryptotis* is "first incisor, 1/1; unicuspids, 4/1; fourth premolar, 1/1; molars, 3/3."

3-S   Examine a beaver (Castor) skull. There is a total of 20 teeth, and only the last premolar is present. Write a dental formula combining the cheek teeth.

_____

Write a standard dental formula.

_____

3-T   Write a dental formula grouping cheek teeth for the porpoise examined in 3-S above.

3-U   Examine a shrew skull. How many unicuspids are present? (You will need a binocular microscope or a hand lens to see them clearly.) Write a dental formula grouping the unicuspids.

Formulas Identifying Missing Teeth

Occasionally a dental formula is written to indicate exactly which teeth have been lost. For *Canis,* this type of formula is:

$$I\frac{123}{123} \; C\frac{1}{1} \; P\frac{1234}{1234} \; M\frac{120}{123} = 42$$

Here each number represents a particular tooth. The zero indicates that the last upper molar is absent. A similar dental formula for the Norway rat is:

$$I\frac{100}{100} \; C\frac{0}{0} \; P\frac{0000}{0000} \; M\frac{123}{123} = 16$$

The last incisors and last molars are usually lost before the first of either of these kinds of teeth, whereas the first premolars are usually lost before the last premolars. Thus, if a mammal has only one incisor or one molar, it is usually I 100/100 or M 100/100 rather than I 003/003 or I 020/020. If a mammal has only two premolars, these are usually P 0034/0034 rather than P 1200/1200 or P 0230/0230. However, this is a usual trend and not a rule! Teeth are sometimes lost from the opposite end or from the middle of a series.

3-V   Write a dental formula for a cat (*Felis* or *Lynx*) that shows which teeth are absent. (Remember that the carnassial pair in adults is always the last upper premolar and the first lower molar.)

Notation for Single Teeth

P^2 is a shorthand method of saying "the second upper premolar," and M_3 is a shorthand method of saying "the third lower molar." This use of the tooth-type initial combined with a superscript or subscript is in common usage in scientific literature and is used throughout this manual. Some authors use a capital letter to represent a tooth in the upper jaw and a lowercase letter to represent a tooth in the lower jaw. By this system, P2 is the second upper premolar, and m3 is third lower molar. A lowercase letter may also be used to refer to a tooth in the deciduous dentition, but the letter "d" usually accompanies such a designation. For example, the fourth upper deciduous premolar could be termed p^4, dp^4, or pd^4. *Note!* If a mammal has a dental formula of:

I 123/123, C 1/1, P 0234/0234, M 120/123 = 38

and an author refers to P^3, he/she may be referring to the third upper premolar present in the specimen, or he/she may be referring to the third upper premolar that is potentially present based upon the primitive dental formula. In this manual, a shorthand note such as P^3 will always refer only to the teeth actually present in the particular species.

Variation in Formulas

The dental formula is generally considered to be a characteristic of a genus or a higher taxon, but there are certain genera in which there is variation in dental formulas. For instance, the gray squirrel, *Sciurus carolinensis,* has a tiny upper premolar, giving a dental formula of:

I 1/1, C 0/0, P 2/1, M 3/3 = 22

whereas the closely related fox squirrel, *Sciurus niger,* lacks this tiny premolar, giving a dental formula of:

I 1/1, C 0/0, P 1/1, M 3/3 = 20

There is also variation of dental formulas within certain species. This variation is frequently associated with secondary sexual differences, as when canines are developed in the male but are absent in the female or with age as with the late eruption of the last molars (wisdom teeth) in humans. In certain other species (usually, though not always, those with a degenerate homodont dentition), the number of teeth may vary among individuals without regard to sex or age.

3-W   Compare the dentition of a gray squirrel with that of a fox squirrel.

3-X   Write and compare the dental formulas of a male (ungelded) and female horse.
    Ungelded
        male horse: I   C   P   M   =
    Female horse: I   C   P   M   =

3-Y   Compare the tooth counts in a series of armadillo and/or black bear skulls.

CHAPTER 4

THE INTEGUMENT

The skin of mammals and other vertebrates is composed of two layers, the dermis and epidermis (Fig. 4.1). In the **epidermis,** only the cells in the two lowest layers, the deepest **stratum basale** (or **stratum germinativum**) and the cells in the layer immediately above, the **stratum spinosum,** are living and dividing (Sokolov 1982). Progressing toward the surface, successive layers of epithelial cells become more flattened and cornified, or keratinized. This surface layer of dead cells, the **stratum corneum,** receives the brunt of environmental wear and tear and continually flakes off the skin and is replaced by growth from below. Thickened portions of this keratinized epithelium form the **tori** (or **pads**) on the feet (Fig. 4.2A) of most mammals and form the **friction ridges** (Fig. 4.2B) on the digits, palms, soles, and naked prehensile-tail pads of primates. A fingerprint is an impression of such ridges. **Calluses,** which form where the skin is subjected to constant friction, are further thickenings of this cornified layer. Epidermal scales, hair, horn, and claws are all modifications of cornified epithelial cells.

The **dermis** lies below the epidermis and is a thick layer of fibrous connective tissue with associated skin glands, specialized muscles, nerves, and sensory structures. Most blood vessels and all specialized sensory receptors associated with the skin are in the dermal layer. Some reptiles (turtles, crocodilians, and many lizards), and some mammals (armadillos) also have bone present in the dermis for armor. Below the dermis, in the deepest layer of the integument, lies the **hypodermis,** which consists of adipose or fat tissue.

4-A Examine a slide of mammal skin under the compound microscope. Differentiate between epidermis, dermis, and hypodermis. Note the change in shape and degree of cornification of epithelial cells from base to surface. In the dermis and hypodermis, differentiate between connective, vascular, muscular, nervous, and adipose tissues.

4-B Examine the tip of your finger under a binocular microscope. Compare the friction ridges of humans with the pads on the foot of a rodent, a carnivore, or both, and with the ischial callosities on the buttocks of a nonhuman primate.

SCALES

Dermal bone is true bone formed within the dermal layer of the integument. This bone formed the protective shells of ancient jawless, fishlike vertebrates, and in modern animals occurs in the scales of elasmobranch fishes (e.g., sharks), the shells of turtles, and the skins of many lizards and all crocodilians. Dermally derived bones contribute portions of the skulls of bony fishes and tetrapods, including mammals (e.g., the frontals and parietals are dermal bones), and teeth are believed to be derivatives of dermal denticles. In many "lower" vertebrates, dermal bones compose much of the pectoral girdle, but in mammals, only the interclavicle (found only in Monotremata) and the clavicle are of dermal origin. Except for these elements, dermal bone is absent in the integument of all living mammals except the armadillos.

Epidermal scales are modifications of the cornified epithelium and are never bony, although they may be intimately associated with underlying dermal bone. Reptiles are usually completely covered with epidermal

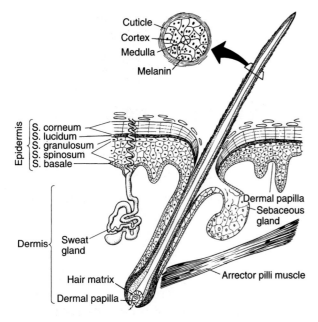

Figure 4.1 Sectional view of skin showing hair and various structures in epidermis and dermis.
(After Kardong 1998:211)

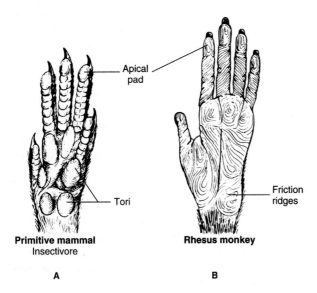

Figure 4.2 (A), Tori on the feet of an insectivore; (B), friction ridges on a primate.
(After Kent and Miller 1997:114)

scales. Birds have epidermal scales on their legs, and their bodies are covered with modified epidermal scales termed feathers. Various species of mammals retain epidermal scales on the tail, feet, or both (e.g. Rodentia: Anomaluridae, Castoridae). However, only two groups of mammals, the armadillos (order Xenarthra) and pangolins (order Pholidota), have a major portion of the body covered with scales of epidermal derivation.

The armadillos (Xenarthra, Dasypodidae) bear bony dermal scales embedded in the skin over the top of the head, the back, and (usually) the sides of the body. Also,

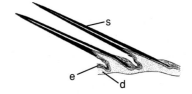

Figure 4.3 Diagram of a section through two scales of a pangolin showing the scales (s), epidermis (e), and dermis (d).
(After Feldhamer et al. 1999:253)

most species have rings of dermal scales encircling the tail. Overlying these dermal scales are thin horny epidermal scales that usually abut one another but do not overlap (except on the movable overlapping bands).

The pangolins (Pholidota) lack dermal scales but have most of the body covered with large, leaf-shaped, imbricate (overlapping) keratinized scales (Figs. 4.3, 18.1 and 18.2) of epidermal origin.

4-C Examine a piece of armadillo shell and an armadillo skin. Note the arrangement of dermal bone and epidermal scales. How do the sizes and shapes of the units of the two layers compare? How are the sutures of the bony shell positioned relative to the lines of contact between the epidermal scales? How is the shell constructed to allow for flexibility?

4-D Examine a pangolin and note the arrangement of scales over the body. What is the function of the scales in pangolins?

4-E Examine the scaly tail of a beaver or rat and note the placement of hairs in relation to the placement of scales. Compare this to the arrangement of hair follicles on the inner surface of a pigskin and the back of your hand on the "little" finger (digit V) side. How do these arrangements support the belief that hair is not a derivative of epidermal scales but is of separate origin?

HAIR ANATOMY

Though hairlike structures may be found on birds, insects, and even plants, true epidermal hair is unique to mammals. In most mammals, it is conspicuous; in others, it is sparse (the naked mole rat, *Heterocephalus glaber;* mysticete whales; and freshwater dolphins) or absent in most adult odontocete whales (in odontocete whales, vibrissae may be present in the embryo but disappear later in development) (Sokolov 1982:289).

A hair first develops as an epidermal thickening that pushes into the dermis to form a **follicle** (see Fig. 4.1). Directly under the follicle is a tiny invagination of blood vessels and connective tissue, the **papilla.** The papilla is richly supplied with blood vessels that nourish the growing hair. The epidermal cells at the base of the follicle proliferate to form a column of dead, keratinized cells (the **hair**), which pushes out through the neck of the follicle. An outgrowth of epidermal cells from the side of the follicle forms a **sebaceous** (oil) **gland,** the secretions of which keep the hair from becoming brittle, and render a degree of water-proofing (see the "Integumentary Glands" section).

The hair follicle and shaft are oriented obliquely in the skin. Associated with each follicle are bundles of smooth muscle fibers of the **arrector pili muscle** (Sokolov 1982:34). Contraction of this muscle erects the hair. In humans, this brings about "goosebumps." In other mammals, it raises the hair, thereby increasing insulating properties or serving as a threat. Echidnas, marsupials, colugos, pinnipeds, and sirenians lack these muscles (Sokolov 1982).

4-F   On a slide of mammal skin, locate a follicle. Identify the papilla, hair, sebaceous gland, and arrector pili muscle.

4-G   The arrector pili muscle raises the hair to increase insulation or to make the animal seem larger in threat postures. In humans, "goosebumps" are usually produced by cold or fear. Is this consistent with what occurs in other mammals?

Each hair normally consists of a central core and two well-defined outer layers (see Fig. 4.1). The central **medulla,** present in all but the smallest hairs, may be continuous or may be interrupted by regularly or irregularly spaced air cavities. In some mammals, the medulla is absent, leaving a continuous central air column. The **cortex** immediately surrounds the medulla and makes up the bulk of a hair. The **cuticle** is a thin outer layer of **cuticular scales** covering the cortex. The scales may form a relatively smooth surface or may overlap in various distinctive patterns (Fig. 4.4). Pigment granules may be located in the medulla, the cortex, or both but are never found in the cuticle, although colored substances produced by skin glands may coat or adhere in flakes to the cuticle. In cross section, a hair may be circular or flattened. Flattened hairs are often curly, and cylindrical hairs are usually straight.

Certain combinations of hair structure characteristics are distinctive for various mammalian groups (Bruner and Coman 1974; Hess et al. 1985). Short (1978) used SEM microscopy to examine hair samples from 7 orders, 18 families, and 48 genera of mammals and found that cuticular scale patterns were less useful than color, size, and shape of the medulla in identifying hair to a particular taxonomic group. Keys can be constructed to aid in identifying mammals to species (e.g., Haffner and Ziswiller 1989, for species of vespertilionid bats) or to

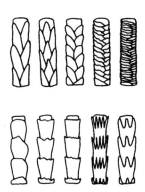

Figure 4.4 The basic types of cuticular scales found on mammalian hair.
(Nason 1948)

identify hairs found in dens, fecal material, owl pellets, etc. (e.g., Benedict 1957; Day 1966; Mayer 1952; Miles 1965). Hair size, the character of the medulla, the amount and distribution of pigments, and the type of cuticular scale patterns are all considered in writing these keys.

4-H   Remove a hair from your head and from at least three other species of mammals. Mount each hair in a drop of water on a clean microscope slide and cover with a coverslip. With a compound microscope, compare the structure of the medulla and the location and relative abundance of pigment granules in each of the hairs.

4-I   Place a sample of each of the kinds of hair used above on a separate microscope slide. Cover the hairs with plastic coverslips and then pass each slide through a flame until the coverslip has melted somewhat. Allow it to cool before removing hairs. Remove the hairs from the impressions made, mount the coverslips on microscope slides, and examine the impressions of the cuticular scale patterns under the compound microscope. See Chapter 34 for additional references on making slide preparations of hair.

CLASSIFICATION OF HAIR

On the basis of growth, hair may be classified as either **definitive** or **angora.** Hair with definitive growth reaches a certain length characteristic for the species and body location, and then growth ceases. These hairs are shed and replaced periodically, as with human eyelashes. Angora hair grows almost continuously and reaches a considerable length before being shed. Some angora hairs, such as a domestic horse's mane, are never shed but continue to grow through the life of the animal. Growth patterns and hair morphology and functions are combined to classify hair into the various types recognized on mammals. Although Sokolov (1982:33–34) divided hair into five basic types, most reference sources divide hair into three main types: vibrissae, guard hairs, and underhairs. The latter two types are often further classified into subgroups.

Vibrissae

Vibrissae are usually long (short in colugos, cetaceans, and, with environmental wear, walruses) stiff hairs with well-innervated bases surrounded by many blood vessels. They primarily serve as tactile receptors. The best known vibrissae are the "whiskers" on a mammal's face (Fig. 4.5), but vibrissae may also be located on the ankles and elsewhere on the body.

Guard Hairs

Guard hairs (or **overhairs**) are the most conspicuous hairs on most mammals. They serve primarily for protection. The three major types of guard hair recognized are spines, bristles, and awns, although intermediates between these conditions are known. **Spines** are greatly enlarged, stiff guard hairs with definitive growth. They serve primarily as defense from predators. The New World porcupines, Erethizontidae, and the tenrec genus *Hemicentetes* (Eisenberg and Gould 1970:49), have **barbs** on the tips of the spines (Fig. 4.6). Once imbedded in a predator's skin, such a spine cannot be easily removed, and the victim's muscular action will actually cause the spine to embed more deeply. Spines of other groups of mammals (e.g., Old World porcupines, Hystricidae; hedgehogs, Erinaceidae; and echidnas, Tachyglossidae) do not have barbs. **Bristles** are long, firm hairs with angora growth (e.g., horse and lion manes). **Awns** are hairs with definitive growth that have a firm, expanded distal portion and a smaller, weaker base. Awns are the most noticeable hairs on most mammals.

Underhairs

Underhairs function primarily for insulation. Three major types are generally recognized, but intermediate types do occur. **Wool** is angora (that is, evergrowing) underhair and is usually long, soft, and curly. **Fur** is fine, relatively short hair with definitive growth that grows densely over the body. **Velli** (or down, fuzz) are very fine, short hairs that are velvety in appearance. The embryonic hair or **lanugo** of humans is a type of vellus.

4-J   Note the location of vibrissae on a variety of mammals. Explain these locations considering the habits and/or habitat of each species. Observe the action of the vibrissae in a live mammal.

4-K   Under a binocluar microscope compare the spines of a New World porcupine with those of an Old World porcupine, hedgehog, or other mammal with spines. How do they differ? Under what circumstances may a porcupine be said to "throw its quills"?

4-L   Examine a variety of mammals. What types of hair are found on each? What is the function of each type of hair?

Hair Replacement

Angora hairs grow continuously for the life of the mammal and are continuously worn away at the tips. Most hair, however, is shed and replaced periodically in a process termed **molting.**

Molts may occur continuously with at least some hairs being replaced at all times, (e.g., human eyelashes). But most mammals, particularly those living in temperate or polar climates, have an **annual molt** during which all hairs are replaced in a short period of time. Such molts usually begin in a specific region or regions of the body and spread in orderly sequential fashion until all hairs have been replaced. The **molt pattern** varies with the species and occasionally with the age of an individual (Fig. 4.7). Some mammals have **seasonal molts** with more than one molt per year. This is most conspicuous in species that change from a brown summer **pelage** (the hair covering of a mammal) to a white winter pelage. In northern populations of the long-tailed weasel, *Mustela frenata,* for example, the

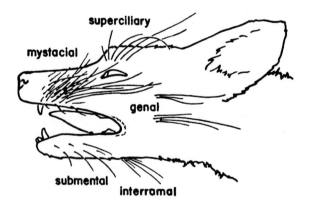

Figure 4.5 Locations of vibrissae on the head of a gray fox (*Urocyon cinereoargenteus*).
(Hildebrand 1952:422)

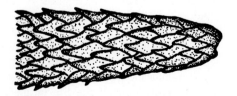

Figure 4.6 Enlarged view of a quill tip of a New World porcupine (*Erethizon dorsatum*). Note presence of barbs.
(Shadle and Chedley 1949:172)

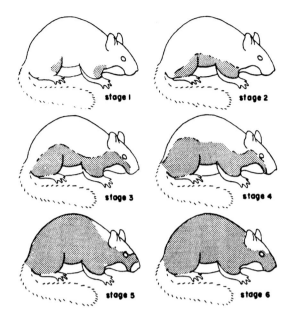

Figure 4.7 Postjuvenal molt pattern in the wood rat, *Neotoma cinerea*. Stippled areas represent appearance of new pelage as the molt progresses.
(Egoscue 1962:335)

spring molt begins along the dorsal surface, and brown hairs replace the white winter hairs over the dorsal parts of the animal. In the fall, the molt pattern is reversed with replacement of brown hairs by white ones progressing dorsally from the perpetually white venter. In the southern parts of the range of this species, the molts occur but do not result in a color change, and the weasel has a brown dorsal pelage throughout the year. The change from winter to summer pelage is influenced by hormones, photoperiod, and temperature.

A distinctly **juvenile pelage** is recognizable in many mammals. This is usually grayer and duller than typical **adult pelage.** The juvenile pelage can also be variously striped or spotted, whereas the adult is more or less uniformly colored. In some groups, a distinctive **subadult pelage** occurs between the juvenile and the adult pelages.

4-M Examine series of skins of several species collected while the animals were in the process of molting. How does the molt pattern differ among the species examined?

4-N Compare winter and summer pelts of northern populations of *Mustela frenata, Lepus americanus,* or some other species that has seasonal alteration of colors. Compare the pelages of specimens collected in spring and fall.

4-O Examine a large series of deer mice (*Peromyscus*) and identify juvenile individuals on the basis of pelage. Compare pelages of juvenile and adult specimens of other species of mammals.

COLOR

The color of an individual hair is affected by numerous factors. Differences in kind, amount, and distribution of pigment granules in a hair produce different effects. In addition, hair surface texture, the thickness of the hair, and the amount of air space in the medulla can all alter the way in which light is reflected by the hair and, therefore, change the apparent color.

The overall coloration of a mammal is determined by the coloring of individual hairs and the relationships among these hairs in the pelage. An animal may, therefore, be red- and brown-speckled because each hair has red and brown color bands or because the pelage has a mixture of red hairs and brown hairs.

There are two main types of pigment in mammalian hair. **Eumelanin** in various concentrations produces blacks and browns. **Pheomelanin** in various concentrations produces reds and yellows. White is the complete lack of pigment. Each hair usually has a series of color bands. An **agouti** hair has a black tip followed by successive bands of pheomelanin and eumelanin.

Some unmolted pelage may show obvious changes of color over time. This can be caused by simple wear or by bleaching owing to sunlight or powerful artificial light. Old study skins may show **foxing,** chemical changes in pigments that lead toward dull reddish brown. Some opossum study skins show various marked color changes over time, apparently owing to breakdown of derivatives of the amino acid tryptophan.

4-P Examine individual hairs from a cottontail rabbit (*Sylvilagus*) or other animal with agouti hair, and note the sequence of color bands.

4-Q Examine the color banding on hairs of at least 10 species of mammals that could be described as brown. How do these compare in number, width, sequence, and hue of the bands? Do all of the hairs in a given body region exhibit the same banding?

4-R Examine a series of skins or study skins from the same general area (e.g., same county) but collected over a number of years (20+). Assess differences that may be due to age, sex, or season. Then, within samples of these subcategories, try to determine if any of the specimens show evidence of "foxing." Which years of collection show the most evidence of foxing in the skins? Which years the least? Would foxing be a factor that a taxonomist must consider in assessing geographic and interspecific differences in mammals?

Mammalian hair coloration and skin coloration serve several functions. Among these are concealment, communication, and protection from ultraviolet (UV) radiation (see Timm and Kermott 1982 for discussion and references). Many predators and many prey have **concealing** or **cryptic coloration** that allows them to blend with their habitat to avoid detection. The primitive agouti pattern provides a coloration that is usually very similar to the color of the earth and dead vegetation. Mammals such as the tigers and okapis are strikingly marked with sharply defined light and dark colors. In these animals' normal habitats, however, these examples of **disruptive coloration** obscure the body contours and cause the animal to blend into the patterns of light and shadow caused by sunlight penetrating the vegetation. Facial stripes, present in many mammals, are also disruptive coloration, usually intended to conceal the eye. Most mammals have a ventral surface that is paler than the dorsum—an arrangement termed **countershading.** It conceals the usual pattern of contour-revealing highlights and shadow, thus making the animal more difficult to distinguish.

Coloration is also important for intra- and interspecific communication. The color pattern typical of a certain species may serve to elicit appropriate intraspecific behavior. In several species, display of distinctive color regions, such as the red genital area of some baboons, is an important part of the courtship ritual. Conspicuously colored **flags,** such as the white underside of the tail in many rabbits and certain deer, may alert others of a group to dangerous conditions. **Warning coloration** is present in some species that have special means of defense. The striking black and white patterns of skunks and the convergently patterned (Fig. 4.8) African zorilla, *Ictony striatus,* are examples.

4-S Examine skins of all of the mammal species known from your state or province. Which exhibit concealing coloration? Do any areas of the body have disruptive coloration patterns? Which exhibit countershading? Which are equipped with white flags or other color signals? Do any exhibit warning coloration?

4-T Do any mammals other than skunks and zorillas exhibit warning coloration? Examine illustrations in Nowak (1999), Burt and Grossenheider (1964), van den Brink (1967), Dorst and Dandelot (1970), etc.

4-U Which mammals other than the tiger and okapi mentioned above exhibit a marked degree of disruptive coloration? Check the same sources listed in 4-T above.

Prolonged exposure to UV radiation from the sun can cause burns and in other ways be deleterious to an animal's health. The melanin pigments in hair and skin are known to filter this harmful UV radiation from sunlight (see references in Timm and Kermott 1982).

Terrestrial vertebrates inhabiting arid regions are usually paler in coloration than closely related forms inhabiting more humid regions. This phenomenon is known as **Gloger's Rule.** Although the rule links color and aridity, the influencing factor is most likely background color of the habitat. As the habitat becomes more arid, vegetation becomes more sparse, and soil color becomes generally lighter. Thus light-colored pelages are adaptations for concealment. In desert areas where there are large expanses of black volcanic rock, the small mammals are usually black like the rock rather than pale as Gloger's Rule would dictate.

4-V Examine a number of rabbits (*Sylvilagus*) or woodrats (*Neotoma*) including specimens collected in the eastern deciduous forests, the Great Plains, and the southwestern deserts. Do differences in color correlate with annual precipitation? If possible, compare these with specimens collected from an arid lava field.

Albinism, the complete lack of integumentary pigments, is a genetic trait that has been observed in many species of mammals. Because albinos are not well adapted for camouflage, communication, or UV radiation protection, they generally do not become established as an appreciable portion of a wild population. But geographically localized populations containing numerous albinos have become established in several species. True **albinos,** which lack pigments in the irises and thus are pink-eyed (due to the red blood visible through the transparent, colorless tissues of the eye), have been propagated in captivity, and albino rats and rabbits, in particular, are common domestic animals. **Piebald** individuals, having patches of white on the body, are not as radically different from normal and are, therefore, better able to survive in the wild. The white patches are often caused by somatic mutations and thus cannot be passed on to the offspring.

Melanism, a tendency toward completely black coloration, is also a genetic trait sometimes encountered in wild populations. Although melanistic individuals may

Figure 4.8 Warning coloration in the zorilla, *Ictonyx striatus,* an African mustelid that has a pelage similar to that found in North American striped skunks of the genus *Mephitis* and *Spilogale.*
(After Feldhamer et al. 1999:264)

differ from the normal members of their species with respect to concealment and communication, they do have ample protection against UV radiation. In some species, melanistic individuals are common. Examples of these are the silver and cross varieties of the red fox, *Vulpes vulpes;* black fox squirrels, *Sciurus niger;* black gray squirrels, *Sciurus carolinensis;* and black leopards ("black panthers"), *Panthera pardus.*

4-W   Examine a live albino animal (e.g., laboratory mouse, *Mus musculus*). Note color of hair, skin, and eyes. Compare the color of an individual hair with one from a normally colored individual of the same species. Is the albino really white?

4-X   Examine mammal skins exhibiting unpigmented areas. (Such irregular white patches are common in the eastern mole, *Scalopus aquaticus,* and the Mexican freetailed bat, *Tadarida brasiliensis.*)

4-Y   Examine a number of squirrel or fox skins showing the typical color for the species and others showing a variety of melanistic shades. Compare several hairs from the melanistic specimens with one from a normally colored individual of the same species. Does each hair have fewer or smaller red bands? Is each hair completely black? Are all the hairs black?

A few species of mammals have coloration that does not result from pigmentation. Sloths, for example, have coarse overhair with numerous external grooves. Algae grow in these grooves and often give the animals a greenish color that allows them to better blend in with their forest environment. Some mammals, (e.g., golden moles, Chrysochloridae) show iridescence of the pelage, caused by structural characteristics rather than by pigment.

4-Z   Prepare a temporary wet mount of a sloth hair and examine under the microscope. Are grooves evident? Are algae visible?

4-AA Examine a specimen of a mole of the genus *Talpa* or *Scalopus* (Talpidae) and a golden mole (Chrysochloridae). Alternately move each of these specimens in different directions under an intense source of light (sunlight or illuminator). Do the colors of the specimens change under the light source as they are moved about? Which specimen exhibits iridescence? Why does one of the specimens not show dramatic color changes as it is moved under the light source?

INTEGUMENTARY GLANDS

There are two basic types of glands in the skin of mammals, sweat glands and sebaceous glands. All other integumentary glands are considered to be—or hypothesized to be—modifications of one of these two types.

Sweat glands are found only in mammals, although several kinds of mammals—echidnas, megachiropteran bats, sirenians, elephants, lagomorphs, and rodents—have none. Sweat glands consist of two basic types, sudoriferous and eccrine. **Sudoriferous (=apocrine sweat) glands** are highly coiled and empty their secretions into the cavity of the hair follicle (Sokolov 1982). Sudoriferous glands produce the odorous component of perspiration. In humans, apocrine sweat glands are concentrated in the axillae, navel, anogenital areas, nipples, and ears. **Eccrine sweat glands** are also highly coiled (see Fig. 4.1) but open directly onto the skin surface, independent of the hair follicles. Eccrine sweat glands are responsible for most of the fluid portion of sweat and also excrete some metabolic wastes and salts. Evaporation of sweat from the surface of the skin is a cooling mechanism for the body, and perspiration can also improve tactile sensitivity and grip when secreted onto the palms and soles. The wax-producing glands of the external auditory meatus, the **glands of Moll,** are modified apocrine sweat glands that protect the tympanic membrane from becoming dry and losing flexibility.

Sebaceous glands (see Fig. 4.1), which are usually associated with hair follicles, serve primarily to keep the hair from becoming too dry and brittle. These glands are absent in elephants and sirenians (Sokolov 1982). In many mammals, they are also important in waterproofing the pelage. Sebaceous secretions in the hair of otters and fur seals, for instance, keep cold water from penetrating the fur and contacting the skin, thereby retarding heat loss. Some sebaceous glands, as in the upper lip, nose, and upper cheek areas of humans, open directly onto the skin surface rather than into hair follicles. Mammals' **Meibomian glands,** located on the eyelid, and **Hartner's glands,** located behind the eyeball, are modified sebaceous glands that lubricate the eyelid and nictitating membrane, respectively.

4-BB Examine prepared slides of mammal skin. Locate and compare the structures of a sebaceous gland and sweat gland. Where is the sebaceous gland situated with respect to the hair follicle? What portion of the skin (dermis or epidermis) houses the bulk of the eccrine sweat gland?

Scent glands (Fig. 4.9) are complex odor-producing glands of variable composition. They may be predominately sebaceous or sudoriferous or divided about equally between these two types of glands. Although much remains to be learned about the function and significance of scent glands and their secretions (Doty 1976; Eisenberg and Kleiman 1972; Ralls 1971; Sokolov 1982), the functions can be divided into three general categories: defense, marking of territory, and other social interactions.

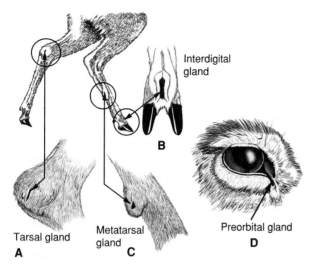

Figure 4.9 Scent glands in white-tailed deer, *Odocoileus virginianus*. (A) Tarsal gland; (B) interdigital gland; (C) metatarsal gland; (D) preorbital gland.
(After Feldhamer et al. 1999:107)

In defense, skunks (Mephitinae) will discharge a mercaptan-based **musk** from anal glands. Wolverines (*Gulo*) and peccaries (*Tayassu*) are examples of other mammals that emit a musk when they are in danger.

Many mammals (e.g., canids and felids) establish scent trails in their territories to guide their travels. This process of labeling an area with scent is one kind of **marking.** An area marked may even be recognizable as a visible signpost (see Chapter 28). The interdigital glands of deer (Cervidae [e.g., *Odocoileus*]) and the anal musk glands of American badgers (*Taxidea*) apparently provide the scent for trail marking in these mammals. In rodents and some primates, urine and preputial gland secretions are important for marking territories.

A **pheromone** is an odor or musk (or odorless secretion) that has a behavioral or physiological effect on another individual or individuals of the same species. For example, pheromones in the urine of some rodents are known to influence the onset of estrus and other phenomena of the reproductive cycle. An **alarm pheromone** is released when an animal is in danger. The secretions of the metatarsal glands of deer are thought to be alarm pheromones.

The secretions of scent glands are sometimes utilized by humans in commercial enterprises. Some scent glands of mammals are removed by trappers and used to prepare scent baits or attractants (see Chapter 30). The musk, or **civet,** produced by the anal glands of certain civets (Viverridae, principally *Viverra, Viverricula,* and *Civettictis*) is utilized (as are glandular substances from other mammals) as a base in the manufacture of fine perfumes.

4-CC Examine preserved material of a variety of mammals. How many scent glands can you find? Where on the animals' bodies are these located?

4-DD Examine a demonstration dissection of the anal glands of a skunk (e.g., *Mephitis*). How are these animals able to propel their musk such great distances?

Mammary Glands

Mammary glands, unique to mammals, may be derived from sweat or sebaceous glands, although the precise origin is not clear. These glands develop from two ridges of tissue, termed **milk lines,** in the integument (Kent and Miller 1997). Mammary glands are present in both sexes in eutherians but normally only reach their full size and development in females. Secretions of **milk** from these glands nourish young during the early stages of their lives. In monotremes, there are two mammary glands that secrete milk into abdominal depressions. The monotreme hatchlings then suck milk (Griffiths 1968) from hairs associated with these mammary glands (Fig. 4.10A). All female (and many male) eutherian mammals have mammary glands that are equipped with either nipples or teats. **Nipples** are found in most mammals and have numerous small glandular ducts that exit from the tip of a small, fleshy projection (Fig. 4.10B). **Teats,** such as those found in artiodactyls, have ducts that lead from the glands into a common reservoir or **cistern** that in turn is connected to the exterior through a single duct (Fig. 4.10C).

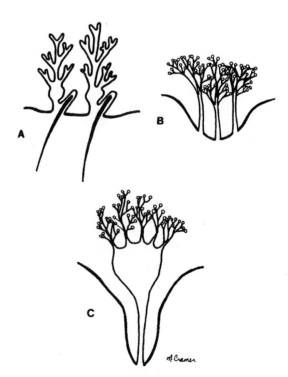

Figure 4.10 Lateral view (diagrammatic) of mammary gland of monotreme (A), and structure of nipple (B) and teat (C).
(Mary Ann Cramer)

The number and placement of mammary glands vary greatly and are usually correlated with the typical litter size for a species. Marsupials frequently have a somewhat circular arrangement of nipples. The opossum *Didelphis virginiana*, for instance, has 13 nipples, 12 arranged in a large U shape and the thirteenth centrally located. Eutherian mammals usually have the nipples arranged in two longitudinal ventral rows (certain pouchless marsupials with many nipples—up to 27 in *Monodelphis sorex*—show a similar pattern). In species such as the hog, *Sus scrofa,* which has a relatively large litter size, each row extends from a point between the pectoral limbs to a point between the pelvic limbs. "Higher" primates, which usually have a single young at a time, have a single pectoral pair of mammae, whereas horses, which have the same litter size, have a single abdominal pair. Some rodents (e.g., species of *Mastomys*, which have an extremely large litter size) have the nipples extending out onto the backs of the thighs. Some hystricomorph rodents (e.g., *Myocastor*) have the nipples located relatively high on the sides, apparently as an adaptation for nursing the large, precocial young of this group of mammals.

4-EE Examine a demonstration dissection or illustrations of the mammary tissue of the inner surface of skin that shows well-developed mammary tissue. How is the mammary tissue arranged? Are there differences in the arrangement of the mammary tissue in rodents, carnivores, and bovids?

4-FF Examine preparations of a teat and a nipple. How do these structures compare?

Horns and other integumental derivatives on the head and body are discussed in Chapter 5. Claws, nails, and hoofs, which are composed of keratinized epithelial cells, are discussed in Chapter 6.

CHAPTER 5

Horns and Antlers

In modern mammals, horns and antlers are confined to the ungulate orders Artiodactyla and Perissodactyla. However, the fossil record includes horned mammals in other orders, even the Rodentia. Head **excrescences** or outgrowths in living mammals may be divided on the basis of structure and method of formation into five major types.

True Horns

True horns, which occur only in the family Bovidae (buffaloes, sheep, goats, cattle, antelopes, etc.), are unbranched and permanent. Each is composed of an inner bony core that is an extension of a frontal bone and an outer layer of true **horn,** formed from keratinized epidermis (Fig. 5.1). *Note!* "Horn" can refer either to the entire structure (e.g., a cow's horn) or to the keratinized material that forms the sheath, or to the sheath itself. A true horn grows from its base throughout the adult life of the animal, but neither the bony core nor the keratinized portion is shed. Portions of the sheath are often worn away, and in some species parts or layers of the sheath may regularly break away (O'Gara and Matson 1975), but the entire sheath is not shed. In many horned bovids, each season's growth produces a ring at the base of the sheath (Fig. 5.2), and counts of these annual rings have proven useful in determining the age of wild sheep and certain other species.

Horns may be present on both sexes or may occur only on males. When present on both sexes, they are usually larger on the males. A few breeds of domestic bovids (e.g., Aberdeen Angus, polled Hereford) are hornless. Horns usually occur as a single pair; however, one living bovid, the four-horned antelope, *Tetracerus quadricornis,* has four well-developed horns (Fig. 5.3).

5-A Examine horns and horn cores of a variety of bovids. Note differences in size, length, and curvature.

Pronghorns

Pronghorns are found among modern mammals only on the North American pronghorn, *Antilocapra americana,* the single living species of the family Antilocapridae. Their basic structure (Fig. 5.4) is similar to that of the bovid horn, consisting of a permanent, unbranched, bony core that is part of the frontal bone, and an epidermal horny sheath. However, in pronghorns, the horny sheath is shed annually and is branched, having a small anterior projection or prong. When the sheath is about to be shed, it becomes loose, and a new one begins to form on the bone core. O'Gara and Matson (1975) have described this process in detail. Female pronghorns are sometimes hornless and frequently lack prongs. The horns of the males are larger than those of the females.

5-B Compare the horns (both cores and sheaths) of a pronghorn with those of the bovid series examined above. How do the horn cores differ? (Be sure to use a goat in your comparisons.)

Antlers

Antlers occur on males of all deer except the Chinese water deer (*Hydropotes inermis*) of the family Cervidae, and are found in both sexes of the genus *Rangifer,* the reindeer and caribou. Fully developed antlers are entirely

Horns and Antlers   31

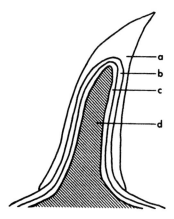

Figure 5.1 Diagrammatic section of a bovid horn: a, horn or keratinized epidermis; b, epidermis; c, dermis; d, bone.
(L.P. Martin)

Figure 5.3 Skull of the four-horned antelope, *Tetracerus quadricornis,* the only living bovid with more than one pair of horns.
(Owen 1868:625)

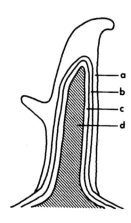

Figure 5.4 Diagrammatic section of a pronghorn: a, keratinized epidermis; b, epidermis; c, dermis; d, bone.
(L.P. Martin)

Figure 5.2 Head of Grant's gazelle, *Gazella granti.* Note the growth rings at the bases of the horns.
(Flower and Lydekker 1891)

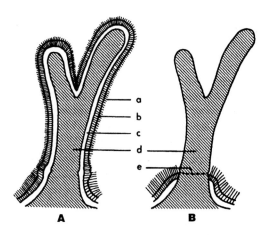

Figure 5.5 Diagrammatic section of antler with (A) and without (B) velvet: a, hair; b, epidermis; c, dermis; d, bone (or antler); e, abscission line at region of burr.
(L.P. Martin)

bony structures (Fig. 5.5B) that are branched in older adults of most species and are shed periodically (annually in the temperate zones). While the antler is growing, the bone is covered with skin, the **velvet** (Fig. 5.5A), which carries blood vessels and nerves supplying the growing bone. When the bone is fully ossified, the velvet is shed. After each mating season in the temperate zones, the bony antler is shed, and in the spring a new set begins to grow (Fig. 5.6).

The antler forms from the **pedicel,** an extension of the frontal bone. A **burr** marks the point of separation between the permanent pedicel and the deciduous antler.

Figure 5.6 Stages in the growth of antlers of the red deer, *Cervus elaphus*.
(Sokolov 1959)

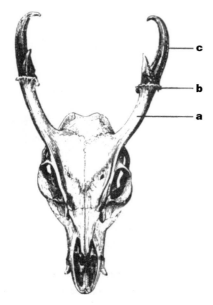

Figure 5.7 Skull of a male muntjac, *Muntiacus muntjak,* a deer with unusually long pedicels: a, pedicel; b, burr; c, antler.
(Owen 1866)

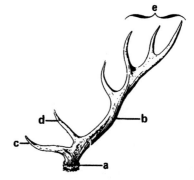

Figure 5.8 Red deer, *Cervus elaphus,* antler: a, burr; b, beam; c, brow tine; d, bez tine; e, crown.
(Sokolov 1959)

The pedicel is usually very short, but in the muntjacs, *Muntiacus,* it is as long as or longer than the antler (Fig. 5.7). The antler usually consists of a main stem, the **beam,** with a variable number of branches or **tines.** The first tine to arise from the beam, immediately over the forehead, is termed the **brow tine,** and the second is termed the **bez tine.** The points of the summit of the antler are collectively termed the **crown** (Fig. 5.8). The pair of antlers together are termed the **rack.** Commonly, all of the branches of the antler are essentially cylindrical, but in some species, as in the moose, *Alces alces,* they are more or less expanded and flattened. Such flattened antlers are termed **palmate** antlers (Fig. 5.9A).

In the white-tailed deer, *Odocoileus virginianus,* the first indication of future antler growth begins at about nine months when small, paired bulges first appear on the frontal bones and rapidly develop into the pedicels. The first antlers start to grow from the pedicels at about 18 months. These begin to harden at the base, and the process continues toward the tip. The growth is very rapid and may be completed in 14 weeks. By late summer or early fall, growth is completed, blood circulation in the velvet becomes sluggish, and the skin dies. The antlers are rubbed against trees and brush until the velvet hangs in shreds and falls off. After the mating season, absorption just under the burr results in a plane of weakness, and the antlers fall off. In the spring, the cycle begins again with new growth.

The number of points or tines displayed by an individual increases with age until the largest number for the individual is reached when the animal is in its prime. Antler development is influenced by several factors other than age (genetics, nutrition, hormone levels, etc.); thus, it is never possible to determine the absolute age of any animal merely by counting the number of tines (see Chapter 33). Nutrition plays an important role in antler development, and undernourished individuals will never have as many tines or as well-developed racks as properly nourished individuals of the same age. In most deer, antlers are secondary sex characters; thus, their formation is also controlled by

Figure 5.9 Relative sizes and configurations of antlers, Cervidae. (A) Moose, *Alces alces,* with palmate antlers; (B) caribou, *Rangifer tarandus,* where both sexes have antlers; (C) white-tailed deer, *Odocoileus virginianus;* (D) pudu, *Pudu puda,* with small spike antlers; (E) Père David's deer, *Elaphurus davidianus.*
(Feldhamer et al. 1999:337)

Figure 5.10 Skull with deformed antlers from a castrated male white-tailed deer, *Odocoileus virginianus.*
(Michael Gilliland)

5-C Compare skulls or racks of *Odocoileus* collected in different seasons. What has caused the small grooves visible on a mature antler?

5-D Examine racks of *Odocoileus* of different ages. What factors influence the differences in antler complexity? Which, if any, can you age absolutely?

5-E Locate the burr, beam, brow tine, bez tine, and crown on a white-tailed deer, mule deer, wapiti, moose, and caribou. How do the general size and arrangement of these racks vary?

male hormones. Injury to the testes or other factors influencing hormonal production may result in stunted or deformed antlers (Fig. 5.10). In very old males, poor nutrition or reduced hormonal production will frequently result in antlers consisting of a single spike or a rough burr. Deformed antlers are frequently seen. Although these may be the result of hormonal or nutritional deficiencies, they are frequently due to mechanical injury to the antler while it is in velvet and still growing.

GIRAFFE "HORNS"

The primary head protuberances of giraffes and male okapis, Giraffidae, consist of a pair of short, unbranched, permanent, bony processes that are situated over the sutures between the frontal and parietal bones and are permanently covered with skin and hair (Fig. 5.11). They ossify from distinct centers, the **ossicones,** and then fuse to the skull (Goss 1983:67). Thus, they are not projections of the frontal bones as are the head excrescences of

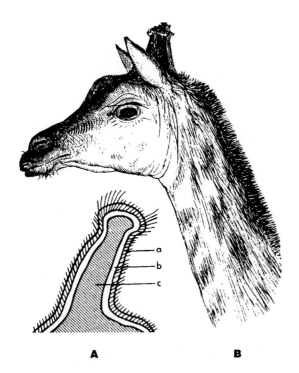

Figure 5.11 (A) Giraffe "horn"; a, hair-covered epidermis; b, dermis; c, bone. (In younger animals, a suture is present between the frontal bone and the "horn," as shown in Figure 27.11.) (B) Head of giraffe, *Giraffa camelopardalis*.
(A, L.P. Martin; B, Giebel 1859:369)

other types of artiodactyls. "Horns" are present in both sexes of giraffes and even in newborn animals. Anterior to the paired "horns" of giraffes, a median protuberance of the frontal bone is frequently present in some populations (see Fig. 27.1). This "third horn" increases in size with the age of the individual.

5-F Examine a "horn" of a giraffe (Fig. 5.11) or okapi and compare the structure to that of a true horn and of an antler in velvet (see Fig. 5-5A).

Rhinoceros "Horns"

The only living nonartiodactyls to possess keratinized hornlike structures on the head are the rhinoceroses of the order Perissodactyla (see Fig. 26.4). The rhino "horn" (sometimes called a *hair horn,* although the fibers making up the horn are not formed like true hair) does not have a distinct core and sheath but is a solid mass of hardened epidermal cells that are formed from a cluster of long **dermal papillae** (Ryder 1962). The cells formed around each papilla constitute a distinct horny fiber resembling a thick hair. These fibers are cemented together by a mass of epidermal cells that grow up from the spaces between the fibers. They differ from true hairs in growing from a dermal papilla that extends up into them, rather than from an epidermal follicle extending down into the dermis (Fig. 5.12A). The skin bearing the

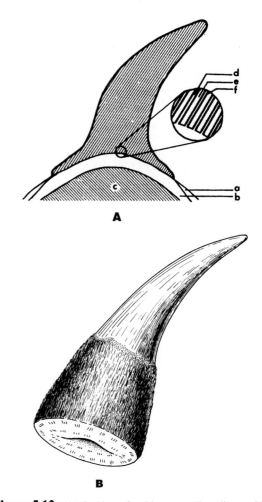

Figure 5.12 (A) Section of a rhinoceros "horn": a, epidermis; b, dermis; c, bone; d, dermal papilla; e, matrix of epidermal cells; f, fiber. (B) Rhino "horn" showing surface features.
(A, L.P. Martin; B, Kent and Miller 1997:112)

horn is situated over the fused nasal bones. These bones are generally enlarged and have a rough surface that ensures firm seating of the skin and horn (see Fig. 26.4). In species with two horns, the second is positioned over the frontal bones. The horns are conical and frequently curve posteriorly. In some species, they may reach a length of 4 feet (1.2 meters).

5-G Examine a rhino "horn" (see Fig. 5.12B) and note its fibrous texture. What is the reason for the deep pores in the base?

5-H Examine a rhinoceros skull (see Fig. 26.4). Where is/are the horn(s) situated in relation to the skull bones?

Function

Horns, antlers, and other excrescences from the head region of mammals serve a number of functions. The primary function of these structures is apparently to better enable males to carry out combat among themselves and thereby gain access to mates (see Ewer 1968; Geist 1966; Goss 1983; Schaffer and Reed 1972). Secondarily, they may serve as display structures, as indicators of social status, as protection against predators, and as devices to deposit secretions from glands onto objects (e.g., trees and rocks) in their environments. Removal of tines or otherwise reducing the size of an antler or horn often causes a decrease in the social status of the affected individual (Goss 1983 and Suttie 1980, for review). Some scientists have suggested that the antlers of the Cervidae could serve as thermoregulatory devices by facilitating heat loss. Goss (1983), however, did not think that the thermoregulation argument was convincing because species from tropical and warm environments tend to have small antlers and those from colder climates have the largest antlers.

CHAPTER 6

THE POSTCRANIAL SKELETON

The skeleton is the bony framework that supports the body and protects many of the internal organs (Fig. 6.1). The bones composing the skeleton are grouped in three categories. The **axial skeleton** is the skeleton of the midline of the body proper, including the skull (which has already been discussed in detail), the vertebrae, and the bones of the thoracic cavity or rib cage. This group also includes some miscellaneous, often unpaired bones to be discussed later. The postcranial **dermal skeleton**, in mammals, is usually rudimentary or absent with the notable exception of the armadillos. Dermal bones were discussed in Chapter 4. The **appendicular skeleton** is the skeleton of the paired appendages, the forelimbs and the hind limbs, as well as associated pectoral and pelvic girdles.

POSTCRANIAL AXIAL SKELETON

Posterior to the skull is the vertebral column, a series of articulated bones that provide the suspension system from which the rest of the body is hung and that encase and protect the spinal cord. Immediately posterior to the skull are the **cervical**, or neck, vertebrae. The first cervical vertebra is termed the **atlas**, and the second is the **axis**. In virtually all mammals from the smallest mouse or bat to the tallest giraffe, seven cervical vertebrae are present. The only exceptions to this are among the sloths (Xenarthra, Chapter 17) and the sirenians (Chapter 25). Next come the **thoracic vertebrae**. These are the vertebrae that articulate with the ribs and help form the thoracic cavity, or rib cage. Each **rib** has a head that articulates with the body of a thoracic vertebra and another articulation point that touches the transverse process of an adjacent rib. The distal ends of the more anterior ribs articulate directly with the **sternum**, a series of bony segments that provide the midventral completion of the thoracic cavity. The more posterior ribs are often **costal ribs**, connecting with each other, and eventually the sternum, indirectly through lengths of cartilage. One or more pairs of **free ribs** may also be present; they are always the most posterior ribs and do not connect with the sternum directly or indirectly.

Posterior to the thoracic vertebrae are the **lumbar vertebrae**. These form the spine in the lumbar region of the lower back and are the arch supporting the muscular-walled abdominal cavity. Next come the **sacral vertebrae**. These are the vertebrae articulating with the pelvic girdle, often somewhat fused together and sometimes fused to the pelvic bones as well. The fifth type of vertebrae are the tail, or **caudal, vertebrae**, which may be reduced to a few small fused rudiments in a supposedly tailless species such as humans, but at least some are usually present.

6-A Examine the mounted, articulated skeleton of a cat. Identify each of the five types of vertebrae. How many of each type are present? Compare with the mounted articulated skeleton of a human and any other mammal specimen available. How do the numbers of each kind compare?

6-B Using the same demonstration material, examine how the ribs articulate with both the thoracic vertebrae and the sternum.

All carnivores, most primates (but not humans), most rodents and bats, and some insectivores have a bone, the

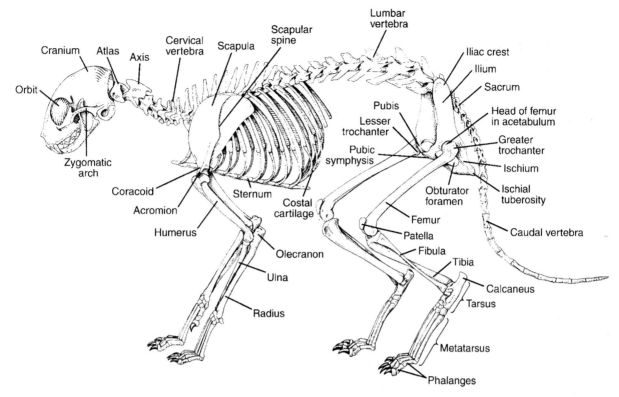

Figure 6.1 Skeleton of a cat.
(Hickman et al. 1997)

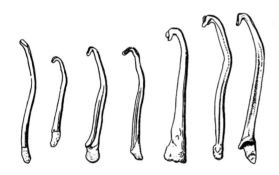

Figure 6.2 Bacula of seven species of *Mustela*, Mustelidae.
(Novikov 1956)

baculum (Fig. 6.2) or **os penis,** within the penis. The structure of the baculum varies greatly from group to group and is frequently used as a major taxonomic character. A much smaller bone, the **os clitoris,** is found in the glans clitoris of females of most species in which the males possess a baculum.

THE APPENDICULAR SKELETON

Mammals are tetrapods, and with few exceptions (cetaceans and sirenians) have four conspicuous locomotor appendages—an anterior pectoral pair and a posterior or pelvic pair. Like the skull and teeth, the limbs of mammals have a consistent general anatomy but show considerable variation in size and proportions because of numerous adaptations for different ways of life in diverse habitats.

The most proximal portion of the **pectoral limb,** or forelimb, is the shoulder girdle or **pectoral girdle** (see Fig. 6.1). In most mammals, this is composed of two elements: the **scapula** and **clavicle.** The scapula, or shoulder blade, is a large platelike bone embedded in muscles dorsal and/or lateral to the ribs and has no direct articulations with any of the bones of the axial skeleton. The clavicle extends from the glenoid fossa, the shoulder socket in the scapula, to the sternum and provides a firm base for the anterior limb. The **scapular spine** is a ridge of bone extending vertically for much of the length of the scapula, this ridge provides additional surface areas for muscle attachment. Many mammals adapted to run on hard ground have the clavicle reduced or absent. In these mammals, there is no direct contact between the pectoral and axial portions of the skeleton, so that the shock of the feet striking the ground is absorbed by the muscles and other soft tissues. The **coracoid,** the third main bone of the pectoral girdles of other vertebrates, is rudimentary and fused with the scapula in marsupials and placentals. Monotremes have a pectoral girdle that is much more reptilian in structure, with well-developed coracoids, precoracoids, and an interclavicle.

The skeleton of the pectoral limb, or foreleg, is composed of a series of articulating elements. The single proximal element, the **humerus,** has a large head that articulates in the glenoid fossa of the scapula, forming a ball-and-socket joint. This type of joint generally allows great mobility of the limb in various planes. Distally, the humerus articulates with the two bones of the lower forelimb, the **radius** and **ulna.** This joint is termed the **elbow.** The ulna articulates with the humerus as a hinge joint, allowing movement in only one plane. The olecranon process of the ulna extends proximally beyond the humerus and serves as the short arm of the lever for attachment of the muscles that extend the forearm. In most mammals, this process prevents the forelimb from being completely straightened. The radius, the more medial of the two forearm bones and the one that aligns with the first digit, articulates at both ends in a manner that allows the two bones to rotate somewhat around each other. This is the action that allows humans to turn their hands either to a **supine** (i.e., palm up) or **pronate** (i.e., palm down) position.

Distal to the radius and ulna are a group of small bones, the **carpals,** or wrist bones, which provide great yet sturdy flexibility to the wrist (Fig. 6.3). Distal to these are the bones of the **manus, hand,** or **forefoot.** Primitively, mammals are pentadactyl, or five-digited. Although reduction in the number of digits frequently occurs, an increase in the number of digits is very rare and does not normally occur in any mammalian group. Following the carpals are five elongate **metacarpals,** one for each digit. The metacarpals are enclosed within the body of the forefoot or hand. A series of **phalanges** (singular, **phalanx**) extends from the distal end of each metacarpal to form each manual digit. The first or most medial digit on the forefoot is termed the **pollex** or **thumb,** with two phalanges. In some mammals, including humans, the pollex is opposable. The second through fifth digits typically have three phalanges each. With modifications of the limb for various environments and ways of life, the number of phalanges may be reduced, or, in a few groups, increased.

The most proximal element of the **pelvic limb** or hind limb, is the **hip girdle** or **pelvic girdle** (see Fig. 6.1), a single, unitary structure formed by the fusion of three pairs of bones. The dorsal **ilia** (singular, **ilium**) articulate with the sacral vertebrae. The **ischia** (singular, **ischium**) are directed posteriorly and form the bony part of the rump. The paired **pubic bones** project anteriorly and ventrally and are joined on their distal ends. These three pairs of bones, together with the sacral vertebrae, form a ring through which the digestive, urinary, and reproductive tracts all exit from the body. In females of some species, the junction between the pubic bones, the **pubic symphysis,** is somewhat elastic to allow for passage of a large fetus.

The skeletal structure of the hind limb is similar to that of the forelimb. At the point where the three pelvic bones meet, a large socket, the **acetabulum,** receives the **head of the femur,** the proximal element of the pelvic limb. The **femur** is followed by a pair of bones, the **fibula** and **tibia.** The more medial of the two, the tibia, is larger in most mammals, and the fibula is often reduced greatly. Whereas the elbow bends to project the distal segments of the forelimb anteriorly, the **knee** bends to project the distal segments of the hind limb posteriorly. A sesamoid bone, the **patella** or kneecap, develops in a tendon on the anterior side of this joint.

Distal to the tibia and fibula is a complex of **tarsal** bones or ankle bones that correspond to the carpals of the forelimb (Fig. 6.4). The largest of the tarsals, the **calcaneum** or heel bone, extends posteriorly from the joint with the tibia and serves for attachment of the **Achilles tendon.** The lever arrangement is similar to that of the olecranon process of the ulna. The **astragalus** is the large tarsal bone adjacent and medial to the calcaneum. Five elongate **metatarsal** bones extend from the tarsals and correspond to the metacarpals of the forelimb. A series of phalanges extends from each metatarsal and forms each **pedal digit,** or **toe.** The first or most medial digit, the **hallux,** consists of two phalanges. The remaining digits have three each. The hind foot, including the tarsals, metatarsals, and phalanges, is termed the **pes.**

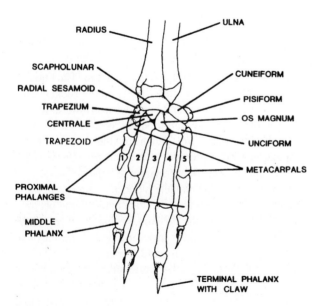

Figure 6.3 Skeleton of the left manus of a rat, *Rattus norvegicus.*
(Chiasson 1969)

6-C Examine the mounted, articulated skeleton of a raccoon, an opossum, or another plantigrade mammal. Locate each of the bones or structures

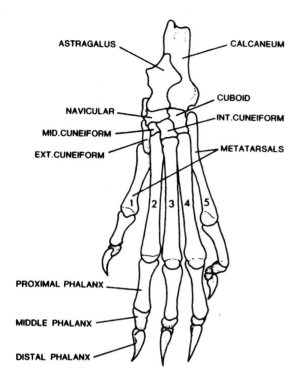

Figure 6.4 Skeleton of the left pes of a rat, *Rattus norvegicus*.
(Chiasson 1969)

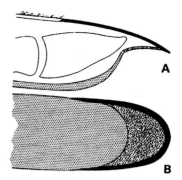

Figure 6.5 Digit with a claw. (A) Lateral section; (B) ventral view; unguis shaded, subunguis stippled, and pad dotted.
(R. Roesener)

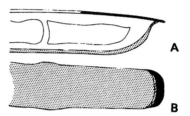

Figure 6.6 Digit with a nail. (A) Lateral section; (B) ventral view: unguis shaded, subunguis stippled, and pad dotted.
(R. Roesener)

given in boldface type above. Skeletons of a cat or human may be used, but each of these has modifications from the primitive ambulatory limb.

CLAWS, NAILS, AND HOOFS

The extremities of the digits of mammals are protected by plates or sheaths formed by keratinized epidermal cells. Only the cetaceans and most sirenians lack digital keratinizations.

In their most primitive form, these structures are **claws** (Fig. 6.5), similar to those of birds and most reptiles. A claw is composed of a dorsal, scalelike plate, the **unguis,** and a ventral plate, the **subunguis.** The unguis, the better developed and harder of the two, is curved longitudinally and transversely and encloses the subunguis between its lower edges. The subunguis is continuous with the **pad** at the end of the digit. The claw encases the last phalanx of the digit. In addition to protection for the ends of the digits, claws serve several other functions. In some running mammals, they aid in increasing traction and stability. Mammals that dig have long stout claws to aid in excavation. Arboreal species frequently have sharp claws that enable them to scamper up the side of a tree. The sloths have long curving claws from which they hang suspended in trees. Many carnivorous mammals use their claws to help hold or kill prey. Claws are usually fixed in position, but most cats and some viverrids have retractile terminal phalanges so that claws may be drawn back into protective bony sheaths. This is an arrangement that minimizes wear on the sharp tips.

6-D   Examine the claws of a tree squirrel, cat, dog, and badger or mole. Locate the unguis and subunguis on each. What is the main function of the claws in each of these mammals?

A **nail** (Fig. 6.6) is a modified claw that covers only the dorsal surface of the end of the digit. In nails, the unguis is broad and flattened, and the subunguis is reduced to a small remnant that lies under the tip of the nail. Nails provide less protection to the ends of digits than do claws, but they also allow for greater precision in manipulations of objects and for increased tactile perception.

6-E   Examine your fingernail and locate the unguis and subunguis. Compare with the nails of other primates.

In the **hoof** (Fig. 6.7) of an ungulate mammal, the unguis curves almost completely around the end of the digit and encloses the subunguis within it. The pad lies just behind the hoof and is called the **frog.** Because the unguis is harder than the subunguis, it wears away more slowly, and a rather sharp edge is maintained. Ungulates have only their digital keratinizations (hoofs) in contact with the ground. These durable structures provide good traction and protect the digits from wear.

6-F Examine the hoofs of a horse and bovid (cow or sheep). Locate the unguis, subunguis, and frog.

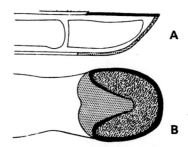

Figure 6.7 Digit with a hoof. (A) Lateral section; (B) ventral view: unguis shaded, subunguis stippled, and pad (frog) dotted.
(R. Roesener)

CHAPTER 7

Locomotor Adaptations

The generalized limb structure discussed in Chapter 6 is well suited for walking or **ambulatory** locomotion. The metacarpals and metatarsals are unmodified, the pectoral and pelvic limbs are about equal in length, and the joints allow movement of the limb in several planes. The feet are **plantigrade** (Fig. 7.1A), and five digits are usually present. This primitive ambulatory limb structure has been variously modified in ways to produce the diversity of limb types found in mammals.

Terrestrial Locomotion

Most mammals with the ambulatory limb are capable of running, or **cursorial,** locomotion. Species that rely heavily upon running to catch prey or escape predators have specially adapted limbs. As the length of limb increases, so does the length of stride. When walking, humans are plantigrade, resting their weight on the toes and the entire sole of the foot; but when sprinting, humans, in effect, lengthen their leg by raising their heels off the ground and placing their weight only on the digits and ball of the foot. Many cursorial mammals are permanently **digitigrade** (Fig. 7.1B) with the metatarsal and metacarpal portions of the foot never contacting the surface during locomotion. Digitigrade mammals frequently exhibit reduction in the number of toes, and elongation of the metacarpals and metatarsals.

Most digitigrade mammals have limbs capable of moving in several planes (e.g., cats). This freedom of movement allows great agility and allows the limbs to aid in the capture of prey. Several digitigrade carnivores also have great suppleness of the back, which allows the hindfeet to be placed well in front of the forefeet when the animal is in full gallop. This capability is found, for example, in the cheetah, the fastest living mammal.

Hoofed animals illustrate the ultimate in cursorial limb structure, **unguligrade** limbs (Fig. 7.2). These have the phalanges elevated so that only the hoofs, the modified digital keratinizations, are in contact with the substrate. The proximal portions of the limbs are shortened and heavily muscular. The radius is usually fused with the ulna, and the fibula with the tibia. These bones, as well as the metacarpals, metatarsals, and phalanges, are usually greatly elongated, and the number of digits is usually greatly reduced. The ultimate of reductions is reached in the equids, which all run on the tip of a single digit. Most of the "cloven-hoofed" mammals, order Artiodactyla, have two functional digits. The pairs of metacarpals and metatarsals associated with these two digits may be fused to form a **cannon bone.** Unguligrade mammals generally have limbs capable of little lateral motion. The resulting restricted agility is usually compensated for by a long neck with a large head, which may be used as a counterweight in maintaining balance.

7-A   Examine the limbs of a mounted skin and a skeleton of a cat or other digitigrade mammal. Compare the lengths and relative positions of the various limb elements with those of a plantigrade mammal. Compare the placement of pads on the feet of a cat with that of a raccoon or other plantigrade mammal.

7-B   Examine the limbs of mounted unguligrade animals and their skeletons. Compare the lengths and arrangements of limb elements with those of digitigrade and plantigrade mammals. What

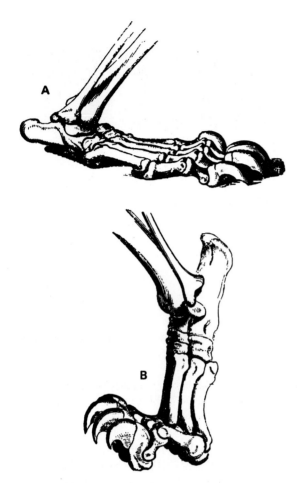

Figure 7.1 Skeletons of (A) the plantigrade right hindfoot of a bear, Ursidae; and (B) the digitigrade right hindfoot of a lion, Felidae.
(Owen 1866)

Figure 7.2 The skeleton of the unguligrade pectoral limb and girdle of a horse, Equidae.
(Owen 1866)

Figure 7.3 The hindfoot of *Salpingotus,* a small rodent inhabiting arid, sandy areas of Asia.
(Vinogradov 1937)

causes the restriction of lateral movement in the unguligrade limb?

Mammals that normally walk on particularly soft surfaces such as snow, sand, and mud have special limb modifications. For example, the lemmings of the genus *Dicrostonyx,* and the hare, *Lepus americanus,* inhabit areas that have snow much of the year. These and several other mammals that occupy similar habitats have feet that are larger than those of their relatives in more temperate climates. Arctic mammals also frequently have the soles of their feet well covered with hair, which offers insulation and further increases the functional size of the foot. The enlarged foot distributes the weight of the animal over a greater surface area and functions in the same manner as snowshoes. The jerboas (e.g., *Jaculus*) of the African and Asian deserts are examples of mammals that have similarly hairy feet for getting about on soft sand (Fig. 7.3). The camels of the same region have broad, tough pads on the bottom of the feet for the same purpose. The swamp deer, *Blastocerus dichotomus,* and waterbuck, *Kobus ellipsiprymnus,* are examples of ungulates that inhabit wet areas. Their elongated and widely spread hoofs facilitate movement on mud.

7-C Compare the feet of small mammals that inhabit regions of snow with those from areas of loose sand. Why are they so similar?

7-D Compare the hoofs of a reindeer or caribou, *Rangifer,* with those of a similarly sized deer from temperate habitats. How do you explain the difference in terms of habitat?

7-E Compare the feet of a jerboa with those of a camel. How do these different feet serve the same function?

Several mammals are **saltatorial;** they progress by a series of leaps with the hindlimbs providing the main propulsive force. Some of these, such as the rabbits and

Locomotor Adaptations   43

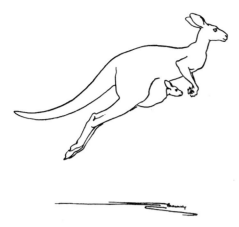

Figure 7.4 A kangaroo exhibiting ricochetal locomotion.
(Brazenor 1950)

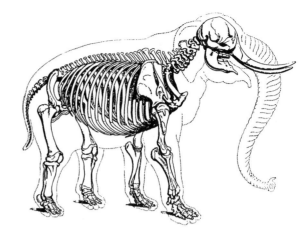

Figure 7.6 The skeleton of an Indian elephant, *Elephas maximus*.
(Owen 1866)

Figure 7.5 Skeleton of a ricochetal rodent, the three-toed jerboa, *Dipus sagitta*.
(Owen 1866)

hares, are quadrupedal. Their leap is termed a **spring,** and all four feet are involved.

Many only distantly related saltatorial mammals (kangaroos, jerboas, kangaroo rats, jumping mice, etc.) are bipedal. Their leap is termed a **ricochet,** and the forefeet are not utilized (Fig. 7.4). Ricochetal species have greatly elongated and very muscular hind limbs and reduced forelimbs. The pes is particularly long (Fig. 7.5). The tail, which is usually long and frequently tufted at the tip, functions as a counterbalance when the animal is moving and provides support when the animal is resting in a bipedal position. Ricochetal locomotion is particularly useful for rapid movement over soft substrates where vegetation is sparse, and is common in sand-dwelling desert rodents.

7-F Examine mounted specimens and skeletons of a kangaroo, a kangaroo rat, or a jumping mouse. Compare the structures of hind- and forelimbs and compare these with those of plantigrade and digitigrade mammals.

7-G Observe rapid locomotion of a live kangaroo rat, *Dipodomys,* or a jumping mouse, *Zapus.* How does the tail aid in maneuvering?

7-H Examine a mounted or live rabbit or hare and a rabbit or hare skeleton. How does the limb structure compare with that of an ambulatory or cursorial species? With that of a ricochetal species?

Some terrestrial mammals (e.g., the elephants) are extremely heavy-bodied and have a limb structure that is dictated by the great weight of the animal. This **graviportal** limb (Fig. 7.6) is essentially columnar, with each element situated directly above the one below it. The digits radiate out to form a series of arches that are incorporated in the flesh to form the base of a massive column. The bottom of the foot has a thick cushioning pad (Fig. 7.7).

7-I Examine specimens or illustrations of an elephant or hippopotamus leg. Compare the leg skeleton with the complete leg. Was the animal digitigrade or plantigrade?

Fossorial Adaptations

Many species of mammals have habits that involve digging. These range from species that dig only to obtain food and seldom go entirely beneath the surface, to species that feed, live, and breed beneath the ground and seldom, if ever, come out on the surface. Tendencies associated with life underground include reduction of external body projections, development of valvular body

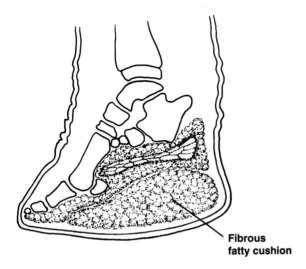

Figure 7.7 Forefoot of an African elephant, showing the fibrous fatty cushion under the digitigrade foot skeleton.
(Feldhamer et al. 1999)

Figure 7.8 A golden mole, *Chrysochloris sp.*, a fully fossorial mammal.
(Flower and Lydekker 1891)

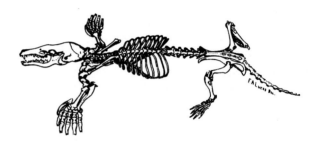

Figure 7.9 Skeleton of a mole, *Scalopus aquaticus*, a fully fossorial mammal.
(Hatt 1964)

openings, reduction of vision, increase in number and sensitivity of tactile receptors, enlargement of the forefeet and claws, and reduction in length of the tail and neck.

Semifossorial mammals are those that burrow into the ground but also spend a great deal of time moving around on the surface. Ground squirrels, marmots, prairie dogs, kangaroo rats, grasshopper mice, pygmy rabbits, and many other rodents and lagomorphs live in burrows that they themselves dig, but they eat or gather food primarily on the surface. As prey species, they are constantly alert, and their eyes are frequently positioned high on the head so that they may observe their surroundings before exposing themselves. Badgers, some armadillos, and many other mammals both den and gather food below the surface but still spend a great deal of time above ground.

Fully **fossorial** mammals live underground and only rarely move about on the surface. In all of these animals, the body is very compact; the tail is reduced or rudimentary; the neck is very short; pinnae are tiny or absent; eyes are frequently vestigial (Fig 7.8); and the forefeet, pectoral girdle, and associated musculature are large and powerful (Fig. 7.9).

7-J Compare the limb structure and general body shape of a badger with those of a raccoon. What modifications for semifossorial life are present in the badger?

7-K Compare the position of the orbits in skulls of a raccoon and a marmot. Explain the reasons for the difference observed.

7-L Examine a skinned mole carcass and a mole skeleton. Compare the structures of the hind- and forelimbs.

7-M Compare the forefeet of a mole (Talpidae), golden mole (Chrysochloridae), pocket gopher (Geomyidae), molerat (Bathyergidae), and marsupial mole (Notoryctidae). Use photographs when necessary. How are limbs on each of these distantly related fossorial mammals modified to serve the same function? Compare their eyes, external ears, and tails.

7-N Examine a skin of a mole (Talpidae). Brush the hairs in various directions. What is the advantage of this type of pelage to a fossorial mammal?

7-O Examine a specimen or photograph of the burrowing rodent, *Heterocephalus glaber* (see Fig. 23.38C). How do you explain this unusual pelage?

7-P The lemmings, *Dicrostonyx*, live in the far north where they must spend a considerable portion of the year burrowing through snow. Examine the claws on the forefeet of a winter-caught specimen. How do these compare with the claws of a winter-caught vole (*Microtus*) from a temperate region?

Aquatic Adaptations

Mammals with special adaptations for life in the water range from those that swim frequently to ones that never leave the water.

Semiaquatic mammals, such as the beavers, otters, water shrews, platypus, polar bear, etc., occur in several orders. They all exhibit some degree of aquatic adaptation. With the exception of the sea otter, all of these spend a good deal of their lives out of water either on land or in dens. Many have **webbed feet,** but none other than the otters come close to having true flippers. Some have a fringe of stout hair along the edge of the foot to increase the surface area (Fig. 7.10). Almost all have a generally fusiform body streamlined for easy movement through the water. The neck and tail are often thick at the base so that there is no sudden change of body diameter where they join the body. The tail is frequently flattened or equipped with a keel of stiff hairs (Fig. 7.11) to assist in propulsion, maneuvering, or both. The neck may be elongated to provide a counterbalance or shortened to provide a compact torpedo shape. The ears and nostrils are usually **valvular,** and each eye is protected by a nictitating membrane. Pinnae and other external projections are reduced.

7-Q Examine a platypus, water opossum, water shrew, otter shrew, beaver, muskrat, river otter, and sea otter for the above modifications. Use photos when necessary. In what ways are these distantly related forms very similar in structure? How are the feet of each modified to increase the surface area for swimming?

7-R The hippopotamus and the polar bear are both also semiaquatic mammals. How does their structure compare with those of other semiaquatic species? In what ways is each similar to the other species? In what ways are they different?

7-S The platypus digs nest burrows into the banks of streams where it feeds underwater. Examine the forefeet (Fig. 7.12) and explain.

Pinniped mammals, the seals, sea lions, and walruses, spend a large part of their lives in the water but bear their young ashore. The forelimbs are modified into **flippers** that, in otariids and odobenids, can be used for locomotion on land. Hair seals, Phocidae, have hindflippers that are permanently rotated backwards, whereas sea lions and fur seals, Otariidae, and walruses, Odobenidae, have hindflippers that can be rotated forward during locomotion. The main propulsive force of hair seals in water is produced by a lateral undulation of the posterior portion of the body, including the hindlimbs. The neck is shortened to provide a torpedo shape, and the forelimbs are relatively small and used primarily for maneuvering (Fig. 7.13). Otariids and walruses use their relatively large forelimbs for propulsion. The neck is long to provide a counterbalance.

Figure 7-10 Hindfoot of a water shrew, *Neomys fodiens*. Note the fringe of stiff hairs that increase the surface area of the foot.
(Stroganov 1957)

Figure 7.11 The tail of a water shrew, *Neomys fodiens*. Note the ventral keel of stiff hairs.
(Stroganov 1957)

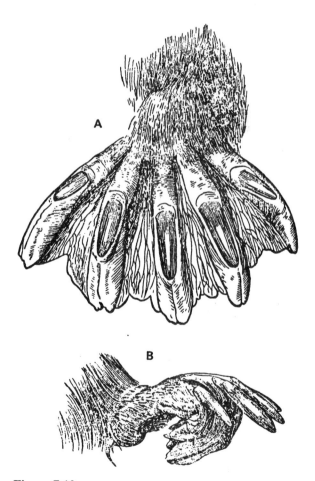

Figure 7.12 Forefoot of a platypus, *Ornithorynchus anatinus*. (A) with webbing extended; (B) with the membrane folded back.
(Ognev 1951)

7-T Examine the ears, nostrils, and flippers of a hair seal and a sea lion or fur seal. Compare the positions and sizes of the flippers and the lengths of the necks.

7-U Study photographs or films of phocids and otariids resting and moving on land and in the water. Compare the methods of propulsion and maneuvering.

7-V Compare photographs or specimens of polar bears with those of other bears. How do the neck lengths compare? Do these indicate the method of propulsion in water?

The fully **aquatic** mammals, the cetaceans and sirenians, do not regularly come onto land. Pinnae are absent. The body is extremely fusiform with a very short, thick neck and an even taper from the trunk to the tip of the tail. The forelimbs are modified as flippers, and the hindlimbs are absent externally (Fig. 7.14). The tail tip is laterally expanded and dorsoventrally flattened to form a paddle-shaped structure (manatees) or **flukes** (dugong and cetaceans). These mammals swim by undulating the posterior part of the body in a vertical plane. The horizontally expanded tail tips provide increased surface area for propulsion, and the pectoral appendages are used primarily for maneuvering. A **dorsal fin,** present in many cetaceans, aids in stabilization. The paddle, flukes, and fin are composed of fibrous connective tissue without bony support (Fig. 7.14).

7-W Examine models or photos of cetaceans and sirenians (live mammals if possible). Note the structure of the tail, fin, and pectoral appendages. Locate the nostril(s).

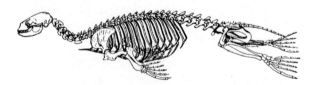

Figure 7.13 Skeleton of a hair seal, Phocidae.
(Hatt 1946)

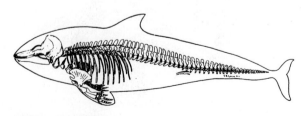

Figure 7.14 Skeleton and body outline of a porpoise.
(Hatt 1946)

7-X Examine the skeletons of the flippers of a cetacean and a pinniped. How many digits are present? How many phalanges are present in each digit? How do these flippers compare with the feet of terrestrial mammals?

7-Y Locate the pelvic rudiments on cetacean and sirenian skeletons. Which bone elements remain? What skeletal support is present in the flukes?

7-Z Compare the tail structure of a cetacean with that of a fish. The fish undulates laterally. Of what advantage is dorsoventral undulation to an air-breathing mammal?

7-AA Compare the placement of the orbits of several aquatic and semiaquatic mammals with those of several terrestrial and fossorial ones. Is orbit placement in aquatic forms more similar to that of terrestrial ones or the fossorial ones? Why?

ARBOREAL ADAPTATIONS

Locomotion in and between trees has taken diverse forms, ranging from the rapid scampering of squirrels to the slow motion of sloths. Modifications of the limbs and the tail are many and diverse. **Prehensile** tails have evolved in several groups of arboreal mammals and are particularly common in the Neotropical Region. The mammals that move rapidly through trees, particularly those such as the gibbons and flying squirrels that swing or jump from point to point, have well-developed binocular vision and depth perception. Some animals that commonly move about in trees, such as the tree hyraxes, *Dendrohyrax,* have no striking adaptations unique to their habitat, but all hyraxes have amazingly adhesive foot structure and good balance. They also move carefully.

Scansorial mammals, such as the tree squirrels (e.g. *Sciurus*) show little obvious modification for arboreal life. However, they have sharp, strong claws that allow them to scamper up vertical surfaces and also have long fluffy tails that aid in balance.

Arboreal mammals, such as arboreal mice, arboreal monkeys, and arboreal opossums, cling to branches by prehensile, opposable digits, prehensile tails, or both.

Brachiating mammals, such as the gibbons, swing through the trees with their hands (Fig. 7.15). This movement may be described as an inverted bipedal walk with the arms. The olecranon process of the ulna is very small and thus allows the arm to be extended perfectly straight to support the weight of the body. The pollex of brachiating mammals is frequently reduced or absent, and the fingers are very long.

Sloth movement, found in sloths and colugos, is an inverted quadrupedal walk by an animal hanging sus-

Locomotor Adaptations   47

Figure 7.15 White-handed gibbons, *Hylobates lar*, brachiating.
(Flower and Lydekker 1891)

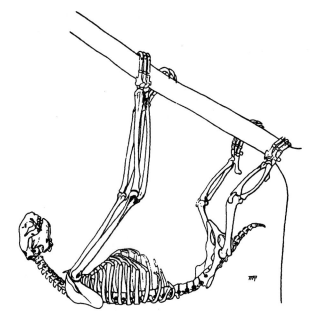

Figure 7.16 Three-toed sloth skeleton, Bradypodidae.
(Hatt 1946)

pended from all four limbs. Sloths hang from strongly curved claws on all four limbs (Fig. 7.16).

The various flying squirrels, most anomalurids, colugos, and certain diprotodont marsupials have a thin flap of skin, a **patagium,** extending between the forelimbs and hindlimbs. The colugos have exceptionally well-developed additional patagia extending from the neck to forelimbs and from the hindlimbs to the tip of the tail (Fig. 7.17). These flaps offer maximum surface area for gliding through the air from tree to tree. Those "extra" patagia are either absent or not as well developed in the other gliding mammals, but their long tails sometimes have hairs arranged to provide an overall structure that is dorsoventrally flattened and that further increases the surface area (Fig. 7.18). Lifting and lowering the limbs and tail facilitate maneuvering in midair.

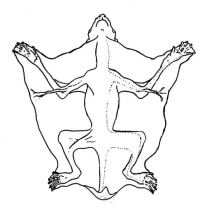

Figure 7.17 Ventral view of a colugo, *Cynocephalus sp.*, with hair removed, showing the extent of the gliding patagia.
(Pocock 1926)

7-BB   Examine the limbs and tails of a tree squirrel, arboreal opossum, prehensile-tailed cebid monkey, arboreal cercopithecid monkey, marmoset, gibbon, tree hyrax, raccoon, arboreal mouse, sloth, flying squirrel, and colugo. Into which of the above categories (scansorial, etc.) does each of these fit? Use photos where necessary.

7-CC   How do the gliding patagia and tails of a flying squirrel and a colugo compare? How do the foot structures of a sloth and a colugo compare?

7-DD   Examine a series of skulls of arboreal mammals. Which have the orbits directed anteriorly

Figure 7.18 A "flying" squirrel, *Hylopetes*.
(Hsia et al. 1963)

to provide overlapping visual ranges and, therefore, binocular vision? What is the selective advantage of depth perception in these forms?

7-EE Prehensile tails have evolved in diverse mammalian groups. List mammals having a prehensile tail and the continent(s) on which each is found.

AERIAL ADAPTATIONS

True flight has developed in only one mammalian order, the Chiroptera. The wing is made up of naked or usually naked patagia that extend between the greatly elongated manual digits, body, the hindfoot and, often, the hindlegs and tail (see Fig. 14.1). The major skeletal modifications for flight include: elongated distal portions of the arm; elongated metacarpals and phalanges of the second, third, fourth, and fifth digits; in most species, a keeled sternum to which powerful flight muscles are attached; a knee that is directed outward and back; and a cartilaginous spur, the **calcar,** on the ankle to help support the uropatagium. All long bones are very slender and light in weight (Figs. 7.19, 14.1).

Associated with the nocturnal habits of one of the two suborders of bats, the Microchiroptera, is the ability to echolocate. For this activity, the ears of these bats are generally enlarged. A structure inside the pinna, the **tragus,** is very much enlarged in many bats and is believed to aid in echolocation. Elaborate structures are found on the noses of many kinds of bats, and these are also believed to aid in echolocation.

Bats differ greatly in their habits and habitats. Many are relatively sedentary, spending their entire lifetime in a few square miles of cave and forest. Others migrate great distances annually. Due to these factors and others, there are different requirements for general wing structure. Long narrow wings provide swift, sustained flight at the expense of lift and maneuverability, whereas broad, relatively short wings provide greater lift and maneuverability at the expense of speed and sustained flight.

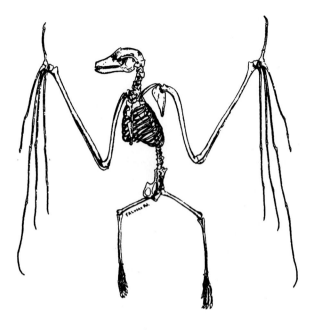

Figure 7.19 Skeleton of a fruit bat, Pteropodidae.
(Hatt 1946)

7-FF Examine a bat skeleton and note the modifications for flight mentioned in this chapter. How does the structure of a bat wing differ from those of the wings of the other flying vertebrates, the birds and pterosaurs?

7-GG Examine a series of microchiropteran bats of the families Vespertilionidae, Phylostomidae, Rhinolophidae, Molossidae, and Emballonuridae. Compare the structures of the ears, noses, and tails of these forms. Do you find any relationship between size of tragus and development of nasal foliation?

7-HH Compare the extended wing of a freetailed bat (*Tadarida*) with that of a vespertilionid bat such as *Plecotus, Myotis, Pipistrellus,* or *Eptesicus.* Does this reveal anything about their respective habits?

CHAPTER 8

KEYS AND KEYING

In biology, a key is a tool for the identification of specimens to phylum, order, family, genus, or species. It generally consists of a series of pairs of mutually exclusive statements (Fig. 8.1). Each pair of statements is termed a couplet, and the couplets are usually numbered or lettered consecutively on the left side of the page.

The person using the key reads both parts of the first couplet and judges which of the two statements best describes the specimen he or she is trying to identify. Once a decision is made, the number at the right-hand margin indicates the next couplet to be considered. In Figure 8.1, for instance, if the user decides that the second set of statements (1') in the first couplet best matches the specimen, the user then proceeds to couplet 3. If the specimen best matches the first part of this couplet (3), the user is told that the specimen is a mole of the genus *Parascalops*. However, if this specimen best matches the second part (3'), the user is directed to couplet 4. The user thus proceeds, by the process of elimination, to identify the specimen.

Although the above paragraph may seem too basic for inclusion in an advanced manual, it is essential that practicing biologists be able to use a key. No one can be an expert on the identification of all of the kinds of living organisms in any given area, let alone in the world as a whole. All biologists, including taxonomists, ecologists, physiologists, ethologists, etc., must identify the organisms with which they are working. A key makes the task of identification much simpler.

The essentials of key construction were known to Aristotle (Mayr 1969), but keys were not widely used for identification purposes until the seventeenth and eighteenth centuries (Voss 1952). Modern keys usually serve solely for straightforward and accurate identification of specimens. They are not a **synopsis** (summarized descriptions of particular taxonomic groups (Metcalf 1954) and need not reflect relationships among the groups included). This last statement cannot be overemphasized. The characters used in a key should be those most easily observed by the user. Although occasionally these will be the same characters as those indicating relationships between various groups, it should never be assumed that this is the case.

SELECTION OF KEY CHARACTERS

Once the taxonomic and geographic limitations of the key have been defined, the author's first step is the selection of the key characters that will be used to distinguish the organisms to be covered. The characters selected should be valid for both sexes and all age classes; if this is not possible, the limitations of the key with respect to gender and age should be clearly stated.

The author of the key must keep in mind the portion(s) of the organism that the reader of the key will be able to examine. For example, if the key is intended for use by a field ecologist or ethologist who will be identifying live animals, an internal structure such as length of intestine or presence or absence of an interparietal bone is useless. Similarly, a behavioral or physiological character is useless to a taxonomist trying to identify a museum skin or skull.

Key characters should be easily observable with a minimum of manipulation and should be described or stated clearly and concisely. In addition, the characters selected should exhibit little individual variation. Absolute conditions such as the presence or absence of a structure are best. Meristic or discontinuous characters that compare

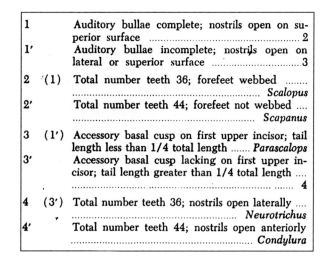

Figure 8.1 Example of a dichotomous bracket key using some of the characters in Table 8.1.

counts of discrete objects (e.g., two incisors versus three incisors) are useful as long as individual variation does not result in overlapping of the counts. Continuous characters (e.g., width of zygomatic arch) should be used only when there is no possibility of overlap between the groups being separated or when these characters offer support to other mutually exclusive characters.

Characters that call for a value judgment on the part of the user are least desirable. Judgment against a generally recognized standard may be used when all else fails. For instance, the alternatives may be "blond pelage" versus "brown pelage." Most people have an idea of what blond and brown mean, but who can decide precisely when "dark blond" becomes "light brown"? Comparative judgments (e.g., "size large" versus "size small") should never be used unless they are tied to extremes of measurements (e.g., "size large, over 100 mm" versus "size small, never exceeding 75 mm").

Ideally, each statement in the first part of a couplet is completely contradicted by a corresponding statement in the second part of the couplet. In practice, however, it is sometimes not possible for *each statement* to be contradicted, but it is essential that the two parts of a couplet be mutually exclusive. Thus, if each part of a couplet contains three statements, at least one must exhibit no overlap. The other two might exhibit minimal overlap that will render them useless with some specimens but will be highly useful with others. If, for instance, the following couplet were given and you had only a skull, you could identify the specimen as belonging to Taxon A if it had two upper incisors or as belonging to Taxon B if it had no canines. But if it had three upper incisors and had canines, you could not identify it without also examining the skin.

1 Hair present on soles of feet; incisors 2–3/3; canines present .... **Taxon A**

1' Hair absent on soles of feet; incisors 3/3; canines present or absent .... **Taxon B**

Occasionally it is impossible to use any single character to distinguish between two groups and it is necessary to write a couplet, such as the following:

1 Rostrum long and pointed; cheek teeth bunodont .... **Taxon C**

1' Rostrum short and broadly rounded, or if long and pointed, cheek teeth lophodont .... **Taxon D**

Specimens identified as Taxon C must have both a long pointed rostrum and bunodont teeth. But specimens identified as Taxon D may have either a short and rounded rostrum and any kind of teeth or a long pointed rostrum and lophodont teeth.

Once the group of characters to be used has been decided upon, it is helpful to summarize them in a table (Table 8.1). The table can then be used to organize the series of couplets in the keys. Methods have been devel-

TABLE 8.1 | **Some Key Characters of Five Genera of Moles (Talpidae)**

Genus	Total Number Teeth	Auditory Bullae	Accessory Basal Cusp on First Upper Incisor	Tail Length	Nostril Opens	Forefeet
Scalopus	36	Complete	Absent	< 1/4 total length	Superiorly	Webbed
Scapanus	44	Complete	Absent	< 1/4 total length	Superiorly	Not webbed
Parascalops	44	Incomplete	Present	< 1/4 total length	Laterally	Not webbed
Condylura	44	Incomplete	Absent	> 1/4 total length	Anteriorly	Not webbed
Neurotrichus	36	Incomplete	Absent	> 1/4 total length	Laterally	Not webbed

Based on data in Jackson (1915).

oped for programming data on key characters into a computer and using the computer to organize these data into a key (Hall 1970; Morse 1971, 1974; Pankhurst 1971; Willcox et al. 1973).

KEY CONSTRUCTION

Thus far, we have described **dichotomous** keys that are composed of couplets. Each couplet has only two parts (e.g., 1 and 1'). Keys have been written with three or more alternatives (e.g., 1, 1', and 1") in each couplet, but the writing of such keys is discouraged.

Each couplet should be written with two series of parallel and mutually exclusive statements. The following example is *not* an acceptable couplet:

1   Dental formula 1/3 1/1 3/3 2/3 = 34; tail long and bushy .... **Taxon E**

1'   Cheek teeth unicuspid; tail sharply bicolored .... **Taxon F**

Although a dental formula of 1/3 1/1 3/3 2/3 = 34 may be exclusive to Taxon E, and unicuspid teeth may be exclusive to Taxon F, this fact is not made clear by the key. The two parts of the couplet are not parallel. In order for them to be parallel, the couplet must be rewritten so that the condition of each key character is clearly stated in each part of the couplet. Thus, the above couplet could be rewritten as follows:

1   Dental formula 1/3 1/1 3/3 2/3 = 34; cheek teeth bunodont; tail long, bushy, and uniformly colored .... **Taxon E**

1'   Dental formula 1/1 1/1 3/3 2/2 = 28; cheek teeth unicuspid; tail short, sparsely haired, and sharply bicolored .... **Taxon F**

A key is usually written in telegraphic style (i.e., unnecessary articles and verbs are eliminated) in order to economize on text space and to simplify reading. The most important or most easily used character is presented first with the remaining characters arranged in the same order in each portion of the couplet. The alternative condition for a character (e.g., canines absent) is given in the second member. Such an arrangement allows for a more efficient and rapid use of the key.

There are several ways of arranging couplets into a key. (Metcalf 1954 and Mayr 1969 present reviews of the types of keys used in biology.) Figure 8.1 illustrates a **bracket key** format in which the two portions of each couplet are presented in immediate succession to one another. All of the keys in the following chapters of this manual are bracket keys.

The second frequently used format is the **indented key** (Fig. 8.2). This style of organization can, and fre-

```
A.  Auditory bullae complete; nostrils open on superior
    surface
    B.  Total number teeth 36; forefeet webbed ........
        .................................................. Scalopus
    BB. Total number teeth 44; forefeet not webbed ....
        .................................................. Scapanus
AA. Auditory bullae incomplete; nostrils open on lateral
    or superior surface
    B.  Accessory basal cusp on first upper incisor; tail
        length less than 1/4 total length .... Parascalops
    BB. Accessory basal cusp lacking on first upper in-
        cisor; tail length greater than 1/4 total length
        C.  Total number teeth 36; nostrils open laterally
            .............................................. Neurotrichus
        CC. Total number teeth 44; nostrils open anter-
            iorly .......................................... Condylura
```

Figure 8.2 Example of an indented key using some of the characters in Table 8.1. Note that the characters utilized are identical to those in Figure 8.1.

quently does, result in wide separation between the two parts of each couplet. We do not recommend it.

Occasionally a key is written as a flowchart (Fig. 8.3) with lines indicating the appropriate succession of couplets. Flowchart keys are useful only for very short keys and are most commonly found in popular or semi-popular works. Sometimes flowchart keys are prepared using illustrations rather than printed descriptions. Construction of such picture keys is not feasible if a character is variable in form, because a word description can fit a variety of conditions more easily than can a figure.

Occasionally, it will be necessary for the user to move backwards in the key in order to locate the point at which an incorrect decision was made. By the nature of their construction, the indented and flowchart keys are easily reversible. To facilitate reverse use of a bracket key, the number of the referring couplet may be placed in

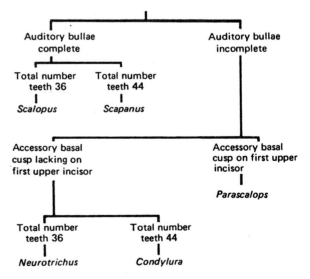

Figure 8.3 Example of a flowchart key using characters in Table 8.1.
(R.E. Martin)

parentheses after the number of the couplet. In Figure 8.1, the number 3 followed by "(1')" indicates that it was "1'" that referred the user to couplet 3.

Before finalizing the text of a key, it is useful to try it out on zoologically naive subjects, utilizing specimens of all the contained taxa. This is the best way to pick up errors and ambiguities.

8-A Study Table 8.1 and Figures 8.1, 8.2, and 8.3. Why are some characters omitted in preparing the keys? Which key is easiest to use? Why? Which would be easiest to use if the key were longer and continued for several pages? Which key is easiest to use in reverse? Which key is most economical of space? Which least? Do you see other ways that the keys could be constructed using any of the characters in the table? How would you proceed?

8-B Prepare a dichotomous bracket key using the information present in Table 8.2. Are your couplets mutually exclusive and parallel? Rearrange your couplets into an indented key.

8-C Assemble a selection of at least 10 different "taxa" of common office supplies including various kinds of rubber bands, thumbtacks, staples, paper clips, etc. Write a key that can be used to identify these "taxa."

8-D Using an assortment of at least 10 different kinds of identified mammal skulls, skins, or both, write a dichotomous key.

The Keys in This Manual

It is relatively easy to say how a key is ideally constructed, but if you have tried the above exercises, you will know that application of these principles is not always easy. In writing the keys in the following chapters, we have attempted to practice what we preach.

The keys in the next 19 chapters are designed to identify to order (Chapter 9) or family (Chapters 10 through 27) extant mammals of the world. All adult mammals should be identifiable, but immature specimens without fully erupted dentition or without fully ossified cranial sutures, or with juvenile pelage may or may not key out correctly.

External features are included in the keys wherever possible, but none of the keys is designed to work with skins alone. Naturally, it is impossible for any author or team of authors to examine the complete range of individual variation that occurs in any species or the complete range of species variation in many families. If you find species or specimens that are not identifiable, we would appreciate hearing from you.

Table 8.2 | Key Characters of Eight Imaginary Taxa of Rodents

Taxon	Dominant Color of Dorsum	Tail Pattern	Number Toes on Hindfoot	Cheek Tooth Formula	Grooves on Incisors	Postorbital Process	Paraoccipital Process
A	Black	Unicolored	4 or 5	4/3	None	Present	Present
B	Black or brown	Unicolored	4	3/3	1	Absent	Present
C	Gray	Bicolored	5	3/3	None	Absent	Present
D	Brown	Unicolored	5	4/4	None	Present	Absent
E	Brown	Bicolored	4 or 5	3/3	2	Absent	Present
F	Brown	Unicolored	3	4/4	None	Absent	Absent
G	Brown and white	Bicolored	4	3/3	1	Absent	Absent
H	Gray	Bicolored	4	3/3	None	Absent	Present

Key to the Orders of Living Mammals

CHAPTER 9

THE ORDERS OF LIVING MAMMALS

Mammals are today among the dominant animals on earth. They range in size from the smallest shrews and bats, weighing a few grams, to the blue whale, which at 112,500 kg is the largest animal ever known to have lived. Mammals run over the earth's surface, burrow underground, swim in the waters, climb in the trees, and fly and glide through the air. They feed upon the larger zooplankton, vegetation, fruit, nectar, pollen, mollusks, annelids, arthropods, fish, and other vertebrates. Some provide the power to plant humans' crops, and others steal the food from their granaries. Some provide meat, milk, hides, and/or serve as beasts of burden for human use, while others feed upon humans. One mammal, *Homo sapiens*, has developed the power to drastically alter the earth's environments and to destroy much, if not all, life on earth.

The class name Mammalia refers to one factor uniting all living species: The females have **mammary glands,** which produce a secretion, milk, that nourishes the young (Fig. 9.1). The extant members of the class Mammalia are grouped into two subclasses, one with two infraclasses, based largely upon reproductive anatomy and reproductive modes: subclass **Prototheria,** the **oviparous** or egg-laying mammals, and the subclass **Theria,** the **viviparous** mammals, which give birth to live young. Subclass Theria is, in turn, divided into two infraclasses based largely upon the method by which the young are nourished prior to and following birth. The infraclass **Metatheria** contains the marsupials, which give birth to only partially developed young that are then often lodged in a pouch, or marsupium, where they receive milk from nipples while they complete development. The infraclass **Eutheria** contains the placental mammals in which the young are nourished within the uterus by means of a **placenta,** and birth occurs only after sufficient development allowing the young to survive without constant attachment to the mother's nipples. *Note!* Use of the adjective *placental* to describe this group should not imply that this organ is absent in marsupials. Some metatherians also have a placenta, but it is not used as an endocrine organ during prolonged gestation as it is in the Eutheria.

These reproductive modes are the way we generally think about these three major groups of extant mammals. But many marsupials have no marsupium, and some do have a placenta, and there are other overlaps and exceptions. More important for defining each of the groups are details of the reproductive anatomy, such as those illustrated in Figure 9.2.

The Prototheria contains only one living order with two families and only three species total, all of which are confined to the Australian Faunal Region. The Metatheria includes approximately 270 living species, which, with one exception, are confined to the Australian and Neotropical faunal region. The exception is one species of opossum that extends from the Neotropical well into the Nearctic. Eutheria includes approximately 94% of the extant species of mammals, and these range over virtually all of the earth's surface.

Division of the class Mammalia into the various orders has varied significantly over the years and at any one time. The following list of orders is used in this text. It follows the classification of Wilson and Reeder (1993) and is consistent with that used by Feldhamer et al. (1999).

CLASS MAMMALIA

Subclass Prototheria
  Order Monotremata — Platypus, echidnas

Subclass Theria
 Infraclass Metatheria
  Order Didelphimorphia — Opossums
  Order Paucituberculata — Caenolestids
  Order Microbiotheria — Monito del monte
  Order Dasyuromorphia — Numbat, marsupial "mice," native "cats," Tasmanian "wolf," devil, etc.
  Order Peramelemorphia — Bandicoots, bilbies
  Order Diprotodontia — Koala, wombats, cuscuses, possums, gliders, kangaroos, wallabies
  Order Notoryctemorphia — Marsupial "moles"

 Infraclass Eutheria
  Order Insectivora — Shrews, moles, desmans, hedgehogs, tenrecs, solenodons, etc.
  Order Dermoptera — Colugos
  Order Chiroptera — Bats
  Order Scandentia — Tupaiids
  Order Primates — Primates
  Order Xenarthra — Sloths, armadillos, anteaters
  Order Pholidota — Pangolins
  Order Carnivora — Carnivores
  Order Cetacea — Whales, dolphins, porpoises
  Order Macroscelidea — Macroscelideans
  Order Lagomorpha — Rabbits, hares, pikas
  Order Rodentia — Squirrels, rats, mice, and other rodents
  Order Tubulidentata — Aardvark
  Order Proboscidea — Elephants
  Order Hyracoidea — Hyraxes
  Order Sirenia — Manatees, dugong
  Order Perissodactyla — Odd-toed ungulates: horses, rhinos, tapirs
  Order Artiodactyla — Even-toed ungulates: pigs, camels, deer, antelope, cattle, goats, sheep, etc.

Figure 9.1 A domestic horse, *Equus caballus,* Equidae, nursing her foal.
(F & W Publications 1997)

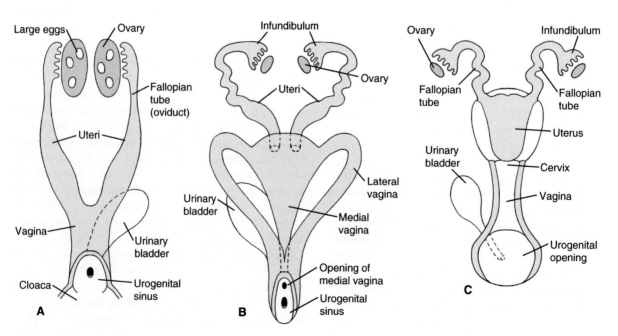

Figure 9.2 The female reproductive tracts from A, a prototherian; B, a metatherian; and C, a eutherian.
(Feldhamer et al. 1999)

Key to the Orders of Living Mammals

This key is based primarily upon skull characters. In some instances, external features are included as supplementary information.

1	Teeth totally absent	**2**
1'	One or more teeth present	**6**
2 (1')	Greatest length of skull more than 500 mm	**3**
2'	Greatest length of skull less than 500 mm	**4**

3 (2) Skull essentially symmetrical; nasal bones form part of roof of nasal passages (Figs. 20.6, 20.8); two external nostrils; baleen present in living specimens .... **Cetacea** (in part)
suborder **Mysticeti**
baleen whales

3' Skull usually asymmetrical, particularly in the region of the external nares; nasal bones form no part of the roof of the nasal passages (Figs. 20.13, 20.17); one external nostril; baleen absent .... **Cetacea** (in part)
suborder **Odontoceti** (in part)
some female, and old male, toothed whales

4 (2') Zygomatic arch complete (Figs. 10.3, 10.4); body with spines (Fig. 10.2) or well furred; if with fur only, then feet webbed (Fig. 7.12) ......
....**Monotremata**
echidnas, platypus

4' Zygomatic arch incomplete (Figs 17.4, 18.2); body with large, overlapping scales or body well-haired; feet never webbed **5**

5 (4') Bony palate with conspicuous medial, longitudinal depression (Fig. 18.2); body covered with large overlapping scales (Fig. 18.1) ....
....**Pholidota**
pangolins

5' Bony palate without conspicuous medial, longitudinal depression; body furred or haired, never with large scales .... **Xenarthra** (in part)
anteaters

6 (1) Incisors 1/1, both upper and lower incisors large and curving; canines absent, wide diastema present between incisors and cheek teeth **7**

6' Incisors more than 1/1, or fewer than 1/1; canines present or absent, diastema present or absent; or if incisors 1/1, then canines present **9**

7 (6) Orbit and temporal fossa separated by a postorbital bar (Fig. 16.4); foramen magnum opens ventrally .... **Primates** (in part)
aye-aye

7' Orbit and temporal fossa confluent; postorbital bar usually absent; if present, then foramen magnum opens posteriorly **8**

8 (7') Angular process of dentary inflected (Fig. 11.17); limbs subequal, hallux vestigial, tail vestigial (Fig. 11.17) ....
....**Diprotodontia** (in part)
wombats

8' Angular process of dentary absent, or, if present, not inflected; limbs various, hallux present or absent, tail usually present ....
.... **Rodentia**
squirrels, rats, mice, and other rodents

9 (6') Incisors present or absent, if present, never numbering more than 3/3, lower incisors never diprotodont **10**

9' Incisors 3–5/2–4, if upper incisors are 3, then lower incisors diprotodont, with I_1 much larger than I_2 and often projecting nearly horizontally forward (Fig. 11.2) **35**

10 (9) Incisors 1/2; upper incisors triangular in cross section (Fig. 25.4) .... **Hyracoidea**
hyraxes

10' Incisors not 1/2, or if 1/2, then upper incisors not triangular in cross section **11**

11 (10') Incisors 2/1; first pair of incisors large and strongly curved; the upper, second pair small, peglike and situated immediately behind first pair (Fig. 22.3); anterior portion of the maxilla perforated .... **Lagomorpha**
rabbits, hares, pikas

11' Incisors not 2/1, or if 2/1, then size and arrangement not as above, and anterior portion of maxillae not perforated **12**

12 (11') Upper incisors forming large curving tusks (Fig. 25.1); cheek teeth present; limbs columnar for terrestrial locomotion ....
.... **Proboscidea** (in part)
most elephants

12'   Upper incisors, if present, not tusklike, or if tusklike, tusk is long and straight with a spiraling twist (Fig. 20.15), other teeth are absent; limbs are modified for marine locomotion **13**

13 (12') External nares displaced posteriorly, opening at, or behind the anterior margins of the orbits; nasal bones reduced or absent; pelvic limbs absent externally **14**

13'   External nares not, or only slightly, displaced posteriorly, opening well anterior to the anterior margins of the orbits; nasal bones usually well-developed; pelvic limbs present externally **15**

14 (13) Skull usually somewhat asymmetrical, particularly in the region of the external nares (Fig. 20.13); rostrum long and pointed or broadly rounded (Figs. 20.24, 20.26) teeth usually conical .... **Cetacea** (in part)
most toothed whales

14'   Skull not asymmetrical; rostrum short and blunt (Figs. 25.6, 25.7); at least some cheek teeth not conical, may be peglike, or have multiple cusps or ridged lophs .... **Sirenia**
dugong, manatees

15 (13') Orbit and temporal fossa separated by postorbital bar or postorbital plate **16**

15'   Orbit and temporal fossa confluent, neither postorbital bar nor postorbital plate present **24**

16 (15) Upper incisors (or incisiform teeth) absent **17**

16'   Upper incisors (or incisiform teeth) present **18**

17 (16) Greatest length of skull less than 100 mm; upper tooth-row starts at anterior edge of rostrum; neither horns nor antlers present; manus with five functional, prehensile, digits .... **Primates** (in part)
some individual lemurs

17'   Greatest length of skull usually more than 100 mm; pronounced interval between anterior end of rostrum and first tooth (Figs. 27.11, 27.16); horns or antlers frequently present; two or four functional digits, with hoofs ....**Artiodactyla** (in part)
ruminants and some individual camelids

18 (16') Incisors 2/1, 2/2, or 2/3 .... **19**

18'   Incisors 1/3 or 3/3 .... **21**

19 (18) Incisors 2/3 (Fig. 15.2); W-shaped cusp pattern present on occlusal surface of molars ....**Scandentia**
tupaiids

19'   Incisors 2/1 or 2/2; molars with cusps present or absent; if present, not arranged in W-shaped pattern **20**

20 (19') Cheek teeth lack cusps, have sharp lateral edges and a medial longitudinal furrow; if cusps present, then the canine is bicuspid (Fig. 14.5B); forelimbs modified as wings (Fig. 14.1) ....**Chiroptera** (in part)
some Megachiroptera

20'   Cheek teeth cuspidate, never as above; canine not bicuspid; forelimbs not modified as wings ....**Primates** (in part)
most primates

21 (18') Incisors 1/3, molars selenodont; limbs with two digits bearing rudimentary hoofs ....**Artiodactyla** (in part)
most camelids

21'   Incisors 3/3; molars zalambdodont, secodont, or with complex infolded pattern of cusps and ridges, never strictly selenodont; digits one, or four or more, never two, and hooves, if present, well-developed **22**

22 (21') Molars zalambdodont (Fig. 12.10B); canines quite small but present, evenly spaced with other teeth (Fig. 11.29); a small fossorial mammal, molelike in general appearance (Fig. 11.29) .... **Notoryctemorphia**
some individual marsupial "moles"

22'   Molars secodont or with complex infolded pattern of cusps and ridges, never zalambdodont; canines present or absent, if present they are either much larger than adjacent teeth, or if small are separated from the cheek teeth by a wide diastema; body form varies, but is never molelike **23**

23 (22') Molars secodont, carnassial pair well-developed (Fig. 19.3); canines present and much longer than adjacent teeth; digits number four or five, hoofs absent .... **Carnivora** (in part)
felids and most viverrids, and herpestids

23'   Molars with complex folded pattern of cusps and ridges (Fig. 3.7); canines present or absent,

if present not longer than adjacent incisors (Fig. 26.3) .... **Perissodactyla** (in part)
asses, zebras, and other horses

24 (15') First two lower incisors pectinate (Fig. 13.3); patagium extends from side of neck to manus to pes to side of tail (Fig. 13.1) ...... **Dermoptera**
colugos

24' Incisors never as above; patagium usually absent, if present, forelimb modified as a wing **25**

25 (24') Postcanine teeth homodont, never with complex folds and ridges **26**

25' Postcanine teeth heterodont or, if homodont, possessing complex folds and ridges **28**

26 (25) Canines, or caniniform teeth, conspicuously longer than other teeth; one to three teeth in each premaxilla (Figs. 19.5, 19.8, 19.12) ....**Carnivora** (in part)
walrus, seals, sea lions, aardwolf

26' Canines, or caniniform teeth, absent, or not conspicuously longer than other teeth (Figs. 17.5, 24.3); or, if long caniniform teeth are present, no teeth are present in the premaxillae (Fig. 17.6) **27**

27 (26') Incisors and canines absent (Fig. 24.3); some cheek teeth constricted medially so as to form distinct anterior and posterior lobes (Fig. 24.4); each tooth composed of numerous hexagonal prisms of dentine surrounding tubules in the pulp cavity (Fig. 24.4); (hand lens or binocular microscope required) .... **Tubulidentata**
aardvark

27' Incisors and/or canines or caniniform teeth present or absent (Figs. 17.5, 17.6); cheek teeth cylindrical or platelike, never bilobed; never having dentine prisms as above .... **Xenarthra** (in part)
armadillos, sloths

28 (25') Size large, greatest length of skull more than 200 mm **29**

28' Size moderate to small, greatest length of skull less than 200 mm **32**

29 (28) Cheek teeth secodont (Figs. 19.2, 19.3) with carnassial pair usually well-developed; limbs various but never unguligrade or graviportal, digits with claws .... **Carnivora** (in part)
certain large carnivores

29' Cheek teeth not secodont, carnassial pair never developed; limbs unguligrade or graviportal, digits with hoofs, or modified hooves **30**

30 (29') Incisors and canines absent, one to two large cheek teeth (Fig. 25.2) present in each jaw quadrant; skull massive, forehead broad and highly domed; limbs graviportal with modified hooves .... **Proboscidea** (in part)
some female Indian elephants

30' Incisors and/or canines often present; cheek teeth more than two per jaw quadrant; skull various but forehead not broad and highly domed, limbs unguligrade, with well-developed hooves **31**

31 (30') Lower canines more or less triangular in cross section, usually with sharp edges; upper canines large, often curving upward and outward (Figs. 27.6, 27.7, 27.9); pes with two, three, or four digits; if three, two of about equal size, the third considerably smaller ....**Artiodactyla** (in part)
suborder **Suiformes**
pigs, peccaries, hippos

31' Canines, if present, not triangular in cross section, not sharp-edged; upper canines, if present, small, never curving upward or outward (Figs. 26.4, 26.6); pes with three digits, central digit larger than the two lateral digits, which are about equal in size (Fig. 26.2A & B) ....**Perissodactyla** (in part)
tapirs, rhinos

32 (28') Upper incisors 0–2, the total number of teeth never exceeding 38; forearm adapted for sustained flight; phalanges of manus greatly elongated, second through fifth digits wholly enclosed in patagium (Fig. 14.1) .... **Chiroptera** (in part)
most bats

32' Upper incisors 0–3, the total number of teeth various, up to 48 or 50; forearm not adapted for flight; phalanges of manus not greatly elongated, and digits never wholly enclosed in a patagium **33**

33 (32') Canines large and clearly differentiated from premolars; incisors always much smaller than canines or cheek teeth; carnassial pair may be well-developed; zygomatic arch and auditory bullae both complete ...... **Carnivora** (in part)
certain smaller carnivores

33'   Canine or caniniform tooth (most anterior tooth in maxilla) small and not clearly differentiated from premolars; incisors small or greatly enlarged; carnassial pair never developed; zygomatic arch often incomplete, auditory bullae complete or incomplete **34**

34 (33')   Jugal absent or vestigial, maxilla extending into medial wall of orbit .... **Insectivora**
solenodons, shrews, moles, desmans, hedgehogs, tenrecs, etc.

34'   Jugal well-developed, maxilla not extending into medial wall of orbit .... **Macroscelidea**
macroscelideans

35 (9')   Lower incisors diprotodont (Fig. 11.2B); I_1 very large, usually projecting nearly horizontally forward, I_2 much smaller or absent; canines may be absent **36**

35'   Incisors polyprotodont (Fig. 11.2A); I_1 not particularly larger than I_2, does not protrude horizontally forward; canines present **37**

36 (35)   Dental formula 4/3–4 1/1 3/3 4/4 = 46–48 (Fig. 11.9); no syndactylous digits ....
....**Paucituberculata**
caenolestids

36'   Dental formula various but upper incisors always fewer than four, and C^1 never present; second and third pedal digits syndactylous (Figs. 11.3, 11.4) ...... **Diprotodontia** (in part)
koala, cuscuses, possums, gliders, kangaroos, wallabies, etc.

37 (35')   Dental formula 4–5/3 1/1 3/3 4/4 = 46–48; upper incisors roughly uniform in size; C^1 widely separated from both last incisor and first premolar (Fig. 11.11); second and third pedal digits syndactylous; tail never prehensile ....
.... **Peramelina**
bandicoots, bilbies

37'   Dental formula diverse; I^1 frequently larger than other upper incisors; C^1 not widely separated from both last incisor and first premolar; no syndactylous digits; tail frequently prehensile **38**

38 (37')   Incisors 5/4 (Fig. 11.7); hallux well-developed and opposable; tail prehensile **39**

38'   Incisors 4/3 (Figs. 11.13, 11.29); hallux reduced or absent, never opposable; tail not prehensile **40**

39 (38)   Size small, condylobasal length less than 30 mm; first upper premolar and last upper incisor approximately equal in width (Fig. 11.10) ......
....**Microbiotheria**
monito del monte

39'   Size variable, condylobasal length more than 30 mm, or if less, then first upper premolar at least twice as wide as last upper incisor (Fig 11.7) ..
....**Didelphimorphia**
opossums

40 (38')   Canines small, no larger than last incisor and smaller than first premolar; skull blunt and squarish in shape (Fig. 11.29); body shape adapted for fossorial existence, molelike in general appearance (Fig. 11.29) ....
....**Notoryctemorphia**
most individual marsupial "moles"

40'   Canines larger than last incisor and first premolar; skull tapering and conical (Figs. 11.11, 11.13); not fossorial, not molelike in general appearance .... **Dasyuromorphia**
numbat, native "cats," marsupial "mice," Tasmanian "wolf," and devil, etc.

COMMENTS AND SUGGESTIONS ON IDENTIFICATION

The characters used in this key are not necessarily those used to define an order. This key was based primarily upon cranial characters that can be easily seen in the lab by students. Mammalian orders may be based on cranial characters but also on limb structure, methods of reproduction, or various other characters or combinations of characters.

    Some comments follow that will, with practice, help in rapidly identifying a specimen to order. How do you tell the difference between a golf ball and a tennis ball? Between a football and a soccer ball? Do you need a key? No, of course not. You just know what each one should look like. This is exactly the kind of understanding you should try to develop for skulls and skins of the various orders of mammals. When you look at a rodent skull, or a carnivore skull, or any of the others, you should be able to recognize it because of characteristics you do not even have to think about. The following hints will help you on your way to this kind of gestalt understanding.

    Monotremes are mammals that lay eggs. Of course, this cannot be observed in a study skin or skull. There are two basic types of monotremes, the platypus and the echidnas: learn what the skin and skull of each looks like.

    Marsupials are a diverse group of several closely related orders. They all have an inflected angle of the ramus, which is the best characteristic to immediately identify the skull of any metatherian. The various orders are further identified by the presence or absence of diprotodont dentition and the presence or absence of syndactylous digits. In the three Neotropical orders, Didelphimorpia is the most diverse and widely distributed and has neither diprotodont dentition nor syndactylous digits, and the monotypic Microbiotheria has the same pair of characters but is a unique tiny animal. The Paucituberculata have diprotodont dentition but no syndactyly. In the Australian orders, the Dasyuromorphia are highly diverse but have neither diprotodont dentition nor syndactylous digits. The Notoryctemorphia, the marsupial "moles," also lack these two specializations but are very molelike in structure and easily identified in this way. The Peramelemorphia have polyprotodont dentition and syndactylous digits, whereas the very diverse group of Diprotodontia have both diprotodont dentition and syndactylous digits.

Insectivora are very difficult to characterize. They are generally small with long, pointed snouts and numerous, cuspidate teeth. They can best be identified by eliminating other orders.

Dermopteran (colugo) skulls are recognized by the wide pectinate lower incisors and by the skins by the extensive patagium, which runs from the side of the neck to the manus, to the pes, to the side of the tail.

Chiroperans (bats) are identified by a forelimb modified as a wing. The skulls are, however, highly diverse. Some could be confused with those of insectivores or with those of small carnivores. Only practice will help.

Tupaiiads are a small group of scansorial and arboreal mammals, often resembling long-nosed squirrels. The zygomatic arch is perforated, and there is a complete postorbital bar.

Primates are usually monkeylike in general external appearance, but some of the more primitive forms (e.g., some lemurs and galagos) may seem squirrel-like. The skulls always have a postorbital bar or plate separating the orbit and temporal fossa. No other small mammals have a postorbital plate—a few have a postorbital bar, but these generally can be identified by other characteristics.

Xenarthrans include three very different-appearing groups. The anteaters have long, conical, toothless skulls. Although the three anteater genera differ greatly in size, they resemble each other in general shape and proportions. The sloths have squarish skulls with cylindrical, homodont cheek teeth. Externally, they are easily identified by their general appearance. Armadillos have conical skulls with numerous homodont cheek teeth. Externally, they are easily identified by their characteristic armor.

Pholidotes (pangolins) are identified externally by their armor of overlapping epidermal scales. The skull is conical and toothless, but, with practice, can be distinguished readily from those of anteaters or echidnas.

Carnivores almost always have large, well-developed canines. Fissiped carnivores often have well-developed carnassials, and pinniped carnivores have homodont postcanine dentition. Externally, carnivores are diverse, but many types (dogs, bears, cats, otters, weasels, skunks, hyenas, raccoons, seals, sea lions, etc.) are familiar to most students.

Cetaceans, whether the great blue whale or the smallest dolphin, are externally unique and easily identified. Their skulls are just as unique and cannot be easily confused with anything else.

Macroscelideans are a small group of quite small mammals with greatly elongated hindfeet and long, slender, flexible snouts. Externally, they would be most easily confused with some of the small saltatorial rodents, and cranially they might be confused with Insectivora.

Lagomorphs have incisors 2/1, with the first pair long like those of rodents, and the second pair of upper incisors are small pegs situated directly behind the first incisors. This arrangement is unique and distinctive. Externally, there are two main body forms: that of the pikas and that of the rabbits and hares. The latter is familiar to all students.

Rodents have incisors 1/1 and lack canines. The incisors are long, usually strongly curved, and frequently pigmented on the anterior surface. Unfortunately, a few other mammals have similar incisors (e.g. wombats and the aye-aye). Learn these exceptions. Externally, rodents are diverse, but there is a basic resemblance.

The *aardvark* has a long conical skull with flat-crowned, cylindrical, or bilobed teeth composed of numerous hexagonal prisms of dentine. Because there is only one species, all aardvarks closely resemble one another.

Proboscideans (elephants) are identified readily both externally and by skulls. Once you have seen an elephant or an elephant skull, they cannot be confused with anything else.

Hyracoideans (hyraxes) are rabbit-sized mammals that superficially resemble rodents. Externally, the species bear close resemblance to each other. The unique foot structure is diagnostic. The upper incisors are long and somewhat resemble those of rodents, but they are distinctly triangular in cross section.

Sirenians have distinctive skulls with posteriorly displaced external nares. Manatees and the dugong lack external hindlimbs and could be confused only with dolphins. However, their blunt snouts distinguish them from these small cetaceans.

Perissodactyls are hoofed mammals that have the axis of weight passing through the central digit. They usually have an odd number of toes (one or three), but even when four toes are present (on the forelimbs of tapirs), one toe is larger than the others. Skull and skin characters readily separate this order into three groups (the equids, the tapirs, and the rhinos), each of which may be easily recognized once the basic appearance is learned.

Artiodactyls are hoofed mammals that have the axis of weight passing between the third and fourth digits. They almost always have an even number of toes (two or four). (On most peccaries, a rudimentary third toe is present on each hindleg.) Externally, the various body forms (pig, camel, deer, cow, sheep, antelope, etc.) are familiar to most students. Horns or antlers are frequently present.

CHAPTER 10

Key to Living Families

THE MONOTREMES
Order Monotremata

ORDER MONOTREMATA

The name Monotremata, meaning "one-holed ones," refers to the cloaca, a common chamber into which the digestive, excretory, and reproductive tracts open and from which the products of these tracts leave the body. This is only one of the primitive characters found in these unique mammals.

Living monotremes are very primitive, but highly specialized, mammals. The three extant genera are the only living representatives of the subclass **Prototheria.** They possess many features that are typically reptilian. The most striking of these is the structure of their reproductive systems. Monotremes are the only living mammals that lay eggs.

The two living families differ in many ways. The duckbilled platypus (Fig. 10.1), Ornithorhynchidae, is both semiaquatic and semifossorial. The feet are webbed, and the tail is dorsoventrally flattened. The tail and body are covered with dense fur, but the large ducklike beak is hairless. The platypus eats aquatic invertebrates, which it locates by means of electroreceptors in the beak. The food is stored temporarily in cheek pouches. It digs burrows and dens in the banks of streams and ponds. The one to three eggs are laid in a den and incubated until they hatch. Upon hatching, the young feed on the thick milk secreted from the mammary glands.

The echidnas (Fig. 10.2), Tachyglossidae, are terrestrial and also semifossorial. They eat termites, ants, other insects, and earthworms, and they frequently dig to obtain these foods. When frightened, they can dig rapidly into the ground to escape predators, and they are known to excavate burrows. A temporary pouch develops on the abdomen of the female during the breeding season. A single egg (rarely two to three) is laid directly into this pouch, where it is carried until it hatches. Initially, the young remains in the pouch and drinks the mother's milk. When the young becomes too large, the pouch disappears, but the mammary glands continue to secrete, and the young continues to drink. The echidnas have a pelage of coarse hair and heavy, sharp spines, and are sometimes called spiny "anteaters."

DISTINGUISHING CHARACTERS

Teeth are absent in adults of all living species. Tooth buds form in embryonic Ornithorhynchidae, and some cheek teeth may erupt in young animals, but all traces of teeth disappear soon after birth. These temporary teeth are small, with little enamel and numerous roots. The teeth are replaced by continually growing horny plates (Marshall 1984).

The mandibular fossa is located entirely within the squamosal. The lower jaws are reduced. Only a vestige of the coronoid process remains, and no true angular process is present. Lacrimal bones and auditory bullae are absent. The jugals are reduced or absent, but the zygomatic processes of the maxilla and squamosal meet to form complete zygomatic arches.

Large precoracoids, coracoids, and an interclavicle are present in the pectoral girdle. Cervical ribs are present, as are large epipubic bones. The digits have large claws. Males of all three species, and some female echidnas as well, have a large, hollow spur on each ankle. A venom-secreting gland is located at the base of each spur in Ornithorhynchidae. The jaws are covered with rubbery, hairless skin. Vibrissae are lacking.

Figure 10.1 The duckbilled platypus, *Ornithorhynchus anatinus,* Ornithorhynchidae.
(Brazenor 1950)

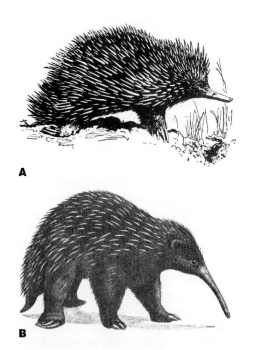

Figure 10.2 (A) The short-nosed echidna, *Tachyglossus aculeatus,* and (B) the long-nosed echidna, *Zaglossus bruijni.* Both Tachyglossidae.
(A, Brazenor 1950; B, Cabrera 1919)

The uteri are completely unfused and poorly developed. Shell glands are present, and the mammary glands lack nipples. The penis is bifurcate at the tip, attached to the ventral wall of the cloaca, and used only for the passage of semen. The testes are permanently abdominal, and a baculum is not known.

Living Families of Monotremata

Ornithorhynchidae contains only one species, the duckbilled platypus, *Ornithorhynchus anatinus,* which occurs in streams and lakes in eastern Australia, including Tasmania. Tachyglossidae contains two monotypic genera of echidnas. *Tachyglossus aculeatus* occurs in Australia and on the island of New Guinea. *Zaglossus bruijni,* is found only on New Guinea. (Groves 1993).

Key to Living Families of Monotremata

1 Rostrum with widely flaring premaxillae (Fig. 10.3) and snout broad, "duckbilled" (Fig. 10.1); tail well-developed; pelage of soft hair, external ear pinnae absent .... **Ornithorhynchidae**
duckbilled platypus

1' Rostrum and snout slender, terete (Fig. 10.4); tail vestigial; pelage of coarse hair and spines (Fig 10.2); external ear pinnae well-developed .... **Tachyglossidae**
echidnas

Comments and Suggestions on Identification

The platypus and echidnas are distinctive mammals and, except for the echidnas' skulls, should present no identification problems. The skulls of the echidnas resemble those of the smaller anteaters and pangolins in being toothless and cone-shaped. However, they have a more elevated braincase and, in the more common genus, *Tachyglossus,* the tips of the premaxillae are bent slightly upwards. The anteaters possess well-developed lacrimal

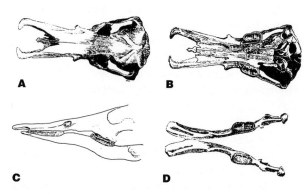

Figure 10.3 Skull of the platypus. (A and B) Dorsal and ventral views, respectively, of the skull. (C) Lateral view of the skull and lower jaw in occlusion. (D) Dorsal view of lower jaw.
(Giebel 1859)

bones, that are absent in the echidnas and pangolins, but the presence or absence of these bones is not always easily determined. Pangolin skulls are more robust than those of the others, but their lower jaws are weaker and lack an angular process.

Several groups of small mammals (e.g., hedgehogs, tenrecs, and certain rodents) possess spines and in this way superficially resemble echidnas. Check the snout, the presence of teeth, and the rest of the skull.

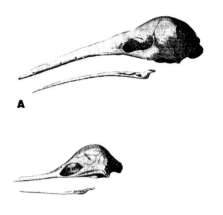

Figure 10.4 Skulls of the echidnas. (A) *Zaglossus bruijni*, and (B) *Tachyglossus aculeatus*.
(Cabrera 1919)

CHAPTER 11

THE MARSUPIALS

Order Didelphimorphia *Order Peramelemorphia*
Order Paucituberculata *Order Diprotodontia*
Order Microbiotheria   *Order Notoryctemorphia*
Order Dasyuromorphia

Keys to Living Families

The supraordinal name **Marsupialia** refers to the abdominal pouch or **marsupium** (Fig. 11.1) in which many newborn marsupials complete what could, with justification, be called larval development. About a third of the species lack a pouch, however, and the great majority of the New World forms have none. Although the idea sounds nonparsimonious, it is clear that pouchlessness is the primitive condition in marsupials and that marsupia have evolved independently several times.

The Marsupialia was long regarded as a single, extremely diverse order including insectivorous, carnivorous, omnivorous, nectarivorous, browsing, and grazing animals. They occupied diverse habitats in the Australian and Neotropical regions, ranging from the deserts of central Australia and Chile to the wet tropics of New Guinea and Brazil. Marsupials have evolved fossorial, semifossorial, ambulatory, cursorial, saltatorial, semiaquatic, arboreal, and gliding forms.

The number of incisors in the upper jaw is higher than the number in the lower jaw in all living forms except Vombatidae. The primitive marsupial dental formula, with premolars 3/3 and molars 4/4, is the reverse of the "late primitive" premolar and molar numbers in placentals. The total number of teeth often exceeds the basic eutherian number of 44. In 11 families, the incisors have been modified to form the **diprotodont** condition (Fig. 11.2). In these, the lower jaw is shortened, and the first pair of lower incisors is greatly enlarged and elongated. The upper incisors may be enlarged also but are usually unspecialized. The canines and the first premolars are frequently incisoriform. In potoroids, each anterior premolar above and below is **plagiaulacoid** (see Fig. 11.20A), meaning that it is an elongated bladelike tooth with vertical grooves, similar to that of †*Plagiaulax,* a member of the extinct order Multituberculata. The burramyid *Burramys* has such a tooth in the maxilla. In marsupials, the only tooth replacement is in the third premolar position above and below.

The brain is often relatively small, and a corpus callosum is absent. The auditory bullae, when present, are formed principally or entirely by the tympanic process of the alisphenoid. Large palatal vacuities are often present. The angular process of the dentary is inflected, and the jugal contributes to the mandibular fossa in all except *Tarsipes,* and it is only weakly developed in *Phascolarctos* and in *Myrmecobius.*

Limb structure varies considerably among the marsupials. Most are plantigrade, but some are digitigrade. Microbiotheriidae, Phalangeridae, Burramyidae, Petauridae, Phascolarctidae, Acrobatidae, Pseudocheiridae, Tarsipedidae, and many Didelphidae are well-adapted for arboreal life, and one species of didelphid, *Chironectes minimus,* has webbed hindfeet and is semiaquatic. Vombatidae and some Dasyuridae and Peramelidae are semifossorial, and Notoryctidae is fully fossorial. All Peramelemorphia hop, and most Macropodidae are highly modified for ricochetal locomotion (some are arboreal, however). In arboreal forms, the hallux (which is clawless in all marsupials) is typically opposable, but in other forms it is frequently reduced or absent. In the Peramelemorphia and Diprotodontia, the second and third digits of the hindfoot are syndactylous (Figs. 11.3 and 11.4). The toes are fused so that the skeletal elements of the two toes are encased within a single skin sheath. Two claws, one for each toe, project from the end of this syndactylous digit (see Fig. 11.4).

A simple yolk sac placenta is present in most forms. A chlorioallantoic placenta, similar to that of eutherians

Figure 11.1 A swamp wallaby, *Wallabia bicolor*, Macropodidae, with a "joey" in its marsupium.
(Brazenor 1950)

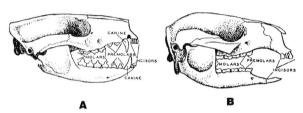

Figure 11.2 Polyprotodont (A) and diprotodont (B) marsupial skulls. A spotted-tailed quoll, Dasyurus *maculatus*, Dasyuridae (A), and a wallaby, Macropodidae (B).
(Brazenor 1950)

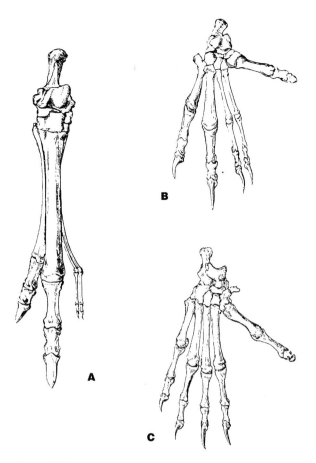

Figure 11.3 Skeletons of the right hindfeet of three marsupials. (A) A kangaroo, Macropodidae, and (B) a koala, *Phascolarctos cinereus*, Phascolarctidae, both with syndactylous digits. (C) Foot skeleton of an opossum, *Didelphis virginiana*, Didelphidae, without syndactylous digits.
(Flower and Lydekker 1891)

but lacking villi, is found in the Peramelemorphia. The gestation period is short, and the young are undeveloped at birth in comparison to those of placental mammals. There is a marsupium on the abdomen of about two-thirds of the species. This pouch contains the young as they complete development after birth. Epipubic bones are present in both sexes of all species (Fig. 11.5) but are vestigial in Notoryctidae and Thylacinidae. The uteri are completely separate, and there are two vaginal canals. These are for copulation only, and young are born through a temporary median birth canal (this becomes permanent after the first parturition in kangaroos and the noolbenger). The penis is often bifurcate at the tip and lacks a baculum. The scrotum is anterior to the penis. A shallow cloaca is present in young, but, except in Microbiotheriidae, the adults typically have separate urogenital and anal openings. Many marsupials have pelage that fluoresces vividly under ultraviolet light.

Several species of marsupials are of some economic importance. The North American opossum, *Didelphis virginiana*, for instance, is of minor importance as a furbearer and as food. Kangaroos are hunted for hides

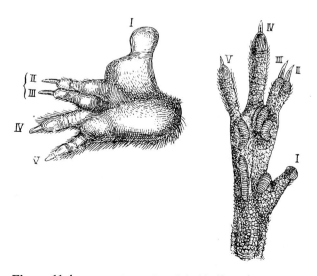

Figure 11.4 Views of the soles of the hindfeet of two marsupials with syndactylous digits.
(Weber 1928)

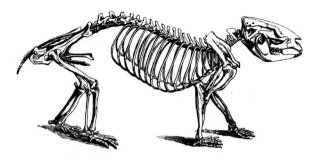

Figure 11.5 Skeleton of a wombat, Vombatidae. Note the epipubic bones projecting anteriorly from the pelvis.
(Owen 1866)

and for meat, which is used to some extent for human consumption but mostly for pet food. The brush-tailed possum, *Trichosurus vulpecula,* of Australia, is an important furbearer and also a nuisance where it lives as a semicommensal with humans, and especially in New Zealand where it has been introduced.

Many species of marsupials are used for food by Australian aborigines and by native peoples of New Guinea. Many Australian species, particularly the carnivorous forms and the smaller, less conspicuous herbivores, are endangered or extinct, probably due to competition with and/or predation by introduced placental species such as rabbits, dingoes, cats, foxes, etc., and/or due to destruction of habitat. Other species were apparently never very common. Leadbeater's possum, *Gymnobelideus leadbeateri,* was known from only five specimens collected between 1867 and 1909 in Victoria, Australia, and was considered to be extinct until a small population was discovered living at Marysville, Victoria, in 1961. The genus *Burramys* was known only as a Pleistocene fossil until a representative was discovered alive at a ski lodge near Mt. Hotham, Victoria, in 1966.

ORDER DIDELPHIMORPHIA

The Didelphimorphia consist exclusively of the opossums, family Didelphidae. They have varied food habits. Some eat other vertebrates on occasion, many are highly insectivorous, and some are frugivorous. Most are probably omnivorous. They occupy terrestrial and arboreal habitats, one is semiaquatic (Fig. 11.6), and some are probably semifossorial.

DISTINGUISHING CHARACTERS

Incisors 5/4, small and peglike. Canines well-developed. Polyprotodont. Toes didactylous. Hallux clawless, well-developed, and more or less opposable. Each foot with five digits. Marsupium absent more often than present. (The larger species mostly have pouches, whereas all of the smaller ones lack them.) Tail prehensile to semi-prehensile. Stomach simple, caecum small.

LIVING FAMILY OF DIDELPHIMORPHIA

Didelphimorphia contains a single living family, **Didelphidae,** the opossums, with 15 genera and 63 species (Gardner 1993). All species are confined to the Neotropical region except for *Didelphis virginiana,* which ranges widely into the Nearctic.

COMMENTS AND SUGGESTIONS ON IDENTIFICATION

The skulls of Didelphimorphia are distinctive (Fig. 11.7) and not easily confused with those of other mammals. Aside from the Microbiotheria and most Peramelemorphia, no other mammals have five upper incisors. Microbiotheria have large and inflated auditory bullae, quite unlike anything seen in Didelphimorphia, and Peramelemorphia have three, rather than four, lower incisors.

The skins of Didelphimorphia can resemble those of Microbiotheria and certain Dasyuromorphia, Pauci-

Figure 11.6 A didelphid opossum, the yapok or water opossum, *Chironectes minimus.*
(Duncan 1877)

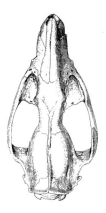 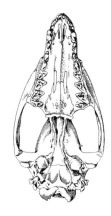 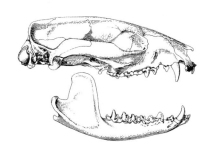

Figure 11.7 Skull of a didelphid, the coligrueso, *Lutreolina crassicaudata*.
(Marshall, L. G. 1978 *Lutreolina crassicaudata*. Mammalian Species, 91: 1–4, fig. 1. Copyright 1978 American Society of Mammalogists. Reprinted with permission.)

tuberculata, Diprotodontia, Insectivora, Primates, and Rodentia. The combination of five separate toes on each hindfoot with a clawless/nailless, often obviously offset, hallux, frequently associated with a naked, ratlike tail, will, generally, identify Didelphimorphia. Paucituberculata have characteristic flaps on their lips and have more uniformly dark pelage and smaller eyes than most Didelphimorphia. Microbiotheria combine the size of a mouse with a tail about as long as the body, and that is furred to the tip, a combination not found in any known Didelphimorphia.

ORDER PAUCITUBERCULATA

The caenolestids or flaplips are small and rather shrew-like in appearance (Fig. 11.8). They live in humid areas and feed on insects, earthworms, vegetation, fungi, and seeds (Barkley and Whitaker 1984; Meserve et al. 1988; Patterson and Gallardo 1987).

DISTINGUISHING CHARACTERS

Diprotodont incisors, usually 4/3 in number (Fig. 11.9). A preorbital vacuity is present between the nasal, maxillary, and frontal bones. The mastoid is large and broadly exposed laterally. Both the upper and lower lips are provided with external membraneous flaps on each side. A marsupium is always absent. The tail is long and haired to the tip. Stomach divided into three distinct parts. The limbs are subequal, with the hindfoot didactylous. Caenolestids share, with didelphids, pairing of the spermatozoa, a trait unknown in Australian marsupials or in placentals.

LIVING FAMILY OF PAUCITUBERCULATA

The Paucituberculata contains a single living family, **Caenolestidae,** the caenolestids or flaplips, with three genera and five species (Gardner 1993). The order is confined to western South America where two genera, *Caenolestes* and *Lestoros,* live in high, wet, cool areas on the western slope of the northern Andes. *Rhyncholestes* is known from only a few specimens from Chiloé Island and adjacent coastal areas of central Chile (Meserve et al. 1982).

Figure 11.8 A caenolestid or flaplip, *Caenolestes fuliginosus*.
(Osgood 1921)

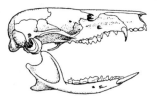

Figure 11.9 Skull of a caenolestid, *Caenolestes fuliginosus*.
(Osgood 1924)

COMMENTS AND SUGGESTIONS ON IDENTIFICATION

The diprotodont dentition separates the caenolestid skull from those of all marsupials except the Diprotodontia. Caenolestids are much smaller than most diprotodonts and differ from them in having four upper incisors. The skins resemble those of mice or shrews and some tenrecids but can be distinguished from these by the presence of a clawless/nailless hallux and flaps on the lips.

ORDER MICROBIOTHERIA

The mouse-sized monito del monte, *Dromiciops australis,* is the only extant species of microbiotherian. It is very opossumlike in appearance, is semiarboreal or scansorial, lives in dense, humid forests, and apparently feeds primarily on insects and other invertebrates.

DISTINGUISHING CHARACTERS

Polyprotodont incisors 5/4. First incisors and canines not particularly prominent. Large, inflated auditory bullae. Toes didactylous. Hallux opposable. Marsupium present. Cloaca present. Tail prehensile, equal in length to body, furred throughout its length except for narrow strip ventrally on terminal portion.

LIVING FAMILY OF MICROBIOTHERIA

The order Microbiotheria contains one living family, **Microbiotheriidae,** which contains a single living species, *Dromiciops australis,* the monito del monte (Gardner 1993). The geographic range is restricted to central Chile, including Chiloé Island, and barely into adjacent Argentina.

COMMENTS AND SUGGESTIONS ON IDENTIFICATION

The five upper incisors and four lower incisors, combined with large, inflated bullae (Fig. 11.10), each about a third as wide as the braincase, separate the microbiotherian skull from those of all other mammals. The skin resembles those of certain opossums, but its being mouse-sized in combination with a tail as long as the body and furred to the tip, separates it from those. The naked strip under the tail distinguishes it from all dasyuromorphians and from the very small Madagascan primates. All diprotodonts are syndactylous rather than didactylous. Most rodents lack an opposable hallux, and none we know of has a clawless/nailless one.

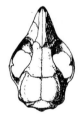

Figure 11.10 Skull of the living microbiotherian, the monito del monte, *Dromiciops australis.*
(Osgood 1943)

ORDER DASYUROMORPHIA

The Dasyuromorphia include marsupials which feed mostly upon animal material. Some eat other vertebrates, and many are largely insectivorous. One, the numbat, *Myrmecobius fasciatus* (Fig. 11.11), feeds almost exclusively on termites. Most are terrestrial, but a few are arboreal, and some dig burrows, although none is truly fossorial. The presumably extinct Tasmanian "wolf," *Thylacinus cynocephalus,* showed remarkable convergence toward members of the carnivoran family Canidae.

DISTINGUISHING CHARACTERS

Polyprotodont. Incisors 4/3. Canines well-developed. Toes didactylous. Hallux, if present, clawless. Marsupium present or absent or conspicuous only seasonally. Tail nonprehensile. Stomach simple, caecum absent.

LIVING FAMILIES OF DASYUROMORPHIA

The Dasyuromorphia contains three families that were living early in the twentieth century (Groves 1993). All are restricted to the Australian Region. **Dasyuridae** contains 15 genera and 61 species of basically carnivorous marsupials which occur widely in Australia, New Guinea, and adjacent islands. **Myrmecobiidae** contains only the numbat, *Myrmecobius fasciatus,* which is found only in southwestern Australia. **Thylacinidae** had only a single living species during historic times, the Tasmanian "wolf," *Thylacinus cynocephalus,* which is almost certainly extinct and occurred only in Tasmania in historic times.

2'   Size smaller, condylobasal length less than 120 mm, usually considerably less, if 120 mm or more, then dental formula 4/3, 1/1, 2/2, 4/4 = 42 (Fig. 11.13); dorsal pelage plain or spotted (Fig. 11.14) but not striped as above ....
....**Dasyuridae**
phascogales, quolls (native "cats"), dunnarts (marsupial "mice"), Tasmanian devil

Figure 11.11 The numbat or banded anteater, *Myrmecobius fasciatus,* Myrmecobiidae; feeding, and lateral view of skull.
(Whole animal, Flower and Lydekker 1891; skull, Gregory 1910)

KEY TO LIVING FAMILIES OF DASYUROMORPHIA

1   Postcanine teeth 7–8/8–9, for a total of 48 to 52, reduced and widely separated (Fig. 11.11); dorsal pelage dark with six to nine transverse pale bands (Fig. 11.11); total length less than 500 mm .... **Myrmecobiidae**
numbat, *Myrmecobius fasciatus*

1' Postcanine teeth 6–7/6–7 for a total of 42 to 46, well-developed for chopping insects or shearing flesh, not widely separated (Figs. 11.12, 11.13); dorsal pelage without bands, or if with bands then with more than seven dark bands on a paler general pelage color (Fig. 11.12); total length various ....,............................. 2

2 (1) Size larger, condylobasal length greater than 120 mm; dental formula 4/3, 1/1, 3/3, 4/4 = 46 (Fig. 11.12); dorsal pelage with several dark transverse stripes over the rump (Fig. 11.12) ..
....**Thylacinidae**
Tasmanian "wolf," *Thylacinus cynocephalus*

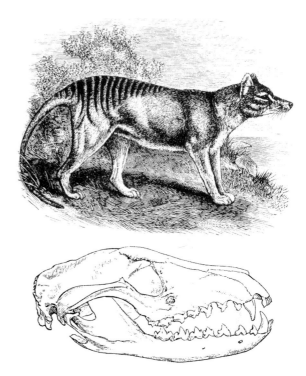

Figure 11.12 The presumably extinct Tasmanian "wolf" or thylacine, *Thylacinus cynocephalus,* Thylacinidae, and a lateral view of its skull.
(Flower and Lydekker 1891)

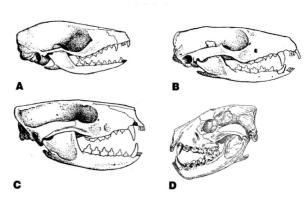

Figure 11.13 Skulls of representative Dasyuridae. (A) The fat-tailed dunnart, *Sminthopsis crassicaudata,* (B) the brush-tailed phascogale, *Phascogale tapoatafa,* (C) the spotted-tailed quoll, *Dasyurus maculatus,* (D) the Tasmanian devil, *Sarcophilus lanarius.* Not all to same scale.
(A–C, Brazenor 1950; D, Gregory 1910)

Figure 11.14 Representative Dasyuridae. (A) The common dunnart, *Sminthopsis murina*, (B) the brush-tailed phascogale, *Phascogale tapoatafa*, (C) the eastern quoll, *Dasyurus viverrinus*.
(Brazenor 1950)

Comments and Suggestions on Identification

The thylacinid skull and those of the larger dasyurids look like those of Carnivora at first glance. The numbat skull and smaller dasyurid skulls resemble those of certain Insectivora. Check the number of upper incisors and the angular process of the ramus.

The skins of Dasyuromorphia frequently resemble those of Didelphimorphia, Microbiotheria, Insectivora, Carnivora, or Rodentia. As for the skins, the nonprehensility of the tails of dasyuromorphians, signified by the concomitant lack of any truly naked portion of the tail, should generally separate dasyuromorphians from didelphimorphians and microbiotherians. The clawless/nailless hallux should be useful in most cases for separating dasyuromorphians from members of the eutherian orders just mentioned.

ORDER PERAMELEMORPHIA

The bandicoots are terrestrial, primarily insectivorous, Australian marsupials, some of which have been compared to rabbits in size and appearance (Fig. 11.15). The bandicoots have sometimes been considered to be phylogenetically intermediate between the Dasyuromorphia and the Diprotodontia. They share the primitive nondiprotodont dentition with the Dasyuromorphia and derived syndactylous toes with the Diprotodontia.

Distinguishing Characters

Dental formula 4–5/3, 1/1, 3/3, 4/4 = 46–48. Incisors with flattened, not pointed, crowns, and the canines widely spaced from both the last incisor and first premolar. Rostrum elongate and skull conical in general shape. Hindlimbs noticeably longer than forelimbs. Only the

Figure 11.15 A peramelid, a barred bandicoot, *Perameles*. Some Peramelidae have shorter ears than members of this genus, but bilbies (*Macrotis*) have much longer ears.
(Brazenor 1950)

medial two or three manual digits well-developed and clawed—lateral digits on manus rudimentary or absent. First pedal digits rudimentary or absent, second and third pedal digits slender and syndactylous, fourth pedal digit largest, and fifth reduced in size but usually functional. Marsupium always present, opening to the rear. Stomach simple, moderate-sized caecum present. With chorioallanatoic placenta unlike eutherian ones in lacking villi. Clavicle rudimentary or absent.

LIVING FAMILIES OF PERAMELEMORPHIA

Peramelemorphia contains two living families. The **Peramelidae**, the dry-country bandicoots and bilbies, contains four genera and 10 species (Groves 1993). They are found throughout mainland Australia, Tasmania, southern and eastern New Guinea (one species), and some adjacent islands. The **Peroryctidae**, the rainforest bandicoots, containing four genera and 11 species (Groves 1993), are found only in New Guinea and nearby islands, with one species in northeastern Australia.

KEY TO LIVING FAMILIES OF PERAMELEMORPHIA

1 Molar crowns squarish, main palatal vacuities almost touching (often fused in adults), mesopterygoid fossa narrowed in middle (Groves *in litt.*) .... **Peramelidae** dry-country bandicoots

1' Molar crowns more or less triangular or trapezoidal, main palatal vacuities showing no tendency to fuse, mesopterygoid fossa broad and parallel-sided (Groves *in litt.*) .... **Peroryctidae** rainforest bandicoots

COMMENTS AND SUGGESTIONS ON IDENTIFICATION

Bandicoot skulls (Fig. 11.16) resemble those of some Insectivora, certain Macroscelidea, or Scandentia, but they can be readily distinguished from these by the number of upper incisors and the inflected angular process of the ramus. The skins can resemble those of large rodents, insectivores, or certain macroscelideans, but they differ from all of these in the presence of the syndactylous digits on the hindfeet.

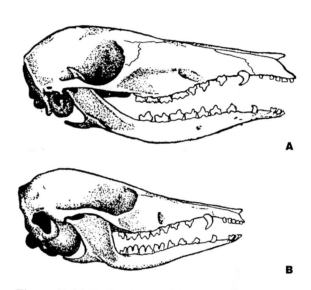

Figure 11.16 Skulls of peramelids. (A) The western barred bandicoot, *Perameles bougainville*, (B) the lesser bilby, *Macrotis leucura* (probably extinct).
(Brazenor 1950)

ORDER DIPROTODONTIA

The ordinal name refers to the diprotodont dentition that this group shares with the Paucituberculata. The Diprotodontia include, among others, the primarily herbivorous marsupials of the Australian Region. They vary considerably in body size and form, ranging from small mouselike creatures to large kangaroos. They occupy diverse habitats and include terrestrial, semifossorial, and arboreal animals.

DISTINGUISHING CHARACTERS

Diprotodont incisors may be 3/2–3, 3/1, 2/1, or 1/1 in number, second and third lower incisors are minute when present. Marsupium always present, opening anteriorly or posteriorly. Stomach simple in most forms, complex in Macropodidae, in which it functions much like that of ruminants. Caecum absent (Tarsipedidae) to very long and complex (Phascolarctidae). Second and third digits of hindfoot syndactylous.

LIVING FAMILIES OF DIPROTODONTIA

A list of living families or Diprotodontia, and their contents, is given in Table 11.1. The Diprotodontia are confined to the Australian Faunal Region. The **Phalangeridae** range throughout the Australian Region from Sulawesi east to the

TABLE 11.1	Living Families of Diprotodontia*			
		Number of		
Family	Common Name	Genera	Species	Distribution
Pseudocheiridae	Ring-tailed possums, rock possum, greater gliding possum	5	14	Australian
Phalangeridae	Cuscuses, brush-tailed possums, scaly-tailed possum	6	18	Australian
Burramyidae	Pygmy possums	2	5	Australian
Petauridae	Striped possums, Leadbeater's possum, lesser gliding possums	3	10	Australian
Potoroidae	"rat"-kangaroos	5	9†	Australian
Macropodidae	Kangaroos and wallabies	11	54	Australian
Phascolarctidae	Koala	1	1	Eastern Australia
Vombatidae	Wombats	2	3	Southeastern Australia
Tarsipedidae	Noolbenger/honey possum	1	1	Southwestern Australia
Acrobatidae	Feather-tailed possum, feathertail glider	2	2	Australian

* Groves 1993.
†1 extinct, 1 possibly extinct.

Solomon Islands and south to Tasmania. One species of phalangerid, *Trichosurus vulpecula,* was introduced into New Zealand, where it has become a pest. The **Macropodidae** and **Pseudocheiridae** are both known from mainland Australia, Tasmania, New Guinea, and some adjacent islands. Macropodids have also been introduced into New Zealand, Hawaii, and elsewhere. **Petauridae** live in New Guinea, mainland Australia, Tasmania, and various islands in the region. The **Burramyidae** occur in New Guinea, mainland Australia, and Tasmania. **Potoroidae** are found in eastern and southern mainland Australia, nearby islands, and Tasmania, whereas **Acrobatidae** live only in New Guinea and the eastern coast of mainland Australia. **Vombatidae** are restricted to Tasmania and southeastern and southcentral mainland Australia, and **Tarsipedidae** occur only in southwestern Australia. **Phascolarctidae** are found only in eastern Australia, but fossils are known from the southwestern portion of the country.

KEY TO LIVING FAMILIES OF DIPROTODONTIA

1 Dental formula 1/1, 0/0, 1/1, 4/4 = 24; incisors resemble those of rodents (Fig. 11.17); limbs subequal; hallux vestigial; tail vestigial (Fig. 11.17) .... **Vombatidae** wombats

1' Dental formula various but never as above; upper incisors number 2 or 3, not like those of rodents; limbs subequal or with hindlimbs noticeably longer than forelimbs; hallux absent or well-developed, or if vestigial, hindlimbs larger than forelimbs; tail long, or if vestigial, then hallux well-developed and opposable 2

Figure 11.17 The coarse-haired wombat, *Vombatus ursinus,* Vombatidae, and a skull of the same species.
(Whole animal, Brazenor 1950; skull, Gregory 1910)

2 (1') Incisors 2/1; canines 1/0; cheek teeth peglike and vestigial (Fig. 11.18), never more than 3/3; lower jaw and zygomatic arch extremely slender; tail long and sparsely haired; except for clawless hallux and clawed syndactylous toes, all digits with nails (Fig. 11.18); dorsal pelage with three dark longitudinal stripes (Fig. 11.19) ...... **Tarsipedidae** noolbenger or honey possum, *Tarsipes spenserae*

2' Incisors 3/1–3; canines 0–1/0; cheek teeth never peglike or vestigial, usually more than 3/3; lower jaw and zygomatic arch well-developed; tail long and fully haired for at least part of its length, long and naked but scaly, or short and vestigial; only claws present on digits, nails absent; dorsal pelage coloration variable, if dorsal pelage with three longitudinal dark stripes, then tail bushy **3**

3 (2') Dental formula 3/1–2, 0–1/0, 0–2/0–2, 3–4/3–4 = 30–34; upper incisors approximately equal in size, lower incisor extremely procumbent; masseteric fossa deep, masseteric canal present but not always obvious (Fig. 11.20); hindlegs (including excessively long feet) much larger than forelegs (including feet); tail long and usually tapering from a stout base (Figs. 11.1, 11.21) **9**

3' Dental formula various but if incisors 3/1 then I^1 larger than I^2 and I^3, and I_1 less procumbent (Fig. 11.22); masseteric fossa shallow, masseteric canal absent (Figs. 11.22, 11.24, 11.25, 11.26); limbs subequal; tail short and vestigial or long and slender, never tapering from a stout base (Figs. 11.23, 11.28) **4**

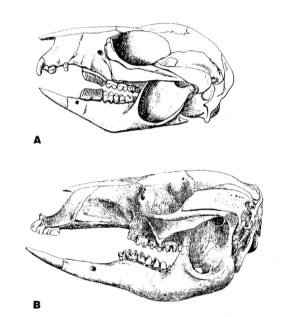

Figure 11.20 Skulls of a potoroid and a macropodid. (A) A burrowing bettong, *Bettongia lesueur* (note the bladelike, plagiaulacoid premolars), (B) a red-necked wallaby, *Macropus rufogriseus* (note the deep masseteric fossa and the masseteric canal).
(Flower and Lydekker 1891)

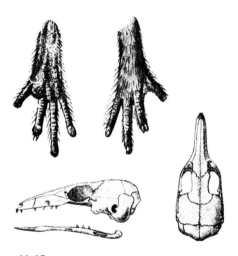

Figure 11.18 Dorsal and plantar views of the right hindfoot and lateral and dorsal views of the skull of the noolbenger or honey possum, *Tarsipes spenserae*, Tarsipedidae.
(Cabrera 1919)

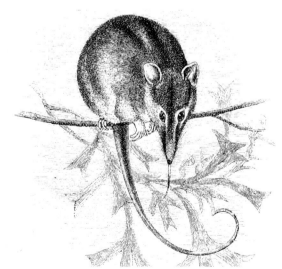

Figure 11.19 The noolbenger or honey possum, *Tarsipes spenserae*, Tarsipedidae.
(Flower and Lydekker 1891)

Figure 11.21 The long-nosed potoroo, *Potorous tridactylus*, Potoroidae.
(Brazenor 1950)

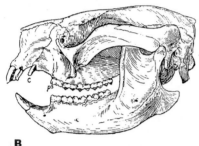

Figure 11.22 (A) The koala, *Phascolarctos cinereus*, Phascolarctidae, and (B) a lateral view of its skull.
(Whole animal, Kingsley 1884; skull, Gregory 1910)

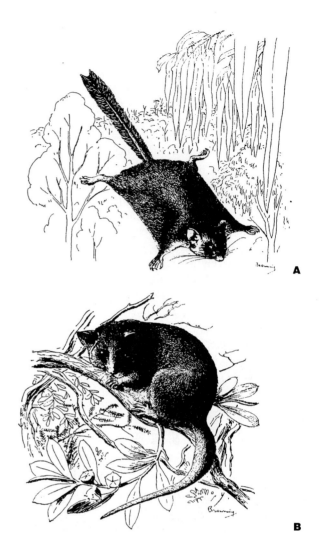

Figure 11.23 An acrobatid and a burramyid. (A) The feathertail glider, *Acrobates pygmaeus*, and (B) the eastern pygmy possum, *Cercatetus nanus*.
(Brazenor 1950)

4 (3')   Incisors 3/1 (Fig. 11.22); molars with crescent-shaped ridges; tail rudimentary; marsupium opens posteriorly .... **Phascolarctidae**
koala, *Phascolarctos cinereus*

4'   Incisors usually 3/2–3, second and third lower incisors tiny or sometimes absent (Figs. 11.24, 11.25, 11.26); molars with rounded cusps or crescent-shaped ridges; tail long (Figs. 11.23 11.28); marsupium opens anteriorly **5**

5 (4')   One upper premolar plagiaulacoid or not. Mouse-sized, head and body length 60–130 mm; tail either furred at base and essentially naked for remainder of length or covered with very short hair throughout its length (Fig. 11.23B); tail never with lateral fringes of hair; never with gliding membranes .... **Burramyidae**
pygmy possums

5'   No plagiaulacoid upper premolar. Rat-sized or larger, head and body length more than 130 mm, or if mouse-sized, then tail with lateral fringes of hair; tail heavily furred, mostly naked, or in between; gliding membranes present or absent **6**

6 (5')   First lower incisor greatly enlarged, maximum exposure approaching length of rest of toothrow, procumbent (Fig. 11.24B); upper molars diminish progressively in size posteriorly; tail fully furred to tip or underside at or near extreme tip naked; back of ears not excessively hairy; fourth manual digit may be markedly elongated, in which case with three bold dark brown or black dorsal stripes ...... **Petauridae**
striped possums, lesser gliders, Leadbeater's possum

6'   First lower incisor usually not as greatly enlarged and procumbent as described above; upper molars not diminishing progressively in size posteriorly; underside of tail with at least substantial portion toward tip naked (Fig. 11.27) unless with gliding membranes and/or tail with lateral fringes of hair; back of ears may be excessively hairy; fourth manual digit not markedly elongated; without three bold dark dorsal stripes **7**

7 (6') Molars 3/3 (Fig. 11.25), tail with prominent lateral fringes of hair (Fig. 11.23A); mouse-sized, head and body length = 65–132 mm ...... **Acrobatidae**
feather-tailed possum, and feathertail glider

7' Molars 4/4, tail without prominent lateral fringes of hair, rat-sized or larger, head plus body length more than 132 mm **8**

8 (7') Last lower premolar enlarged, "swollen," somewhat triangular, more-or-less unicuspid, projecting above level of molar crowns, often somewhat offset or "splayed outwards" (Fig. 11.26); last upper premolar without multiple cusps, sometimes similar in appearance to last lower premolar but in less exaggerated form; never with gliding membranes or with large ears with extremely hairy backs (Fig. 11.27) ....
.... **Phalangeridae**
cuscuses, brush-tailed possums, scaly-tailed possum

8' Last upper and lower premolars not projecting beyond level of molar crowns, elongated, with multiple cusps (Fig. 11.24A); gliding membrane absent (Fig. 11.28B) or present (Fig. 11.28A), if present then animal with backs of rather large ears extremely hairy .... **Pseudocheiridae**
ring-tailed possums, rock possum, greater gliding possum

Figure 11.25 Skull of an acrobatid, the feathertail glider, *Acrobates pygmaeus*.
(Brazenor 1950)

Figure 11.26 Skull of the gray cuscus, *Phalanger orientalis*, Phalangeridae.
(Flower and Lydekker 1891)

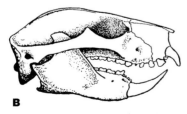

Figure 11.24 Skulls of a pseudocheirid and a petaurid. (A) The ring-tailed possum, *Pseudocheirus peregrinus*, and (B) the striped possum, *Dactylopsila trivirgata*.
(Brazenor 1950)

Figure 11.27 Two representatives of the Phalangeridae. (A) The common brush-tailed possum, *Trichosurus vulpecula*, and (B) the spotted cuscus, *Spilocuscus maculatus*.
(Brazenor 1950)

Comments and Suggestions on Identification

The families Vombatidae, Phascolarctidae, and Tarsipedidae are distinctive, unique in many ways. The wombat and koala skulls could be mistakenly identified as of rodents, but the inflected angular process of the ramus identifies them as marsupials. The *Tarsipes* skull somewhat resembles that of nectarivorous bats, but the combination of otherwise greatly reduced dentition coupled with a pair of markedly elongated lower incisors is unique in this mammal. The skulls of the Macropodidae, Potoroidae, and the remaining phalangeroid families can be distinguished from those of all other marsupials except Caenolestidae by the procumbent lower incisors. These animals are usually larger than caenolestids and differ from them in the number of upper incisors. The macropodids and potoroids can be differentiated from the other phalangeroids by the masseteric fossa and canal. Potoroids, unlike macropodids, have plagiaulacoid premolars. Burramyid skulls are the size of those of mice, and one species has a plagiaulacoid upper premolar. Acrobatid skulls are also mouse-sized and have molars 3/3. Petaurid skulls show very large, rodentlike first lower incisors, and the upper molars diminish in size posteriorly. Phalangerids have enlarged, triangular last premolars, and pseudocheirids have nonprojecting, elongated, multicusped last premolars.

The skins of various possums look like those of some didelphids, various rodents, and some primates. The syndactylous toes of the possums should distinguish them from these kinds of mammals. Petaurids may have gliding membranes inserting anteriorly on the fifth digit, are rat-sized or larger, have three bold dark dorsal stripes, or have a tail the bushiness of which increases posteriorly and appears somewhat laterally compressed. Burramyids are mouse-sized, with scantily haired tails. Acrobatids have tails with lateral hairy fringes, which make the tails appear featherlike. The gliding pseudocheirid, *Petauroides*, has its gliding membranes inserting on the elbows and has rather large, very furry ears.

Figure 11.28 Two representatives of the Pseudocheiridae. (A) The greater gliding possum, *Petauroides volans* (gliding animal's front limbs shown in incorrect posture—elbows should be out but front feet tucked in by head), and (B) the ring-tailed possum, *Pseudocheirus peregrinus*.
(Brazenor 1950)

9 (3)   Premolars 1/1, plagiaulacoid (Fig.11.20A) ......
....**Potoroidae**
"rat"-kangaroos

9'   Premolars 2/2, not plagiaulacoid (Fig. 11.20B)
.... **Macropodidae**
kangaroos and wallabies

ORDER NOTORYCTEMORPHIA

The Australian marsupial "moles" are among the most fossorially adapted mammals. They live in sandy deserts and feed on insects.

DISTINGUISHING CHARACTERS

Dental formula 3–4/3, 1/1, 2/2–3, 4/4 = 40–44. Skull conical. Skin with horny rostral shield. Claws of third and fourth digits of forefoot enormously enlarged (Fig. 11.29), forming spadelike digging unit. Pinnae and externally visible eyes absent. Fur silky, pale, and iridescent. Tail short, naked, conical, and annulated. Marsupium present. Epipubic bones vestigial.

LIVING FAMILY OF NOTORYCTEMORPHIA

The order Notoryctemorphia has a single living family, **Notoryctidae,** the marsupial "moles," with one genus and two species (Groves 1993), which are both confined to central and western portions of Australia.

COMMENTS AND SUGGESTIONS ON IDENTIFICATION

The skulls of notoryctemorphians are probably most reminiscent of those of various Insectivora. The presence of the inflected jaw angle and four molars (Fig.11.29) should identify the notoryctemorphian skull, however.

The skins of marsupial "moles" superficially resemble those of true moles (Insectivora: Talpidae), certain fossorial Rodentia (Muridae: *Spalax, Nannospalax*; some Bathyergidae), and, especially, those of golden-moles (Insectivora: Chrysochloridae). True moles have five subequal-sized digits on the forefeet. They and the fossorial rodents also lack the horny rostral shield found in notoryctids. The golden moles have a short rostral pad but have no external tail.

Figure 11.29 A marsupial "mole," *Notoryctes typhlops,* Notoryctidae.
(Whole animal, Beddard 1902, skull, Cabrera 1919)

CHAPTER 12

Key to Living Families

THE INSECTIVORES
Order Insectivora

ORDER INSECTIVORA

The name Insectivora refers to the diet of most members of this order. Although a diet of insects and other small invertebrates is common among the diverse mammals included in Insectivora, the ordinal name does not indicate a diagnostic character. Some members of the order Insectivora are quite omnivorous or carnivorous, whereas various species in several other orders have diets that are almost exclusively insectivorous. Insectivores are generally small, rather primitive mammals. The order has been used as a "wastebasket taxon" into which many kinds of living and fossil mammals of dubious relationships have been placed. Most species are terrestrial, but fossorial and/or semiaquatic forms are not uncommon, and semiarboreal forms exist.

Some insectivores (e.g., some moles) are considered pests in some parts of their range, but none are of major economic importance. The Russian desman, *Desmana moschata,* and several other species in Talpidae, were once commercially important for their fur. One species of shrew, *Suncus murinus,* is a commensal with humans through much of the Orient.

DISTINGUISHING CHARACTERS

No single character or simple combination of characters can be given that will distinguish all insectivores from all other mammals. They are small animals ranging in size from some of the smallest known mammals, the shrews *Suncus etruscus* and *Sorex hoyi,* to species the size of a rabbit. The pelage usually consists of only one kind of hair other than vibrissae, though some forms have spines as well. The feet are usually plantigrade and with five toes. The dentition is generally simple, and the cheek teeth are zalambdodont in some groups.

The olfactory capsules are longer than the brain and largely interorbital. The maxillae extend into the orbital walls and separate the lacrimal bones from the palatines. The jugals are reduced or absent, and the zygomatic arches are sometimes incomplete. Postorbital bars are never present. A baculum has been reported in some insectivores and may exist in others.

LIVING FAMILIES OF INSECTIVORA

For a list of families and their contents see Table 12.1.

Insectivores range over most of the world's land surface except for the Australian and southern Neotropical Regions, Antarctica, and most oceanic islands. The European hedgehog, *Erinaceus europaeus,* has been introduced into New Zealand; a tenrec, *Tenrec eacaudatus,* has been introduced onto Réunion, Mauritius, and islands in the Comoro and Seychelles groups in the Indian Ocean; and one commensal shrew, *Suncus murinus,* has followed modern humans into New Guinea, Guam, Madagascar, and various Old World continental seaports.

TABLE 12.1	Living Families of Insectivora*			
Family	Common Name	Number of Genera	Species	Distribution
Solenodontidae	Solenodons	1	2	Cuba, Hispaniola
Tenrecidae	Tenrecs, otter shrews	10	24	Madagascar (tenrecs) and westcentral Africa (otter shrews)
Chrysochloridae	Golden moles	7	18	Central and southern Africa
Erinaceidae	Hedgehogs, gymnures	7	21	Ethiopian, Palearctic, Oriental
Soricidae	Shrews	23	312	Holarctic, Ethiopian, Oriental, northern Neotropical
Talpidae	moles and desmans	17	42	Holarctic, Oriental

*The families listed here, and the numbers of genera and species, follow Hutterer (1993), except that the completely extinct family Nesophontidae and one extinct species of Solenodontidae are omitted from this table and from discussion in this chapter.

KEY TO LIVING FAMILIES OF INSECTIVORA

1 I^1 large, protruding forward and hooked, small cusp present behind main cusp (Fig. 12.2); teeth may or may not be pigmented; small animals with short dense fur (Fig. 12.1) ...... **Soricidae** shrews

1' I^1 may be large, but if hooked, no accessory cusp present; teeth never pigmented; size and pelage various, frequently spiny **2**

2 (1') Zygomatic arch incomplete **3**

2' Zygomatic arch complete **4**

3 (2) Dental formula 3/3 1/1 3/3 3/3 = 40; I^1 large (Fig. 12.3); I_2 large with longitudinal, lingual groove (Fig. 12.4); pelage long and lax; tail long and scaly (Fig 12.5) .... **Solenodontidae** solenodons

3' Dental formula various; I^1 small or large; I_2 may be large but never grooved (Fig. 12.6); pelage various, usually very short or with spines (Fig. 12.7); tail various .... **Tenrecidae** tenrecs and otter shrews

4 (2') Crowns of upper molars quadrate (Fig. 12.8); eyes and pinnae large, pelage may include spines (Fig. 12.9) .... **Erinaceidae** hedgehogs and gymnures

4' Crowns of upper molars triangular (zalambdodont) or with cusps in a W-shaped pattern (dilambdodont) (Fig. 12.10A); eyes and pinnae small or absent, pelage never spiny **5**

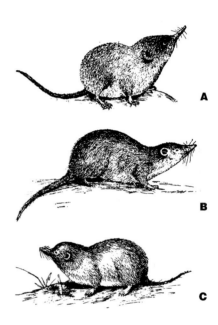

Figure 12.1 Representatives of the three most widespread genera of Soricidae. (A) *Sorex araneus,* (B) *Suncus murinus,* and (C) *Crocidura suaveolens.* Not all to same scale.
(Hsia et al. 1964)

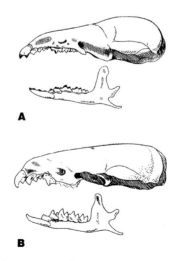

Figure 12.2 Skulls of a pigmented-toothed shrew. (A) *Sorex araneus,* and (B) an unpigmented-toothed shrew, *Crocidura suaveolens.* Both Soricidae.
(Stroganov 1957)

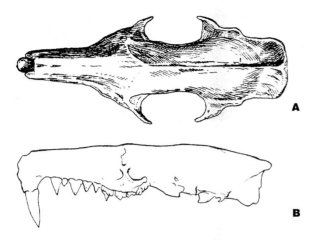

Figure 12.3 Dorsal (A) and lateral (B) views of the skull of a solenodon, *Solenodon paradoxus,* Solenodontidae.
(A, Cabrera 1925; B, John D. Whitesell)

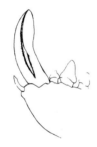

Figure 12.4 A lingual view of the anterior portion of the mandible of a solenodon.
(John D. Whitesell)

Figure 12.5 The Cuban solenodon, *Solenodon cubanus,* Solenodontidae.
(Flower and Lydekker 1891)

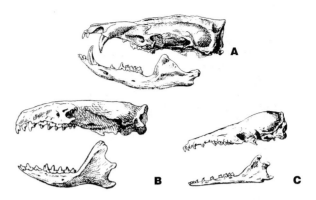

Figure 12.6 Skulls of three tenrecids illustrating some of the diversity that occurs in this family. (A) *Tenrec ecaudatus,* (B) *Setifer setosus,* and (C) *Hemicentetes semispinosus.*
(Cabrera 1925)

Figure 12.7 Representative tenrecids. (A) An otter shrew, *Potamogale velox,* and (B) a tenrec, *Tenrec ecaudatus.*
(Cabrera 1925)

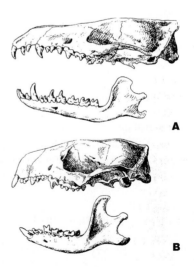

Figure 12.8 Skulls of representative Erinaceidae. (A) A gymnure, the moon "rat," *Echinosorex gymnura,* and (B) a hedgehog, *Atelerix algirus.*
(Cabrera 1925)

5 (4') Upper molars zalambdodont (V-shaped) (Fig. 12.10B); dental formula 3/3 1/1 3/3 2/2 or 3/3 = 36 or 40; I^1 enlarged; skull a short cone, zygomatic arches broad (Fig. 12.11); forefoot with four digits and with two central claws much larger than others; tail rudimentary (Fig. 12.12) ....**Chrysochloridae**
golden moles

5' Upper molars dilambdodont (Fig 12.10A); dental formula various; I^1 may (Fig. 12.13) or may not (Fig. 12.14) be enlarged; skull generally long and conical, zygomatic arches generally weak; forefoot with five digits; claws on digits two and three not particularly larger than others (Fig. 12.15), tail length various ...... **Talpidae**
moles and desmans

Figure 12.9 Representative Erinaceidae, (A) A hedgehog, *Atelerix algirus,* and (B) a gymnure, *Hylomys suillus.*
(Cabrera 1925)

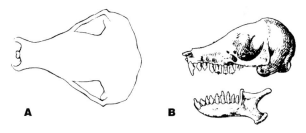

Figure 12.11 Skull of a golden mole, *Chrysochloris asiatica* Chrysochloridae. (A) Diagrammatic dorsal view and (B) lateral view.
(A, John D. Whitesell; B, Cabrera 1925)

Figure 12.12 A golden mole, *Chlorotalpa leucorhina,* Chrysochloridae.
(Dekeyser 1955)

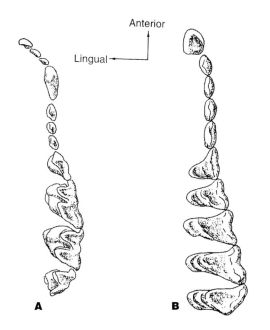

Figure 12.10 Representative occlusal surfaces of the teeth of Insectivora. (A) Occlusal surfaces of the left upper tooth row from a European mole, *Talpa europaea,* Talpidae, with a dilambdodont (W-shaped) cusp pattern on the molars. (B) Left upper tooth-row from a giant otter shrew, *Potamogale velox,* Tenrecidae, with a zalambdodont (V-shaped) cusp pattern on the molars.
(Feldhamer et al. 1999)

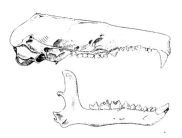

Figure 12.13 Skull of a Russian desman, *Desmana moschata,* Talpidae.
(Gromov et al. 1963)

Figure 12.14 Skull of the European mole, *Talpa europaea*, Talpidae.
(Cabrera 1925)

Figure 12.15 Representative species of Talpidae. (A) A mole, *Scalopus aquaticus,* and (B) a Russian desman, *Desmana moschata.*
(A, Kingsley 1884; B, Flower and Lydekker 1891)

COMMENTS AND SUGGESTIONS ON IDENTIFICATION

Unfortunately, there is no single good characteristic distinguishing all insectivores from all other mammals. Familiarize yourself with the general appearance of each family and the other groups with which it might be confused. Only the Tenrecidae has great diversity within the family, and except for the subfamily Potamogalinae, the otter shrews, tenrecids are confined to Madagascar. Also keep in mind that:

1. Moles, golden moles, and marsupial "moles" resemble each other. Check and compare forefeet, skulls, and localities collected.
2. Hedgehogs, tenrecs, and echidnas resemble each other. Check skulls and localities. There are no hedgehogs in Madagascar, and spiny tenrecids are confined to that island.
3. Certain shrews and certain moles resemble each other. Check the skulls for the distinctive first upper incisor of shrews.

CHAPTER 13

THE COLUGOS
Order Dermoptera

ORDER DERMOPTERA

The ordinal name, Dermoptera, literally means "skin winged ones" and refers to the extensive patagia of these animals (Fig. 13.1). Dermopterans are called colugos or "flying lemurs," but *flying* is inaccurate because these animals glide but do not fly, and *lemur* is incorrect because true lemurs are in the order Primates. Colugo is the preferred common name.

Colugos are herbivorous, feeding upon the young leaves, buds, and fruits of trees in which they live. Adept gliders, they can cover great distances with little loss of altitude. When moving about in trees, they hang from their large curved claws and move in a slow, suspended quadrupedal fashion similar to that of sloths.

In some parts of their range, colugos are hunted for their meat and skins, but they are of little economic importance. No zoo has yet kept one alive for more than a few months (Schultze-Westrum 1975).

DISTINGUISHING CHARACTERS

The dental formula is 2/3 1/1 2/2 3/3 = 34. The first upper incisors are small and widely separated (Fig. 13.2). The second upper incisors are caniniform. The first two lower incisors are pectinate (Fig. 13.3), each with from five to 20 long, slender cusps resembling the teeth of a comb. The third lower incisor has five or six cusps. The canines are incisiform, and the cheek teeth are brachyodont. The skull is broad and dorsoventrally flattened.

Well-developed and completely furred patagia extend from the sides of the neck, to the manual phalanges, to the pedal phalanges, and to the tail (see Figure 7.17).

Figure 13.1 A colugo, *Cynocephalus*, Cynocephalidae.
(Ognev 1951)

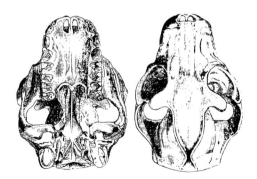

Figure 13.2 Ventral and dorsal views of the skull of a colugo, *Cynocephalus volans*.
(Giebel 1859)

Figure 13.3 The lower incisors of a colugo, *Cynocephalus variegatus*.
(Cabrera 1925)

LIVING FAMILY OF DERMOPTERA

There is one living family, **Cynocephalidae,** the colugos, with one genus, *Cynocephalus,* and two species (Wilson 1993). Colugos range through the tropical forests of southern Thailand and Vietnam, Malaysia, Indonesia, and the southernmost Philippine Islands (Wilson 1993; Nowak 1999).

COMMENTS AND SUGGESTIONS ON IDENTIFICATION

The colugo skin and skull are distinctive and not easily confused with those of any other mammals. Certain primates have pectinate lower incisors, but in these primates each tooth of the "comb" is a single incisor, whereas in colugos each tooth of the comb is only a single cusp of a large incisor. Colugos are the only gliding mammals with gliding membranes extending completely around the body from the sides of the neck to the tip of the tail.

Key to Living Families

CHAPTER 14

THE BATS
Order Chiroptera

ORDER CHIROPTERA

The ordinal name Chiroptera, literally meaning "hand-winged ones" refers to bats' most distinctive characteristic. Bats are the only mammals that have their forelimbs modified as wings, and they are the only mammals capable of sustained flight.

Most bats are insectivorous. Many eat flying insects that they capture in the air. Others glean insects from foliage, from the ground, or from the surface of water. In tropical and semitropical areas, insectivorous bats exist together with species specialized for diets of fruit, nectar, pollen, fish, other small vertebrates, or blood. The three families of bats that range widely into temperate regions, Rhinolophidae, Vespertilionidae, and Molossidae, are insectivorous. Because their food supply is generally unavailable in winter, these temperate species either hibernate or migrate seasonally to warmer climates.

Most bats are nocturnal, and most navigate with the aid of echolocation. Ultrasonic sounds are emitted through the nose or mouth, and the reflected sounds are received by the ears. The detailed mechanism of this varies (Henson 1970; Novick 1977). Many groups have simple to elaborate fleshy flaps, ridges, or other projections associated with the nose, the ears, or both. These structures are believed to aid in echolocation.

Bats affect humans in many ways. Many nectar-feeding and pollen-feeding bats provide service in pollination of plants, and many fruit bats aid in seed dispersal. Deposits of bat guano (feces) in caves are, or have been, mined in many parts of the world for use as fertilizer. Insectivorous bats, particularly in temperate regions, may play an important role in insect control. Vampire bats are responsible for the deaths of many domestic animals from disease (e.g., rabies). Some species are eaten by humans and may be endangered as a result.

DISTINGUISHING CHARACTERS

The forelimb is modified to form a wing. The metacarpals and phalanges of digits two through five are elongated and interconnected by a thin web of skin (Fig. 14.1). The **propatagium** extends from the shoulder to the wrist anterior to the upper arm (brachium) and forearm (antebrachium). The **dactylopatagia** fill the spaces between the digits, with the dactlopatagium minus between digits 2 and 3, the dactylopatagium longus between digits 3 and 4, and the dactylopatagium latus between digits 4 and 5. The **plagiopatagium** extends from digit five to the side of the body and the hindleg. The **uropatagium** extends between the hindlegs and usually incorporates the tail, if one is present (Fig. 14.2). A cartilaginous spur, termed the **calcar,** is sometimes present. It extends medially from the ankle region of each hindfoot and helps support the uropatagium (Fig. 14.2A–E). The first manual digit is clawed, the second bears a claw only in most Pteropodidae, and the remaining digits are always clawless. The pedal digits all bear claws. The clavicle is well-developed, and the sternum is frequently keeled.

The **tragus,** a fleshy lobe projecting from the lower medial corner of the inner margin of the pinna, is present in most microchiropterans (Fig 14.3A–C), but is absent in the megachiropterans (Fig. 14.3E). A flap, termed the **antitragus,** on the lower edge of the outer margin of the

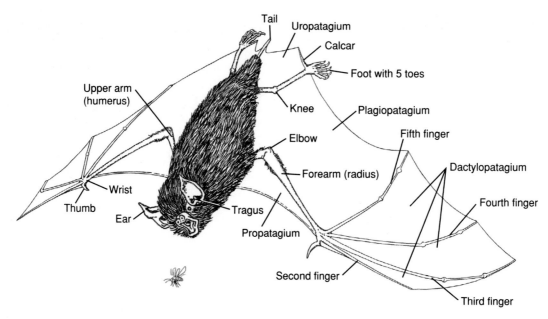

Figure 14.1 The major external features of bats.
(Feldhamer et al. 1999)

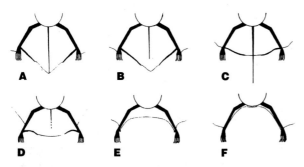

Figure 14.2 Diagrams of various tail and uropatagium arrangements found in bats. (A and B) Tail enclosed in uropatagium, (C) tail extending beyond posterior margin of uropatagium, (D) tail protruding from dorsal surface of uropatagium, (E) tail absent, uropatagium short, and (F) tail absent and uropatagium essentially absent.
(R. Roesener)

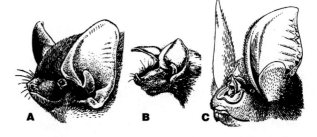

Figure 14.4 Bat heads showing varying degrees of development of flaps and projections on the nose. (A) Plain nose of a *Pipistrellus*, (B) simple leaf of a *Rhinopoma*, and (C) the complex noseleaf of *Rhinolophus*. Not all to same scale.
(A, Kuzyakin 1950; B, Harrison 1964, used with permission; C, Gromov et al. 1963)

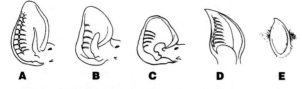

Figure 14.3 The ears of some representative bats. (A–D) are of Microchiroptera, (E) is of a Megachiropteran. (A–C), all vespertilionids, show differences in tragus size and shape. (D), a rhinolophid, lacks a tragus but has the antitragus enlarged. (E), a pteropodid, has a continuous inner ear margin and completely lacks both tragus and antitragus. Not all to same scale.
(A–D, Barrett-Hamilton, 1910; E, R. Roesener)

pinna is particularly well-developed in microchiropterans that lack a tragus (Fig. 14.3D). The noses of many microchiropteran bats are ornamented with noseleaves that vary from simple flaps of skin to highly complex structures (Fig. 14.4).

LIVING FAMILIES OF CHIROPTERA

Living families and their contents are listed in Table 14.1.

Bats occur over most of the earth's land surface except the permanently cold polar regions and a few small oceanic islands. Even such remote islands as New Zealand and the Hawaiian Islands have endemic bats.

TABLE 14.1 | Living Families of Chiroptera*

Family	Common Name	Number of Genera	Species	Distribution
Suborder Megachiroptera				
Pteropodidae	Old World fruit bats	42	166	Australian, Oriental, Ethiopian, south Palearctic, some Oceanic Islands
Suborder Microchiroptera				
Rhinopomatidae	Mouse-tailed bats	1	3	North Ethiopian, south Palearctic, west Oriental
Craseonycteridae	Bumblebee bat	1	1	Thailand
Emballonuridae	Sheath-tailed bats	13	47	North Neotropical, Ethiopian, south Palearctic, Oriental, Australian
Nycteridae	Hollow-faced bats	1	12	Ethiopian, Oriental, south Palearctic
Megadermatidae	Old World false vampire bats	4	5	Ethiopian, Oriental, Australian
Rhinolophidae	Horseshoe bats	10	130	Most of Eastern Hemisphere
Noctilionidae	Bulldog bats	1	2	North Neotropical
Mormoopidae	Leaf-chinned bats	2	8	Neotropical, south Nearctic
Phyllostomidae	New World leaf-nosed bats, vampire bats	49	141	Neotropical, south Nearctic
Natalidae	Funnel-eared bats	1	5	North Neotropical
Furipteridae	Smoky bats	2	2	North Neotropical
Thyropteridae	Disk-winged bats	1	3	North Neotropical
Myzopodidae	Sucker-footed bats	1	1	Madagascar
Vespertilionidae	Vespertilionid bats	35	318	Worldwide
Mystacinidae	New Zealand short-tailed bats	1	2	New Zealand
Molossidae	Free-tailed bats	12	80	All regions except northern Holarctic

*List of families and number of genera and species in each follow Koopman (1993) except for the number of species of Thyropteridae which is based upon Pine (1993).

KEY TO LIVING FAMILIES OF CHIROPTERA

1 Second finger retaining evident degree of independence from third, usually with a claw present; postorbital process well-developed (Fig. 14.5), occasionally forming postorbital bar; ears simple, the inner margin forming a continuous ring, tragus and antitragus absent (Fig. 14.3E) ....
.... Suborder **Megachiroptera**
Pteropodidae
Old World fruit bats

1' Second finger scarcely, if at all, independent from third, never clawed; postorbital process usually absent (if present, rostrum short); ears may be complex, inner margin never a complete ring, tragus usually present, or if absent, antitragus well-developed (Fig. 14.3A–D) ....
.... Suborder **Microchiroptera 2**

2 (1') Postcanine teeth usually fewer than 4/4; upper incisors enlarged to form blades (Fig. 14.6); mandible with pit to receive large upper incisors; no tail, uropatagium greatly abbreviated .... **Phyllostomidae** (in part)
Desmodontinae, vampire bats

2' Postcanine teeth 4/4 or more; upper incisors not large and bladelike; mandible without pit; tail present or absent, development of uropatagium various 3

3 (2') Incisors 2/2, 2/1, or 2/0 4

3' Incisors not as above 6

4 (3) Incisors 2/1; noseleaf absent (Fig. 14.7) ....
.... **Noctilionidae**
Bulldog or hare-lipped bats

4'   Incisors 2/2, 2/1, or 2/0; noseleaf present or absent, if absent, incisors 2/2 ...... **5**

5 (4')   Noseleaf absent; chinleaf present, extending ventrally from lower lip (Fig. 14.8); tail protruding from dorsal surface of uropatagium; rostrum tilted up anteriorly giving a dorsal profile of skull that is strongly concave in lateral view (Fig. 14.9); dental formula 2/2 1/1 2/3 3/3 = 34 ...... **Mormoopidae** leaf-chinned bats

5'   Noseleaf usually present (Fig. 14.10); chin leaf absent; tail, if present, various but never protruding from dorsal surface of uropatagium; dorsal profile of skull usually convex or flat, not strongly concave; dental formula various, including above ...... **Phyllostomidae** (in part) all New World leaf-nosed bats except Desmodontinae

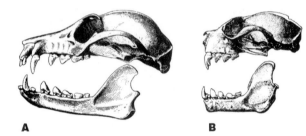

Figure 14.5 Skulls of representative Pteropodidae. (A) *Pteropus*, and (B) *Nyctimene*.
(A, Miller 1907; B, Andersen 1912)

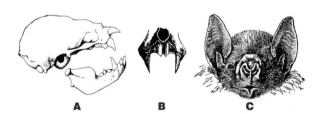

Figure 14.6 A vampire bat, *Desmodus rotundus*, Phyllostomidae. (A) Lateral view of the skull, (B) anterior view of upper jaw showing incisors, and (C) face.
(A, Ruschi 1953; B, Dobson 1878, C, Goodwin and Greenhall 1961)

Figure 14.7 Face of *Noctilio leporinus*, Noctilionidae.
(Goodwin and Greenhall 1961)

Figure 14.8 Faces of mormoopids. (A) *Pteronotus*, and (B) *Mormoops*.
(Goodwin and Greenhall 1961)

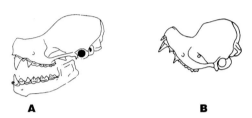

Figure 14.9 Lateral views of the skulls of mormoopids. (A) *Pteronotus*, and (B) *Mormoops*.
(Smith 1972)

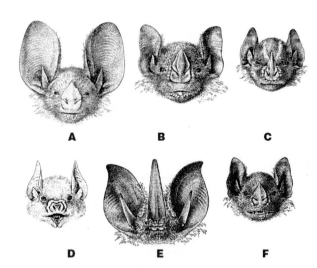

Figure 14.10 Faces of representative phyllostomids. (A) *Tonatia saurophila*, (B) *Artibeus jamaicensis*, (C) *Platyrrhinus helleri*, (D) *Brachyphylla sp.*, (E) *Lonchorhina aurita*, and (F) *Sturnira lilium*.
(D, Dobson 1878; all others, Goodwin and Greenhall 1961)

6 (3')   Dental formula 1/1 1/1 2/2 3/3 = 28; basal talon present on claws on thumb (Fig. 14.11) and toes; short tail projects from midpoint of dorsal surface of uropatagium .... **Mystacinidae** New Zealand short-tailed bats

6'   Dental formula not as above, incisors not 1/1, or if 1/1, total number of teeth 26; claws without basal talon; tail, if any, various **7**

Figure 14.11 Wrist and thumb of *Mystacina tuberculata*, Mystacinidae, showing the basal talon on the claw.
(Dobson 1876)

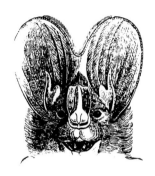

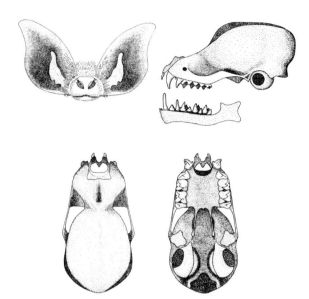

Figure 14.13 Face and skull of *Craseonycteris thonglongyai*, Craseonycteridae.
(Hill 1974) (A new family, genus and species of bat [Mammalia: Chiroptera]. Bull. British Museum [Nat. Hist.]. 32920:29–43, figs. 1–4. Copyright 1974 Trustees of the British Museum [Natural History]. Reprinted with permission.)

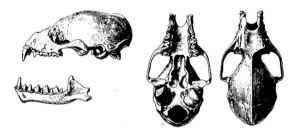

Figure 14.12 Representative megadermatids, face of *Megadermy lyra* and skull of *M. spasma*.
(Face, Finn 1929; skull, Miller 1907)

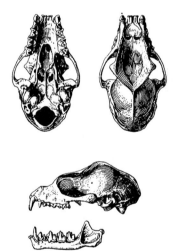

Figure 14.14 Skull of a hollow-faced bat, *Nycteris javanica,* Nycteridae.
(Miller 1907)

7 (6') Premaxillae and upper incisors absent (Fig. 14.12); nose leaf present, tragus bifurcated (Fig. 14.12); tail absent .... **Megadermatidae**
Old World false vampire bats

7' Premaxillae and upper incisors present; nose leaf present or absent; tragus, if present, not bifurcated; tail present or absent **8**

8 (7') Premaxillae form a complete ring around the external nares (Fig. 14.13); dental formula 1/2 1/1 1/2 3/3 = 28; tail absent, uropatagium well-developed .... **Craseonycteridae**
bumblebee bat

8' Premaxillae not forming a complete ring around external nares; dental formula various; tail present, uropatagium various **9**

9 (8') Postorbital processes present, sometimes incorporated into a broad supraorbital shelf (Figs. 14.14, 14.16); tail completely enclosed in uropatagium or protruding from dorsal surface of uropatagium **10**

9' Postorbital processes and supraorbital shelves absent; tail various but never protruding from dorsal surface of uropatagium **11**

10 (9') Premaxillae with palatal branches only, nasal branches absent; skull with a deep concavity in the frontal area (Fig. 14.14); tip of terminal tail vertebra attached to T-shaped or Y-shaped cartilage in hind edge of uropatagium (Fig. 14.15) .... **Nycteridae**
hollow-faced bats

10' Premaxillae with nasal branches only, palatal branches absent; usually no frontal concavity (Fig. 14.16); tail usually protrudes from dorsal surface of uropatagium ...... **Emballonuridae**
sheath-tailed or sac-winged bats

11 (9') Nasal branches of premaxillae absent, palatal branches fused to each other medially, but widely separated from the maxillae laterally (Fig 14.17); complex noseleaf present, tragus absent, antitragus well-developed (Fig. 14.18) .... **Rhinolophidae**
horseshoe bats

11' Nasal branches of premaxillae present, palatal branches, if present, not as above; nose leaf absent, or is a simple flap or projection; tragus present **12**

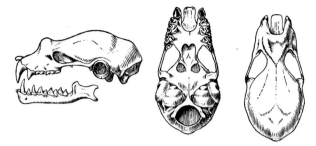

Figure 14.17 Skull of a rhinolophid, *Rhinolophus euryale*.
(Kuzyakin 1950)

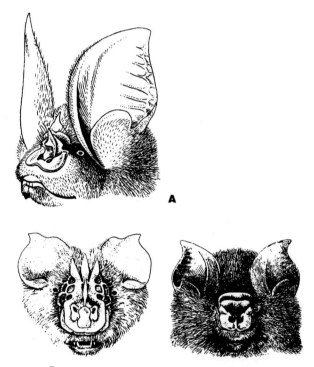

Figure 14.18 Faces of representative Rhinolophidae. (A) *Rhinolophus mehelyi,* (B) *Trianops persicus,* and (C) *Hipposideros caffer.*
(A, Gromov et al. 1963; B, Dobson 1878; C, Cabrera 1932)

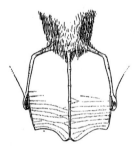

Figure 14.15 Tail and uropatagium of *Nycteris arge,* Nycteridae.
(Lang and Chapin 1917)

Figure 14.16 Skull of a sac-winged bat, *Saccopteryx bilineata,* Emballonuridae.
(Miller 1907)

12 (11') Dental formula 1/2 1/1 1/2 3/3 = 28; nasal region of skull inflated (Fig. 14.19); tail long and slender, almost as long as head plus body; uropatagium short, less than 1/3 the length of the tail (Fig. 14.20) .... **Rhinopomatidae**
mouse-tailed bats

12' Dental formula various, teeth usually number more than 28; nasal region of skull not inflated; tail usually less than 2/3 the length of head plus body; uropatagium extends for more than 1/3 of the length of the tail **13**

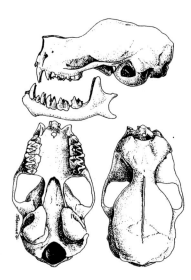

Figure 14.19 Skull of a mouse-tailed bat, *Rhinopoma hardwickei,* Rhinopomatidae.
(R. Roesener)

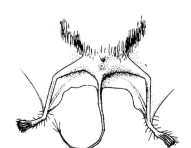

Figure 14.20 Posterior end of a mouse-tailed bat, Rhinopomatidae.
(Lang and Chapin 1917)

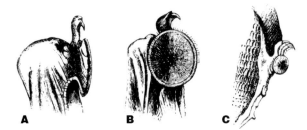

Figure 14.21 Sucking disks of a disk-winged bat, *Thyroptera tricolor,* Thyropteridae. (A and B) Wrist, and (C) foot.
(Dobson 1876)

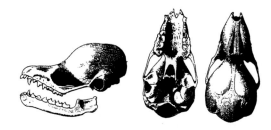

Figure 14.22 Skull of *Thyroptera discifera,* Thyropteridae.
(Miller 1907)

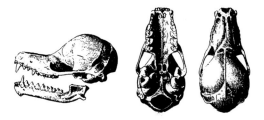

Figure 14.23 Skull of *Natalus stramineus,* Natalidae.
(Miller 1907)

13 (12') Posterior bases of pterygoids connected by a distinct ridge on the basisphenoid; dental formula 2/3 1/1 3/3 3/3 = 38 .... **14**

13' Posterior bases of pterygoids not connected by a ridge; dental formula various **15**

14 (13) Full length of third and fourth toes and claws fused; sucking disks present on feet and thumbs (Fig. 14.21); anterior palatal emargination, if present, V-shaped, incisive foramina absent, rostrum tapered anteriorly (Fig 14.22) .... **Thyropteridae**
disk-winged bats

14' Third and fourth toes and claws not fused; no sucking disks present; anterior palate not deeply notched, and two small incisive foramina usually present, rostrum squared or rounded off anteriorly (Fig. 14.23).......... **Natalidae** (in part)
funnel-eared bats

15 (13') Palate with anterior emargination and no incisive foramina (Figs. 14.25, 14.27) **16**

15' Palate entire, or with only very shallow anterior emargination, two small incisive foramina present (Fig. 14.23) or absent (Fig. 14.24) **17**

16 (15) Incisors 1–2/2–3, first upper incisors widely separated at bases and tips (Fig. 14.25), teeth number from 28 to 38; tail attached to uropatagium for all, or almost all, its length; ears usually more or less upright; tragus usually elongate (Fig. 14.26) .... **Vespertilionidae**
vespertilionid bats

16' Incisors 1/2–3 (Fig. 14.27), first upper incisors widely separated at base but closer together at tips, teeth number from 28 to 32; tail extending 1/3 to 2/3 of its length beyond the posterior edge of the uropatagium (Fig. 14.28); ears frequently directed nearly horizontally forward; tragus usually short and rounded (Fig. 14.28) .... **Molossidae** (in part)
free-tailed bats

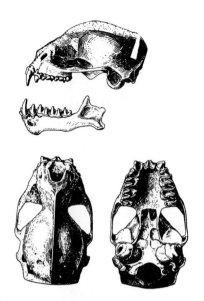

Figure 14.24 Skull of a molossid bat that has no palatal emargination, *Molossus ater.*
(Miller 1907)

Figure 14.25 Skull of *Myotis myotis,* Vespertilionidae.
(Popov 1956)

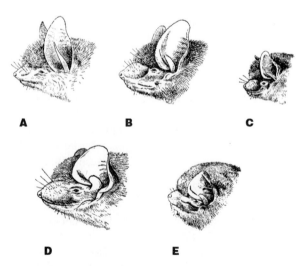

Figure 14.26 Heads of representative Vespertilionidae. (A) *Myotis nattereri,* (B) *Eptesicus serotinus,* (C) *Pipistrellus pipistrellus,* (D) *Nyctalus noctula,* and (E) *Miniopterus schreibersi.*
(Cabrera 1914)

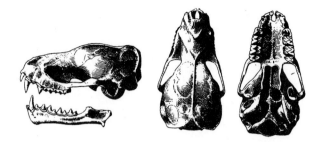

Figure 14.27 Skull of a molossid bat with an anterior palatal emargination, *Nyctinomops macrotis.*
(Miller 1907)

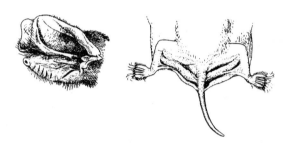

Figure 14.28 Head and posterior end of the free-tailed bat, *Tadarida teniotis,* Molossidae.
(Head, Cabrera 1914; tail, Lang and Chapin 1917)

Figure 14.29 Wrist region of the smoky bat, *Furipterus horrens,* Furipteridae.
(Goodwin and Greenhall 1961)

17 (15') Dental formula 1/1–3 1/1 1–2/2 3/3 = 26 to 32; tail extending 1/3 to 2/3 of its length beyond the posterior edge of the uropatagium (Fig. 14.28) .... **Molossidae** (in part) free-tailed bats

17'   Dental formula 2/3 1/1 2–3/3 3/3 = 36 or 38; tail attached to uropatagium for all, or almost all, of its length **18**

18 (17') Dental formula 2/3 1/1 2/3 3/3 = 36; thumb reduced, enclosed in patagium to base of minute claw (Fig. 14.29) .... **Furipteridae** smoky bats

18'   Dental formula 2/3 1/1 3/3 3/3 = 38; thumb free of patagium or, if enclosed, a large claw present **19**

19 (18') Posterior palatal emargination reaches anteriorly to last molar; canine in contact with adjacent incisor; rostrum broader than long; unique mushroom-shaped structure in ear partially blocking auditory canal (Fig. 14.30); thumb free; sucker pads on wrists and ankles .... **Myzopodidae**
sucker-footed bat

19' Palate extending posteriorly well beyond last molar (Fig. 14.23); a distinct space present between canine and adjacent incisor; rostrum as long as or longer than broad; no mushroom-shaped structure in ear; thumb enclosed in patagium; sucker pads absent ...... **Natalidae** (in part)
funnel-eared bats

COMMENTS AND SUGGESTIONS ON IDENTIFICATION

Externally, bats are easily recognized by their wings. The two suborders are usually easily separated by size and shape of ears, eye size, and overall body size, but keep in mind that some megachiropteran species are smaller than some of the largest microchiropteran species.

Figure 14.30 Head of *Myzopoda aurita,* Myzopodidae. (Thomas 1904)

The microchiropteran families Craseonycteridae, Mystacinidae, Rhinolophidae, Rhinopomatidae, Molossidae, Noctilionidae, Nycteridae, Megadermatidae, Emballonuridae, Furipteridae, Myzopodidae, Thyropteridae, and Natalidae are each recognized by one or two distinctive characteristics. The family Vespertilionidae is very large and relatively homogeneous but lacks any one good distinctive character. The family Phyllostomidae is large and very heterogeneous, probably more so cranially than any other family of mammals. Become familiar with the many different types.

CHAPTER 15

THE TUPAIIDS
Order Scandentia

ORDER SCANDENTIA

Scandentia means "the climbing ones," from the Latin. The English name for this group, "tree shrews," is misleading because some species are terrestrial; nor are these animals in any way shrewlike or related to shrews. Some tupaiids are arboreal, some semiarboreal, and others fully terrestrial. Their habitats include forested areas and brushy, rocky slopes. Depending on the species, they shelter in holes between rocks or roots or in tree hollows. Except for the arboreal pen-tailed tree "shrew," *Ptilocercus lowii*, which is nocturnal, members of the order are diurnal. These mammals are omnivorous, with insects and fruits making up major portions of their diet.

The Scandentia are small, active, and alert mammals, some of which resemble long-nosed squirrels, or small mongooses (Fig. 15.1). The eyes are considerably larger than those of the Insectivora. Facial vibrissae are short, and the pinnae are low and rounded except in *Ptilocercus*.

The tupaiids were previously often included in the Insectivora "wastebasket" and sometimes treated as the most primitive Primates. As stated before, the tupaiids are not closely related to shrews, nor to any other insectivores. These mammals are of little direct economic importance.

DISTINGUISHING CHARACTERS

The dental formula is 2/3 1/1 3/3 3/3 = 38. The upper incisors are often large and sometimes caniniform (Fig. 15.2A); the canines are small and premolariform. The upper molars are dilambdodont. The palate lacks large perforations, a postorbital bar is present (Fig. 15.2), and the zygomatic arch is complete and perforated (Fig. 15.2).

LIVING FAMILY OF SCANDENTIA

The Scandentia contains one living family, **Tupaiidae,** the tree "shrews," which includes five genera and 19 species (Wilson 1993), ranging through most of the Oriental Region.

COMMENTS AND SUGGESTIONS ON IDENTIFICATION

Except for *Ptilocercus*, tree shrews resemble squirrels externally but lack the long vibrissae of the latter, and have a much more pointed face than most. *Ptilocercus*, with its terminally featherlike tail, is unmistakable. The complete postorbital bar, coupled with a perforate zygomatic arch, distinguishes all tupaiid skulls.

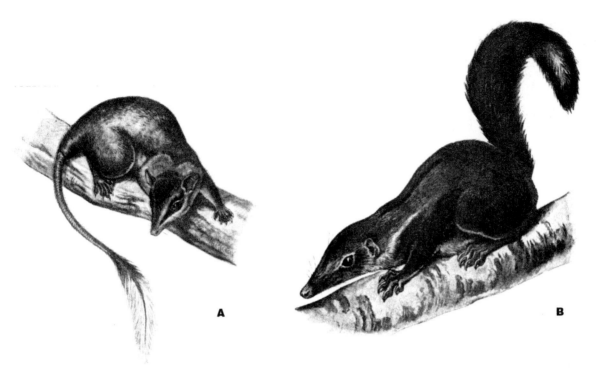

Figure 15.1 Representative tree "shrews." (A) The pentail tree "shrew," *Ptilocercus lowii,* and (B) the large tree "shrew," *Tupaia tana.*
(Cabrera 1925)

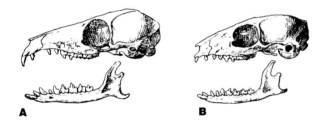

Figure 15.2 Skulls of two tupaiids. (A) Common tree "shrew," *Tupaia glis,* and (B) the Indian tree shrew, *Anathana ellioti.*
(Cabrera 1925)

CHAPTER 16

THE PRIMATES
Order Primates

Key to Living Families

ORDER PRIMATES

The ordinal name Primates means the "first" or "primary" ones. Although this name reflects the anthropocentrism of humans, it is of no help in defining the group of mammals included in it.

Primates are mainly arboreal mammals (Fig. 16.1), but terrestrial forms occur in several groups. Examples of these are the ring-tailed lemur, *Lemur catta;* baboons, *Papio, Mandrillus,* and *Theropithecus;* the gorilla, *Gorilla gorilla;* and humans, *Homo sapiens.* Humans are the only completely bipedal primates. Most primates are omnivorous, eating a higher percentage of plant than animal material, and usually softer foods, such as fruits, buds, and soft-bodied insects, are favored. The gorilla, the leaf-eating monkeys (Asian langurs, African colobus monkeys, Neotropical howler monkeys) and the Madagascan indrid and megaladapid lemurs are almost entirely herbivorous in the wild. The aye-aye, the galagos, and the tarsiers are largely insectivorous.

Most primates are pentadactyl, although certain fingers are shortened in some Loridae, the thumb is reduced in several brachiating forms, and it is completely absent in the spider monkeys, *Ateles;* some muriquis, *Brachyteles arachnoides;* and the African genus *Colobus.* In most primates, the digits are prehensile, and the pollex (thumb) and/or the hallux (big toe) are more or less opposable. Napier and Napier (1967) discussed the opposability of the thumb and gave the following classification:

Nonopposable thumb: Tarsiidae, Callitrichidae

Pseudo-opposable thumb: Strepsirhini, Cebidae

Opposable thumb: Catarrhini

These prehensile and opposable digits produce a hand or foot with great dexterity—the animal is better able to grasp and manipulate objects. A prehensile tail is present in some Cebidae but occurs in no other primate family.

In general, the sense of smell in primates becomes less acute as the hand becomes better adapted for manipulation (Napier and Napier 1967), but this should not be interpreted as cause and effect. Along with the loss of olfactory sensitivity, primates have developed very good vision. The field of vision has a large area of overlap between the fields of

Figure 16.1 A capuchin monkey, *Cebus apella,* Cebidae. (Gray 1865)

the two eyes, resulting in precision in depth perception. Only some nocturnal primates (some lemurs, the Galagonidae, Loridae, Tarsiidae, and the cebid genus *Aotus*) have retinas composed entirely of rods (Napier and Napier 1967). Other primates have both rods and cones in the retinas. Most Strepsirhini and all Tarsii are nocturnal, and most Platyrrhini and all Catarrhini are diurnal.

Primates are eaten in various parts of the world, and the skins of some species (e.g., *Colobus guereza*) are valuable. Some species do a great deal of damage to crops, and some, primarily baboons, can be dangerous to human infants and toddlers (Fiedler et al. 1975). Primates are major attractions at zoos, and many species are kept as pets. The rhesus monkey, *Macaca mulatta*, and several other species are widely used in medical research. Many species are now endangered.

Distinguishing Characters

The dentition varies, but the cheek teeth are bunodont and brachydont. The orbits are directed more or less anteriorly, and either a postorbital bar or a postorbital plate is present. In more "advanced" forms, the braincase is large, and the rostrum is usually shortened. The auditory bulla is formed by the petrosal.

The gait is plantigrade, and the hands/feet are usually pentadactyl. In some forms, the pollex or the second manual digit is reduced or absent. A nail is always present on the pollex, and most species have nails (as opposed to claws) on the other digits.

The uterus is either bicornuate (Strepsirhini and Tarsii) or simplex (Platyrrhini and Catarrhini), and the penis is usually pendant. The testes are scrotal. So far as is known, bacula are lacking in only four cebid species, in the tarsiers, and in humans (Hershkovitz 1977).

Living Families of Primates

A list of living families of Primates, and their contents, is given in Table 16.1.

Primates are essentially tropical mammals. Five families are confined to Madagascar (two species also occur on the nearby Comoro Islands), and two are found only in the tropical portions of the Neotropical Region. One family is confined to the tropical portions of the Ethiopian Region, two are confined to the tropical parts of the Oriental Region, and two, plus the great apes, occur in both of these areas. Several species of Cercopithecidae are native to various areas in the southern Palearctic. Humans are worldwide in distribution.

TABLE 16.1 | Living Families of Primates*

Family	Common Name	Number of Genera	Species	Distribution
Suborder Strepsirhini				
Cheirogaleidae	Dwarf lemurs, mouse lemurs	4	7	Madagascar
Megaladapidae	Sportive lemurs	1	7	Madagascar
Lemuridae	Lemurs	4	10	Madagascar, Comoro Islands
Indridae	Indrid lemurs (avahi, sifakas, indri)	3	5	Madagascar
Daubentoniidae	Aye-aye	1	1	Madagascar
Loridae	Lorises, potto, angwantibos	5	7	Oriental, Ethiopian
Galagonidae	Galagos or bushbabies	4	11	Ethiopian
Suborder Haplorhini				
Infraorder Tarsii				
Tarsiidae	Tarsiers	1	5	Indonesia, Philippines
Infraorder Platyrrhini				
Callitrichidae	Marmosets, tamarins	4	26	Neotropical
Cebidae	New World monkeys	11	58	Neotropical
Infraorder Catarrhini				
Cercopithecidae	Old World monkeys	18	81	Ethiopian, Oriental, southern Palearctic
Hylobatidae	Gibbons	1	11	Oriental
Hominidae	Gorilla, chimpanzees, orangutan, humans	4	5	Central Ethiopian, Sumatra, Borneo, humans cosmopolitan

*Numbers of genera and species based on Groves (1993), except numbers for Loridae based on Nowak (1999).

Key to Living Families of Primates

1 Postorbital bar present,[1] orbit and temporal fossa broadly confluent; some digits with nails and others with claws; nostrils are crescentic lateral slits associated with a moist glandular "nose" or rhinarium (Fig. 16.2) **2**

1' Postorbital plate present,* orbit and temporal fossa completely or mostly separated; all digits with nails or some digits with nails and others with claws; nostrils less slitlike, no moist rhinarium (Fig. 16.3) **7**

2 (1) Teeth number 18 to 20; incisors 1/1, large and similar to those of a rodent, with enamel on only anterior surfaces; canines absent (Fig. 16.4); third manual digit much more slender than adjacent digits (Fig. 16.5); hallux with a nail, other digits with claws; tail long and bushy .... **Daubentoniidae** the aye-aye, *Daubentonia madagascariensis*

2' Teeth number 30 or more; incisors never 1/1, and never resemble those of a rodent; C^1 present, C_1 usually present; third manual digit not appreciably more slender than other digits; distribution of nails and claws varies; tails of various sorts **3**

[1]The skulls of Tarsiidae could be interpreted by the student as having either a postorbital bar or a postorbital plate. Therefore, the key is constructed to allow Tarsiidae to be keyed to by either selection in this couplet. However, the nostrils and "nose" of Tarsiidae are consistent only with the condition described in 1'.

Figure 16.2 A sifaka, *Propithecus verreauxi coronatus*, Indridae.
(Flower and Lydekker 1891)

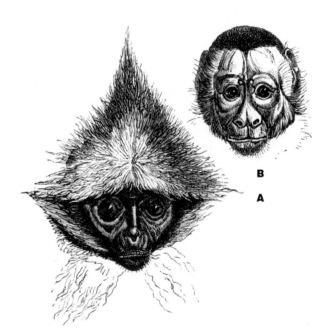

Figure 16.3 Faces of a catarrhine monkey (A) and a platyrrhine monkey (B).
(Duncan 1877)

Figure 16.4 Skull of an aye-aye, *Daubentonia madagascariensis*, Daubentoniidae.
(Flower and Lydekker 1891)

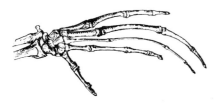

Figure 16.5 Skeleton of the left manus of an aye-aye, *Daubentonia madagascariensis,* Daubentoniidae.
(Duncan 1877)

Figure 16.7 Skull of the Philippine tarsier, *Tarsius syrichta,* Tarsiidae.
(Mary Ann Cramer)

Figure 16.6 Skull of an indrid lemur, the indri, *Indri indri,* Indridae.
(Duncan 1877)

Figure 16.8 Tarsiers, *Tarsius spectrum,* Tarsiidae.
(Kingsley 1884)

3 (2')   Dental formula 2/1 1/1 2/2 3/3 = 30; P_1 caniniform (Fig. 16.6); basal portion of toes at least partly webbed .... **Indridae**
indrid lemurs

3'   Teeth number 32 or more; P_1 varies; toes never webbed **4**

4 (3')   Dental formula 2/1 1/1 3/3 3/3 = 34; I^1 considerably larger than I^2 (Fig. 16.7); rhinarium absent, area around nostrils haired; tail long and naked or only sparsely haired for most of its length (Fig. 16.8) .... **Tarsiidae**
tarsiers

4'   Dental formula 0–2/3 1/1 3/3 3/2–3 = 32–36; I^1 and I^2, if present, essentially equal in size; rhinarium present, area around nostrils naked; tail short, or if long, well-haired **5**

5 (4')   Upper incisors 0 or 2; rostrum long, braincase elongate (Fig. 16.9B); ventral surface of pes naked with coarsely ridged pads **14**

5'   Upper incisors 1 or 2; rostrum short, braincase essentially spherical (Fig. 16.10); ventral surface of pes haired at heel (Fig. 16.11) **6**

6 (5')   Skull with prominent temporal ridges (Fig. 16.12); tail short or absent (Fig. 16.13) ....
.... **Loridae**
lorises, potto, angwantibos

6'   Skull without temporal ridges (Fig. 16.10); tail long and bushy (Fig. 16.14) ...... **Galagonidae**
galagos or bushbabies

7 (1')   Dental formula 2/1 1/1 3/3 3/3 = 34; orbit and temporal fossa confluent ventrally (Fig. 16.7); two tarsal bones greatly elongated, tips of digits with large circular pads (Fig. 16.8) .... **Tarsiidae**
tarsiers

7'   Dental formula 2/2 1/1 2–3/2–3 2–3/2–3 = 32 or 36; orbit and temporal fossa separated ventrally by postorbital plate (Fig. 16.15); no tarsal bones elongated; tips of digits without enlarged pads..
.... **8**

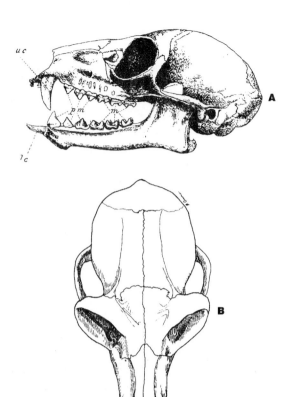

Figure 16.9 Skull of a ring-tailed lemur, *Lemur catta*, Lemuridae.
(Lateral view, Flower and Lydekker 1891; dorsal view, Mary Ann Cramer)

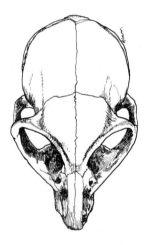

Figure 16.10 Skull of a greater bushbaby, *Otolemur crassicaudatus*, Galagonidae.
(Mary Ann Cramer)

Figure 16.11 Plantar view of right pes of a greater bushbaby, *Otolemur crassicaudatus*, Galagonidae.
(Duncan 1877)

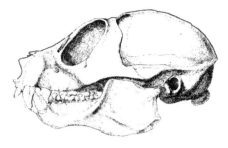

Figure 16.12 Skull of a slender loris, *Loris tardigradus*, Loridae.
(Giebel and Leche 1874)

Figure 16.13 A slow loris, *Nycticebus coucang*, Loridae.
(Beddard 1902)

8 (7')   Dental formula 2/2 1/1 3/3 2/2 = 32; third postcanine tooth more like second than fourth in shape and size; second incisors pointed (Fig. 16.15); tail long; hallux with a nail, other digits clawed; pollex present but not opposable .... **Callitrichidae** (in part)
marmosets, tamarins

8'   Dental formula 2/2 1/1 3/3 3/3 = 36, or 2/2 1/1 2/2 3/3 = 32; if postcanine teeth number five, then third postcanine tooth more like fourth than second in shape and size; second incisors chisel-shaped or pointed; tails of various sorts; all digits with nails; pollex absent or, if present, at least partly opposable; or if claws and pollex as in 8 above, then with six postcanine teeth **9**

9 (8') Dental formula 2/2 1/1 3/3 3/3 = 36; nostrils laterally directed (Fig. 16.3B); tail present, sometimes prehensile **10**

9' Dental formula 2/2 1/1 2/2 3/3 = 32; nostrils open forward or down (Fig. 16.3A); tail present or absent, never prehensile **11**

10 (9) Suture between maxilla and premaxilla well-defined; hallux with a nail, all other digits clawed, pollex unopposable; tail not prehensile; color black .... **Callitrichidae** (in part)
Goeldi's monkey, *Callimico goeldii*

10' Maxilla and premaxilla fused in adults, suture not distinguishable; all digits with nails; pollex absent or, if present, at least partly opposable; tails of various sorts, sometimes prehensile (Fig. 16.1); color varies ....
.... **Cebidae**
New World monkeys

11 (9') Greatest length of skull less than 150 mm ......
.... **12**

11' Greatest length of skull more than 150 mm ....
.... **13**

12 (11) Anterior root of P_1 sloping forward into dentary and exposed for much of its length (Fig. 16.16), wears against upper canine in occlusion; P_1 distinctly different from P_2; tail present or absent; pollex variable ....
..............................**Cercopithecidae** (in part)
most Old World monkeys

12' Anterior root of P_1 entering dentary vertically or at only a slight anterior angle and unexposed or barely exposed (Fig. 16.17), P_1 similar to P_2; tail absent; pollex reduced ....
....**Hylobatidae**
gibbons

Figure 16.14 A greater bushbaby, *Otolemur crassicaudatus*, Galagonidae.
(Duncan 1877)

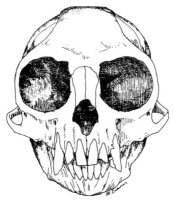

Figure 16.15 Skull of a marmoset, *Callithrix humeralifer*, Callitrichidae.
(Mary Ann Cramer)

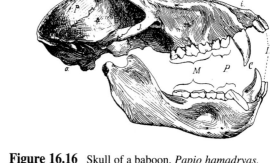

Figure 16.16 Skull of a baboon, *Papio hamadryas*, Cercopithecidae, showing the distinctive lower premolar.
(Weber 1928)

Figure 16.17 The mandible of the siamang gibbon, *Hylobates syndactylus,* Hylobatidae, showing a first lower premolar characteristic of this family and of Hominidae. (Mary Ann Cramer)

13 (11') Anterior root of P_1 sloping forward into dentary and exposed for much of its length (Fig. 16.16); longitudinal groove frequently present on anterior surface of C^1; ischial callosities present .... **Cercopithecidae** (in part)
baboons

13 Anterior root of P_1 entering dentary vertically (Fig. 16.17) or at only a slight anterior angle, unexposed or only barely exposed; canine never grooved; ischial callosities absent .... **Hominidae**
great apes, humans

14 (5) Upper incisors absent in adults; head plus body length 300–350 mm; never with definite markings or bold color pattern ...... **Megaladapidae**
sportive lemurs

14' Upper incisors present in adults; head plus body length 125–460 mm; sometimes with definite markings or bold color patterns **15**

15 (14') Upper incisors small and peglike (Fig. 16.9), first upper incisors separated by a wide space; head plus body length 280–460 mm; sometimes with definite markings or bold color patterns .. .. **Lemuridae**
lemurs

15' Upper incisors not small and peglike, first upper incisors not separated by a wide space; head plus body length 125–285 mm; without definite markings or bold color patterns unless with middorsal stripe that splits in two anteriorly to pass across dorsum of head and then reunite at rhinarium .... **Cheirogaleidae**
mouse lemurs, dwarf lemurs

COMMENTS AND SUGGESTIONS ON IDENTIFICATION

Primates as an order are distinctive. Some of the smaller forms may at first glance be confused with marsupials, carnivores, or rodents, but close examination should reveal their correct placement. Most primate families are rather homogeneous in basic characteristics, though they may include a large size range. Some possible sources of confusion are listed here.

The skulls of Daubentoniidae resemble those of rodents. Check for a postorbital bar and examine the animal externally.

The marsupials that resemble the smallest primates can be told from them by means of such characters as dental formulas, ventral naked strips on the tail, syndactyly, and clawless/nailless pollex. The rodents that resemble these small primates will have typical rodent dentition, shorter legs, and will generally lack an opposable pollex and/or hallux.

To tell cebids from cercopithecids, check the tooth number, geographic origin, and characteristics of the nose, third postcanine tooth, and P_1. All prehensile-tailed monkeys are members of the family Cebidae, but only about half of the species of Cebidae have prehensile tails.

Key to Living Families

CHAPTER 17

THE XENARTHRANS: ANTEATERS, ARMADILLOS, AND SLOTHS
Order Xenarthra

ORDER XENARTHRA

Xenarthra contains four families and three strikingly different groups of living mammals (Fig. 17.1): the true anteaters, Myrmecophagidae; the armadillos, Dasypodidae; and the two families of sloths, Bradypodidae and Megalonychidae. This order was formerly called Edentata, which means "ones without teeth." Although the diverse groups included in Xenarthra share many defining characteristics, being edentulate is not one of them. Only the anteaters completely lack teeth, a condition they share with several mammals in other orders. The name Xenarthra refers to the accessory "xenarthrous" articular surfaces between vertebrae, primarily in the lumbar area of the spine (Fig. 17.2). These xenarthrous articulations are found in all living members of the order and in most fossil forms as well.

Myrmecophagidae, the anteaters, feed primarily upon ants and termites. It includes three living genera. The giant anteater, *Myrmecophaga tridactyla*, is a large (30–35 kg) terrestrial mammal. Its tongue is about 60 cm long and can be protruded and withdrawn into the mouth up to 160 times per minute (Moeller 1975). The tamanduas, *Tamandua tetradactyla* (Fig. 17.1B) and *T. mexicana* are medium-sized anteaters (3–5 kg) that are both terrestrial and arboreal. The pygmy or two-toed anteater, *Cyclopes didactylus,* is small (500 g) and completely arboreal. Both *Tamandua* and *Cyclopes* have prehensile tails. All of the anteaters have long, slender rostra, very small mouth openings, and long vermiform tongues. Their large foreclaws are used to tear open ant and termite nests.

The Dasypodidae, the armadillos, are terrestrial to fossorial mammals possessing a carapace over much of the body (Fig. 17.1C). This shell of dermal bony scutes overlain by epidermal scales (see section on scales in Chapter 4) ranges from full dorsolateral armor that completely protects the animals when they roll into a ball (*Tolypeutes*) to one in which the flanks are unarmored, and most of the dorsum is covered only with a thin, loose mantle of soft, highly flexible, armor attached by only a fragile, translucent, middorsal membrane (*Chlamyphorus truncatus*). Most armadillos are only sparsely haired, but the lesser pichiciego, *C. truncatus*, has dense hair, even under its mantlelike dorsal armor. The rare *Dasypus pilosus* has such dense hair over the trunk that its full body armor is concealed. Most armadillos are primarily terrestrial, but all are well-equipped for digging for food and burrowing. The pichiciegos are the most fully fossorial. Armadillos have cylindrical, homodont, ever-growing cheek teeth and no incisors or canines. Deciduous teeth have been recorded only in the genus *Dasypus*. Most armadillos feed exclusively or primarily upon insects, but other invertebrates, small vertebrates, carrion, eggs, fungi, berries, and other plant material may be eaten (Bolkovic et al 1995; Davis 1974; Moeller 1975).

The sloths are arboreal animals with long limbs (see Fig 17.1A), syndactylous toes, and large curved claws used to hang from tree limbs. They are covered by coarse hair that harbors algae growing in grooves in the cuticle. This gives the slow-moving animals a greenish tinge providing protective coloration. Sloths are strictly vegetarian, eating leaves, buds, flowers, and fruit. Incisors and canines are absent, and the cheek teeth are cylindrical, ever-growing, and essentially homodont. The tail is rudimentary. Sloths

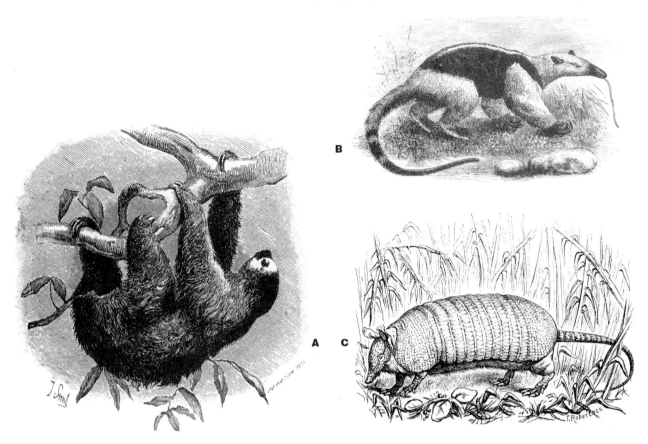

Figure 17.1 Representatives of three of the living families of Xenarthra. (A) A two-toed sloth, *Choloepus hoffmanni,* Megalonychidae; (B) a tamandua, *Tamandua tetradactyla,* Myrmecophagidae; and (C) the nine-banded armadillo, *Dasypus novemcinctus,* Dasypodidae.

(A, Sclater and Sclater 1899; B and C, Flower and Lydekker 1891)

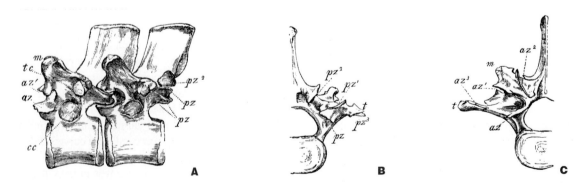

Figure 17.2 Vertebrae of the giant anteater, *Myrmecophaga tridactyla.* (A) Side view of twelfth and thirteenth thoracic vertebrae. (B) Posterior view of second lumbar vertebra. (C) Anterior surface of third lumbar vertebra. az = anterior zygapophysis; az^1, az^2, az^3 = additional (xenarthrous) anterior articular facets; cc = facet for capitulum of rib; m = metapophysis; pz = posterior zygopophysis; pz^1, pz^2, pz^3 = additional (xenarthrous) posterior articular facets; t = transverse process; tc = facet for articulation of tubercle of rib.

(Flower 1885)

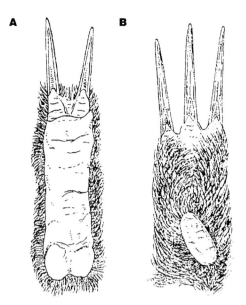

Figure 17.3 Ventral views of the right forefoot of (A) a two-toed sloth, *Choloepus,* Megalonychidae, and (B) a three-toed sloth, *Bradypus,* Bradypodidae.
(Pocock 1924b)

and manatees (Sirenia: Trichechidae) are the only mammals that possess cervical vertebrae that number more or fewer than seven. Individual two-toed sloths may have five, six, seven, or occasionally, eight cervical vertebrae. The three-toed sloths have eight or nine vertebrae in their long, very flexible necks.

The extant Bradypodidae are three species of *Bradypus,* the three-toed sloths. These have three large claws (Fig. 17.3B) on each of the four limbs. The cheek teeth are essentially homodont. There are two living species of Megalonychidae, the two-toed sloths, *Choloepus*. These have only two claws on the forefeet (Fig. 17.3A) and three on the hindfeet. The anterior tooth in each jaw is caniniform, larger, and more pointed than the teeth posterior to it.

Living Families of Xenarthra

A listing of the living families of Xenarthra and their contents is given in Table 17.1.

Living xenarthrans are confined to the Neotropical Region except for one species. The nine-banded armadillo, *Dasypus novemcinctus,* ranges into the south-central Nearctic.

Key to Living Families of Xenarthra

1   Teeth absent (Fig. 17.4); body covered with thick coat of hair, no dermal bony armor, tail long, prehensile in three of the four species .... **Myrmecophagidae**
   anteaters

1' Teeth present; dermal bony armor may be present; tail various, never prehensile **2**

2 (1') Skull conical, rostrum usually long and tapering (Fig 17.5); four or five digits on each foot, none syndactylous; most of body, or at least back, covered with bony dermal armor overlain with epidermal scales (in some species only epidermal scales on back) .... **Dasypodidae**
   armadillos

2' Skull cuboid, rostrum blunt, not tapered (Fig. 17.6); two or three digits on each foot, all syndactylous; dermal armor and epidermal scales absent **3**

3 (2') Anterior tooth in each jaw caniniform, considerably larger and more pointed than teeth behind it (Fig. 17.6); forefoot with two syndactylous toes and two large claws ....**Megalonychidae**
   two-toed sloths

3' All teeth in each jaw approximately equal in size and shape; forefoot with three syndactylous toes and three large claws ...... **Bradypodidae**
   three-toed sloths

TABLE 17.1 | **Living Families of Xenarthra***

Family	Common name	Number of Genera	Species	Distribution
Bradypodidae	Three-toed sloths	1	3	Neotropical
Megalonychidae	Two-toed sloths	1	2	Neotropical
Dasypodidae	Armadillos	8	20	Neotropical, southern Nearctic
Myrmecophagidae	Anteaters	3	4	Neotropical

*List of families and numbers of genera and species based upon Gardner (1993).

Figure 17.4 Skull of the giant anteater, *Myrmecophaga tridactyla*, Myrmecophagidae. Skulls of *Tamandua* and *Cyclopes* have proportionally, and absolutely, shorter rostra than this.
(Owen 1866)

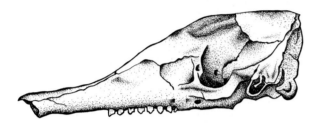

Figure 17.5 Cranium of the armadillo. *Dasypus novemcinctus*, Dasypodidae.
(Michael Gilliland)

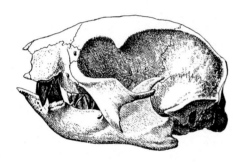

Figure 17.6 Skull of a two-toed sloth, *Choloepus*, Megalonychidae.
(Weber 1928)

COMMENTS AND SUGGESTIONS ON IDENTIFICATION

The three groups of xenarthrans are very distinctive and cannot be confused with each other. Sloths are unique in both external and cranial features, and the two families are easily distinguished from each other by the dentition and number of manual digits. The bony shells of armadillos are unique as skin features. Confusion could arise between skulls of large armadillos and those of aardvarks, but close examination of teeth easily separates these two groups. The toothless skulls of anteaters could be confused with the toothless skulls of echidnas and pangolins. Methods of distinguishing these three groups are discussed in the chapter on Monotremata.

CHAPTER 18

THE PANGOLINS
Order Pholidota

ORDER PHOLIDOTA

The ordinal name, Pholidota, literally means "scaly ones" and points out the major diagnostic characteristic of this group. The pangolins, or scaly anteaters, are covered with large epidermal scales (Fig. 18.1) that are usually described as being formed by agglutinated hairs. However, Rahm (1975a:183–84) stated that

The scales are not "glued hairs," as was once believed; they are two-sided symmetrical elevations of the skin which are flattened from the back toward the stomach, and which face the back of the animal. The epidermis, as it grows into horn material, leads to scale formation. The horn scales, which are lost through wear, are constantly replaced by the skin base. Throughout its life, the pangolin always has the same number of scales. The pattern, quality, shape, and size of the scales differ from species to species, and can also differ slightly within a species, depending on the part of the body they cover.

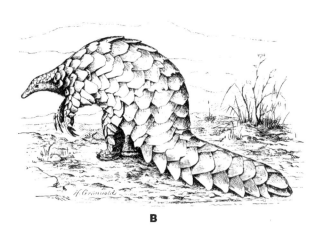

Figure 18.1 Representative pangolins. (A) An arboreal form, *Manis tricuspis*, and (B) a terrestrial form, *Manis temminckii*.
(A, Flower and Lydekker 1891; B, Sclater 1901)

Some pangolin species have a long prehensile tail and are arboreal (see Fig. 18.1A); others have a short, rather blunt tail and are fully terrestrial (see Fig. 18.1B). Pangolins are insectivorous, gathering ants or termites with their long, vermiform, sticky tongues. In the four African species, the xiphisternum is greatly elongated, passing between the peritoneum and lower abdominal wall ventral and posterior to the abdominal cavity. It then curves back dorsally to the region of the kidneys where it forms a support for the base of the greatly elongated tongue (Emry 1970). The tongue itself, in these species, extends well back into the chest cavity when retracted (Rahm 1975a). Teeth are absent, and insects are crushed in the stomach where a horny, layered epithelium with horny teeth replaces the usual mucous membrane lining.

Distinguishing Characters

Teeth are completely absent. The cranium is robust and conical, and the mandible is slender. The zygomatic arch is incomplete (Fig. 18.2). The limbs are plantigrade, and pentadactyl, and posses large claws. The tail is prehensile in arboreal species. The top of the head, top and sides of the neck and body, most parts of the limbs, and all parts of the tail are covered with large, overlapping, epidermal scales. The undersides of the head, neck, and body are scaleless and hairy. The arboreal forms have a scaleless spot ventrally at the tip of the tail. The uterus is bicornuate, and the testes are inguinal but never scrotal. Bacula have not been reported.

Living Family of Pholidota

There is a single living family, **Manidae,** the pangolins or scaly anteaters, with one genus, *Manis,* and seven species (Schlitter 1993). Four species occur in the Ethiopian Region, and three in the Oriental Region. Both arboreal and terrestrial forms occur in each region.

Comments and Suggestions on Identification

Pangolin skins are unique and cannot be confused with those of any other mammals. The toothless skulls could be confused with those of echidnas or Neotropical anteaters. See this section in the chapter on Monotremata for comments on distinguishing these three groups.

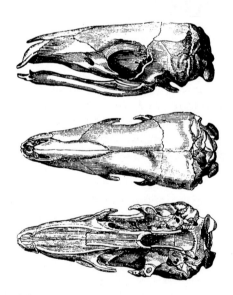

Figure 18.2 Skull of a pangolin, *Manis* sp.
(Giebel 1859)

Key to Living Families

CHAPTER 19

THE CARNIVORES
Order Carnivora

ORDER CARNIVORA

Carnivora means "flesh eaters" (Jaeger 1955), and this is an at least partially accurate description of most members of this order. The polar bear (Fig. 19.1), the leopard seal, and the lion are examples of Carnivora that live almost exclusively on other vertebrates. Most otariids and phocids feed primarily on fish, and several members of other families (e.g., the raccoon dog, the brown bear, the American mink, the otters, and the fishing cats and flat-headed cats) include a significant amount of fish in their diets. The walrus and sea otter feed exclusively, or almost exclusively, on marine invertebrates, and many carnivores (e.g., the sloth bear, the banded mongoose, the meerkat, and the aardwolf) are primarily insectivorous. A few members of the order Carnivora (e.g., the spectacled bear, the giant panda, and the red panda) are almost exclusively vegetarian.

Regardless of their ordinal name, most Carnivora eat significant quantities of both plant and animal matter. The dietary habits of particular species may change drastically from season to season as various kinds of foods change in abundance and ease of gathering. Berries, nuts, fruits, eggs, carrion, and invertebrates are eaten by almost all terrestrial Carnivora. Guggisberg (1960) reported lions eating termites, grass, and fruits; Doutt (1967) reported polar bears eating grass and berries; and Ewer (1973) mentioned giant pandas eating meat in captivity. The subject of carnivore diets and feeding behavior was discussed by Ewer (1973), and the above is based primarily upon her review.

The English word *carnivore* is often used for any member of the order Carnivora, regardless of diet, but is also commonly used as a descriptive term for various predatory or partially predatory animals regardless of order (or even class or phylum). In order to avoid confusion, some zoologists employ the term *carnivoran* to designate a member of the order Carnivora.

Carnivores (in the sense of carnivorans) exhibit a great range in size, from the American least weasel with a head and body length of 135–185 mm and a weight of 35–70 grams, to the huge brown bears of Kodiak Island, Alaska, which reach a length of over 3 m and a weight up to 780 kg (polar bears reach equal or greater weights), and the adult male southern elephant seals, which can exceed 6 m in length and weigh over 3600 kg (Coffey 1977).

Carnivores occur in nearly all possible mammal habitats. They range from the ice floes of the Arctic

Figure 19.1 A polar bear, *Ursus maritimus,* Ursidae, one of the most carnivorous members of the order Carnivora.
(Zerov and Pavlovskii 1943)

Ocean to the Antarctic continent, and from the driest deserts and highest mountains into the oceans. Most are essentially terrestrial, but many are variously adapted for arboreal, fossorial, or aquatic existences. Arboreal adaptations range from infrequent tree climbing to the nearly continuous arboreal existence of some procyonids and viverrids. Many carnivores dig for part of their food, and several excavate dens, but the badgers, of the family Mustelidae, are probably the most fossorial. All carnivores can, and most readily do, swim. The river otters, of the family Mustelidae, are well-adapted for a semiaquatic existence, but the sea otter and all members of the families Phocidae, Otariidae, and Odobenidae are much better adapted to life in the water than on land. Compared to them, only the cetaceans and sirenians are more highly modified for aquatic life.

Economically, carnivores form one of the most important groups of mammals. One carnivore, the dog, *Canis familiaris,* was the first domesticated animal. The dog and another carnivore, the domestic cat, *Felis catus,* are the most popular nonhuman companions of modern humans. The larger species of wild carnivores are important game animals in many parts of the world, and many smaller species (e.g., coyote) are considered to be pests, although they are important predators of many rodents and other small animals. One family, Mustelidae, includes many of the world's most important furbearers (e.g., mink, ermine, sable, sea otter), and most other families include furbearers of commercial importance. Some carnivores (e.g., polar bear, tiger) are endangered, due to uncontrolled or inadequately controlled hunting, "pest" control, and pelt seeking. Many species (e.g., wolf, American black bear, and puma) have declined in number with the increase in human population, whereas other species (e.g., coyotes, one species of raccoon, and some mongooses) have thrived in company with humans. By contrast, the Caribbean monk seal has been driven to extinction.

DISTINGUISHING CHARACTERS

The canine teeth are usually large and conical (Fig. 19.2). Three lower incisors are present in all **fissipeds**[1] except the sea otter, *Enhydra lutris,* which has only two. The number of lower incisors in Otariidae and Phocidae ranges from 0 to 2. The postcanine teeth of fissipeds are usually secodont, and in adults of many species the last upper premolar and the first lower molar are elongated and bladelike (Fig. 19.2). These are termed the carnassial pair and serve for

---

[1] The term *fissiped* as used in this chapter refers to all Carnivora other than the Phocidae, Otariidae, and Odobenidae: the pinnipeds. Neither fissiped or pinniped is intended to necessarily have any phylogenetic/taxonomic connotation.

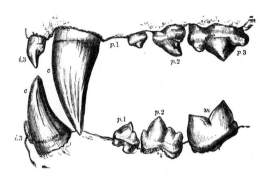

Figure 19.2 Labial view of the dentition of a lion, *Panthera leo,* Felidae. In this species, the third upper postcanine tooth is the last upper premolar, and the third lower postcanine is the first lower molar. Together, these teeth form the canassial pair in adults.
(Duncan 1877–83)

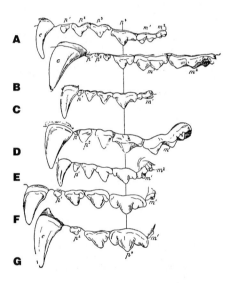

Figure 19.3 The maxillary dentition of representative fissiped carnivores showing varying development of premolars and molars. The vertical line runs through the upper carnassial, the last upper premolar. (A) A dog, Canidae; (B) a bear, Ursidae; (C) a marten, Mustelidae; (D) a badger, Mustelidae; (E) a mongoose, Herpestidae; (F) a hyena, Hyaenidae; and (G) a lion, Felidae.
(Weber 1928)

shearing flesh. In immatures of these species, the carnassial pair is formed by the next to last upper premolar and the last lower premolar. The premolars and molars anterior and posterior to the carnassials are variously developed, reduced, or lost (Fig. 19.3).

In other fissipeds, the postcanine teeth may be modified for crushing (e.g., the sea otter, Fig. 19.4) or may be greatly reduced (e.g., the aardwolf, see Fig. 19.12). In the pinnipeds, the postcanine teeth are always essentially

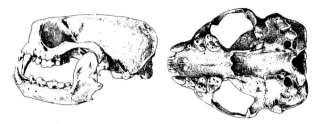

Figure 19.4 Skull of the sea otter, *Enhydra lutris*, Mustelidae.
(Novikov 1956)

Figure 19.6 A walrus, *Odobenus rosmarus*, Odobenidae.
(Gromov et al. 1963)

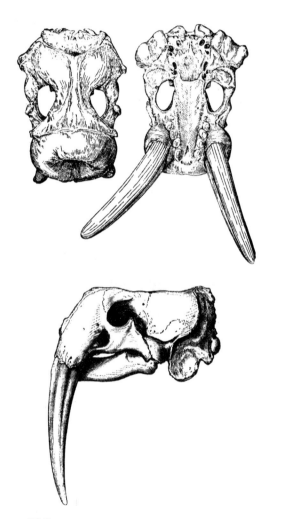

Figure 19.5 Three views of the skull of a walrus, *Odobenus rosmarus*, Odobenidae.
(A, Gromov et al. 1963; B, Duncan 1877–83)

homodont and frequently unicuspid (Figs. 19.5, 19.8, and 19.10). Carnassials are never present in the **pinnipeds**. The zygomatic arches of Carnivora are well-developed, and a sagittal crest is frequently present. The auditory bullae are usually fully ossified and are frequently large.

All present-day carnivores have the carpals designated as the centrale, scaphoid, and lunar fused. The feet of fissipeds are plantigrade or digitigrade and have four or five toes. Each toe ends with a large, sharp, curved claw. The pinnipeds have their knees and elbows included within the contour of the body, the metatarsals and metacarpals are elongate, and the five digits on each foot are fully webbed (Figs. 19.6, 19.9, and 19.11). These feet are termed flippers.

The stomach is simple, and the caecum is usually small or absent. The uterus is bicornuate, and the testes may be internal or scrotal. A baculum is present and is very well-developed in many species (Fig. 6.2).

LIVING FAMILIES OF CARNIVORA

A list of living families of Carnivora, and their contents, is given in Table 19.1.

Carnivores range from Arctic ice floes through all continents to the Antarctic continent. Though most oceanic islands lack fissipeds, pinnipeds are found on many, including Hawaii and New Zealand. Until early human immigrants (presumably) brought dingoes with them into the Australian Region, there were no Carnivora on that side of Wallace's Line. Modern humans have introduced other fissipeds, such as the domestic dog and cat and the red fox, to Australia, to New Zealand, and to many other areas where no terrestrial carnivores had existed before.

Most fissipeds are confined to continental land masses and fresh waters. The polar bear, *Ursus maritimus,* the sea otter, *Enhydra lutris,* and the South American marine otter, *Lontra felina,* are marine, or partially so. Conversely, most pinnipeds are marine, with exclusively landlocked species only in the saline Caspian Sea and in the fresh water of Lake Baikal in Siberia.

TABLE 19.1 | Living Families of Carnivora*

Family	Common Name	Number of Genera	Number of Species	Distribution
Canidae	Dogs, wolves, foxes, jackals	15	33	Holarctic, Neotropical (one species, the Falkland Islands "wolf," extirpated in 1800s), Ethiopian, Oriental, introduced worldwide
Ursidae	Bears, pandas	6	9	Holarctic, Oriental, Northwest Neotropical
Otariidae	Sea lions, fur seals	7	15	Colder coastlines of Pacific, South Atlantic, and Indian Oceans
Odobenidae	Walrus	1	1	Arctic Ocean and adjacent far northern marine waters
Procyonidae	Raccoons and allies	6	18	Nearctic, Neotropical
Mustelidae	Badgers, skunks, otters, weasels, and allies	25	65	Holarctic, Neotropical, Ethiopian, Oriental
Phocidae	True seals	10	19	All oceans and seas (recently extirpated in Caribbean), Lake Baikal, Caspian "Sea," other lakes
Viverridae	Civets, genets, linsangs, fossa	20	34	Ethiopian, Madagascar, Oriental, southern Palearctic
Herpestidae	Mongooses	18	37	Ethiopian, Madagascar, Oriental, southern Palearctic, widely introduced
Hyaenidae	Hyenas, aardwolf	4	4	Ethiopian, southcentral Palearctic
Felidae	Cats: lion, tiger, lynxes, tabby, cheetah, etc.	18	36	Holarctic, Neotropical, Ethiopian, Oriental, introduced worldwide

*List of families and numbers of genera and species based upon Wozencraft (1993).

KEY TO LIVING FAMILIES OF CARNIVORA

1. Lower incisors 0, 1, or 2; cheek teeth essentially homodont and usually unicuspid; carnassials never present; limbs modified to form flippers; tail very short **2**

1'. Lower incisors 3 (2 in one species); cheek teeth usually heterodont and multicuspid; dentition frequently secodont, and carnassials often well-developed; limbs usually not modified as flippers; tail various; *if* lower incisors number 2, *then* cheek teeth very broad and flat crowned (Fig. 19.4) and tail of medium length (Fig. 19.27C) **4**

2. Teeth number 24 or fewer; lower incisors absent; upper canines very long, forming tusks (Fig. 19.5); extending well beyond the lips (Fig. 19.6) .... **Odobenidae**
walrus, *Odobenus rosmarus*

2'. Teeth number 26 or more; lower incisors present; upper canines not tusks, normally concealed behind lips **3**

3 (2'). Dental formula 3/2, 1/1, postcanines 5–7/5 = 34–36; alisphenoid canal present (Fig. 19.7); postcanine teeth usually unicuspid; postorbital process of frontal usually well-developed (Fig. 19.8); hindlimbs capable of being rotated forward (Fig. 19.9); small pinnae present; pelt is usually uniform in color, and underfur may be present .... **Otariidae**
sea lions and fur seals

3'. Dental formula 2–3/1–2, 1/1, postcanines 4–6/4–6 = 26–36; alisphenoid canal absent; postcanine teeth usually not unicuspid; postorbital process of frontal usually absent (Fig. 19.10); hindlimbs not capable of being rotated forward, always point posteriorly (Fig. 19.11); pinnae absent; pelt often spotted or ringed, underfur never present in adults .... **Phocidae**
true seals

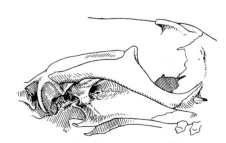

Figure 19.7 The midsection of an otariid cranium showing the alisphenoid canal (arrow).
(Gromov et al. 1963)

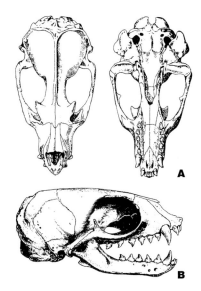

Figure 19.8 Cranium and skull of two members of the Otariidae. (A) Dorsal and ventral views of the cranium of the California sea lion, *Zalophus californianus;* (B) lateral view of the skull of the New Zealand fur seal, *Arctocephalus forsteri.*
(A, Gromov et al. 1963; B, Flower and Lydekker 1891)

Figure 19.9 Two members of the Otariidae. (A) A male northern sea lion, *Eumetopias jubatus;* (B) a male northern fur seal, *Callorhinus ursinus.*
(Gromov et al. 1963)

4 (1')	Canines well-developed, postcanine teeth vestigial, widely spaced, unicuspid (Fig. 19.12); with vertical dark stripes on body and a mane from back of head to base of tail (Fig. 19.12); head and body length 550–800 mm .... **Hyaenidae** (in part) aardwolf, *Proteles cristatus*
4'	Postcanine teeth rarely vestigial, never unicuspid, if greatly reduced in size, then canines poorly developed (Fig. 19.13); markings various, with or without mane; sizes various **5**
5 (4')	Premolars 2–4/2–4, molars 2/3, last upper molar elongated anteroposteriorly—is the longest upper postcanine tooth (Fig. 19.14) unless tied in length with second upper molar, all molars without sharp cusps and/or shearing ridges (Fig. 19.14); tail about as long as hindfoot or shorter (Figs. 19.1, 19.15); all feet with five toes side-by-side; size large: skull length more than 200 mm; head and body length 1,000–3,000 mm .... **Ursidae** (in part) bears and giant panda
5'	Premolar and molar numbers various, last upper molar not elongated anteroposteriorly, one or more molars with or without sharp cusps and/or shearing ridges; tail usually manifestly longer than hindfoot; numbers and arrangements of toes various; skull length usually much less than 200 mm; head and body length usually much less than 1,000 mm **6**
6 (5')	Postcanine dental formula 4/3 1/1 or 4/4 2/3; each foot with only four toes (check for possible presence of inconspicuous fifth digit high up on foot); size large, head and body length 750–1,700 mm **7**
6'	Postcanine dental formula not 4/3 1/1 or 4/4 2/3, or, if so, then forefeet with at least five toes (one may be a pollex located high on foot and thus separate from other toes) and/or head and body length less than 750 mm **8**

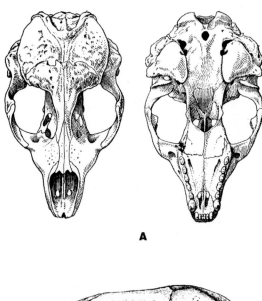

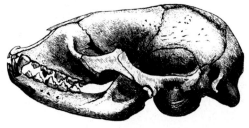

Figure 19.10 Skull of the harbor seal, *Phoca vitulina*, Phocidae. (A) Dorsal and ventral views; (B) lateral view.
(A, Gromov et al. 1963; B, Flower and Lydekker 1891)

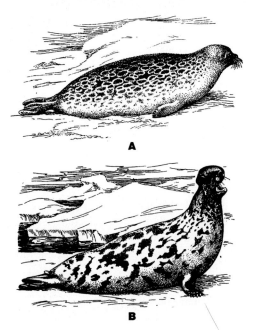

Figure 19.11 Representative true seals or hair seals, Phocidae. (A) The ringed seal, *Phoca hispida;* (B) a male hooded seal, *Cystophora cristata,* with its "hood" inflated.
(Gromov et al. 1963)

Figure 19.12 The aardwolf, *Proteles cristatus,* Hyaenidae, a termite-eating hyaenid.
(Whole animal, Giuliani 1993; skull, Flower and Lydekker 1891)

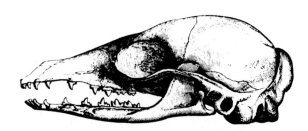

Figure 19.13 Skull of a falanouc, *Eupleres goudotii,* Viverridae, a viverrid that feeds mostly on earthworms.
(Flower and Lydekker 1891)

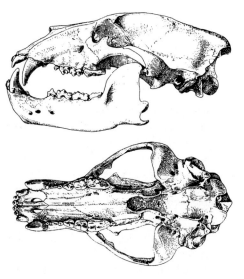

Figure 19.14 Skull of a brown bear, *Ursus arctos,* Ursidae.
(Novikov 1956)

The Carnivores 115

Figure 19.15 The Asiatic black bear, *Ursus thibetanus*.
(Finn 1929)

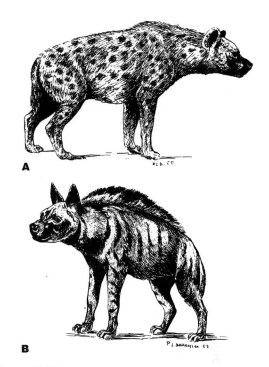

Figure 19.17 The spotted hyena, *Crocuta crocuta* (A), and the striped hyaena, *Hyaena hyaena* (B), both Hyaenidae.
(Dekeyser 1955)

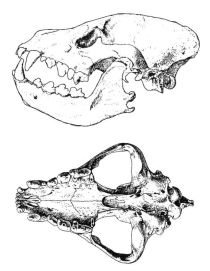

Figure 19.16 Skull of a striped hyena, *Hyaena hyaena*, Hyaenidae.
(Novikov 1956)

Figure 19.18 An African hunting dog, *Lycaon pictus*, Canidae.
(Giuliani 1993)

7 (6)   Postcanine dental formula 4/3 1/1 (Fig. 19.16); color pattern of small dark spots (Fig. 19.17A) or with dark brown crossbars on legs in combination with vertical dark stripes (Fig. 19.17B), or in combination with long, shaggy, brown fur on most of body .... **Hyaenidae** (in part)
hyenas

7'   Postcanine dental formula 4/4 2/3; color pattern irregular blotching and mottling of black, tan, and white; fur never long and shaggy (Fig. 19.18) .... **Canidae** (in part)
African hunting dog, *Lycaon pictus*

8 (6')   Postcanine dental formula 4/4 2–4/3–5, carnassials usually well-developed (Fig. 19.3A, 19.19) but if not then molars 3-4/4-5; forefoot with five toes including nonfunctional vestigial pollex widely separated from functional toes, four toes on hindfoot; tail bushy (Figure 19.20) ....
.... **Canidae** (in part)
most dogs, wolves, foxes, etc.

8'   Postcanine dental formula not 4/4 2–4/3–5, lower postcanine teeth fewer than seven; number and arrangement of toes various; tail bushy or not **9**

9 (8')   Postcanine dental formula 2–3/2 1/1, carnassials well-developed—lower molar shaped somewhat like a "two-toothed saw blade," forming an M-shaped outline (Figs. 19.2, 19.21); five toes on forefoot including pollex separated from other toes, four toes on hindfoot; muzzle short and blunt, not pointed; tail well-haired, more-or-less cylindrical in appearance, not usually manifestly bushy (Fig. 19.22) .... **Felidae**
cats: tabby, lynxes, lion, tiger, cheetah, etc.

9'   Postcanine formula various—if lower molar shaped like "two-toothed saw blade," then premolars 3–4/3 and all feet with five toes; number of toes various; muzzle often pointed; nature of tail various **10**

Figure 19.20 The red fox, *Vulpes vulpes,* Canidae.
(Gromov et al. 1963)

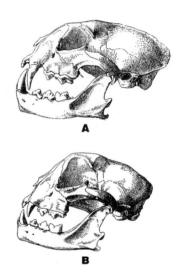

Figure 19.21 Skulls of representative felids, Felidae.
(A) The lynx, *Lynx lynx;* and (B) the cheetah, *Acinonyx jubatus.*
(Gromov et al. 1963)

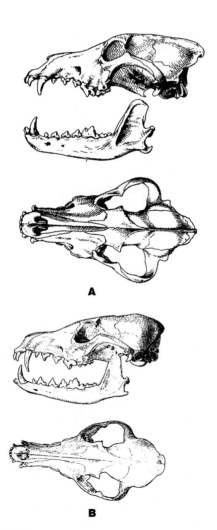

Figure 19.19 Skulls of representative canids, Canidae. (A) The wolf, *Canis lupus;* (B) a red fox, *Vulpes vulpes.*
(A, Stroganov 1962; B, Gromov et al. 1963)

10 (9')   Postcanine dental formula 3/4 2/2; mastoid process approximately as large as or larger than paroccipital process (Fig. 19.23); alisphenoid canal present (Fig. 19.7); size large, head and body length 500–650 mm; muzzle whitish, with a dark band from each eye to corner of mouth, tail ringed (Fig. 19.24) ....
.... **Ursidae** (in part)
red panda, *Ailurus fulgens*

10'   Postcanine dental formula various; if 3/4 2/2, then mastoid process absent or much smaller than paroccipital process; alisphenoid canal present or absent; head and body length usually less than 500–650 mm; markings various
.... **11**

The Carnivores 117

Figure 19.22 Representative felids, Felidae. (A) The lynx, *Lynx lynx;* (B) the leopard, *Panthera pardus.*
(Gromov et al. 1963)

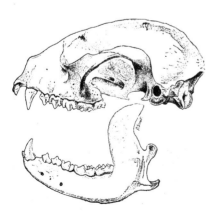

Figure 19.23 The skull of a red or lesser panda, *Ailurus fulgens,* Ursidae.
(Flower and Lydekker 1891)

Figure 19.24 The red or lesser panda, *Ailurus fulgens,* Ursidae.
(Hsia et al. 1964)

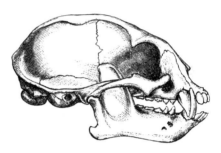

Figure 19.25 Skull of a kinkajou, *Potos flavus,* Procyonidae.
(Duncan 1877)

11 (10') Postcanine dental formula 4/4 1/2 or 4/4 2/2, carnassials well-developed; alisphenoid canal present (Fig. 19.7); forefoot with five toes including vestigial pollex widely separated from functional toes, four toes on hindfoot; size large, head and body length 580–1150 mm; doglike .. .. **Canidae** (in part)
dhole, *Cuon alpinus,* and bush dog, *Speothos venaticus*

11' Postcanine dental formula various, carnassials well-developed or not; alisphenoid canal present or absent; all limbs usually with five toes; head and body length usually less than 580–1150 mm; not doglike **12**

12 (11') Postcanine dental formula 3/3 2/2, canines with two or more grooves on labial (Fig. 19.25) and lingual surfaces, last upper molar essentially circular in occlusal outline; bare, glandular area present at each corner of the mouth; fur short and woolly, overall coloration a more or less uniform yellowish, brownish, or reddish   **Procyonidae** (in part)
kinkajou, *Potos flavus*

12' Postcanine dental formula various, canines without grooves on labial and lingual surfaces, occlusal shape of last upper molar various; no bare glandular area at corner of mouth; fur and color pattern various **13**

13' (12') Postcanine dental formula 2–4/2–4 1/1–2, carnassials often distinctly developed (Fig. 19.3C and D), upper molar either dumbbell-shaped (**Fig**. 19.26A) and oriented transversely, or squarish in occlusal outline—may be greatly enlarged to form many-cusped crushing surface (Fig. 19.4); alisphenoid canal absent; all feet with five digits; body form often elongate with short legs (Fig. 19.27); pinnae usually low, rounded, and not pointed; often more or less unicolored; tail never ringed .... **Mustelidae**
weasels, otters, skunks, badgers, martens, wolverine, etc.

13' Postcanine dental formula various, carnassial development variable, last upper cheek tooth not dumbbell-shaped or dramatically modified for crushing; alisphenoid canal present or absent; numbers of toes varies; body form various; shape of pinnae various; often with spots or stripes; tail may be ringed **14**

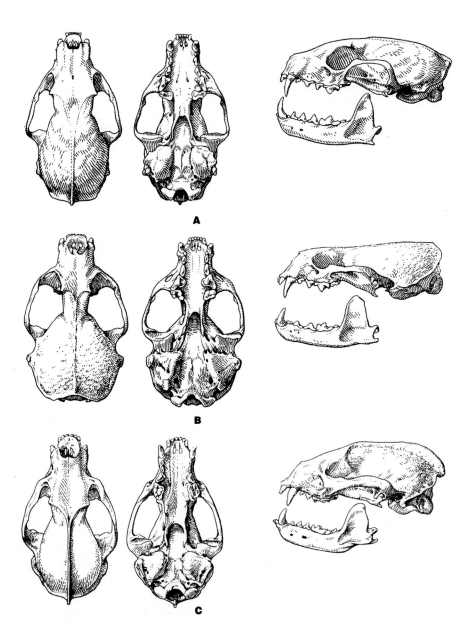

Figure 19.26 Skulls of representative mustelids, Mustelidae. (A) The European pine marten, *Martes martes;* (B) the Palearctic river otter, *Lutra lutra;* (C) the Palearctic badger, *Meles meles.*
(Stroganov 1962)

The Carnivores 119

Figure 19.27 Representative mustelids, Mustelidae. (A) The European pine marten, *Martes martes;* (B) the American badger, *Taxidea taxus;* (C) the sea otter, *Enhydra lutris;* and (D) the spotted-necked otter, *Lutra maculicollis.*
(A C, Gromov et al. 1963; B, Giuliani 1993)

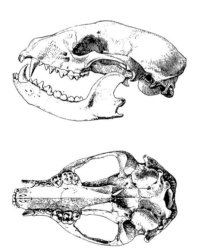

Figure 19.28 Skull of a raccoon, *Procyon lotor,* Procyonidae.
(Novikov 1956)

Figure 19.29 A raccoon, *Procyon lotor,* Procyonidae.
(Davis 1964)

14 (13') Postcanine dental formula 4/4 2/2, carnassials not at all developed (Fig. 19.28); alisphenoid canal absent; five digits on all feet, sole of foot (pes) naked for at least a considerable distance proximal to ball of foot; tail usually with obvious rings but back and/or sides of body never with spots, bands, or stripes (Fig. 19.29) .... **Procyonidae** (in part)
raccoons, coatis, olingo

14' Postcanine dental formula various, carnassials generally developed; alisphenoid canal present or absent; number of digits varies, soles of feet various; tail may be ringed and sometimes with spots and/or bands and/or stripes on back and/or sides of body 15

15 (14') Postcanine dental formula 3–4/3–4 2/2, pinnae usually low and rounded, appearing to be more on the side of the head; some with five toes on all feet, some with only four on hindfeet, some with only four on all feet; dorsum rarely spotted but may be striped or transversely barred; if tail ringed, then with no spots on dorsum and paler tail rings not white or whitish; fur usually somewhat harsh, not soft; form usually more weasel-like rather than catlike or foxlike; tail

often bushiest at base, then tapers evenly to tip (Fig. 19.30C) .... **Herpestidae**
mongooses

15' Postcanine formula 3–4/3–4 1–2/1–2 (Fig. 19.31); pinnae often relatively tall, appearing "perked"; all feet usually with five toes; dorsum and/or sides of body often spotted (Fig. 19.30A, B), usually combined with ringed tail; fur usually relatively soft; form more catlike or foxlike (Fig. 19.30A, B) than weasel-like; tail various **16**

16 (15') Postcanine dental formula 4/4 2/2; alisphenoid canal absent; five digits on all feet; tail long and with conspicuous black and white or with blackish and dirty white rings, dorsum of body more or less solid tan, brownish, or grayish, without spots and/or bands and/or stripes except for possible ill-defined middorsal stripe ....
.... **Procyonidae** (in part)
ringtails, *Bassariscus*[2]

16' Postcanine dental formula various; alisphenoid canal present or absent; sometimes without five digits on all feet; tail various, but if ringed then with spots and/or bands and/or stripes on dorsum and/or sides of body—if dorsal and lateral markings appear to be absent and tail is ringed, then paler rings not white or whitish (Fig. 19.30A and B) .... **Viverridae**
civets, genets, linsangs, fossa, etc.

Comments and Suggestions on Identification

A fair number of the carnivores are familiar to most people and easily recognizable to group. The canids, bears, and felids have characteristic basic body shapes that are more-or-less consistent. Otariids, odobenids, and phocids are easily distinguished from other mammals by general body form and from each other by ability or inability to rotate the hindlimbs forward, presence of pinnae, presence of tusks, and the relative lengths of the necks. Mustelidae, Procyonidae, and Viverridae each contain considerable diversity in body shape, but with practice, each of these groups is recognizable.

The skulls of most families of Carnivora will also, with practice, "look right." But even though the basic shape of members of a family may remain fairly consistent, size can vary considerably—for example, the skulls of a house cat and a lion are quite similar in shape, but the student used to thinking of Felidae as being the size of a tabby will often glance at the lion skull and identify it as of a bear without bothering to check the details.

Figure 19.30 Representative viverrids, Viverridae. (A) The African civet, *Civettictis civetta;* (B) the African lingsang or oyan, *Poiana richardsonii;* and (C) a herpestid, Herpestidae, the banded mongoose, *Mungos mungo.*
(Dekeyser 1955)

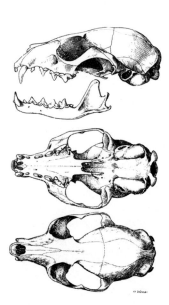

Figure 19.31 Skull of a viverrid, Viverridae, the African linsang or oyan, *Poiana richardsonii.*
(Allen 1924)

[2]Depending on the student's interpretation of characters, it is possible for the species *Bassariscus sumichrasti* to come out here or earlier at 14; in either event, it will be identified as a procyonid.

Keys to Living Families

CHAPTER 20

THE WHALES, DOLPHINS, AND PORPOISES
Order Cetacea

ORDER CETACEA

The word Cetacea, derived from the Greek word meaning "whale," is the name used for an order of mammals with two well-defined suborders, the Mysticeti or baleen whales, and the Odontoceti, containing the toothed whales, dolphins, and porpoises. The two groups are very similar in many ways, but there are also striking and basic differences. Both suborders are fully aquatic, with fusiform bodies and tails provided with a horizontal pair of large flukes made of fibrous connective tissue (Fig. 20.1). A dorsal fin supported by similar connective tissue is present in all but a few species in each suborder, but the height of this fin ranges from 22.5% of the body length in the killer whale, *Orcinus orca,* to 1.5% in the sperm whale, *Physeter catodon,* and 1% in the blue whale, *Balaenoptera musculus* (Mörzer Bruyns 1971).

Posterior limbs are absent externally, but remnants of the pelvic girdle and major leg bones may be present internally (Fig. 20.1). The anterior limb is enclosed in the body contour to the wrist, and the exposed manus is termed a flipper. The humerus, radius, and ulna are greatly shortened, and the phalanges of at least digits two and three greatly exceed the usual mammalian number (Fig. 20.1). The nostrils, termed blowholes, are located high on the dorsal surface of the head (Figs. 20.1, 20.14), and the external nares are located at the proximal end of the rostrum of the skull. To allow for these posteriorly displaced nasal openings, the nasal, maxillary, and frontal bones may be telescoped and come to overlap the parietals (Fig. 20.2).

The neck is very short. There are seven cervical vertebrae, but they may be very thin antero-posteriorly and fused together. The necks of most whales are virtually inflexible; that of the white whale or beluga, *Delphinapterus leucas,* is a notable exception.

The skin is essentially hairless. A few vibrissae are present on the snouts of mysticetes and on the heads and bodies of certain freshwater odontocetes, (e.g., *Platanista gangetica*); other odontocetes have vibrissae only during embryonic development (Sokolov 1982). A thick layer of subcutaneous fat (blubber) provides insulation. The eyes are small, and pinnae are absent. The uterus is bipartite, and the urethra and vagina open separately to the exterior. The penis is completely retractile into the body contour, and the testes are permanently internal.

As air-breathing, fully marine animals, the whales have numerous anatomical, physiological, and behavioral specializations that allow them to dive to great depths and to remain underwater for long periods, to move rapidly through water, to travel great distances, and to communicate and reproduce in the vast oceans.

Whales were of major economic importance in the eighteenth and nineteenth centuries, but their commercial importance has declined over the last several decades. Whales have been hunted by coastal peoples in many parts of the world, and a commercial industry developed in western Europe, North America, Japan, Russia, and some parts of South America. Whale meat was eaten in many parts of the world, and is still a food source in

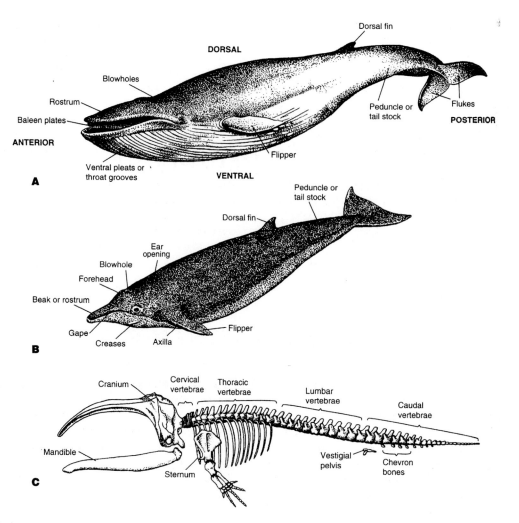

Figure 20.1 Whale body morphology. Lateral view of the general body plan and terminology for (A) baleen whales and (B) toothed whales. (C) Skeleton of a mysticete whale. Note the vestigial pelvis and the chevron bones on the caudal vertebrae. (Feldhamer et al. 1999)

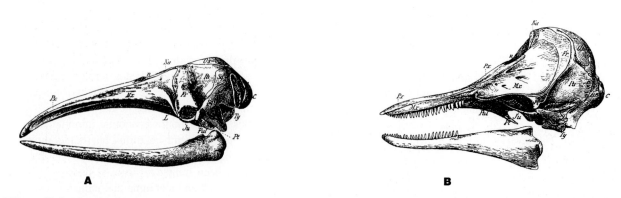

Figure 20.2 Skulls of (A), the northern right whale, *Eubalaena glacialis,* Balaenidae, Mysticeti; and (B), the saddleback dolphin, *Delphinus delphis,* Delphinidae, Odontoceti. C, occipital condyle; Fr, frontal; Ju, jugal; Mx, maxilla; n, external nares; Na, nasal; Oe, exoccipital; Os, supraoccipital; Pa, parietal; Pal, palatine; Pt, pterygold; Px, premaxilla; Sq, squamosal; Ty, tympanic bulla. (Weber 1928)

Japan and the Arctic. Whale oil was an important source of illumination until the development of kerosene, and eventually electricity replaced it.

Modern whaling began with the invention of the harpoon gun that enabled whalers to hit the fast-swimming whales such as the fin and blue, and, later, factory ships were built that allowed complete processing of whales at sea. These developments, together with faster and bigger ships and advanced techniques for locating whales, have had a serious effect on population levels of the "great" whales (the sperm whale and most baleen whales). These whales have become rare, and some, such as the right and blue whales, are near the verge of extinction. In addition, tuna fishermen net huge schools of fish and the dolphins that accompany them, which results in the death of many of these small cetaceans.

The United States has prohibited the importation of all whale products and has instituted some controls for the tuna fishing industry. Some other nations have joined in international agreements protecting cetaceans, but the whaling practices of Japan are still being strongly criticized, and more effective international conservation practices still are needed.

The cetaceans have frequently been described as the most intelligent of nonprimate mammals, and some people claim that they are even more intelligent than the anthropoid apes. They have a large brain in proportion to their body size, and they demonstrate complex patterns of behavior and communication. Stories of dolphins aiding humans by helping with fishing, driving away attackers (e.g., sharks), and saving people who were drowning date from Greek mythology and continue into the present.

SUBORDER MYSTICETI

The subordinal name, Mysticeti, is derived from the Greek words for "mystic whale" (Jaeger 1955), and it is not difficult to understand how these marine giants could seem mysterious to men in small boats. The Mysticeti include the blue whale, *Balaenoptera musculus,* which at 27.5 to 33.5 m, and weighing 150 tons (Mörzer Bruyns 1971), is not only the largest living creature but is also the largest known ever to have lived. It weighs as much as 25 elephants or 1,600 men (Slijper and Heinemann 1975). The smallest mysticete, the pygmy right whale, *Caperea marginata,* is about 6 m long and weighs about five tons (Mörzer Bruyns 1971).

The taking of right, gray, blue, and humpback whales is now prohibited to member nations of the International Whaling Commission, but not all whaling is conducted by countries that are members of the Commission. The remaining mysticetes, together with one odontocete, the sperm whale, *Physeter catodon,* are the mainstays of the whaling industry.

Distinguishing Characters

Present-day baleen whales lack teeth in both upper and lower jaws. From 130 to 400 baleen plates are suspended from each side of the upper jaw (Fig. 20.3). Each plate is composed of longitudinal strands of horny epithelial material embedded in a less resistant matrix. On the lingual edge, the matrix is worn away, producing a fringe of the tougher strands (Fig. 20.4). The plates are so arranged that the fringes of adjacent plates overlap to produce a continuous strainer-like network. The baleen whales feed by taking in huge mouthfuls of sea water (Fig. 20.5).

Then they partially close the mouth and use their large tongues to force the water out between the frayed fringes of the baleen plates. The fringes strain small organisms from the water, and this food is then swallowed.

The skull is bilaterally symmetrical, and the nasal bones extend anteriorly over the nasal passage (Fig. 20.2A). The two nasal passages exit as separate, adjacent blowholes.

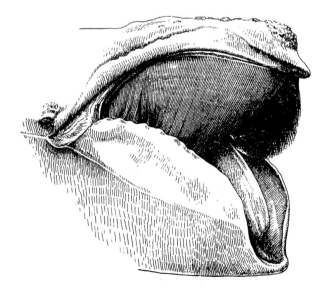

Figure 20.3 The head of a northern right whale, *Eubalaena glacialis,* Balaenidae, showing the arrangement of baleen plates.
(Gromov et al. 1963)

Figure 20.4 A single plate of baleen.
(Duncan 1877)

LIVING FAMILIES OF MYSTICETI

A list of living families of Mysticeti, and their contents, is given in Table 20.1.

Mysticetes are found in all oceans, but they are more common in the far north and far south than in tropical waters. Little is known about the movements of the pygmy right whale and of Bryde's whale. All other mysticetes are known to be migratory, moving between the cooler water areas that they inhabit in the summer and the warmer water areas that they winter in.

KEY TO LIVING FAMILIES OF MYSTICETI

1 Posterior border of nasals and premaxillae anterior to supraorbital process of frontals; rostrum long, slender, and may be very highly arched (Fig. 20.6); throat not grooved (Fig. 20.7) or with only two faintly developed grooves; middle baleen plates on each side considerably longer than anterior and posterior plates **3**

1' Nasals and nasal processes of premaxillae extending posteriorly beyond anterior borders of supraorbital processes of frontals (Fig. 20.8); rostrum broader, less arched; throat grooved (Fig. 20.5); baleen plates all approximately the same length **2**

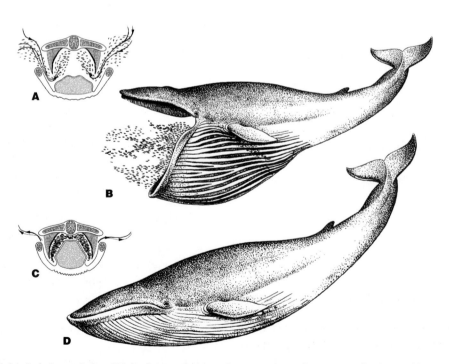

Figure 20.5 Gulping in baleen whales. (A) As the mouth opens, huge amounts of water pour in along with vast quantities of plankton and necton, with the (B) throat grooves allowing for expansion of the oral cavity. (C) This water is then expelled through the filterlike baleen mat as (D) the throat contracts, trapping the food, which is scraped off by the tongue and swallowed.
(Feldhamer et al. 1999)

TABLE 20.1 | Living Families of Mysticeti*

Family	Common Name	Number of Genera	Number of Species	Distribution
Balaenidae	Right whales	2	3	All oceans except in Southern Hemisphere tropics
Eschrichtiidae	Gray whale	1	1	North Pacific coasts (extirpated in North Atlantic in historic times)
Balaenopteridae	Rorquals and humpback whale	2	6	All oceans
Neobalaenidae	Pygmy right whale	1	1	Oceans of Southern Hemisphere in cold to temperate waters

*Based upon Mead and Brownell (1993).

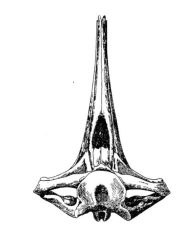

Figure 20.6 Skull of a northern right whale, *Eubalaena glacialis,* Balaenidae.
(Cabrera 1914)

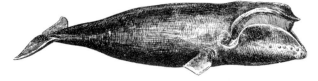

Figure 20.7 A northern right whale, *Eubalaena glacialis,* Balaenidae.
(Cabrera 1914)

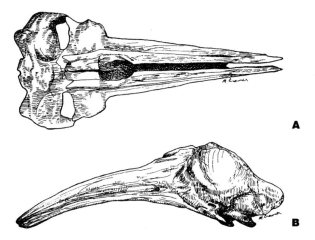

Figure 20.8 Skull of a gray whale, *Eschrichtius robustus,* Eschrichtiidae. (A) Dorsal view; (B) lateral view.
(Mary Ann Cramer)

2 (1')   Nasals large, frontals broadly exposed on vertex of skull; rostrum arched (Fig. 20.8); mandibles not conspicuously bowed outwards; throat with only a few short grooves; dorsal fin absent, but with a series of middorsal bumps posteriorly (Fig. 20.9) .... **Eschrichtiidae**
gray whale, *Eschrichtius robustus*

2'   Nasals reduced, frontals scarcely or not at all exposed on vertex of skull (Fig. 20.10); rostrum not arched (Fig. 20.10); mandibles conspicuously bowed outwards; numerous parallel grooves covering entire throat and chest region (Figs. 20.5 and 20.11); dorsal fin present (Fig. 20.11) .... **Balaenopteridae**
rorquals and humpback whale

3 (1)   Dorsal fin present; two poorly defined throat grooves present; skull length not exceeding 2 m .... **Neobalaenidae**
pygmy right whale, *Caperea marginata*

3'   Dorsal fin absent; no indication of throat grooves (Fig. 20.7), skull length of adults greatly exceeding 2 m .... **Balaenidae**
right whales

Figure 20.9 A gray whale, *Eschrichtius robustus,* Eschrichtiidae. The small raised point on the dorsum is the first and largest of a series of middorsal bumps. The more posterior ones are poorly indicated in this drawing.
(Giuliani 1995)

Figure 20.11 The size range in rorquals and humpback, Balaenopteridae. (A) Blue whale, *Balaenoptera musculus* (27–33.2 m); (B) fin whale, *B. physalus* (20–25 m); (C) sei whale, *B. borealis* (12–18 m); (D) humpback whale, *Megaptera novaeangliae* (± 17 m); (E) minke whale, *Balaenoptera acutorostrata* (7.6–9.2 m). The only species of Balaenopteridae not pictured, Bryde's whale, *B. edeni,* (12–15 m), is about the same size as the sei whale.
(Figures, Gromov et al. 1963; dimensions, Mörzer Bruyns 1971)

COMMENTS AND SUGGESTIONS ON IDENTIFICATION

Skulls of the four mysticete families can generally be distinguished by the degree of arching of the skull (see Figs. 20.6, 20.8, and 20.10). Externally, the gray whale could be confused with a small rorqual, but it has no dorsal fin and only a few throat grooves. A humpback whale could be confused with a balaenid, but the humpback has throat grooves and a dorsal fin, both lacking in the balaenids. The combination of small size with the presence of a dorsal fin and only two poorly defined throat grooves distinguishes the pygmy right whale.

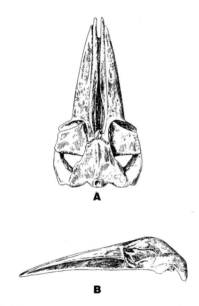

Figure 20.10 Skull of a fin whale, *Balaenoptera physalus,* Balaenopteridae. (A) Dorsal view; (B) lateral view.
(Tomilin 1962)

SUBORDER ODONTOCETI

The subordinal name, Odontoceti, literally means "tooth whales," and the presence of teeth most prominently distinguishes this group from the Mysticeti. The teeth are single-rooted, unicuspid, usually conical, and homodont. In some forms, teeth are lost entirely in the upper and/or lower jaw or are greatly reduced in number (Fig. 20.12A and B). But in most forms, teeth are numerous (Figs. 20.2B, 20.12C and D, 20.26C, 20.30A) and can range up to a total of 220 in the La Plata dolphin, *Pontoporia blainvillei* (Mörzer Bruyns 1971). The toothed cetaceans specialize in utilization of relatively scarce food sources. Different kinds hunt different sorts of fish or invertebrates, and, consequently, they have developed different feeding adaptations. Sight is not well-developed, and olfaction is probably nonexistent, but hearing is extremely well-developed. Most, if possibly not all, odontocetes communicate by sound and sense objects in their environments by sonar.

The sperm whale, *Physeter catodon,* at 20 m and 60 tons for males, and 12 m and 18 tons for females, is the largest odontocete and the only toothed whale that is a major target of the whaling industry. Dolphins are popular zoo animals, and marine exhibits featuring performing delphinids are now common.

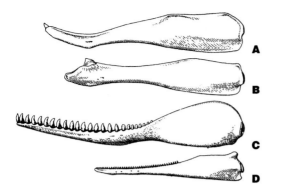

Figure 20.12 Lower jaws of four odontocetes. (A) Male goose-beaked whale, *Ziphius cavirostris;* and (B) Gray's beaked whale, *Mesoplodon grayi,* both Ziphiidae; (C) the sperm whale, *Physeter catodon,* Physeteridae; and (D) a saddleback dolphin, *Delphinus delphis,* Delphinidae.
(Gromov et al. 1963)

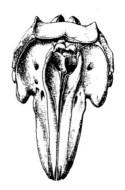

Figure 20.13 Dorsal view of the skull of an odontocete, the killer whale, *Orcinus orca,* Delphinidae, showing asymmetry in the region of the external nares.
(Gromov et al. 1963)

DISTINGUISHING CHARACTERS

Teeth are almost always present; baleen is always absent. The skull is often bilaterally asymmetrical in the area of the external nares (Fig. 20.13), and the nasal bones do not project anteriorly over the nasal passages (Figs. 20.2B, 20.13). Externally, the nostrils are united into a single blowhole (Fig. 20.14).

LIVING FAMILIES OF ODONTOCETI

A list of living families of Odontoceti, and their contents, is given in Table 20.2.

Most odontocetes are marine, but one family, Platanistidae, primarily inhabits fresh water; certain species of Delphinidae are common in fresh water, and at least one species of phocoenid is commonly found there as well.

Odontocetes occur in all oceans and seas, but many species never venture far from coastlines. The sperm

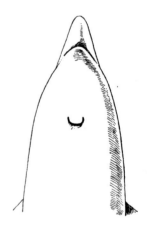

Figure 20.14 Dorsal view of the anterior portion of a bottle-nosed dolphin, *Tursiops truncatus,* Delphinidae, showing the single blowhole.
(Cabrera 1914)

TABLE 20.2 | **Living Families of Odontoceti***

Family	Common Name	Number of Genera	Number of Species	Distribution
Platanistidae	Freshwater dolphins	4	5	Neotropical, Oriental, coastal waters off southeastern South America
Delphinidae	Dolphins, killer whales, etc.	17	32	All oceans and seas, Neotropical, some rivers outside Neotropics
Phocoenidae	Porpoises	4	6	All oceans, some rivers
Monodontidae	Narwhal and beluga	2	2	Arctic Ocean, adjacent seas and large rivers
Physeteridae	Sperm whales	2	3	All oceans
Ziphiidae	Beaked whales	6	19	All oceans

*Based on Mead and Brownell (1993).

whale and many other kinds of Odontoceti, except for the platanistids, are known to be migratory (Rice 1984).

Key to Living Families of Odontoceti

1   Teeth absent (or apparently absent) in lower jaws **2**

1'   Teeth present in the lower jaw; these may be unerupted but form obvious bulges **3**

2 (1)   Rostrum broad, nearly as wide as long; two teeth present in upper jaw, right one in males usually unerupted, both teeth may be unerupted in females, left tooth (and sometimes right one as well) in adult males a long spiraling tusk (Figs. 20.15, 20.16A) .... **Monodontidae** (in part) narwhal, *Monodon monoceros*

2'   Rostrum narrow, much longer than wide (Fig. 20.17); teeth absent in upper jaw, unerupted teeth only may be present in lower jaw .... **Ziphiidae** (in part) those without obvious teeth

3 (1')   Size very large, total length of skull over 2 m, total length of animal 12–20 m; 16 to 30 large conical teeth present in each dentary; no erupted teeth present in upper jaws (rarely a few rudimentary teeth erupted) (Fig. 20.18A); mandibular symphysis at least 35% of mandible length (Fig. 20.18A); head very large and blunt anteriorly (Fig. 20.19A) .... **Physeteridae** (in part) sperm whale, *Physeter catodon*

3'   Size large to small, skull less than 2 m (rarely more than 1 m), total length of animal less than 12 m; teeth present in both upper and lower jaw, or if absent in upper jaw, lower teeth do not number more than 16 per dentary; mandibular symphysis less than 35% of mandible length, or if more than 35%, many well-developed teeth present in upper jaws; head shape various (Fig. 20.20) **4**

Figure 20.16 The Monodontidae. (A) A male narwhal, *Monodon monoceros;* and (B) a beluga, *Delphinapterus leucas*. (Gromov et al. 1963)

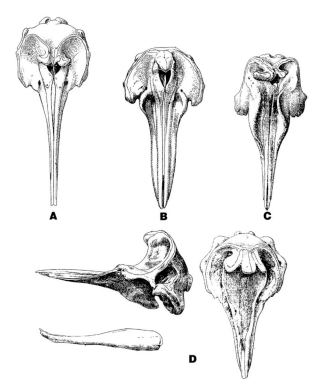

Figure 20.17 Skulls of some beaked whales, Ziphiidae. (A) Sowerby's beaked whale, *Mesoplodon bidens;* (B) Baird's beaked whale, *Berardius bairdii;* (C) northern bottle-nosed whale, *Hyperoodon ampullatus;* and (D) goose-beaked whale, *Ziphius cavirostris*. (Dorsal views, Gromov et al. 1963; lateral view, Cabrera 1914)

Figure 20.15 Skull of a male narwhal, *Monodon monoceros,* Monodontidae. The top of the rostrum has been dissected to show the root of the large left tusk and the small, unerupted, right tusk. (Flower and Lydekker 1891)

The Whales, Dolphins, and Porpoises 129

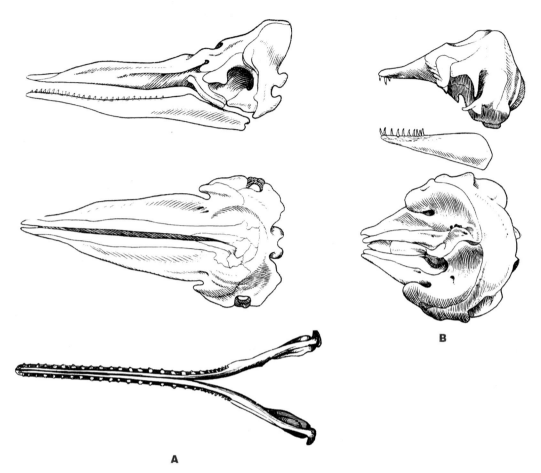

Figure 20.18 Lateral and dorsal views of the skulls of two physeterids. (A) The sperm whale, *Physeter catodon;* and (B) a pygmy sperm whale, *Kogia simus*.
(A, dorsal view of *Physeter* lower jaw, Giebel 1859; A, *Physeter* skull views and B, Bobrinskii et al. 1965)

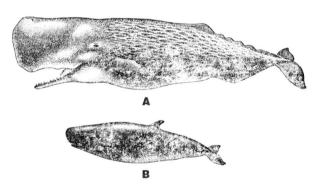

Figure 20.19 Two physeterids. (A) The sperm whale, *Physeter catodon;* and (B) a pygmy sperm whale, *Kogia breviceps*.
(Gromov et al. 1963)

4 (3') Zero to seven teeth present on each side above, or if eight or more on each side above, then anterior one or two lower teeth on each side much larger than upper teeth or posterior lower teeth **5**

4' At least eight well-developed teeth present on each side above and below, anterior lower teeth not appreciably larger than the others **7**

5 (4) One or two teeth in each dentary (Fig. 20.12A, B), usually large—if more than two present, posterior teeth considerably smaller than anterior teeth; rostrum long and narrow (Fig. 20.17); two deep grooves on throat, converging anteriorly (Fig. 20.21); size large, 3.6 to 12 m total length; snout "beaklike" (Figs. 20.20A–D, 20.21) .... **Ziphiidae** (in part)
beaked whales with teeth

5' Three or more teeth in each dentary, posterior teeth not appreciably smaller than anterior teeth (Fig. 20.18B); rostrum short and broad (Figs. 20.18B, 20.22); no grooves on throat; size medium, 3 to 4 m total length; head blunt (Figs. 20.19B, 20.20G) **6**

6 (5') Lower teeth number fewer than eight in each dentary; external nares fully visible in dorsal view of skull (Fig. 20.22); dorsal fin tall, about 42 cm (Fig. 20.23) .... **Delphinidae** (in part)
Risso's dolphin, *Grampus griseus*

6' Lower teeth number eight or more in each dentary; external nares barely visible in dorsal view of skull (Fig. 20.18B); dorsal fin short, about 24 cm high (Fig. 20.19B) .... **Physeteridae** (in part) pygmy and dwarf sperm whales

7 (4') Mandibular symphysis greater than 40% of mandible length; tooth-rows parallel for most of their length; teeth number from 25 to 60 in each quadrant; rostrum very long and very narrow, depth and breadth of rostrum about equal (Figs. 20.24, 20.25) .... **Platanistidae** freshwater dolphins

7' Mandibular symphysis less than 40% of mandible length; tooth-rows diverge, teeth number from five to 52 in each quadrant; rostrum various but never as above, depth always considerably less than breadth **8**

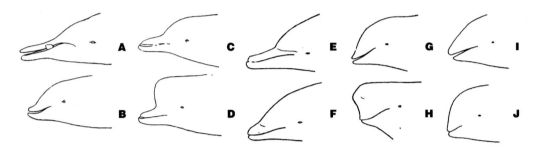

Figure 20.20 Some examples of head shape in the Ziphiidae (A–D), Delphinidae (E–H), and Phocoenidae (I–J). (A) Stejneger's beaked whale, *Mesoplodon stejnegeri;* (B) goose-beaked whale, *Ziphius cavirostris;* (C) a giant bottle-nosed whale, *Berardius;* (D) a bottle-nosed whale, *Hyperoodon;* (E) saddleback dolphin, *Delphinus delphis,* or *Stenella,* or a bottle-nosed dolphin, *Tursiops truncatus;* (F) *Lagenorhynchus,* or a right whale dolphin, *Lissodelphis;* (G) Risso's dolphin, *Grampus griseus;* (H) a pilot whale, *Globicephala;* killer whale, *Orcinus orca;* false killer whale, *Pseudorca crassidens;* (I) a harbor porpoise, *Phocoena,* or Dall porpoise, *Phocoenoides dalli;* and (J) finless porpoise, *Neophocaena phocaenoides.*
(Gromov et al. 1963)

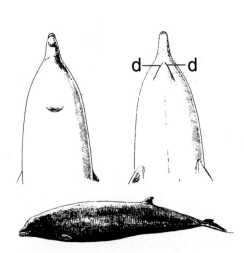

Figure 20.21 Dorsal and ventral views of the anterior end of and a lateral view of an entire goose-beaked whale, *Ziphius cavirostris,* Ziphiidae; d, grooves on the throat.
(Cabrera 1914)

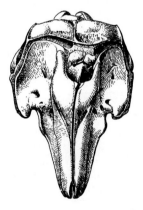

Figure 20.22 Dorsal view of the skull of Risso's dolphin, *Grampus griseus,* Delphinidae.
(Gromov et al. 1963)

Figure 20.23 Risso's dolphin, *Grampus griseus,* Delphinidae.
(Cabrera 1914)

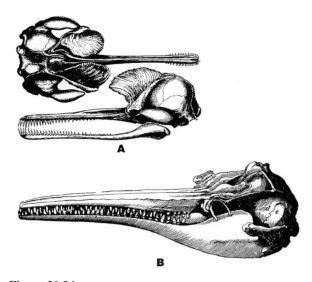

Figure 20.24 Skulls of two platanistids. (A) Dorsal and lateral views of a Ganges river dolphin skull, *Platanista gangetica*. Note the shieldlike processes of bone. These are unique to the genus. (B) Lateral view of the skull of the boto, *Inia geoffrensis*.
(A, Duncan 1877; B, Geibel 1859)

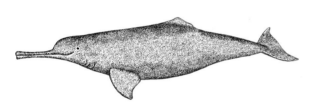

Figure 20.25 A Ganges river dolphin, *Platanista gangetica*, Platanistidae.
(Feldhamer et al. 1999)

9' When skull is viewed in profile, top of rostrum distinctly concave between anterior edge of nares and distal tip of premaxillae (Fig. 20.30); teeth vary in number, usually exceeding 10/8; dorsal fin usually present, color various (Fig. 20.31) .... **Delphinidae** (in part)
all genera except *Grampus*

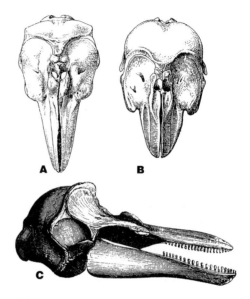

Figure 20.26 Skulls of representative Phocoenidae. (A) Dorsal view of skull of the harbor porpoise, *Phocoena phocoena;* (B) dorsal view of skull of the finless porpoise, *Neophocaena phocaenoides;* and (C) lateral view of a phocoenid skull showing the bosses anterior to the external nares.
(A and B, Gromov et al. 1963; C, Gervais 1855)

8 (7') Premaxillae with prominent bosses (bumps) immediately in front of external nares (Fig. 20.26); teeth laterally compressed and spadelike (Fig. 20.27); head blunt (Figs. 20.20I, J, 20.28) .... **Phocoenidae**
porpoises

8' Premaxillae flat or concave immediately in front of narial openings; teeth generally conical, never laterally compressed and spadelike; head shape various (Fig. 20.20E–I) **9**

9 (8') When skull is viewed in profile, top of rostrum flat or slightly convex between anterior edge of nares and distal tip of premaxillae (Fig. 20.29); teeth 10/8; dorsal fin absent, color white in adults (gray or yellowish in immatures) (Fig. 20.16B) .... **Monodontidae** (in part)
beluga, *Delphinapterus leucas*

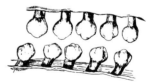

Figure 20.27 The laterally compressed teeth of a porpoise, Phocoenidae.
(Flower and Lydekker 1891)

Figure 20.28 A phocoenid, *Phocoena phocoena*.
(Cabrera 1914)

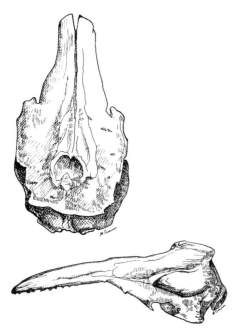

Figure 20.29 Dorsal and lateral views of the cranium of a white whale or beluga, *Delphinapterus leucas,* Delphinidae.
(Mary Ann Cramer from Tomilin 1962)

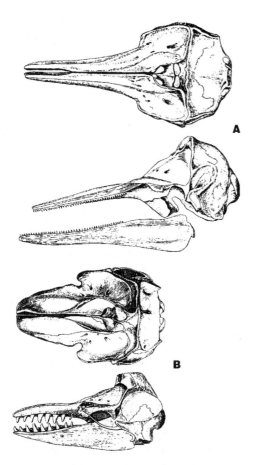

Figure 20.30 Dorsal and ventral views of skulls of two of the many species of Delphinidae. (A) The striped dolphin, *Stenella coeruleoalba;* and (B) the false killer whale, *Pseudorca crassidens.* Not to same scale.
(Tomilin 1962)

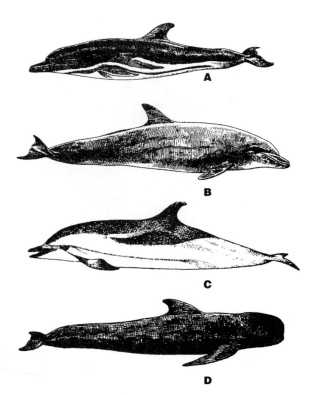

Figure 20.31 Representative Delphinidae. (A) The saddle-back dolphin, *Delphinus delphis;* (B) the bottle-nosed dolphin, *Tursiops truncatus;* (C) the striped dolphin, *Stenella coeruleoalba,* and (D) the long-finned pilot whale, *Globicephala melas.* Not all to same scale.
(A, B, and D, Cabrera 1914; C, Tomilin 1962)

COMMENTS AND SUGGESTIONS ON IDENTIFICATION

Each of the families Monodontidae, Physeteridae, and Ziphiidae is distinctive and, with practice, easily recognizable by either external or cranial characters. The Platanistidae resemble some Delphinidae externally but generally have longer, more slender beaks and lower, more broadly based dorsal fins. Cranially, the rostrum is more slender, and their teeth more numerous than in other families. The members of the Phocoenidae are all similar in appearance, but externally they also resemble several species of Delphinidae. The characters given in the key will identify the skulls, but external identification will require learning to recognize the various species individually.

Many authors have written keys based upon external characteristics of cetaceans. However, unless you are aboard a tuna boat, are scuba diving, or come upon a beached animal, you will rarely have external appearance as a guideline for identification. The only species commonly to be found in marine "zoos" are a few species of Delphinidae, the boto, and the beluga.

CHAPTER 21

THE MACROSCELIDEANS
Order Macroscelidea

ORDER MACROSCELIDEA

The ordinal name Macroscelidea means "large hindlimbed ones." Macroscelideans are small, primarily diurnal mammals that feed mostly on insects. They are found in a variety of terrestrial habitats ranging from forest to semidesert. Some take refuge in burrows made by other mammals, or they may dig their own, or construct temporary surface nests.

The most characteristic feature of these mammals is a long, slender, and highly mobile snout provided with numerous basal vibrissae (Fig. 21.1). This snout is responsible for the English name of these animals: "elephant shrews." The eyes and ears are large, and the lower limbs are slender and elongated, with reduced areas for contact with the ground. The usual mode of progression is a quadrupedal scurry, but when alarmed, they leap or hop on their long, slender, digitigrade hindfeet (Fig. 21.1). The fur is soft, and the long slender tail is covered with scales, which may be largely obscured by short bristles.

Although often referred to as "elephant shrews," this group has no significant relationship to either elephants or shrews. This order was, in the past, often included in the Insectivora "wastebasket."

DISTINGUISHING CHARACTERS

Externally as noted earlier. The hindlimbs are much longer than the forelimbs, the distal portions of the limbs are longer than the proximal portions, and the tibia and fibula are fused.

The dental formula is 1–3/3 1/1 4/4 2/2–3 = 36–42. The fourth premolar is molariform, and the molars are usually four-cusped and zalambdodont. The palate has large perforations (Fig. 21.2). The zygomatic arch is complete and imperforate, and there is no complete postorbital bar. The auditory bullae are well-developed.

LIVING FAMILY OF MACROSCELIDEA

Macroscelidea contains one living family, **Macroscelididae,** the macroscelidids or "elephant shrews," including four genera and 15 species (Schlitter 1993). They range through most of the Ethiopian Region and occur in the Palearctic in North Africa.

Figure 21.1 An elephant shrew, *Elephantulus fuscipes*, Macroscelididae. The stance of the animal in this illustration is misleading. Macroscelideans are neither truly ricochetal nor plantigrade.
(Dekeyser 1955)

Comments and Suggestions on Identification

The slender, mobile snout, relatively large eyes and ears, slender limbs, and small feet are distinctive. No other mammal has this combination of characters. The skulls may be told from superficially similar marsupial skulls by the absence of an inflected angle of the ramus. The combination of a large orbit, complete auditory bulla, complete zygomatic arch, and heavily perforate palate should distinguish macroscelidids from members of the order Insectivora. The imperforate zygomatic arch and lack of a postorbital bar distinguish them from Scandentia and Primates, and poorly developed canines distinguish them from Chiroptera and Carnivora.

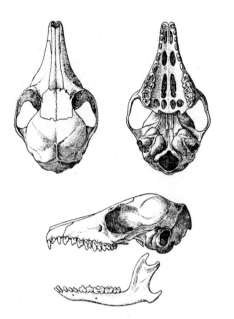

Figure 21.2 Skull of an elephant shrew, *Elephantulus rozeti*, Macroscelididae.
(Cabrera 1932)

Key to Living Families

CHAPTER 22

THE RABBITS, HARES, AND PIKAS
Order Lagomorpha

ORDER LAGOMORPHA

Lagomorph literally means "hare-shaped." Although this name does not describe a diagnostic feature of the order, it does point out the similarity among living species. Lagomorphs are terrestrial mammals, although some species burrow and may be considered semifossorial. Members of the living families have tails that are short to very short to absent and hindfeet that are at least somewhat larger than the forefeet. In rabbits and hares, Leporidae, the hindfeet are considerably larger than the forefeet, the external ears are generally very long, and the tail is short (absent in *Romerolagus diazi*) but usually evident externally (Fig. 22.1). Pikas, Ochotonidae, are smaller than most leporids, the hindfeet are only slightly larger than the forefeet, the ears are relatively short and rounded, and the tail is absent externally (Fig. 22.2).

The term "*rabbit*" is properly applied to those leporids that have **altricial** young (that is, born naked, blind, and helpless). The young of "*hares*" are **precocial** (that is, born furred, sighted, and capable of moving about on their own). In North America, the domesticated rabbit, *Oryctolagus cuniculus,* and the various species of cottontails, *Sylvilagus,* are called "rabbits," whereas the so-called "jack rabbits," arctic hares, and snowshoe rabbits, *Lepus,* are actually "hares."

Lagomorphs are almost totally herbivorous and feed on a wide variety of forbs, grasses, and, to some extent, shrubs. Reingestion of caecotrophic feces is known to occur in most species of lagomorphs and enables them to assimilate more plant nutrients and certain B vitamins that are produced by bacteria in the caecum (Hansen and Flinders 1969).

The social structure in lagomorphs ranges from a dispersed system in many hares (*Lepus*) to that of dominance

Figure 22.1 The mountain hare, *Lepus timidus*, Leporidae.
(Flower and Lydekker 1891:493)

Figure 22.2 A pika, *Ochotona*, Ochotonidae.
(Hsia et al. 1964:24)

hierarchies in the European rabbit, *Oryctolagus cuniculus* (Eisenberg 1966). Certain pikas (*Ochotona*) may establish territories based on the defense of accumulated hay piles, although these territories are not generally defended during the reproductive season (Kawamichi 1976; Lutton 1975). Most species of lagomorphs spend their entire lives above ground, but there are notable exceptions. Montane pikas (*Ochotona*) establish passageways among rock piles, steppe-dwelling pikas live in burrows, and European rabbits (*Oryctolagus cuniculus*) construct extensive underground burrow systems or warrens. Certain other species of rabbits of the genera *Caprolagus, Poelagus,* and *Romerolagus* are also known to construct burrows (Walker et al. 1975).

Wild lagomorphs are of some economic importance as sources of fur and meat and as game animals. In some places, they are pests. This is particularly true in areas such as Australia and southern South America where *Oryctolagus* has been introduced by humans and has greatly multiplied and damaged native ecosystems. The domestic rabbit is raised for meat, for fur, as a pet, and for laboratory research.

DISTINGUISHING CHARACTERS

The dental formula is 2/1 0/0 3/2 2–3/3 = 26–28. The first incisors are large and "rodentlike." The second upper incisors are small, peglike teeth located directly behind the first incisors (Fig. 22.3). The cheek teeth are hypsodont, rootless, and evergrowing. Each maxilla is perforated on the side of the rostrum by a single large opening in the Ochotonidae or by numerous small openings separated by a lattice of bone in the Leporidae (Fig. 22.4).

The forefeet are digitigrade, and the hindfeet are plantigrade. The tail is short to absent. There is no baculum, and the testes descend seasonally into a scrotum located anterior to the penis. The uterus is duplex.

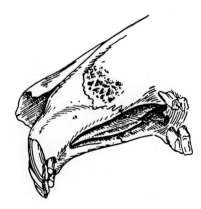

Figure 22.3 Rostrum of lagomorph, showing upper incisors.
(Guryev 1964)

Figure 22.4 Skull of mountain hare, *Lepus timidus,* Leporidae. Note the fenestration of the rostrum.
(Flower and Lydekker 1891:492)

LIVING FAMILIES OF LAGOMORPHA

There are two living families of Lagomorpha. The **Ochotonidae,** the pikas, includes two genera and 26 species (one of these species, †*Prolagus sardus,* is historically extinct from the islands of Corsica and Sardinia) which range through the mountains of the western Nearctic and the mountains and steppes of the Palearctic. The **Leporidae,** the rabbits and hares, includes 11 genera and 54 species and are distributed worldwide except for portions of the Oriental Region and Madagascar. Some species have been introduced into the Australian Region and to many oceanic islands by humans. The number of families, genera, and species (including the extinct †*Prolagus sardus*) is based upon Hoffman (1993).

KEY TO LIVING FAMILIES OF LAGOMORPHA

1   Cutting edge of I^1 with V-shaped notch (Fig. 22.5A); dental formula 2/1 0/0 3/2 2/3 = 26; well-developed supraorbital process of frontal absent (Fig. 22.6A); external tail absent; ears no longer than wide .... **Ochotonidae**
pikas

1'   Cutting edge of each first upper incisor straight (Fig. 22.5B); dental formula usually 2/1 0/0 3/2 3/3 = 28 (2/1 0/0 3/2 2/3 = 26 in *Pentalagus*); supraorbital process of frontal present (Fig. 22.6B); short external tail usually present; ears longer than wide .... **Leporidae**
rabbits and hares

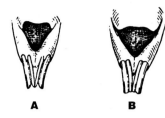

Figure 22.5 Anterior view of the first upper incisors of an ochotonid (A) and a leporid (B).
(Gromov et al. 1963)

COMMENTS AND SUGGESTIONS ON IDENTIFICATION

All rabbits and hares have a similar basic appearance, but some rodents (such as the springhaas) superficially resemble lagomorphs. Pikas could be confused with several kinds of rodents. Be sure to check for the characteristic second upper incisor, which is lacking in all rodents.

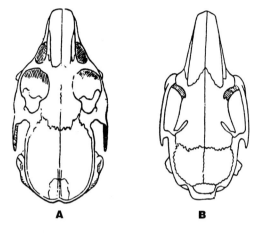

Figure 22.6 Skulls of (A) a pika, *Ochotona,* Ochotonidae, and (B) a domestic rabbit, *Oryctolagus cuniculus,* Leporidae.
(Gromov et al. 1963)

CHAPTER 23

THE RODENTS
Order Rodentia

Key to Living Families of the World
Key to Living Families of North America

ORDER RODENTIA

The order Rodentia, the "gnawing" mammals, contains over 40% of all species in the class Mammalia. A trait shared by all rodents, and only a few other mammals, is an upper and lower pair of arc-shaped, chisel-edged, evergrowing incisors.

Rodents have adapted to most nonmarine habitats open to mammals and include terrestrial, fossorial, saltatorial (Fig. 23.1), arboreal, gliding, and semiaquatic forms. They range in size from the smallest mice (e.g., *Micromys, Baiomys,* and some *Mus*), which weigh only a few grams, up to the largest living rodent, the capybara, *Hydrochaeris hydrochaeris,* that is pig-sized and weighs up to 50 kg.

Most rodents are omnivorous, feeding on bark, grass, other vegetation, seeds, insects, and other animal matter. Some, such as the grasshopper mice, *Onychomys,* also feed on small vertebrates.

Many species of rodents are very important economically. On the negative side, *Rattus norvegicus, Rattus rattus, Mus musculus,* and other species commensal with humans, damage grains stored in unprotected granaries. Under certain conditions, rodents may damage field crops, debark trees in orchards or forests, or establish burrows in areas where humans do not want them (e.g., lawns, dikes). Most of these depredations can be controlled by agricultural management practices or selective trapping.

Rodents also have many beneficial attributes. The burrowing activities of fossorial species aerate the soil and bring mineral nutrients into the topsoil. Insects are eaten by many species. Beavers, muskrats, and ranch stocks of chinchillas are utilized in the fur industry. Medical and zoological research is highly dependent on laboratory-raised stocks of the Norway rat, *Rattus norvegicus;* house mouse, *Mus musculus;* golden hamster, *Mesocricetus auratus;* guinea pig, *Cavia porcellus;* and, more recently, the Mongolian "gerbil," *Meriones unguiculatus.* Many people in various parts of the world utilize rodents for food, especially sciurids and the larger hystricomorphs.

Worldwide, many species of rodents are threatened or endangered, primarily by loss of habitat and, in some cases, by overhunting. The larger species of rodents such as the capybara (*Hydrochaeris hydrochaeris*), paca (*Agouti paca*), and moco (*Kerodon rupestris*) are utilized by humans for food and have been greatly reduced in numbers in parts of their range. The wild chinchillas, *Chinchilla,* of western South America are greatly reduced in numbers due to overhunting for their pelts. Chile has established a national chinchilla reserve to reduce the population decline and help preserve genetic diversity. The nutria, *Myocastor coypus,* has declined in its native Argentina, although it has experienced a population boom in parts of the United States where, as an introduced species, it competes all too successfully with the native muskrat, *Ondatra zibethicus.*

Rodents may live in social groups or be solitary. In solitary species, the young typically disperse after weaning, but in species with family groups, some of the young may join the extended family after they mature. A very unusual form of social organization, similar to the condition in social insects, exists in the naked "mole"-rat, *Heterocephalus glaber.* This is a fossorial species that forms close-knit colonies. In each colony, a

Figure 23.1 Springhaas, *Pedetes capensis,* Pedetidae, one of several species of saltatorial rodents.
(Mary Ann Cramer after photograph in Walker et al. 1975)

single dominant female (sometimes called the "queen") monopolizes breeding and produces the young. The other adult individuals in the colony care for these offspring and the breeding female.

DISTINGUISHING CHARACTERS

A single arc-shaped incisor is present in each jaw quadrant (Fig. 23.2). The distal end of each incisor is chisel-edged, and the sharpness is maintained by wear on the hard enamel (anterior surface) and the softer dentine (bulk of tooth). Canines and most or all premolars are absent, resulting in a wide diastema between the incisors and the remaining cheek teeth. The usual bilateral complement of cheek teeth is 12 to 16, but the range is from four (*Mayermys*) to 28 (*Heliophobius*).

Most rodents are small (less than 300 mm head plus body length), but a few approach or exceed 1 meter in length. With few exceptions (e.g., *Spalax*), an external tail is present, but its form and length vary greatly. All rodents except *Heterocephalus* are well-haired over most of their bodies. Most are quadrupedal, but some arid-land species have greatly enlarged hindlimbs and are capable of ricochetal locomotion. Primitively, the digits are 5/5, but they may be reduced to 4/3.

The uterus is duplex, and the testes may be internal or scrotal. A baculum is present.

IMPORTANT TAXONOMIC CHARACTERS

Cranial and dental characters are those used most often in rodent classification. The cheek teeth show a great diversity in shape and occlusal pattern and may have cusps situated on transverse ridges or **laminae** (see Figs. 23.22 and 23.23) or have isolated **enamel islands** (see Fig. 23.16). An **inner fold** on a cheek tooth is an elongated enamel island (see Fig. 23.20). A **re-entrant fold** is an invagination along the side of a cheek tooth (see Fig. 23.16). A re-entrant fold that is sharply angular is termed a **re-entrant angle** (see Fig. 23.11A).

The zygomatic region of the skull and the infraorbital foramen are variously modified for passage and attachment of different branches of the masseter muscle. In the sewellel, Aplodontidae (Fig. 23.3A), the masseter originates from the lower edge of the zygomatic arch, and only a fascicle of the muscle passes through the infraorbital foramen (Coues, in McLaughlin 1984:267), and this condition is said to be **protrogomorphous.** In the **sciuromorphous** condition, found in "advanced sciuromorphs" (Fig. 23.3B), the middle masseter (*masseter lateralis profundus*) originates high on the zygomatic plate (defined later) at the level of the orbit and with no or slight muscle transmission through the infraorbital foramen. This condition is found in the Sciuridae, Castoridae, Geomyidae, and Heteromyidae. In the **hystricomorphous** condition (Fig. 23.3D), the deep masseter (*masseter medialis pars anterior*) is tremendously enlarged and passes through a correspondingly enlarged infraorbital foramen. This condition is found in caviomorphs and other hystricognaths and in Dipodidae, Anomaluridae, Pedetidae, and Ctenodactylidae. In the **myomorphous** condition (Fig. 23.3C), the deep masseter passes through an infraorbital foramen that is modest in size (intermediate between the sciuromorphous and hystricomorphous conditions). Myomorphy is found in the Muridae and Myoxidae.

The zygomatic process of the maxilla, the **maxillary process,** is variously modified. When the infraorbital foramen is enlarged (myomorphous and hystricomorphous conditions), the process is divided into an **upper maxillary process** and a **lower maxillary process** (see Fig. 23.17A). In many rodents (particularly Muridae), the lower maxillary process may take on a broad, flattened shape, forming a **zygomatic plate** or lower maxillary process (see Fig. 23.17A).

LIVING FAMILIES OF RODENTIA

Living families and their contents are listed in Table 23.1.

Rodents are native to most land areas except some Arctic and oceanic islands, New Zealand, and Antarctica. Some members of the Muridae, Myocastoridae, and Caviidae have been introduced by humans into all parts of the world and have established wild populations.

Because the Rodentia is a very large and complex order, we depart slightly from the usual format and present two family keys. The first is a key to families of the world. The second is a shortened key designed to identify by family all species that occur in North America, including Middle America and the West Indies.

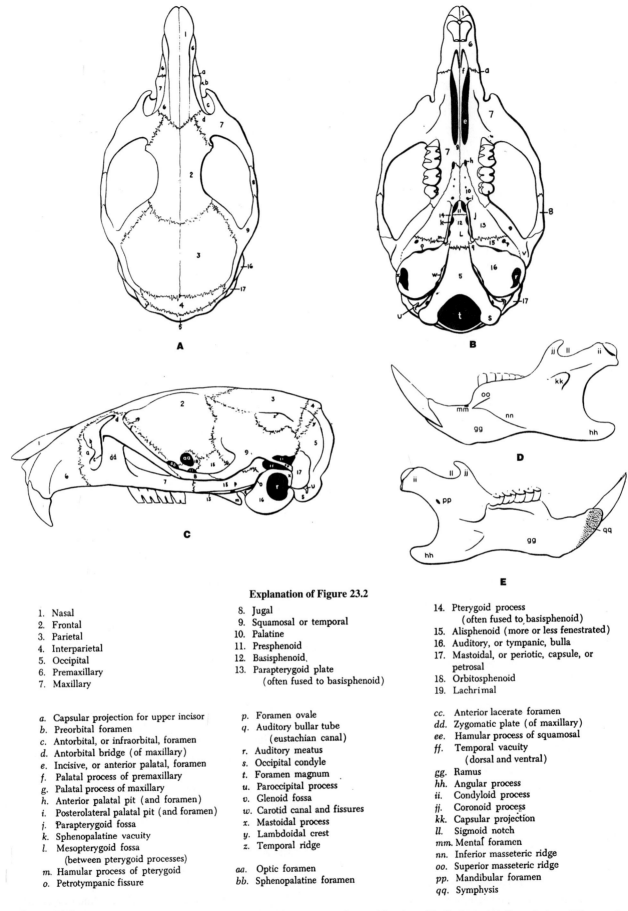

Figure 23.2 Components of the cranium (A–C) and dentary (D–E) of a murid rodent, *Phyllotis* sp. (A) Dorsal view; (B) ventral view; (C–D) lateral views; (E) medial view. See explanation of figure for identification of numbered and lettered portions.
(Modified from Hershkovitz 1962:110–15)

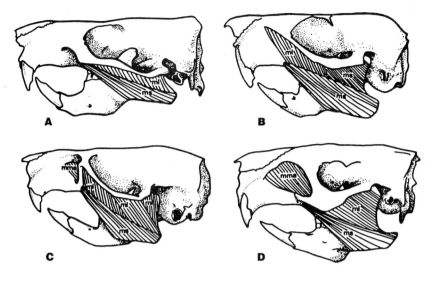

Figure 23.3 Diagrammatic representation of superficial, middle, and deep portions of the masseter muscle in various rodents: (A) **aplodontid,** masseter originates from the lower edge of zygomatic arch; (B) **advanced sciuromorph** condition, middle masseter (masseter lateralis) originates from outer side of skull in front of orbit, infraorbital foramen small, not transmitting any muscle; (C) **myomorph** condition, deep portion of masseter (masseter medialis) pushes up through orbit and passes through a V-shaped, oval, or round infraorbital foramen; (D) **hystricomorph** condition, masseter medialis enormously developed with portion passing through a greatly enlarged infraorbital foramen. **ms,** masseter superficialis; **ml,** masseter lateralis; **mma,** masseter medialis anterior.
(M.A. Cramer)

TABLE 23.1 Living Families of Rodentia[1]

Suborder and Family	Common Name	Number of Genera	Number of Species	Distribution[2]
Suborder Sciuromorpha				
Aplodontidae	Sewellel or mountain "beaver"	1	1	Western Nearctic (California to British Columbia)
Sciuridae	Squirrels and marmots	50	273	Worldwide, except southernmost Neotropical, Madagascar, and Australian
Castoridae	Beavers	1	2	Holarctic
Geomyidae	Pocket gophers	5	35	Southern Nearctic, northwestern Neotropical
Heteromyidae	Kangaroo rats, pocket mice, and kangaroo mice	6	59	Southwestern Nearctic and northern Neotropical
Dipodidae	Jumping mice, birch mice, and jerboas	15	51	Holarctic (except for most of Europe), Northern Ethiopian
Muridae[3]	Rats and mice	281	1326	Worldwide except Antarctica and various oceanic islands (where now introduced)
Arvicolinae	Lemmings, voles, and muskrat	26	143	Holarctic
Cricetinae[3]	Hamsters	9	25	Palearctic and, for *Mystromys,* Ethiopian
Cricetomyinae	African pouched rats	3	6	Ethiopian
Dendromurinae	African climbing mice, gerbil mice, fat mice, and Congo forest mice	8	23	Ethiopian
Gerbillinae	Gerbils	14	110	Ethiopian, southern Palearctic, western Oriental
Lophiomyinae	Maned rats	1	1	Ethiopian
Murinae[3]	Old World rats and mice	112	510	Ethiopian, Palaearctic, Oriental, and Australian; widely introduced elsewhere
Hydromyinae[3]	Australasian water rats and allies	10	19	Australian
Myospalacinae	Zokors	1	7	Palearctic (Siberia) and Oriental
Nesomyinae	Malagasy rats and mice	7	14	Madagascar
Otomyinae	African swamp rats	2	14	Ethiopian
Petromyscinae	Delany's swamp mouse and rock mice	2	5	Southern and eastern Ethiopian
Platacanthomyinae	Spiny dormouse and Chinese pygmy dormice	2	3	Oriental (southern India, southeastern China, and northern Vietnam)
Rhizomyinae	African "mole"-rats and Bamboo rats	3	15	Oriental and Ethiopian (eastern)

TABLE 23.1 Living Families of Rodentia[1] (cont.)

Suborder and Family	Common Name	Number of Genera	Number of Species	Distribution[2]
Suborder Sciuromorpha (cont.)				
Sigmodontinae	New World rats and mice	79	423	Nearctic and Neotropical
Spalacinae	Blind "mole"-rats	3	7	Palearctic (southern)
Anomaluridae	Scaly-tailed squirrels	3	7	Ethiopian (western and central)
Pedetidae	Springhaas	1	1	Ethiopian (southern)
Ctenodactylidae	Gundis	4	5	Ethiopian (northern) and Palearctic (North Africa)
Myoxidae	Dormice and dzhalman	8	26	Ethiopian, Palearctic
Suborder Hystricognathi				
Bathyergidae	Blesmols and naked "mole"-rats	5	12	Ethiopian (south of Sahara)
Hystricidae	Old World porcupines	3	11	Ethiopian, Palearctic (southern), Oriental
Petromuridae	Dassie rat	1	1	Ethiopian (southwestern)
Thryonomyidae	Cane rats	1	2	Ethiopian (south of the Sahara)
Erethizontidae	New World porcupines	4	12	Nearctic and Neotropical
Chinchillidae	Chinchillas and viscachas	3	6	Western and southern Neotropical
Dinomyidae	Pacarana	1	1	Neotropical (Colombia to Bolivia)
Caviidae	Cavies and maras	5	14	Neotropical
Hydrochaeridae	Capybara	1	1	Neotropical (eastern Panama and South America)
Dasyproctidae	Agoutis and acouchis	2	13	Neotropical (tropical middle America and northern two-thirds of South America)
Agoutidae	Pacas	1	2	Neotropical (tropical middle America and northern two-thirds of South America)
Ctenomyidae	Tuco tucos	1	38	Neotropical (southern two-thirds of South America)
Octodontidae	Octodonts and coruro	6	9	Neotropical (parts of Argentina, Bolivia, Chile, and Peru)
Abrocomidae	Chinchilla rats	1	3	Neotropical (parts of Argentina, Bolivia, Chile, and Peru)
Echimyidae	Spiny rats	20	78	Neotropical (southern Central America and northern two-thirds of South America)
Capromyidae	Hutias	8	20	West Indies (most species extinct in historic times)
Myocastoridae	Coypu (or nutria)	1	1	Neotropical (Argentina, Bolivia, Chile); widely introduced elsewhere in world

[1]The list of families follows that of Wilson and Reeder (1993) and Nowak (1993) except that the recently extinct family Heptaxodontidae is not included in the table nor in the keys to families. The numbers of genera and species in the subfamilies of Muridae are adjusted to reflect what is described in footnote 3.

[2]Distribution information based on McLaughlin (1984), Carleton and Musser (1984), Klingener (1984), Woods (1984), and Nowak (1999).

[3]In the Muridae, the list of subfamilies follows that of Musser and Carleton in Wilson and Reeder (1993) with these exceptions: (1) We follow Nowak (1999) in recognizing the Hydromyinae as a distinct subfamily rather than including them in the Murinae; and (2) We follow Nowak (1999) in placing the Calomyscinae (containing *Calomyscus*) and the Mystromyinae (containing *Mystromys*) in the subfamily Cricetinae, although Musser and Carleton (1993) listed these as subfamilies of the Muridae.

KEY TO LIVING FAMILIES OF THE WORLD

1 Infraorbital canal passes through side of rostrum and emerges anterior to zygomatic plate with the infraorbital foramen visible when the skull is viewed laterally (Fig. 23.4) **2**

1' Infraorbital canal and its foramen variable, but if canal emerges anterior to zygomatic plate, then the foramen is not visible when the skull is viewed laterally **3**

2 (1) Infraorbital foramen large, the vacuity extending transversely through the rostrum; zygoma weak and threadlike; tail long, and usually well-haired (Fig. 23.5) .... **Heteromyidae**
kangaroo rats, pocket mice, and kangaroo mice

2' Infraorbital foramen small, the vacuity never extending transversely through rostrum (Fig. 23.5); zygoma robust, never threadlike (Fig. 23.4); tail short, with sparsely distributed tactile hairs (Fig. 23.6) .... **Geomyidae**
pocket gophers

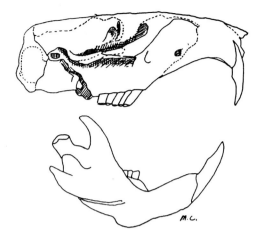

Figure 23.4 Skull of *Geomys pinetis,* a southeastern pocket gopher, Geomyidae, showing infraorbital foramen on side of rostrum.
(Redrawn from Tullberg 1899:pl. 23)

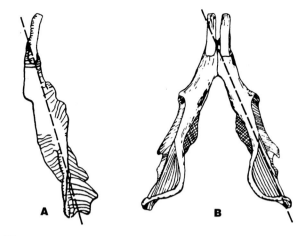

Figure 23.7 Ventral view of sciurognath-type mandible in (A) the American beaver, *Castor canadensis,* Castoridae; (B) the European marmot, *Marmota marmota,* Sciuridae.
(Redrawn: A, Landry 1957; B, Weber 1928)

Figure 23.5 Representative member of the Heteromyidae. Plains pocket mouse, *Perognathus flavescens apache.*
(M.A. Cramer)

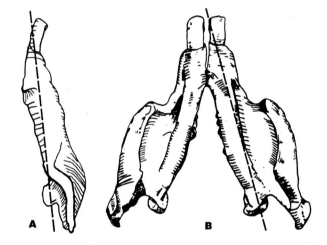

Figure 23.8 Ventral view of hystricognath-type mandible in (A) Old World porcupine, *Hystrix brachyura,* Hystricidae; (B) a blesmol, *Bathyergus suillus,* Bathyergidae.
(Redrawn: A, Landry 1957; B, Weber 1928)

Figure 23.6 Representative member of the Geomyidae, the southeastern pocket gopher, *Geomys pinetis.*
(Merriam 1895:frontispiece)

3 (1') From ventral view, angular process of mandible in line with (Fig. 23.7A) or medial to (Fig. 23.7B) the lateral border of incisive alveolus (sciurognath-type mandible) **4**

3' From ventral view, angular process of mandible lateral to (Fig. 23.8A, B) lateral border of incisive alveolus (hystricognath-type mandible) **32**

4 (3') Postorbital process of frontal present and sharply pointed (Fig. 23.9) .... **Sciuridae**
squirrels and marmots

4' Postorbital process of frontal absent or, if present, small and blunt (Fig. 23.10) **5**

5 (4') Postorbital process of frontal small and blunt (Fig. 23.10); tail flattened dorsoventrally .... **Castoridae**
beavers, *Castor*

5' Postorbital process of frontal absent; tail not flattened dorsoventrally **6**

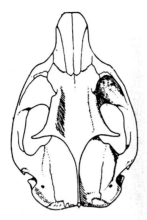

Figure 23.9 Dorsal view of cranium showing sharp-pointed postorbital processes in a member of the Sciuridae, a ground squirrel (large-toothed souslik), *Spermophilus fulvus*.
(Vinogradov and Argiropulo 1941:29)

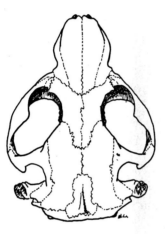

Figure 23.10 Dorsal view of cranium showing blunt postorbital processes in a member of the Castoridae, the European beaver, *Castor fiber*.
(Redrawn from Tullberg 1899)

6 (5')   Each dentary with three or fewer cheek teeth 7

6'   Each dentary with four or more cheek teeth 28

7 (6)   Occlusal surface of cheek teeth with triangles of dentine in all or some of the teeth (Fig. 23.11); teeth high-crowned .... **Muridae** (in part) **Arvicolinae**
lemmings, voles, and muskrats

7'   Occlusal surface of cheek teeth various but lacking triangles in all of the teeth, or cheek teeth low-crowned 8

8 (7')   Occipital crest at (Fig. 23.12A) or slightly behind (Fig. 23.12B) level of zygomatic process of squamosal 9

8'   Occipital crest well behind level of zygomatic process of squamosal (Fig. 23.13) 10

9 (8)   Occipital crest at level of zygomatic process of squamosal (Fig. 23.12A); some upper cheek teeth with S-shaped pattern or isolated enamel lakes; eyes vestigial .... **Muridae** (in part) **Spalacinae**
blind "mole"-rats, *Spalax*, *Nannospalax*

9'   Occipital crest slightly behind level of zygomatic process of squamosal (Fig. 23.12B); upper cheek teeth with sharp-edged re-entrant angles but never forming an S-shaped pattern .... **Muridae** (in part) **Myospalacinae**
zokors, *Myospalax*

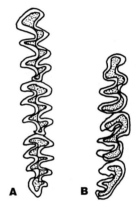

Figure 23.11 Upper cheek teeth in (A) a varying lemming, *Dicrostonyx*, Muridae, Arvicolinae; (B) a zokor, *Myospalax myospalax*, Muridae, Myospalacinae, showing triangles formed by enamel on occlusal surface. Anterior at top.
(Vinogradov and Argiropulo 1941)

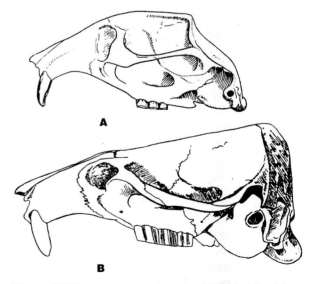

Figure 23.12 Lateral view of crania of (A) blind "mole"-rat, *Spalax microphthalmus*, Muridae, Spalacinae; (B) zokor, *Myospalax myospalax*, Muridae, Myospalacinae.
(Vinogradov and Argiropulo 1941:51, 66)

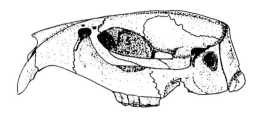

Figure 23.13 Cranium of an African "mole"-rat, *Tachyoryctes* sp., Muridae, Rhizomyinae.
(G.A. Moore)

Figure 23.14 Anterior view of a bamboo rat, *Rhizomys* sp. cranium showing lower border of infraorbital foramina elevated above anterior tip of nasal bones, Muridae, Rhizomyinae.
(M. A. Cramer)

Figure 23.15 A bamboo rat, *Rhizomys sinensis,* Muridae, Rhizomyinae.
(Hsia et al. 1964:44)

11 (10') First two cheek teeth each with deep single inner fold that divides occlusal surface (Fig. 23.16), with upper incisors that project beyond tips of nasal bones; interorbital width narrower than maxillary width .... **Muridae** (in part) **Rhizomyinae** (in part)
African "mole"-rats, *Tachyoryctes*

11' First two cheek teeth and position of incisors of various types, but if single fold present in teeth or if incisors project beyond tip of nasal bones, then width of interorbital region equal to or much greater than maxillary width **12**

12 (11') Zygomatic arch tilted upward anteriorly, with a portion situated above lower border of infraorbital canal (Fig. 23.17); lower maxillary process expanded into a zygomatic plate (Fig. 23.17A) that forms most of lateral wall of infraorbital canal (Fig. 23.17) **13**

12' Zygomatic arch horizontal or nearly so with no portion situated above lower border of infraorbital canal (Fig. 23.18); lower and upper maxillary processes together form lateral wall of infraorbital canal (Fig. 23.18) **25**

13 (12) Two or fewer cheek teeth in each maxilla (Fig. 23.19A); some forms modified for aquatic life (Fig. 23.19B) .... **Muridae** (in part) **Hydromyinae**
Australasian water rats and allies

13' Three or more cheek teeth in each maxilla; body form various **14**

14 (13') Each upper cheek tooth with one, two, or three shallow, basin-shaped depressions (Fig. 23.20A) **15**

14' Upper cheek teeth of various types but never as above **16**

10 (8') Lower margin of infraorbital foramen, in anterior view, even with or above anterior tips of nasal bones (Fig. 23.14) .... **Muridae** (in part) bamboo rats, **Rhizomyinae** (in part) *Rhizomys* and *Cannomys*

10' Lower margin of infraorbital foramen, in anterior view, below anterior tips of nasal bones **11**

Figure 23.16 Left upper cheek teeth in an African "mole"-rat, *Tachyoryctes ruandae,* Muridae, Rhizomyinae.
(Stehlin and Schaub 1951)

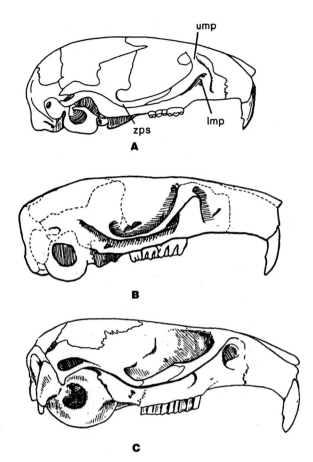

Figure 23.17 Lower maxillary process elevated above lower border of infraorbital foramen in three murids (A) *Mus;* (B) bushy-tailed woodrat *Neotoma cinerea,* Sigmodontinae; (C) a high mountain vole, *Alticola argentatus,* Arvicolinae. **ump,** upper maxillary process; **lmp,** zygomatic plate or lower maxillary process; **zps,** zygomatic process of squamosal.
(A and C, Vinogradov and Argiropulo 1941; B, redrawn from Tullberg 1899:pl. 15–25)

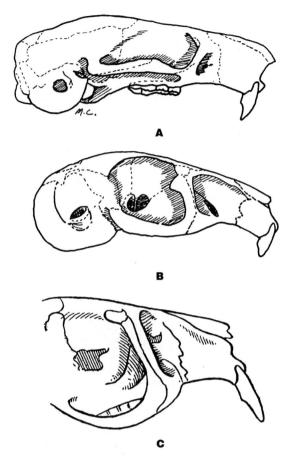

Figure 23.18 Lower maxillary process below infraorbital foramen in (A) a dormouse *Myoxus glis,* Myoxidae; (B) a gundi, *Felovia vae,* Ctenodactylidae; (C) a five-toed jerboa, *Allactaga severtzovi,* Dipodidae.
(A and B, redrawn from Tullberg 1899:pl. 11, 9; C, Vinogradov and Argiropulo 1941:33)

15 (14) Angular process of mandible perforated by foramen; single deep groove on anterior surface of each upper incisor (Fig. 23.20B); size small, greatest length of skull less than 25 mm; tail shorter than head plus body length, covered with short hairs; form not modified for aquatic life (Fig. 23.20C) .... **Myoxidae** (in part)
Selevinia betpakdalensis
dzhalman

15' Angular process of mandible without foramen; anterior surface of upper incisor without groove (if grooved, tail naked or covered with very few short bristles); size medium, greatest length of skull more than 25 mm; tail longer than head plus body length .... **Muridae** (in part)
Murinae (in part)
a few murids

Figure 23.19 Reduced number with cheek teeth basin-shaped (A) in the Australian water-rat, *Hydromys chrysogaster,* Muridae, Hydromyinae, and (B), whole animal.
(Stehlin and Schaub 1951)

Figure 23.21 Right upper cheek teeth in an African swamp rat, *Otomys irroratus*, Muridae, Otomyinae. Anterior at left.
(Redrawn from Tullberg 1899:pl. 29)

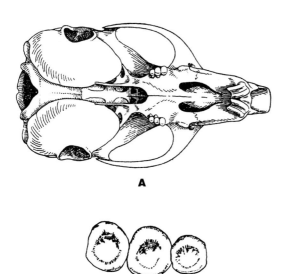

Figure 23.22 Left upper cheek teeth in murids (Muridae) with up to three cusps per lamina. (A) *Rattus* sp., Murinae; (B) a fat mouse, *Steatomys pratensis,* Dendromurinae. Anterior at top; not to same scale.
(A, redrawn from Sokolov 1959; B, redrawn from Stehlin and Schaub 1951)

Figure 23.20 The dzalman, *Selevinia betpakdalensis,* Myoxidae. Cranium (A), cheek teeth (B), and animal (C).
(A, Gromov et al. 1963; B, Afanasev et al. 1953; C, Mary Ann Cramer from photograph in Walker et al. 1975)

16 (14') Occlusal surfaces of cheek teeth consisting of series of transverse ridges and depressions (4 to 9 on last cheek tooth) (Fig. 23.21) .... **Muridae** (in part) **Otomyinae**
Karroo rats (*Parotomys*) and African swamp rats (*Otomys*)

16' Occlusal surfaces of cheek teeth variable; transverse ridges and depressions, if present, number fewer than four on last cheek tooth **17**

17 (16') Cheek teeth with evidence of cusps (Figs. 23.22 and 23.23) **18**

17' Cheek teeth with no evidence of cusps, may possess deep re-entrant folds, transverse noncuspidate laminae, or sharp re-entrant angles **21**

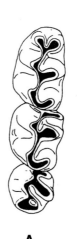 

Figure 23.23 Right upper cheek teeth in murids (Muridae, Cricetinae) with maximum of two cusps per lamina. (A) a rat-like hamster, *Cricetulus migratorius;* (B) a dwarf hamster, *Phodopus roborovskii.*
(A & B, Stehlin and Schaub 1951)

18 (17) Upper surface of skull covered by granular-textured bone; temporal fossae roofed by bone; body densely haired with mane on back and with bushy tail .... **Muridae** (in part)
Lophiomyinae
maned rat, *Lophiomys imhausi*

18' Upper surface of skull of various types but never covered with granular-textured bone; temporal fossae open above, not roofed over by bone; pelage various but not as above **19**

19 (18') Three cusps present on at least one transverse lamina of anterior cheek tooth (Fig. 23.22), although lingual row of cusps may be reduced as to the number of cusps (to as few as one) and in the sizes of the cusps; laminae that bear cusps separated by valleys or pressed together but not interconnected (Fig. 23.22); size and body form variable (Fig. 23.24) .... **Muridae** (in part)
Murinae (in part)
most Old World rats and mice
Cricetomyinae
African pouched rats and African pouched mice
Dendromurinae (in part)
African climbing mice, gerbil mice, fat mice, and Congo forest mice
Petromyscinae
rock mice (*Petromyscus*) and Delany's swamp mouse (*Delanymys*)

19' Maximum of two cusps on each transverse lamina of anterior cheek tooth (Fig. 23.23); laminae that bear cusps interconnected (Fig. 23.23); size and body form variable (Fig. 23.25) **20**

20 (19') Internal cheek pouches present; specimen from Old World (Fig. 23.25D) ...... **Muridae** (in part)
Cricetinae
hamsters

20' Internal cheek pouches absent (extremely small internal pouches may be present in *Ochrotomys*) (Fig. 23.25B & C) ...... **Sigmodontinae** (in part)
some New World rats and mice

21 (17') Auditory bulla greatly enlarged, its greatest dimension more than 1/4 the greatest length of skull (Fig. 23.26) .... **Muridae** (in part)
Gerbillinae
gerbils

21' Auditory bulla not greatly enlarged, its greatest dimension less than 1/4 of greatest length of skull **22**

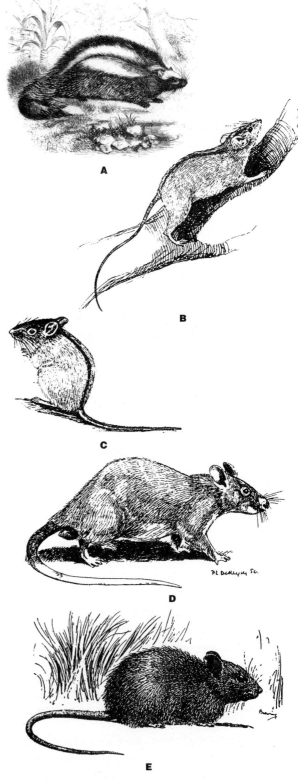

Figure 23.24 Muridae (in part). Representative members of Old World murids, Muridae. (A) the maned rat, *Lophiomys imhausi*, Lophiomyinae; (B) African gray tree mouse, *Dendromus melanotis*, Dendromurinae; (C) Old World striped field mouse, *Apodemus agrarius*, Murinae; (D) an African giant pouched rat, *Cricetomys gambianus*, Cricetomyinae; (E) the Australian bush rat, *Rattus fuscipes*, Murinae.
(A, Flower and Lydekker 1891:460; B, Sclater 1901:32; C, Hsia et al. 1964: 47; D, Dekeyser 1955:213; E, Brazenor 1950:65)

The Rodents 149

Figure 23.25 Muridae (in part). Representative members of New World sigmodontines and of hamsters. (A) Northern pygmy mouse, *Baiomys taylori,* Sigmodontinae; (B) dusky-footed woodrat, *Neotoma fuscipes,* Sigmodontinae; (C) eastern harvest mouse, *Reithrodontomys humulis,* Sigmodontinae; (D) hairy-footed hamster, *Phodopus sungorus,* Cricetinae.
(A–C, Booth 1982:127,122,131; D, Gromov et al. 1963:481)

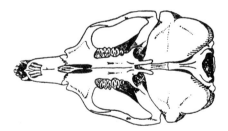

Figure 23.26 Ventral view of cranium of jird, *Meriones* sp., Muridae, Gerbillinae, showing enlarged auditory bullae.
(Vinogradov and Argiropulo 1941:67)

22 (21') Zygomatic plate narrow, its greatest width equal to or less than the length of first upper tooth (Fig. 23.27A); palatine bones with large foramen medial to anterior upper cheek teeth (Fig. 23.27B); distal end of tail bushy (Fig. 23.27C) .... **Muridae** (in part)
Platacanthomyinae
Chinese pygmy dormice (*Typhlomys*) and spiny dormouse (*Platacanthomys*)

22' Zygomatic plate of various types, but greatest width always more than the length of first upper cheek tooth; palatine bone without large foramen medial to anterior upper cheek teeth; distal end of tail of various types, but, if bushy, then palatine bones lack large foramina **23**

23 (22') Length of second upper cheek tooth 80% to 100% length of first upper cheek tooth ....
.... **Muridae** (in part)
Nesomyinae
Malagasy rats
Sigmodontinae (in part)
some New World rats and mice

23' Length of second upper cheek tooth less than 80% length of first upper cheek tooth **24**

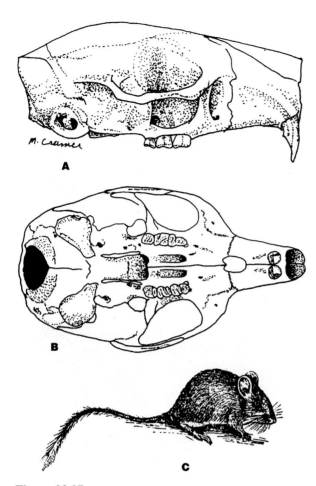

Figure 23.27 Cranium of spiny dormouse, *Platacanthomys lasiurus,* Muridae, Platacanthomyinae, in lateral (A) and ventral (B) views; (C) the chinese pygmy dormouse, *Typhlomys cinereus,* Muridae, Platacanthomyinae.
(A, Mary Ann Cramer; B, G.A. Moore; C, Hsia et al. 1964:43)

24 (23') Occlusal pattern of cheek teeth simple, consisting of two or three noncuspidate laminae on each tooth (Fig. 23.28); specimen from the Old World .... **Muridae** (in part)
Murinae (in part)
some Old World rats and mice

24' Occlusal pattern of cheek teeth variable, but if occlusal pattern that of simple laminae, specimen from the New World .. **Muridae** (in part)
Sigmodontinae (in part)
some New World rats and mice

25 (12') Foramina (one or two) in angular region of mandible (Fig. 23.29); bullae moderately to greatly inflated (Fig. 23.29); body form modified for saltation; head and body length less than 300 mm .... **Dipodidae** (in part)
Cardiocraniinae, Euchoreutinae, Allactaginae, and **Dipodinae**
jerboas

25' No foramen in angular region of mandible; bullae and body form various, but if modified for saltation then head and body length greater than 300 mm 26

Figure 23.28 Left upper cheek teeth in a bandicoot rat, *Bandicota indica,* Muridae, Murinae. Anterior at left.
(Stehlin and Schaub 1951)

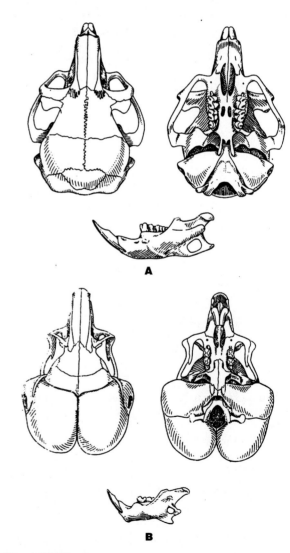

Figure 23.29 Skulls of representative dipodids, Dipodidae. (A) Small five-toed jerboa, *Allactaga elater;* (B) a three-toed dwarf jerboa, *Salpingotus crassicauda.*
(Gromov et al. 1963:260)

26 (25') Cheek teeth evergrowing; each lower cheek tooth with a single deep fold on labial and on lingual surface, resulting in 8-shaped pattern (Fig. 23.30) .... **Ctenodactylidae** (in part)
some gundis

26' Cheek teeth rooted; lower cheek teeth of various types, but never with 8-shaped pattern **27**

27 (26') Anterior surfaces of upper incisors smooth or with one groove (Fig. 23.31); size small, greatest length of skull less than 30 mm; pelage fine to hispid, no white at base of hair ....
.... **Dipodidae** (in part)
Sicistinae and Zapodinae
birch mice and jumping mice

27' Two faint grooves in anterior surface of each upper incisor; size medium, greatest length of skull more than 30 mm; dorsal pelage almost spiny (stiff guard hairs), hair white at base ......
.... **Muridae** (in part)
Dendromurinae (in part)
Congo forest mice, *Deomys ferrugineus*

28 (6') Infraorbital foramen large, and its maximum dimension greater than 10 mm **29**

28' Infraorbital foramen small, and its maximum dimension less than 10 mm **31**

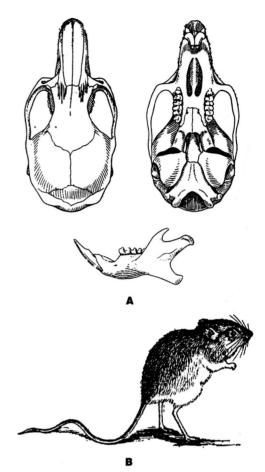

Figure 23.31 Skull (A) of a birch mouse, *Sicista subtilis*, Dipodidae, Sicistinae; (B) the Chinese jumping mouse, *Eozapus setchuanus*, Dipodidae, Zapodinae.
(A, Gromov et al. 1963:376; B, Hsia et al. 1964:40)

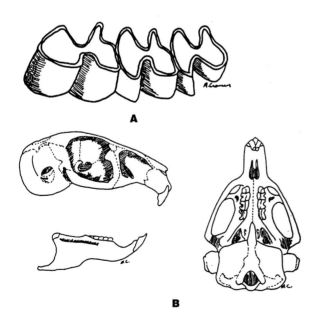

Figure 23.30 Teeth and skull of a gundi, *Ctenodactylus gundi*, Ctenodactylidae. (A) Lower cheek teeth, anterior to left; (B) skull.
(Redrawn from Tullberg 1899:pl 27 and 9)

29 (28) Lower maxillary process projects well forward to a point even with, or anterior to, cutting edge of upper incisor (Fig. 23.32); cheek teeth evergrowing; size large (head and body length > 300 mm), form modified for saltation (Fig. 23.1) ..
.... **Pedetidae**
springhaas, *Pedetes capensis*

29' Lower maxillary process not as above; cheek teeth rooted; size variable, but head and body length < 300 mm and body form not modified for saltation **30**

30 (29') Lower molars 2 and 3 (M_2 and M_3) with one or two deep folds on labial and lingual surfaces, and sometimes resulting in E-shaped pattern (Fig. 23.33); body lacks patagia to permit gliding locomotion; hindfoot with four digits, no platelike scales on ventral surfaces of tail ....
................................**Ctenodactylidae** (in part)
some gundis

30' Lower molars 2 and 3 (M_2 and M_3) with basin-shaped depressions or enamel islands (Fig. 23.34) and never having E-shaped pattern; patagia often present along side of body for gliding locomotion (Fig. 23.35A); hindfoot with five digits, platelike scales on ventral surface of tail (Fig. 23.35B) .... **Anomaluridae**
scaly-tailed squirrels

31 (28') Size large, greatest length of skull more than 50 mm; head and body length greater than 300 mm; portion of angular process of dentary projects inward, medial to mandibular condyle (Fig. 23.36A); no large foramen in angular process of dentary; tail short, less than 15% of head and body length, well-haired (Fig. 23.36B) ....
.... **Aplodontidae**
sewellel or mountain "beaver," *Aplodontia rufa*

31' Size small, greatest length of skull less than 50 mm; head and body length less than 220 mm; angular process of dentary not turned inward, always projecting lateral to mandibular condyle; large foramen in angular process of dentary (Fig. 23.37A); tail long (more than 50% of head and body length), well-haired (Fig. 23.37B) ....
.... **Myoxidae** (in part)
dormice

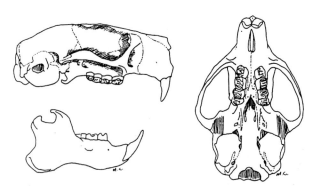

Figure 23.34 Skull of Pel's scaly-tailed flying squirrel, *Anomalurus peli,* Anomaluridae.
(Redrawn from Tullberg 1899:pl. 9)

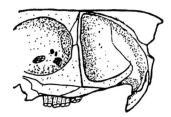

Figure 23.32 Anterior portion of cranium of springhaas, *Pedetes capensis,* Pedetidae, showing anterior projection of lower maxillary process.
(R.E. Martin)

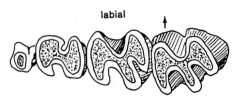

Figure 23.33 Right lower cheek teeth in a ctenodactylid, Speke's pectinator, *Pectinator spekei,* Ctenodactylidae.
(J. Blefeld)

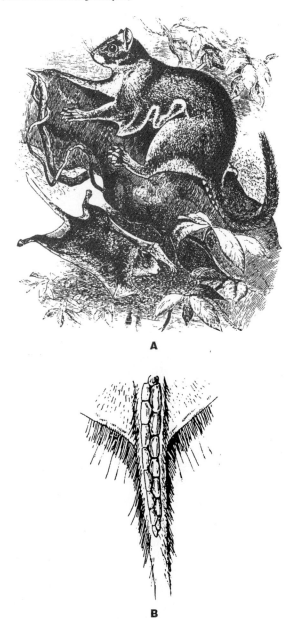

Figure 23.35 Beecroft's scaly-tailed flying squirrel (A), *Anomalurus beecrofti,* Anomaluridae, and (B) ventral view of base of tail.
(A, Flower and Lydekker 1891:449; B, Dekeyser 1955:187)

34' Size smaller, greatest length of skull less than 175 mm; paroccipital process small to moderately elongated (< 25 mm length) and often situated close to auditory bulla (Fig. 23.41A); length of last upper molar less than combined lengths of preceding cheek teeth in each upper tooth row (Fig. 23.41A) .... **Caviidae**
cavies, mara

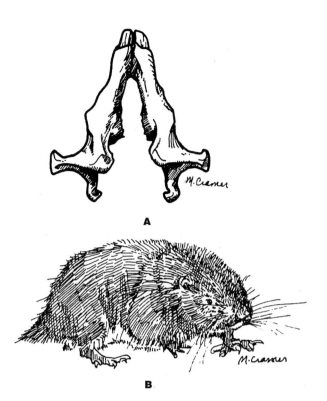

Figure 23.36 Ventral view of mandible (A) of sewellel, *Aplodontia rufa*, Aplodontidae, (B), whole animal.
(A, Mary Ann Cramer; B, M.A. Cramer after Walker et al. 1975:667)

32 (3') Infraorbital foramen smaller than foramen magnum (Fig. 23.38A); body haired (Fig. 23.38B) or nearly hairless (Fig. 23.38C) ....
.... **Bathyergidae**
Blesmols and naked "mole"-rat

32 Infraorbital foramen equal to or larger than foramen magnum; body always well-haired **33**

33 (32') Prominent ridge and groove extending along labial side of dentary, parallel to cheek teeth (Fig. 23.39) **34**

33' No prominent ridge along labial side of dentary **35**

34 (33) Size larger, greatest length of skull more than 175 mm; paroccipital process greatly elongated (> 25 mm length) and projecting almost vertically below auditory bulla (Fig. 23.40); length of last upper molar greater than combined lengths of preceding cheek teeth in each upper toothrow .... **Hydrochaeridae**
Hydrochaeris hydrochaeris
capybara

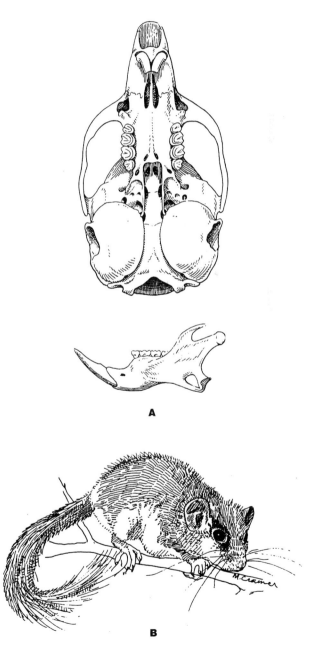

Figure 23.37 Skull (A) of the forest dormouse, *Dryomys nitedula*, Myoxidae, (B) whole animal.
(A, Gromov et al. 1963:364; B, Mary Ann Cramer from photograph in Walker et al. 1975:978)

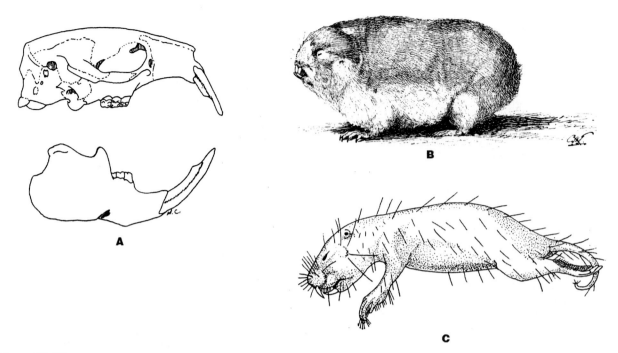

Figure 23.38 Representative members of the Bathyergidae. (A) Skull of a dune "mole"-rat, *Batherygus suillus;* (B) the Cape "mole"-rat, *Georychus capensis;* (C) the naked "mole"-rat, *Heterocephalus glaber.*
(A, redrawn from Tullberg 1899:pl. 2; B, Sclater 1901:75; C, Weber 1928:239)

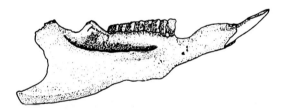

Figure 23.39 Right dentary of dwarf Patagonian cavy, *Dolichotis salinicola,* Caviidae, showing deep groove and ridge along lateral side.
(J. Blefeld)

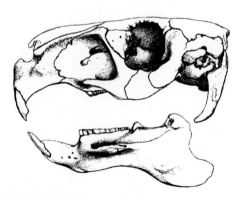

Figure 23.40 Skull of capybara, *Hydrochaeris hydrochaeris,* Hydrochaeridae.
(Flower and Lydekker 1891:48)

35 (33') Auditory canal emerges almost vertically from skull, and most of interior of canal not evident in lateral view (Fig. 23.42); additional foramen leading into tympanic bulla below auditory canal (Fig. 23.42) .... **Chinchillidae**
chinchillas and viscachas

35' Auditory canal emerges laterally from skull, and interior of canal readily evident in lateral view or, if not, then no foramen present below auditory canal (Fig. 23.43) 36

36 (35') Portion of lacrimal canal exposed on side of rostrum (Fig. 23.44) 37

36' No part of lacrimal canal exposed on side of rostrum 38

37 (36) Size large, greatest length of skull more than 70 mm; auditory bullae small, never visible from dorsal aspect of skull; cheek teeth with enamel islands; hindlimbs elongated, digitigrade, pollex vestigial .... **Dasyproctidae**
agoutis, *Dasyprocta*

37' Size small, greatest length of skull less than 70 mm; auditory bullae large, visible from dorsal aspect of skull; each cheek tooth with single re-entrant fold on each side, never forming enamel islands; hindlimbs not elongated, plantigrade, pollex absent (Fig. 23.45) ....
.... **Abrocomidae**
chinchilla rats or chinchillones, *Abrocoma*

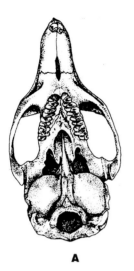

Figure 23.41 Representative species of Caviidae. (A) Ventral view of cranium of a yellow-toothed cavy, *Galea spixii;* (B) Patagonian cavy or mara, *Dolichotis patagonum.*
(A, Mary Ann Cramer; B, Beddard 1902:492)

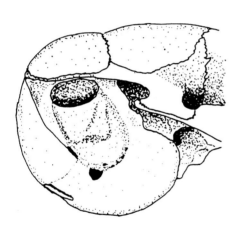

Figure 23.42 Bullar region of a chinchillid, *Chinchilla,* Chinchillidae.
(G. A. Moore)

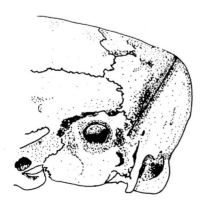

Figure 23.43 Bullar region of the long-tailed porcupine, *Trichys fasciculata,* Hystricidae.
(G.A. Moore)

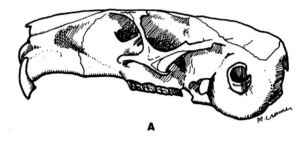
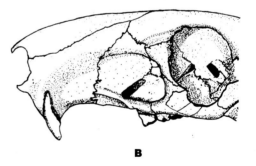

Figure 23.44 Lateral view of cranium of (A) chinchilla rat, *Abrocoma bennetti,* Abrocomidae, and (B) anterior region of cranium of an agouti, *Dasyprocta* sp., Dasyproctidae.
(A, Mary Ann Cramer; B, Janet Blefeld)

38 (36') Additional canal or groove present along ventral, medial wall of infraorbital canal (Fig. 23.46) **39**

38' No canal or groove present along ventral, medial wall of infraorbital canal **43**

39 (38) Upper incisor with three grooves on anterior surface (Fig. 23.47A) .... **Thryonomyidae**
cane rat, *Thryonomys*

39' Upper incisor without grooves **40**

Figure 23.45 Chinchilla rat or chinchillone, *Abrocoma bennetti,* Abrocomidae.
(Mary Ann Cramer from photograph by R.E. Martin)

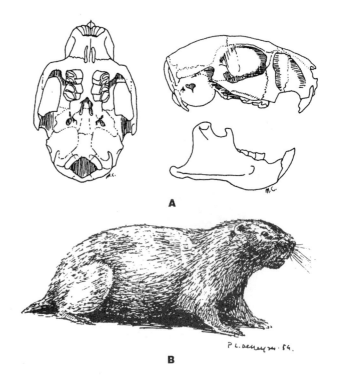

Figure 23.47 Skull (A) of a cane rat, *Thryonomys swinderianus,* Thryonomyidae, and (B) whole animal.
(A, redrawn from Tullberg 1899:pl. 6; B, Dekeyser 1955:174)

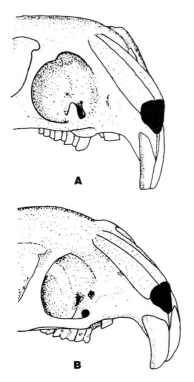

Figure 23.46 Anterior portions of crania of some representative octodontids, Octodontidae. (A) The degu, *Octodon degus,* showing groove, and (B) a rock rat, *Aconaemys fuscus,* showing canal at lower margin of infraorbital foramen.
(G. A. Moore)

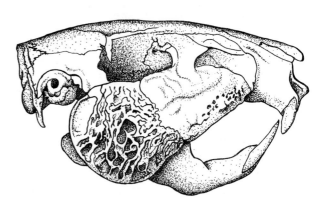

Figure 23.48 Lateral view of paca skull, *Agouti paca,* Agoutidae.
(Michael Gilliland)

40 (39') Size large, greatest length of skull more than 100 mm; inner surface of zygomatic arch with deep depression; outer surface of zygomatic arch pitted (Fig. 23.48) .... **Agoutidae**
pacas, *Agouti*

40' Size small, greatest length of skull less than 100 mm; inner surface of zygomatic arch not modified as above; outer surface of zygomatic arch usually smooth, never pitted **41**

41 (40') Palate long, posterior border behind last upper cheek tooth; cheek teeth terraced, with marked elevation on lingual side of upper cheek teeth (Fig. 23.49) and labial side of lower cheek teeth .... **Petromuridae**
dassie rat, *Petromys typicus*

41' Palate short, posterior border forward of last upper cheek tooth; upper cheek tooth with occlusal surface almost flat and with no marked elevated areas **42**

Figure 23.49 Terraced cheek teeth of the dassie rat, *Petromus typicus*, Petromuridae.
(Hershkovitz 1962)

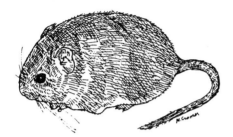

Figure 23.50 Bridges's degu, *Octodon bridgesi*, Octodontidae.
(Mary Ann Cramer from photograph by R.E. Martin)

Figure 23.51 A casiragua, *Proechimys setosus*, Echimyidae.
(Mary Ann Cramer from photograph in Walker et al. 1975)

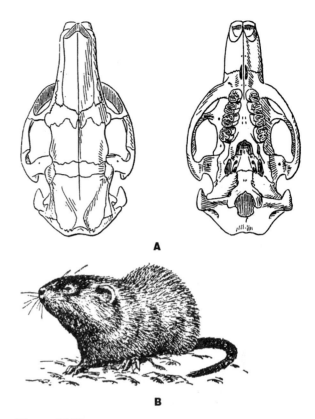

Figure 23.52 Cranium (A) of the nutria, *Myocastor coypus*, Myocastoridae, and (B) whole animal.
(A, Gromov et al. 1963:258; B, Hsia et al. 1964:40)

42 (41') Cheek teeth evergrowing, with single main fold along labial border of each and lacking enamel islands in lake of dentine; pelage without spines (Fig. 23.50) .... **Octodontidae** (in part)
degus and other octodonts except *Spalacopus*

42'   Cheek teeth rooted, with two main folds along labial border of each or, if folds not readily evident, then occlusal surface with one or more enamel islands surrounded by dentine; pelage with weak to moderately developed spines (Fig. 23.51) .... **Echimyidae** (in part)
spiny rats of the genera *Cercomys*, *Euryzygomatomys*, *Hoplomys*, and some *Proechimys*

43 (38') Paroccipital process long, its greatest length more than 15 mm (Fig. 23.52A); hindfoot webbed .... **Myocastoridae**
nutria or coypu, *Myocastor coypus*

43'   Paroccipital process short, its greatest length less than 15 mm; hindfoot not webbed **44**

44 (43') Upper cheek teeth with 8-shaped pattern (Fig. 23.53A); bone enclosing origin of upper incisor forming a prominent ridge (alveolar process) in the orbit (Fig. 23.53B) ....
.... **Octodontidae** (in part)
coruro, *Spalacopus cyanus*

44'   Upper cheek teeth with various patterns, never 8-shaped; usually no prominent ridge in orbit for origin of incisor, but if present, the upper cheek teeth are kidney-shaped **45**

45 (44') Upper cheek teeth with a kidney-shaped pattern (Fig. 23.54) .... **Ctenomyidae**
tuco tucos, *Ctenomys*

45'   Upper cheek teeth with various patterns but never kidney-shaped **46**

46 (45') Paroccipital process curving close to ventral surface of auditory bulla (Fig. 23.55); pelage with weak to moderately developed spines ......
.... **Echimyidae** (in part)
some spiny rats

46'   Paroccipital process projecting almost perpendicular to long axis of skull (Fig. 23.56), never curving close to ventral surface of auditory bulla; pelage with or without spines **47**

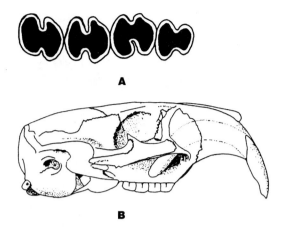

Figure 23.53 Upper left cheek teeth (A) and cranium (B) of the coruro, *Spalacopus cyanus,* Octodontidae. Anterior end of tooth row is to the left.
(A, Reig 1970:596; B, G. A. Moore)

Figure 23.54 Occlusal view of left upper cheek teeth in a tuco tuco, *Ctenomys mendocinus,* Ctenomyidae.
(Reig 1970:599)

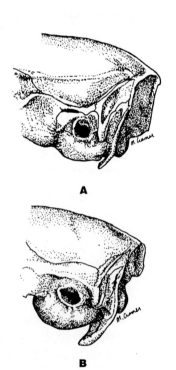

Figure 23.55 Bullar regions in crania of two echimyids, showing paroccipital processes curving toward auditory bullae. (A) The thick-spined rat, *Hoplomys gymnurus,* Echimyidae, Echimyinae; (B) a coro, *Dactylomys boliviensis.*
(G. A. Moore)

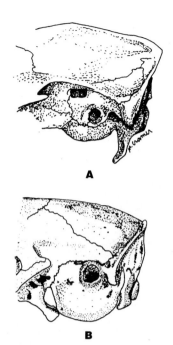

Figure 23.56 Bullar regions in crania of a capromyid and an erethizontid, showing paroccipital processes projecting straight downward and not curving toward auditory bullae. (A) Desmarest's Cuban hutia, *Capromys pilorides,* Capromyidae; (B) North American porcupine, *Erethizon dorsatum,* Erethizontidae.
(A, Mary Ann Cramer; B, G. A. Moore)

47 (46') Palate not constricted at anterior end of tooth row, its greatest breadth at first cheek tooth equal to, or more than, greatest breadth at last cheek tooth (Fig. 23.57A) .... **Hystricidae**
Old World porcupines

47' Palate constricted at anterior end of tooth row, its greatest breadth at first cheek tooth less than greatest breadth at last cheek tooth **48**

48 (47') Paroccipital process long, its distal end extending below most ventral point on auditory bulla (Fig. 23.56A) .... **Capromyidae**
hutias

48' Paroccipital process short, its distal end never extending below auditory bulla (Fig. 23.56B) **49**

49 (48') Coronoid process of mandible absent or vestigial; lower cheek teeth with a series of transverse laminae (Fig. 23.58); pelage with spots; spines absent .... **Dinomyidae**
Dinomys branickii
pacarana

49' Coronoid process of mandible prominent (Fig. 23.59A); each lower cheek tooth with one lingual re-entrant fold, never laminated; pelage lacks spots; spines present (Fig. 23.59B) .... **Erethizontidae**
New World porcupines

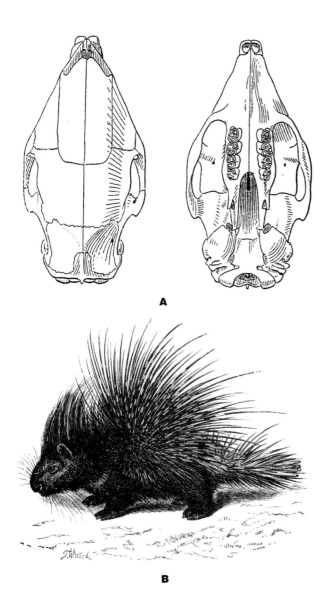

Figure 23.57 Cranium (A) of an Old World porcupine, *Hystrix indica,* Hystricidae, and (B) whole animal.
(A, Gromov et al. 1963:256; B, Finn 1929:165)

Figure 23.58 Right lower cheek teeth (anterior to left) of the pacarana, *Dinomys branickii,* Dinomyidae.
(Janet Blefeld)

Figure 23.59 Representative New World arboreal porcupines, Erethizontidae. (A) Skull of Upper Amazonian porcupine, *Echinoprocta rufescens;* (B) prehensile-tailed porcupine, *Coendou* sp.
(A, redrawn from Tullberg 1899:pl. 7; B, Flower and Lydekker 1891:485)

KEY TO LIVING FAMILIES OF NORTH AMERICA[1]

1 Infraorbital foramen equal to or larger than foramen magnum **2**

1' Infraorbital foramen smaller than foramen magnum **8**

2 (1) Greatest length of paroccipital process more than 15 mm (Figs. 23.40 and 23.52A) **3**

2' Greatest length of paroccipital process less than 15 mm **4**

[1]All of the North American continent south through Panama and including islands of the West Indies.

3 (2)   Anterior face of each upper incisor with single longitudinal groove; tail very short (less than 10% of head and body length); pes with three digits, not webbed or only partially so .... **Hydrochaeridae**
capybara, *Hydrochaeris hydrochaeris*

3'   Upper incisors without grooves; tail long (50% to 80% of head and body length); pes with five digits, manifestly webbed ...... **Myocastoridae**
nutria or coypu (introduced), *Myocastor coypus*

4 (2')   Paroccipital process rudimentary, less than 5 mm in length (Fig. 23.56B); incisors project beyond nasals, visible in dorsal view of skull .. .. **Erethizontidae**
New World porcupines

4'   Paroccipital process never rudimentary, its length always greater than 5 mm; incisors do not project beyond nasals and not visible in dorsal view **5**

5 (4')   Portion of lacrimal canal exposed on side of rostrum (Fig. 23.44B); claws thick and hooflike .. .. **Dasyproctidae**
agoutis, *Dasyprocta*

5'   No part of lacrimal canal exposed on side of rostrum; claws or hooflike digits present **6**

6 (5')   Size large, greatest length of skull more than 100 mm; inner surface of zygomatic arch with deep depression (Fig. 23.48); outer surface of zygomatic arch covered with thick and pitted bone; functional digits 4/3, with thick, hooflike claws; longitudinal rows of white stripes and spots laterally on body .... **Agoutidae**
Agouti paca
paca

6'   Size small to moderate, but if greatest length of skull more than 100 mm, then digits 5–4/5 or 5/4; zygomatic arch not covered with thickened and pitted bone; distal ends of digits with claws, never hooflike; pelage pattern of various types but never with longitudinal rows of white stripes or spots **7**

7 (6')   Paroccipital process peglike and projects almost perpendicularly from long axis of skull, away from surface of auditory bulla (Fig. 23.56A); cheek teeth evergrowing; coronoid process of dentary relatively long (more than twice height of P_4) .... **Capromyidae**
hutias

7'   Paroccipital process bladelike and curves close to surface of auditory bulla (Fig. 23.55A); cheek teeth rooted; coronoid process of dentary relatively short (less than twice height of P_4) .... **Echimyidae**
spiny rats and *Diplomys labilis*

8 (1')   Infraorbital canal passing through side of rostrum, emerging anterior to zygomatic plate with the infraorbital foramen visible when the skull is viewed laterally (Fig. 23.4) **9**

8'   Infraorbital canal and its foramen variable, but if canal emerges anterior to zygomatic plate, the foramen is not visible when the skull is viewed laterally **10**

9 (6)   Infraorbital foramen large, completely perforating rostrum transversely; tail long, usually well-haired (Fig. 23.5) .... **Heteromyidae**
kangaroo rats, kangaroo mice, and pocket mice

9'   Infraorbital foramen small, never completely perforating rostrum transversely; tail short, naked, or sparsely haired (Fig. 23.6) .... **Geomyidae**
pocket gophers

10 (6')   Postorbital processes of frontal present and sharply pointed (Fig. 23.9) .... **Sciuridae**
squirrels and marmots

10'   Postorbital processes of frontal absent or, if present, blunt (Fig. 23.10) **11**

11 (8')   Postorbital processes of frontal present, although blunt and small (Fig. 23.10); tail flattened dorsoventrally (Fig. 23.10) ....**Castoridae**
American beaver, *Castor canadensis*

11'   Postorbital processes of frontal absent; tail not flattened dorsoventrally **12**

12 (11')   Infraorbital foramen a long vertical slit **13**

12'   Infraorbital foramen round or oval **15**

13 (12)   Cheek teeth with cusps (or evidence of cusps) arranged in three longitudinal rows (Fig. 23.24) .... **Muridae** (in part)
Murinae
house mouse, roof rat, and Norway rat (all introduced)

13'   Cheek teeth with cusps (or evidence of cusps) arranged in two longitudinal rows (Fig. 23.23) or with occlusal surfaces having acute triangles or re-entrant folds **14**

14 (13') Occlusal surface of cheek teeth with triangles of dentine in all or some of the cheek teeth (Fig. 23.11A) .... **Muridae** (in part)
Arvicolinae
voles, lemmings, and muskrats

14'   Occlusal surface of cheek teeth with cusps (or evidence of cusps) arranged in two rows or, if pattern that of sharp re-entrant angles, then enamel never enclosing triangles of dentine ....
.... **Muridae** (in part)
Sigmodontinae
New World rats and mice

15 (12') Cheek teeth 5/4; greatest length of skull more than 50 mm; size large (head and body length more than 300 mm), tail short, 5% to 10% of head and body length (Fig. 23.36B) ....
.... **Aplodontidae**
Aplodontia rufa
sewellel or mountain "beaver"

15'   Cheek teeth 3/3 or 4/3; greatest length of skull less than 50 mm; size small (head and body length less than 200 mm); tail long, more than or equal to head and body length ....
.... **Dipodidae** (in part)
Zapus and *Napeozapus*
Zapodinae
jumping mice

COMMENTS AND SUGGESTIONS ON IDENTIFICATION

All rodents possess a single pair of evergrowing, arc-shaped upper incisors. This character is almost unique to the order, and found elsewhere only in the wombats and the aye-aye.

1. Wombats (family Vombatidae, order Diprotodontia). Check angle of ramus and mandibular fossa.
2. Aye-aye (family Daubentoniidae, order Primates). Check postorbital bar, position of foramen magnum, and structure of manus.
3. Rabbits, hares, and pikas (order Lagomorpha) and hyraxes (order Hyracoidea). Check the number and shape of upper and lower incisors.

To help separate rodents into major groups, look first at the characteristics of the infraorbital foramen and the structure of the lower jaw. These features reflect important differences in the ways that the chewing muscles are arranged (see Fig. 23.3) in rodents. All members of the suborder Hystricognathi have a hystricognath-type mandible, and most, Bathyergidae excepted, also have a greatly enlarged or hystricomorphous infraorbital foramen. However, an enlarged infraorbital foramen is also found in certain nonhystricognaths such as the springhaas (Pedetidae), gundis (Ctenodactylidae), dipodids (Dipodidae), and scaly-tailed squirrels (Anomaluridae).

Within suborders of rodents, it is important to look at cusps and other patterns on teeth, at the nature of the paroccipital process, perforations in the mandible, shape and character of the postorbital process, and other characteristics. Inspection of the keys will give you clues concerning which characters are important in separating the different families.

CHAPTER 24

THE AARDVARK
Order Tubulidentata

ORDER TUBULIDENTATA

Tubulidentata is the smallest living order of eutherian mammals. It contains a single living species, the aardvark, *Orycteropus afer*. The ordinal name refers to the unique structure of the teeth, described later. The common name is an Afrikaans term meaning "earth hog," which is appropriate because aardvarks are somewhat piglike in appearance (Fig. 24.1) and are semifossorial. The digits terminate in structures that are intermediate between claws and hoofs (Fig. 24.2). These serve the animal well for burrowing and for tearing open termite mounds. Aardvarks have long, extensible tongues that are used to feed upon termites, ants, and other insects (Melton 1976), but they are also known to feed upon plant materials (Melton 1976; Rahm 1975b), and, in particular, on the fruits of a certain wild cucumber (Patterson 1975).

The flesh and hides of aardvarks are used for food and leather in some areas (Melton 1976). Abandoned aardvark burrows are important to the survival of many other animals that use them as dens and as refuges from bush fires (Rahm 1975b).

Figure 24.1 The aardvark, *Orycteropus afer*.
(Dekeyser 1955)

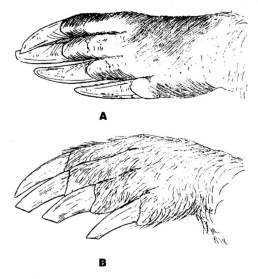

Figure 24.2 (A) Ventral and (B) lateral views of the forefoot of an aardvark, *Orycteropus afer*.
(Pocock 1924a)

Distinguishing Characters

The skull is elongate and conical in shape (Fig. 24.3). Incisors and canines are absent, and the unique cheek teeth usually number 5/5. The ever-growing cheek teeth are oval or 8-shaped, flat-topped, columnar structures (Fig. 24.4) that lack enamel and that are composed of numerous hexagonal prisms of dentine surrounding tubular pulp cavities (Fig. 24.4, detail).

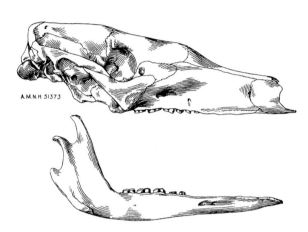

Figure 24.3 Skull of an aardvark, *Orycteropus afer*.
(Hatt 1934)

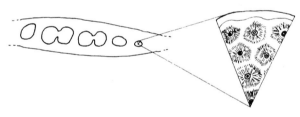

Figure 24.4 Diagram of the occlusal view of the upper tooth-row of an aardvark, Orycteropodidae, and an enlargement of a portion of a tooth surface showing the pulp tubules and surrounding dentine prisms.
(A. F. DeBlase)

The limbs are digitigrade with four manual (see Fig. 24.2) and five pedal digits. The snout is elongate and piglike, and the ears are much longer than wide. The tail is long and tapers gradually from its very thick base. The skin is thick with sparse, bristlelike hairs. Subcutaneous fat is lacking (Rahm 1975b). The caecum is large, especially for an insectivorous mammal (Melton 1976). The testes are never scrotal, and a baculum has not been reported. The uterus is duplex.

Living Family of Tubulidentata

The Tubulidentata contains a single family, **Orycteropodidae,** which contains a single genus and single species, *Orycteropus afer,* the aardvark (Schlitter 1993). It ranges through much of the Ethiopian Region in sub-Saharan Africa except for areas of dense forest. Its pattern of distribution follows that of mound-building termites.

Comments and Suggestions on Identification

Externally, aardvarks are unique and easily identified. The skull is also unique, and on the basis of size and shape alone it could be confused only with that of a giant armadillo. The giant armadillo, however, has numerous tiny teeth that are very different in structure from those of an aardvark.

CHAPTER 25

THE SUBUNGULATES

Order Proboscidea

Order Hyracoidea

Order Sirenia

Key to Living Families

Based upon external appearance, habitat, diet, and nearly every other easily discernible feature, the elephants, hyraxes, and sirenians would have to be regarded as one of the most unlikely assemblages of mammals possible. However, fossil evidence has indicated that these three share a common ancestor in the early Cenozoic of Africa (Carroll 1988; Romer 1966), and these three, together with two completely extinct orders, have been frequently grouped as the subungulates. The extant species are today mere remnants of what were once very diverse and abundant groups.

ORDER PROBOSCIDEA

The ordinal name, Proboscidea, refers to an elephant's most conspicuous structure—its long prehensile proboscis or trunk. This trunk and other unique external features of elephants are distinctive and well-known (Fig. 25.1).

The two extant species of elephants are the largest living land animals. The African elephant, *Loxodonta africana,* is the larger of the two, with large males measuring up to 4 meters at the shoulder and weighing over 7,000 kg (Grzimek 1975). The Asiatic elephant, *Elephas maximus,* is somewhat smaller, rarely reaching 3 meters at the shoulder and weighing up to 5,000 kg (Altevogt and Kurt 1975). The African elephant has much larger ears, a flatter forehead, and a somewhat concave profile of the back (Fig. 25.1), whereas the Asiatic elephant has smaller ears, a more domed forehead, and a convex dorsal profile.

Elephants are browsing mammals that usually live in herds. The Asiatic elephant has been tamed for centuries and used as a beast of burden throughout its range in Asia. The African elephant has been tamed less frequently but was used by Hannibal from 218 to 202 BC (Douglas-Hamilton and Douglas-Hamilton 1975) and is still used in some parts of Africa (Grzimek 1975). Both species have been extensively hunted and, increasingly, poached for their ivory tusks and as trophies. They are endangered in parts of their range because of hunting, illegal poaching, or habitat modification due to agriculture or lumbering.

Figure 25.1 An African elephant, *Loxodonta africana,* Elephantidae.
(Michael Gilliland)

Figure 25.2 Cheek teeth of (A) the Asiatic elephant, *Elephas maximus;* and (B) the African elephant, *Loxodonta africana,* both Elephantidae.
(Flower and Lydekker 1891:424)

DISTINGUISHING CHARACTERS

The incisors, numbering 1/0, are long, ever-growing tusks of solid dentine. These tusks are frequently absent in female Indian elephants. Canines are absent. The cheek teeth, consisting of the second, third, and fourth deciduous premolars and the first, second, and third molars, are hypsodont and lophodont (Fig. 25.2). They are replaced from the back of the jaw, and worn teeth are shed from the front of each tooth row (see Fig. 3.2). Only one or parts of two cheek teeth in a jaw quadrant are functional at any one time (Altevogt 1975). The limbs are graviportal with five digits on each foot. Each digit terminates in a hooflike structure. The upper lip and nose are fused and elongated to form a long, prehensile proboscis with the nostrils at the distal end. The skin is thick and covered with sparse, bristlelike hairs. The caecum is large. Testes are permanently internal, and a baculum is absent.

LIVING FAMILY OF PROBOSCIDEA

There is only a single living family of Proboscidea, **Elephantidae,** which contains two species (Wilson 1993). The African elephant, *Loxodonta africana,* occurs through the Ethiopian Region in sub-Saharan Africa, and in Roman times was found in the Atlas mountains of northern Africa (Grzimek 1975). The Asiatic elephant, *Elephas maximus,* ranged throughout most of the Oriental Region in recent times but is now confined to the continental portion of the Oriental, except for central India, and is found on the islands of Sri Lanka, Sumatra, and Borneo (Altevogt and Kurt 1975). Both species have been eliminated from large areas of their ranges because of human competition for the land and due to hunting and poaching.

COMMENTS AND SUGGESTIONS ON IDENTIFICATION

Both externally and cranially, proboscideans are distinctive and, once seen, are impossible to confuse with any other mammals.

ORDER HYRACOIDEA

The hyraxes are a group of mammals unknown to most people in the areas of the world where they do not exist. Hyraxes (also known as "dassies," and referred to in the King James Bible as "conies") are rabbit-sized animals that look rather like rodents (Fig. 25.3). These herbivorous mammals have a unique foot structure (description follows) that provides a firm grip on the rocks and trees in which they live. The terrestrial forms, in the genera *Heterohyrax* and *Procavia,* live in colonies of six to 50 individuals in areas of jumbled boulders and rock outcrops. The arboreal species, in the genus *Dendrohyrax,* have none of the limb modifications usually associated with arboreal mammals (e.g., opposable digits, sharp claws) but have great agility and remarkably adhesive pads on the feet.

Figure 25.3 A rock hyrax, *Procavia capensis,* Procaviidae.
(Flower and Lydekker 1891)

Distinguishing Characters

The adult dental formula is 1/2 0/0 4/4 3/3 = 34. The long, ever-growing, upper incisors are triangular in cross section and have pointed tips (Fig. 25.4), whereas the lower incisors are chisel-shaped and usually tricuspid. The cheek teeth are somewhat lophodont and are separated from the incisors by a wide diastema. The well-developed postorbital processes usually form postorbital bars. The interparietal is well-developed. The large jugals contribute to the formation of the mandibular fossae.

Limbs are plantigrade. The four manual digits are syndactylous except for their terminal phalanges. The pes has three digits. All digits terminate in short, flat, hooflike nails, except the second pedal digit, which has a long, curved clawlike nail used for grooming (Rahm 1975c). The soles have large, soft, elastic pads that are kept moist by numerous glands. The tail is very short. The testes are internal, and a baculum has not been reported. The uterus is duplex.

Living Family of Hyracoidea

There is a single living family of Hyracoidea, **Procaviidae,** the hyraxes. This family includes three extant genera and six species (Schlitter 1993). Most species are confined to the Ethiopian Region in sub-Saharan Africa; however, *Procavia* extends into the Palearctic, up the Nile valley from the Ethiopian and into Arabia and the Levant.

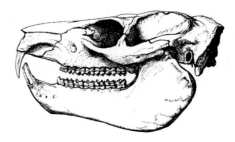

Figure 25.4 Skull of a tree hyrax, *Dendrohyrax dorsalis,* Procaviidae.
(Flower and Lydekker 1891)

Comments and Suggestions on Identification

Hyraxes superficially resemble rodents or large pikas but are quickly identified by their unique foot structure. The skulls may also, at first glance, look like those of rodents. However, the triangular (in cross section) upper incisors and the presence of two lower incisors per side are diagnostic.

ORDER SIRENIA

Sirenians are fully aquatic mammals that lack external hindlimbs and have forelimbs modified to form flippers (Fig. 25.5). Unlike most cetaceans (the only other fully aquatic mammals), sirenians have a short but flexible neck. (The beluga, Cetacea, Monodontidae, has a similarly flexible neck.) The mammae are pectoral, and the female has been said (apparently incorrectly) to float on her back as she clasps her young to her breast to suckle. Some have speculated that the supposedly humanoid appearance of these ungainly animals as they supposedly floated and nursed their young caused sailors, who must have been long at sea, to originate the legends of mermaids. The ordinal name refers to the sirens, sea nymphs who, in Greek mythology, lured mariners to destruction. Columbus recorded seeing three "sirens" (manatees) in an inlet on the island of Hispaniola, and he noted that they were not nearly as beautiful as those described by Horace (Kurt and Wendt 1975).

Sirenians feed on aquatic vegetation, and, in some parts of their range, play an important role in keeping navigation channels free of excess vegetation (Kurt and

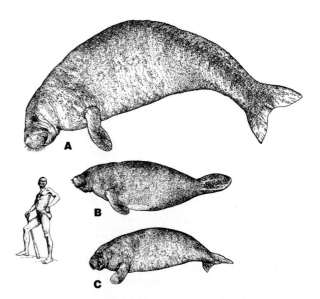

Figure 25.5 Representative sirenians: (A) the recently extinct Steller's sea cow, *Hydrodamalis gigas,* Dugongidae; (B) a manatee, *Trichechus sp.,* Trichechidae, and (C) a dugong, *Dugong dugon,* Dugongidae. These are drawn to scale with each other and with the 6' tall person illustrated.
(Feldhamer et al. 1999)

Wendt 1975). They are hunted for meat, hides, and oil in various parts of their range.

Living sirenians are tropical or subtropical, but one species, the large Steller's sea cow, *Hydrodamalis gigas* (Fig. 25.5A), lived in far northern waters. It was discovered around Bering Island and an adjacent island in the Bering Sea in 1742. For 27 years, this species served as a source of food, oil, and boat-building materials to mariners sailing or marooned in the area, but by 1766, the last of the original population of 1,500 to 2,000 was exterminated (Kurt and Wendt 1975). Steller's sea cow was a member of the family Dugongidae. It differed from living representatives of this family in completely lacking teeth and phalanges and in its much larger size (up to 8 meters long and 4,000 kg in weight) (Kurt and Wendt 1975). This extinct form is not included in the following description or key.

Distinguishing Characters

The external nares are situated high on the skull posterior to the anterior margins of the orbits. The nasal bones are rudimentary or absent. Incisors are absent except in the dugong, which has I 1/0 (Fig. 25.6). Canines are absent. Cheek teeth in the Trichechidae are lost from the front and replaced from the rear as in Proboscidea. In the dugong, most cheek teeth are quickly worn away and may be replaced by horny plates in adults (Kurt and Wendt 1975).

The vestigial pelvic limbs are not visible externally. The pectoral limbs are paddlelike, with their five digits indistinguishable externally. The tail bears a horizontally flattened "fin" that may (Dugongidae, see Fig. 25.5A and C) or may not (Trichechidae, see Fig. 25.5B) be cleft. The dugong has the normal mammalian component of seven cervical vertebrae, but the species of Trichechidae have only six. (Deviation from the basic number of seven cervical vertebrae occurs among living mammals only in Trichechidae and in the two families of sloths in the order Xenarthra.) The ribs are extremely massive and serve as "ballast." Horizontal stability is enhanced by elongated lungs and a horizontal diaphragm.

The nostrils are on the upper surface of the snout. Eyes are small, and pinnae are absent. The lips are large and highly mobile. Numerous stiff vibrissae are present on the upper lip, but the body is otherwise nearly naked. The uterus is bicornuate. Testes are permanently internal, and a baculum has not been reported.

Living Families of Sirenia

There are two living families. **Trichechidae,** the manatees, contains one genus and three species (Wilson 1993). They occur in the Western Hemisphere along the Atlantic coast from North Carolina to southern Brazil and throughout the West Indies. They are found in rivers in Florida and in the Amazon and Orinoco drainages of South America. In the Eastern Hemisphere, they range along the Atlantic coast of Africa from 10° N to 10° S, in Lake Tchad, and throughout the drainages of the Congo, Niger, and several other western and central African rivers (Jones and Johnson 1967; Kurt and Wendt 1975).

The **Dugongidae,** with one living species, *Dugong dugon,* occurs in the Red Sea and ranges throughout coastal waters of the Indian Ocean from Mozambique and Madagascar to northern Australia. It inhabits the Indonesian Region and extends east through the Solomons, north through the Philippines, and along the Chinese coast almost to Japan (Jones and Johnson 1967).

The second species of Dugongidae, *Hydrodamalis gigas,* which became extinct in recent times, was discussed in detail earlier.

The ranges given earlier are being greatly fragmented, and the various sirenian species now are found only in scattered portions of their former ranges.

Key to Living Families of Sirenia

1 Upper incisors present as short tusks in males (Fig. 25.6), unerupted in females; cheek teeth, if present, simple; tail cleft (Fig. 25.5C); upper lip only partly cleft .... **Dugongidae**
  dugong, *Dugong dugon*

1' Incisors absent in adults (Fig. 25.7); cheek teeth with two cuspidate, transverse crests; tail spatulate (Fig. 25.5B); upper lip deeply cleft (Fig. 25.8) .... **Trichechidae**
  manatees

Figure 25.6 Skull of a dugong, *Dugong dugon,* Dugongidae.
(Giebel 1859)

Comments and Suggestions on Identification

Both the external appearances and the skulls of sirenians are distinctive and not easily confused with those of any other mammals. Dental characters and the shapes of the skulls and tails serve to differentiate the two families.

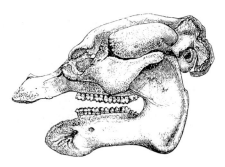

Figure 25.7 Skull of a West African manatee, *Trichechus senegalensis,* Trichechidae.
(Flower and Lydekker 1891)

Figure 25.8 Anterior view of a manatee, *Trichechus manatus,* Trichechidae, showing bilobed upper lip.
(Flower and Lydekker 1891)

Key to Living Families

CHAPTER 26

THE PERISSODACTYLS
Order Perissodactyla

ORDER PERISSODACTYLA

The ordinal name, Perissodactyla, which literally means "odd-fingered ones," points out the major distinctive feature of the order and the one feature that unites three very different extant families. The equids (horses, zebras, and asses, Fig. 26.1), rhinoceroses, and tapirs all have **mesaxonic** limb structure in which a large central digit carries the bulk of the weight of the animal, whereas smaller lateral digits may or may not be present. Hoofs are present on all exposed digits.

All living perissodactyls are herbivorous. Equids and some rhinos are primarily grazing animals inhabiting plains and savannas, whereas tapirs and other rhinos are generally browsing, forest-dwelling mammals. All extant odd-toed ungulates have been used as a source of hides and meat, but only a few species are of major economic importance. The unique horns of rhinos (see Chapter 5) are considered in the Orient to be both a powerful aphrodisiac and a neutralizer of poisons. Rhinos in all parts of their range are slaughtered by poachers to obtain these valuable horns, and today the five living species are considered endangered primarily because of this poaching. The domestic horse, *Equus caballus,* and the domestic donkey, *Equus asinus* (in part), are the major **beasts of burden** over most of the world. The third important equid beast of burden is a sterile hybrid, the mule, produced by crossing a female horse (mare) and a male donkey (jack). The result of a cross between a male horse (stallion) and a female donkey (jenny) is termed a hinny.

DISTINGUISHING CHARACTERS

The limbs are unguligrade and mesaxonic. A large central digit carries the axis of weight, and smaller lateral digits may be present (Fig. 26.2). Manual digits number one (Equidae), three (Rhinocerotidae), or four (Tapiridae). Pedal digits number one (Equidae) or three (Rhinocerotidae and Tapiridae). The skulls tend to be elongated. Canine teeth, if present, are small, and the molars, and usually most

Figure 26.1 The African wild ass, *Equus asinus,* Equidae. (Kingsley 1884)

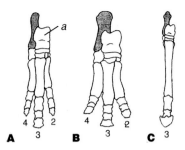

Figure 26.2 The hindfoot skeletons of (A) a tapir, (B) a rhinoceros, and (C) a horse, illustrating their mesaxonic structure. a = astragalus.
(Feldhamer et al. 1999)

of the premolars as well, have a more or less complex pattern of lophs and ridges.

Equids have a dental formula of 3/3 0–1/0–1 3–4/3 3/3 = 36–42. The cheek teeth are hypsodont and have a complex grinding surface of folded enamel ridges. Canines, when present, are small and located in the wide space between the incisors and premolars (Fig. 26.3). The orbit is completely separated from the temporal fossa by a postorbital plate. The body is fully haired.

Rhinos have a dental formula of 0–2/0–1 0/0–1 3–4/3–4 3/3 = 24–34. The cheek teeth are basically lophodont. The orbit and temporal fossa are confluent (Fig. 26.4). One or two "horns," composed of agglutinated dermal papillae, are present on the rostrum (Fig. 26.5). See Chapter 5 for a detailed description of rhino "horn." The body is covered with thick skin that is sparsely haired in most species, but the Sumatran rhinoceros, *Dicerorhinus sumatrensis,* is relatively well-haired.

Tapirs have a dental formula of 3/3 1/1 4/3–4 3/3 = 42–44. A diastema is present between the canines and premolars. Tapirs may be the only mammals in which four premolars all have deciduous precursors. The brachyodont cheek teeth have transverse ridges. The orbit and temporal fossa are confluent (Fig. 26.6). The upper lip and nose are elongated to form a short proboscis. In three of the four living species, the body is covered with a short, sleek coat of hair, but on the mountain tapir, *Tapirus pinchaque,* the coat is soft and woolly (Fradrich 1972).

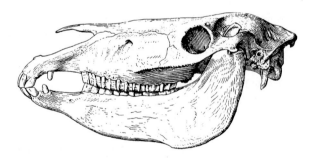

Figure 26.3 Skull of a male Przewalski's horse, *Equus,* Equidae. The canine teeth are usually lacking in females and in males gelded before the adult dentition develops.
(Sokolov 1959)

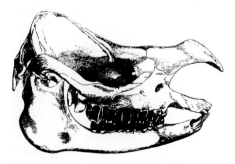

Figure 26.4 Skull of a Javan rhinoceros, *Rhinoceros sondaicus,* Rhinocerotidae.
(Gray 1869)

Figure 26.5 A black rhinoceros, *Diceros bicornis,* Rhinocerotidae.
(Dekeyser 1955)

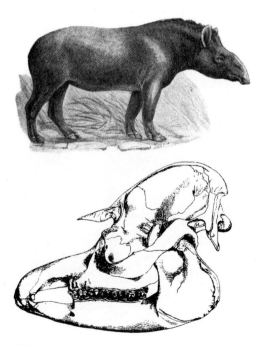

Figure 26.6 A South American tapir, *Tapirus terrestris,* Tapiridae, and a skull of the same species.
(Tapir, Beddard 1902; skull, Gray 1869)

All three families have a simple stomach and large caecum. The uterus is bicornuate. Testes are scrotal in Equidae and internal in Rhinocerotidae and Tapiridae. A baculum is never present.

LIVING FAMILIES OF PERISSODACTYLA

Perissodactyla contains three living families. **Equidae** has a single living genus, *Equus,* with nine species of horses, asses, and zebras. Indigenous equids are found today only in southern and eastern Africa and in the arid regions of southwestern and central Asia (Volf 1975). However, the domesticated forms have been introduced to all parts of the world, and feral populations exist in many areas. **Tapiridae** also has only a single living genus, *Tapirus,* with four species of tapirs, which are found in the tropical portions of the Neotropical, and in the Oriental Region on the Malay Peninsula, and on Sumatra (Fraedrich 1972). The **Rhinocerotidae** contains four living genera and five living species of rhinos. Two live in the central, eastern, and southern portions of the Ethiopian Region, and three range through the Oriental Region from eastern India to Borneo. However, through much of this range, they are now extirpated or very rare and endangered (Lang 1972). (Number of genera and species based upon Grubb 1993.)

KEY TO LIVING FAMILIES OF PERISSODACTYLA

1 Orbit and temporal fossa separated by a postorbital plate (Fig. 26.3); dental formula 3/3 0–1/0–1 3–4/3 3/3 = 36–42; digits 1/1 (Figs. 26.1, 26.2C) .... **Equidae**
horses, asses, and zebras

1' Orbit and temporal fossa confluent; teeth number fewer than 36, or 42 or more; digits 3–4/3 .. .. **2**

2 (1') Dental formula 3/3 1/1 4/3–4 3/3 = 42–44; upper canines well-developed (Fig. 26.6); short muscular proboscis present (Fig. 26.6); no horns .... **Tapiridae**
tapirs

2' Dental formula variable, teeth number 34 or fewer; upper canines absent (Fig. 26.4); proboscis absent; one or two horns on rostrum (Fig. 26.5) .... **Rhinocerotidae**
rhinoceroses

COMMENTS AND SUGGESTIONS ON IDENTIFICATION

Externally, the three families are distinctive and cannot easily be confused with each other or with any other mammals. The general appearance of the skulls of each of the three is also distinctive and easily recognized with a little practice.

CHAPTER 27

THE ARTIODACTYLS
Order Artiodactyla

Key to Living Families

ORDER ARTIODACTYLA

The ordinal name, Artiodactyla, literally meaning "even-digited ones," points out the major feature uniting the families of artiodactyls, the even-toed ungulates, and distinguishing these ungulates from the perissodactyls, the odd-toed ungulates. There are usually either two or four digits, and the limbs are **paraxonic,** with the plane of symmetry passing between the third and fourth digits of each foot, which are about equal in size and which equally share the weight placed on that foot (Fig. 27.1). The second and fifth digits, when present, are smaller than the third and fourth, and in many artiodactyls do not touch the ground unless it forms a very soft substrate.

Most living artiodactyls are herbivorous, but some of the more generalized forms (e.g., the hogs) are omnivorous. The types of plant material eaten by artiodactyls range from lichens on the Arctic tundra to fruits and tubers in tropical forests. Wild artiodactyls are important as sources of meat and hides in many cultures and as game animals in others. By far, the largest number of domesticated animal species comes from this order. Domesticated artiodactyls provide meat, hides, and milk and serve as beasts of burden in nearly all parts of the world. These domestic species include the hog, *Sus scrofa;* llama, *Lama glama;* alpaca, *Lama pacos;* Bactrian (two-humped) camel, *Camelus bactrianus;* dromedary (one-humped camel), *Camelus dromedarius;* reindeer, *Rangifer tarandus;* water buffalo, *Bubalus bubalis;* "cattle," *Bos taurus;* gayal, *Bos frontalis;* domesticated banteng or "Bali cattle," *Bos javanicus;* yak, *Bos grunniens;* goat, *Capra hircus;* and sheep, *Ovis aries.*

DISTINGUISHING CHARACTERS

The third and fourth digits of each foot are subequal in size, with the main axis of weight passing between them. The second and fifth digits are reduced in size or are absent (Fig. 27.2). The first digit is absent. The metapodials (metacarpals or metatarsals) of the third and fourth digits are the largest and may (Tylopoda, Ruminantia) or may not (Suiformes) be fused to form a "cannon bone." All digits terminate in hoofs (modified in Camelidae and Hippopotamidae). A unique characteristic of the order is an astragalus with a pulley-shaped articulating surface both proximally and distally.

Cheek teeth range from bunodont (Suiformes) to selenodont (Tylopoda and Ruminantia) and from brachyodont to hypsodont. The full "late primitive" placental dental formula is present in some forms (most Suidae). The upper incisors are lost (Ruminantia) or reduced in number (Tylopoda) in most species. Upper canines are frequently lost (most Ruminantia) or reduced in size (Tylopoda, some Ruminantia). A complete postorbital bar may (Hippopotamidae, Tylopoda, Ruminantia) or may not (Suidae and Tayassuidae) be present.

In many Ruminantia, an antorbital pit (in which the antorbital gland is located) is conspicuous as a depression on the side of the rostrum just anterior to the orbit (Fig. 27.3). A hole or fenestra is present on the rostrum of many ruminants at the point where the frontal, nasal, lacrimal, and maxillary bones meet (Fig. 27.3). Most male Ruminantia have horns or antlers and so do females of some species. The structure of these is different in

Figure 27.1 A bighorn sheep, *Ovis canadensis*, Bovidae.
(Giuliani 1993)

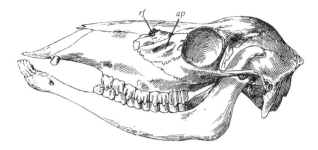

Figure 27.3 Skull of a deer of the genus *Cervus*, Cervidae: ap, antorbital pit; rf, rostral fenestra.
(Sokolov 1959)

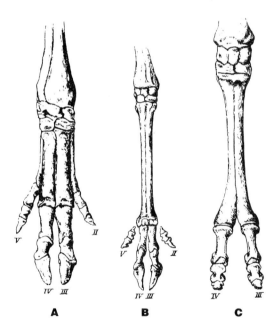

Figure 27.2 Bones of the right forelegs of three representative artiodactyls: (A) a hog, *Sus scrofa*, Suidae; (B) red deer, *Cervus elaphus*, Cervidae; (C) Bactrian camel, *Camelus bactrianus*, Camelidae. Note that in the Suiformes (A), each metacarpal is distinct; in the Tylopoda (C), the metacarpals of digits III and IV are fused for most of their length but separated at their distal end; and in the Ruminantia (B), the two metacarpals are fused for their entire length. Fused metacarpals or metatarsals form what is termed a cannon bone.
(Flower and Lydekker 1891)

Figure 27.4 The African water chevrotain, *Hyemoschus aquaticus*, Tragulidae.
(Flower and Lydekker 1891)

each of the families Cervidae, Giraffidae, Antilocapridae, and Bovidae. See Chapter 5 for the anatomy and development of these structures. Males of the Tragulidae (Fig. 27.4), Moschidae (Fig. 27.16), and the only cervid, *Hydropotes inermis*, which lacks antlers, are equipped with very long upper canines. Two genera of Cervidae, *Muntiacus* and *Elaphodus*, possess both antlers and long upper canines.

The stomach of nonruminant artiodactyls is a relatively simple, two-chambered structure in Suidae, and a three-chambered structure in Tayassuidae and Hippopotamidae. Among those artiodactyls that ruminate, the stomach is a three-chambered structure in Camelidae and Tragulidae,[1] and is a complex, four-chambered structure

---

[1] In Tragulidae, the stomach is four-chambered, but the omasum is rudimentary.

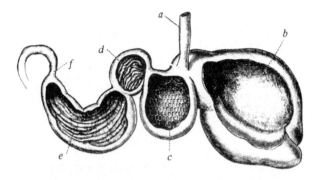

Figure 27.5 A generalized four-chambered "stomach" of a ruminant. a, unmodified portion of esophagus; b, first chamber or rumen; c, second chamber or reticulum; d, third chamber or omasum; e, fourth chamber or abomasum; f, duodenum. (Flower and Lydekker 1891)

(Fig. 27.5) for the breakdown of cellulose by microorganisms, in the remaining five families. The caecum is absent, and the uterus is bicornuate. Testes are scrotal in some species, and a baculum is never present.

LIVING FAMILIES OF ARTIODACTYLA

A list of living families of Artiodactyla, and their contents, is given in Table 27.1.

Artiodactyls are native to all parts of the world's land surface except Antarctica, oceanic islands, and the great majority of the Australian Region. One species of suid occurs in Sulawesi (= Celebes), Indonesia, and on nearby islands, thus barely penetrating the Australian Region. Domestic herds and feral populations of domesticated artiodactyls now occur over much of the Australian Region and on many oceanic islands.

KEY TO LIVING FAMILIES OF ARTIODACTYLA

1   Postorbital bar absent or incomplete; upper incisors present 2

1' Postorbital bar complete[2]; upper incisors present or absent 3

2 (1') Dental formula 2/3 1/1 3/3 3/3 = 38; upper canines relatively straight, point down (Fig. 27.6B); two or three pedal digits .... **Tayassuidae** peccaries

2' Dental formula 3/3 1/1 4/4 3/3 = 44, 1/3 1/1 3/2 3/3 = 34, or 2/3 1/1 2/2 3/3 = 34; upper canines curve either outward, upward, or both (Figs. 27.7 and 27.8); four pedal digits .... **Suidae** swine, hogs, pigs

[2]Incomplete in some specimens of hippos.

TABLE 27.1 | **Living Families of Artiodactyla***

Family	Common Name	Number of Genera	Number of Species	Distribution
Suborder Suiformes				
Suidae	Hogs, pigs, or swine	5	16	Ethiopian, Palearctic, Oriental, Sulawesi in Australian
Tayassuidae	Peccaries	3	3	Neotropical, southern Nearctic
Hippopotamidae	Hippos	2	2	Ethiopian
Suborder Tylopoda				
Camelidae	Camels, guanacos, and allies	3	6	Western and southern South America, eastern Palearctic
Suborder Ruminantia				
Tragulidae	Chevrotains	3	4	Portions of Ethiopian and Oriental
Moschidae	Musk deer	1	4	Eastern Palearctic, northern edge of Oriental
Cervidae	Deer	16	42	Holarctic, Neotropical, Oriental
Giraffidae	Giraffe and okapi	2	2	Ethiopian
Antilocapridae	Pronghorn	1	1	Southwest Nearctic
Bovidae	Cattle, antelopes, sheep, goats, and allies	45	137	Holarctic, Ethiopian, Oriental

*Based upon Grubb (1993).

The Artiodactyls 175

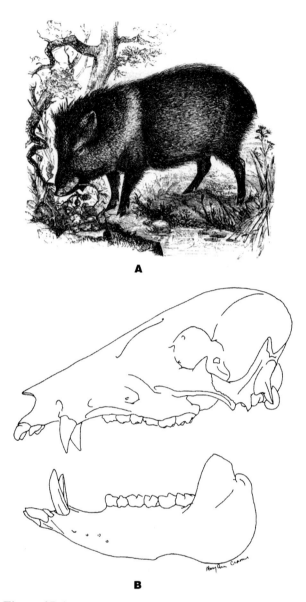

Figure 27.6 The collared peccary (A), *Pecari tajacu*, Tayassuidae, and a skull (B) from the same species.
(A, Flower and Lydekker 1891; B, Mary Ann Cramer)

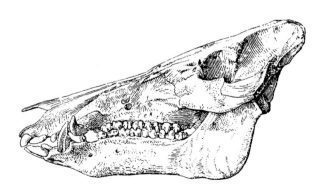

Figure 27.7 The skull of a wild boar, *Sus scrofa*, Suidae. The domestic hog belongs to the same species, but its skull has a more concave dorsal profile.
(Gromov et al. 1963)

Figure 27.8 The babirusa, *Babyrousa babyrussa*, (yes, each of the three names is spelled differently!) a unique hog (Suidae) confined to Sulawesi (= Celebes) and a few smaller islands of Indonesia. The sockets of the upper canines turn up alongside the rostrum so that the teeth grow dorsally through the skin. The upper and lower canines do not come in contact.
(Flower and Lydekker 1891)

3 (1') Lower canine alveoli anterior to alveoli of lower incisors (Fig. 27.9B); lower canines larger than upper canines; skull very massive (Fig. 27.9B); rostrum broader distally than proximally .... **Hippopotamidae**
hippos

3' Lower canine alveoli posterior to alveoli of lower incisors (Fig. 27.10); lower canines, if present, smaller or equal in size to upper canines; skull not particularly massive; rostrum narrower distally than proximally **4**

4 (1') Upper incisors present (Fig. 27.10); hoofs nail-like, with large pads on bottoms of feet posterior to hoofs .... **Camelidae**
camels, guanacos, and allies

4' Upper incisors absent; well-developed hoofs present, pads absent, only hoofs touch ground **5**

5 (4') Horn cores or antler pedicels present **6**

5' No indication of horn cores or antler pedicels **9**

6 (5) Paired "horns" are distinct bones situated over sutures of frontal and parietal bones (Fig. 27.11); a third "horn" sometimes present on midline of skull (Fig. 27.11) .... **Giraffidae**
giraffe and okapi

6' Paired horns or antlers are projections of frontal bone; third medial "horn" never present **7**

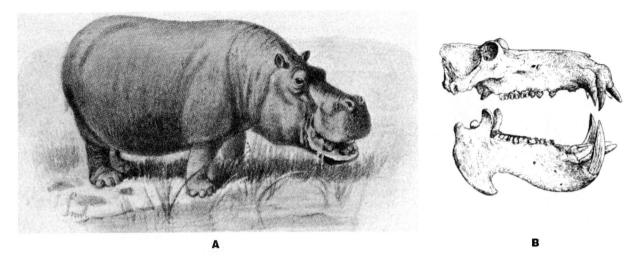

Figure 27.9 A hippopotamus (A) *Hippopotamus amphibius,* Hippopotamidae, and a skull (B) of the same species.
(A, Sclater and Sclater 1899; B, Sclater 1900)

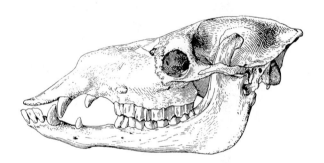

Figure 27.10 The skull of a Bactrian camel, *Camelus bactrianus,* Camelidae.
(Sokolov 1959)

7 (6')   Antlers or antler pedicels present (Fig. 27.12A, B); antorbital pit and rostral fenestra both present (Fig. 27.3) .... **Cervidae** (in part)
deer with antlers

7'   Horns or horn cores present (Figs. 27.13, 27.14, 27.15); presence of antorbital pit and rostral fenestra varies, one or both frequently absent ..
.... **8**

8 (7')   Horn core with sharp anterior edge; one or two large foramina present in frontal at anteromedial base of horn cores (Fig. 27.14A); horn forked, having small anterior projection (Fig. 27.13); digits 2/2 .... **Antilocapridae** (in part)
most pronghorns, *Antilocapra americana*

8'   Horn core with rounded anterior edge, or if anterior edge sharp, no foramina in frontal at base of horn core; horn not forked (Fig. 27.15); digits 2/2 or 4/4 .... **Bovidae** (in part)
horned bovids

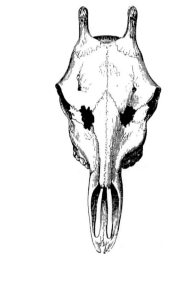

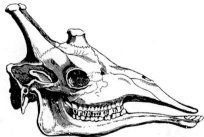

Figure 27.11 Skulls of giraffes, *Giraffa camelopardalis,* Giraffidae. Note that the "horns" (ossicones) are distinct bones and are separated from the frontals by sutures in the specimens figured. They can also completely fuse with the rest of the skull, eliminating obvious sutures.
(Dorsal view, Owen 1866; lateral view, Giebel 1859)

The Artiodactyls 177

Figure 27.12 Representative cervids, Cervidae. (A) Père David's deer, *Elaphurus davidianus;* (B) Sika deer, *Cervus nippon;* and (C) tufted deer, *Elaphodus cephalophus.* Not to same scale.
(Hsia et al. 1964)

10'   Antorbital pit present or absent; rostral fenestra, if present, not as above **11**

11 (10')   Antorbital pit absent; rostral fenestra very small, less than 4 mm in length; greatest length of skull 100 mm or less .... **Tragulidae** (in part)
some female chevrotains

11'   Antorbital pit present or absent, if absent; greatest length of skull more than 100 mm; rostral fenestra usually more than 4 mm in length ......
.... **Bovidae** (in part)
most hornless female bovids

Figure 27.13 Pronghorns, *Antilocapra americana,* Antilocapridae.
(Kingsley 1884)

9 (5')   Either antorbital pit absent or rostral fenestra absent or both absent; upper canine teeth absent (Fig. 27.16) **10**

9'   Both antorbital pit and rostral fenestra present, or if one absent, upper canine teeth present (Figs. 27.17 and 27.18) **12**

10 (9)   Antorbital pit absent; rostral fenestra narrow dorsoventrally and elongate anteroposteriorly (Fig. 27.14B) .... **Antilocapridae** (in part)
hornless female pronghorns, *Antilocapra americana*

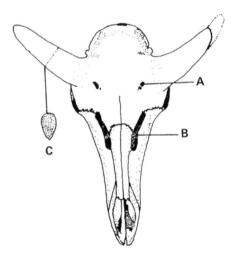

Figure 27.14 The skull of a pronghorn, *Antilocapra americana,* Antilocapridae. (A) Foramen at base of horn core; (B) rostral fenestra; (C) a cross section of horn core.
(G. A. Moore)

Figure 27.15 Representative bovids, Bovidae. (A) The gaur, *Bos frontalis;* (B) the takin, *Budorcas taxicolor;* (C) an ibex, *Capra ibex;* (D) a wild sheep, the argali, *Ovis ammon*. Not to same scale.
(Hsia et al. 1964)

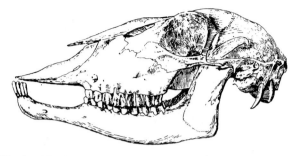

Figure 27.16 Skull of a female gazelle, *Gazella,* Bovidae.
(Sokolov 1959)

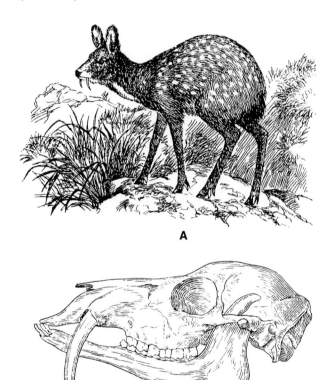

Figure 27.17 A musk deer (A) *Moschus moschiferus,* Moschidae, and its skull (B).
(Gromov et al. 1963)

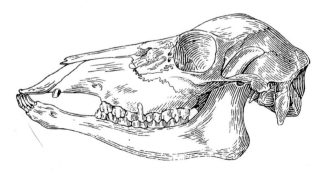

Figure 27.18 Skull of a female sika deer, *Cervus nippon,* Cervidae.
(Gromov et al. 1963)

12 (9') Antorbital pit absent; upper canines present; condylobasal length less than 100 mm ....
.... **Tragulidae** (in part)
most chevrotains

12' Antorbital pit present; upper canines present or absent; condylobasal length usually greater than 100 mm **13**

13 (12) Orifice of lacrimal canal usually single (Fig. 27.19A); upper canines absent ....
.... **Bovidae** (in part)
some female antelopes

13' Orifice of lacrimal canal double (Fig. 27.19B); upper canines present or absent **14**

14 (13) Upper canines present in both sexes, relatively long and pointed (Fig. 27.17); glandular slit (preorbital gland) in front of eye absent; size small, head and body length 700–1000 mm ....
.... **Moschidae**
musk deer, *Moschus*

14' Upper canines present or absent; if present and fang- or tusklike, then preorbital gland present; size usually large, head and body length usually more than 1,000 mm .... **Cervidae** (in part)
deer without antlers (includes females of most species)

COMMENTS AND SUGGESTIONS ON IDENTIFICATION

The structure of the limbs readily distinguishes artiodactyls from members of all other orders. Cranially, the skulls of only perissodactyls could be confused with those of artiodactyls. The large, massive skull and enormous lower canines will serve to identify hippos. Suid skulls may be distinguished from those of tayassuids by the curvature of the upper canines of the former. The rudimentary upper incisors of camelids distinguish them from artiodactyls of other families, and the unique "horns" of the giraffe and okapi identify that family.

Cervids, antilocaprids, and bovids having antlers or horns are easily identified by those structures. However, the "hornless" ruminants are more difficult to distinguish. Their lack of upper incisors identifies them as artiodactyls, but their identification to family is more difficult. Tragulids (and moschids) have large upper canines; however, there are some species of small cervids that also have long upper canines. The Tragulidae have a tiny rostral fenestra and no antorbital pit. The long-canined, antlerless cervids also have a small rostral fenestra but do have an antorbital pit. Unfortunately, this pit is very shallow and frequently difficult to distinguish. Among the larger "hornless" ruminants, the female cervids sometimes possess small upper canines, whereas the antilocaprids and bovids never do. The cervids always have a rostral fenestra and an antorbital pit. The antorbital pit is always lacking in the Antilocapridae. However, the Bovidae may have both the fenestra and the pit, one of these structures, or neither of them.

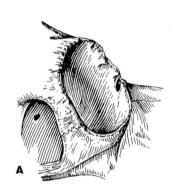
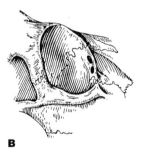

Figure 27.19 Orbital regions of the skulls of a bovid, Bovidae (A) and of a cervid, Cervidae (B). Cervids have a double lacrimal orifice in the orbit, whereas most bovids have a single orifice.
(Gromov et al. 1963)

CHAPTER 28

SIGN AND HABITAT ANALYSIS

Because most mammals are secretive and/or nocturnal, many times they are seldom seen by the casual observer. Their presence, however, is often revealed by tracks, burrows, nests, runways, evidence of feeding, and other **sign** (Fig. 28.1). This sign is often useful in determining the distribution of mammals and learning something of their habits. Learning to recognize and interpret mammal sign is essential to efficient collecting and other field investigation. Techniques to identify sign are covered in the section "Identifying Mammal Sign."

Mammals live in both aquatic and terrestrial habitats. In the terrestrial environment, mammal ecologists use qualitative and quantitative techniques to help define the nature of the habitat that mammals occupy. This information can be useful in understanding how species can coexist in ecological communities. The section "Habitat Analysis" will show techniques that can be used by mammalogists and wildlife professionals to characterize terrestrial habitats.

IDENTIFYING MAMMAL SIGN

Footprints and tail markings can be found in mud along streams and lakes or in any other suitable substrate such as dust, soft soil, sand, or snow. Such tracks can usually be identified and, with experience, are a reliable means for verifying the presence of a particular species in a given locality. The tracks of most larger mammals can be readily identified by their characteristic shapes and arrangements (Fig. 28.2), but it is sometimes difficult to distinguish closely related species. The tracks of small mammals are difficult to distinguish unless one knows which species can be expected in a particular area. Even then, evidence other than tracks is often necessary for positive identification. Olaus Murie (1954), in his *A Field Guide to Animal Tracks,* described and illustrated footprints and other

Figure 28.1 Nest of the hazel dormouse, *Muscardinus avellanarius,* Myoxidae.
(Ognev 1947:542)

Figure 28.2 Tracks of the American beaver, *Castor canadensis*, Castoridae. Forefoot, f; hindfoot, h. Bottom figure shows footprints and tail drag.
(Henderson 1960:54)

28-A Use Murie (1954), Stains (1962), or another book on mammal tracks to answer the following questions. Use casts when available.
  a. How do the footprints of *Felis* and *Canis* differ? What is the reason for this difference?
  b. How, other than by size, can you differentiate between the tracks of a deer and a cow? Between those of a deer and a wild sheep?
  c. What similarities do you note between the footprints of an opossum and a raccoon? How are these two distinguished?
  d. Why does a hog walking on a relatively hard surface leave the impression of two digits per foot whereas on softer ground it registers four digits per foot?

Scats

A mammal's fecal material or **scat** is also frequently species-distinctive and can yield important information on feeding habits, occurrence, and activity. Scats of small mammals can usually be identified only to the generic level, whereas scats of carnivores and ungulates are often species-distinctive (Fig. 28.3). However, the

sign of many North American mammals. This book and several other books, by Perkins (1954) and by Seton (1958) for U.S. species, and by Lawrence and Brown (1967) and by Twigg (1975a) for British species, may be used to identify tracks in the field or casts or photographs of them in the laboratory.

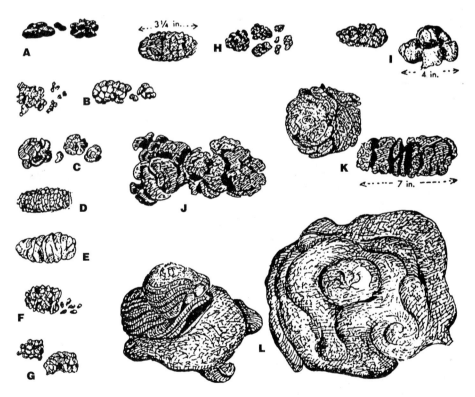

Figure 28.3 Scats of some North American artiodactyls. (A) Collared peccary, *Pecari tajacu;* (B) mule deer, two examples, *Odocoileus hemionus;* (C) white-tailed deer, *Odocoileus virginianus;* (D) Bighorn sheep, *Ovis canadensis;* (E) domestic sheep, *Ovis aries;* (F) mountain goat, *Oreamnos americanus;* (G) domestic goat, *Capra hircus;* (H) caribou, two examples, *Rangifer tarandus;* (I) pronghorn, two examples, *Antilocapra americana;* (J) moose, *Alces alces;* (K) American elk or wapiti, two examples, *Cervus elephus;* (L) American bison, *Bison bison,* two examples.
(Murie, O. 1954. A field guide to animal tracks. Houghton Mifflin Co., Boston, 374 pp., fig. 132. With permission of the publisher.)

shape and appearance of scats vary with the diet and age of the animal.

Examinations of scats can help determine the dietary habits of a species. Relative amounts of different vegetable and animal matter can be determined by indigestible portions, such as hair, feathers, seed coats, and chitin of insect exoskeletons. Microscopic analyses of fecal material (and stomach contents) are usually necessary, however, for quantitative determinations of food habits (see Chapter 34).

28-B Use the illustrations in Murie (1954) or collections of preserved (dried) scats to answer the following questions.
 a. How do you differentiate between the droppings of a deer and of a sheep?
 b. Between those of a deer and of a rabbit?
 c. Between those of a *Peromyscus* and of a *Microtus*?

Figure 28.4 Ridges and mounds made by a mole (Talpidae) with sectional view of some tunnels in lower left of figure. (Silver and Moore 1941:4)

Trails, Runways, and Burrows

Terrestrial mammals of all sizes may form conspicuous trails in their travels to and from feeding sites, water holes, or other areas within their home ranges. Many small rodents and insectivores establish definite **runways** that may be completely open above (e.g., those of ground squirrels, *Spermophilus*), partly covered (e.g., of cotton rats, *Sigmodon*), or entirely roofed over by surrounding vegetation (e.g., of voles, *Microtus*). Some kinds of mammals that usually do not form distinct runways themselves (e.g., deer mice, *Peromyscus*) will frequently use runways formed by others, but other species (e.g., harvest mice, *Reithrodontomys spp.*) apparently disregard runways completely.

The excavations made by individuals of different species of terrestrial mammals are highly variable. Many mammals, such as the hares, *Lepus,* use only shallow depressions, termed **forms,** to rest in. Some species, such as sewellels, *Aplodontia rufa,* and northern pocket gophers, *Thomomys talpoides,* dig elaborate networks of tunnels and nests and storage chambers. Several burrows, each with one or more entrances, may be located within an individual's home range.

The location and shape of the burrow entrance, coupled with a knowledge of the species that potentially inhabit an area, can be used to ascertain the animal responsible. The diameter of the hole places a limit on the size of animal occupying it. Some species leave excavated soil in large mounds at the burrow entrance, whereas others scatter the soil. Individuals of some species dig burrows that enter the ground at slight angles, whereas others dig nearly vertical entrances. Members of some species normally locate burrow entrances at the base of a rock or of vegetation; others select open areas. Many species leave their burrows open, but some place a soil plug in the entrance. Many terrestrial species do not usually dig their own burrows but inhabit those abandoned by other species.

Certain fully fossorial mammals such as moles, pocket gophers, and "mole"-rats have elaborate tunnel systems that may be barely subsurface or very deep. Some moles, Talpidae, usually dig tunnels just below the surface, pushing up serpentine ridges visible above ground, and leaving radially symmetrical eruptions of excavated soil at intervals along the ridges (Fig. 28.4).

Pocket gophers (Geomyidae) and some other burrowing rodents (e.g., *Spalax, Ellobius, Heterocephalus*) excavate tunnels that are several inches underground. The tunnels themselves are not visible at the surface, but mounds of excavated soil indicate their presence. A short tunnel connecting these mounds with the main underground runway system is generally opened to permit removal of excavated soil, to air-condition the burrow system, or to allow the animal to exit and forage aboveground (Fig. 28.5).

28-C Compare illustrations of mole and pocket gopher tunnel systems. What similarities do you see? What differences?

Nests and Dens

Burrowing species usually include nest chambers in their network of tunnels. Frequently, these chambers are lined with dry grass, leaves, or some other cushioning and insulating material. Nonburrowing small mammals may construct nests for retreat, for resting, and/or for the rearing of young. These may be located in small depressions in the ground, in tangled vegetation, in cracks in rocks, or in hollows in trees. Some woodrats or pack rats,

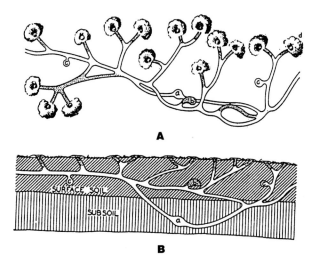

Figure 28.5 Burrow system of pocket gopher (Geomyidae) viewed from above (A) and in diagrammatic section (B). Nest in use (a), old nest filled with dirt (b), chambers (c) for storage of roots and other materials, and (d) one of 15 mounds shown. (Crouch 1933:8)

Neotoma, use twigs, rocks, dried dung, and any other handy material to construct elaborate domed lodges, or dens (Fig. 28.6). Grass-lined and/or fur-lined nests are constructed in chambers within these dens. Beavers, *Castor* (Fig. 28.7) build elaborate houses when appropriate habitat and sufficient construction materials (sticks, limbs, etc.) are available. When these are lacking, they burrow into the side of a pond or into a stream bank.

Arboreal species frequently construct leaf and twig platforms or use cavities in tree trunks. Many tree squirrels inhabit leaf nests during the warm months and move to hollow trees during the winter. Gorillas, orangutans, and chimpanzees (Hominidae) build sleeping platforms in trees for overnight occupancy. The red tree vole, *Arborimus longicaudus,* normally constructs an arboreal nest out of Douglas-fir twigs and resin ducts from the needles. Such a nest may grow to 3 feet in diameter through generations of use.

Hollow trees, caves, protected areas under fallen logs, and other similarly protected spots may be used by mammals as dens for a temporary rest or for hibernation.

Bats may roost in caves, buildings, or hollow trees. Some species select crevices in rocks or spaces under tree bark. Some tropical bats conceal themselves among the leaves of palm trees, and some (e.g., *Artibeus cinereus, Uroderma bilobatum*) modify leaves to construct shelters.

Feeding Residues

Mammals leave many signs of their feeding activities. Species that browse on woody vegetation break twigs and branches and leave conspicuously nipped-off twigs. Rodents of all sizes chew nuts and large seeds and discard the shells. Shrews and rodents eat the soft bodies of

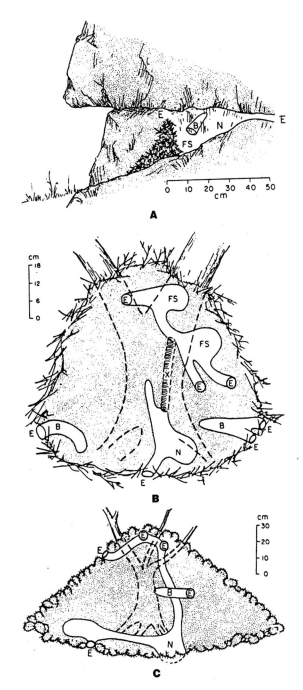

Figure 28.6 Diagrammatic side views of rock (A), stick (B), and cactus (C) houses (dens) made by the desert woodrat, *Neotoma lepida,* Muridae, Sigmodontinae. Blind passageway, B; entrance, E; food cache, FS; nest, N; tree hollow used as passageway, X. Houses may also be made of combinations of these and other materials. (Cameron and Rainey 1972:256, 258, 261)

snails and leave the opened shells. Many species strip bark from trees, and beavers cut down trees for food and construction materials (Fig. 28.7). Hogs, skunks, armadillos, and many other mammals dig and root in the ground for their food and leave evidence of this activity.

Figure 28.7 View of American beaver (*Castor canadensis*, Castoridae) pond and associated sign. (A) Den; (B) feeding sign near shore; (C) trails leading to logging areas; (D) dam; (E) downed tree.
(Henderson 1960:55)

Some bats (e.g., pallid bats, *Antrozous pallidus*) feed upon large insects and/or vertebrates and leave piles of uneaten feathers, wings, and legs under their roosts.

Many mammals, such as squirrels, Sciuridae, woodrats, *Neotoma,* and kangaroo rats, *Dipodomys,* cache quantities of food in or around their nests or dens. Pikas, *Ochotona,* and some mice cut vegetation and allow it to dry. Carnivores may cache their kills by hanging them in trees (e.g., leopard) or by covering them with brush (e.g., tiger) and return to them several times to feed.

28-D   Examine some nuts that have been gnawed by various rodents. How can you tell which were chewed by squirrels and which by mice?

Miscellaneous Sign

Many indications of an animal's presence do not fit into any of the above categories.

Antlers and other bones are frequently well-gnawed by rodents, leaving tooth marks as sign of their presence. Apparently, the animals secure minerals and salts from these objects.

During the fall, male deer in temperate climates lose velvet from their antlers as they rub the antlers against rocks, trees, or brush. Shreds of discarded skin and/or worn spots on the rubbing posts indicate this activity—one that may be partially related to sign-posting. Some carnivores, such as bears and felids, scrape trees with their claws, leaving identifiable marks (Fig. 28.8).

Miscellaneous strands and tufts of hair may be found clinging to brush, barbed-wire fences, or to rocks along a trail where a mammal has brushed against something sharp or rubbed to scratch irritated areas of the body.

Figure 28.8 Tree that has been clawed by a brown bear, *Ursus arctos,* Ursidae.
(Novikov 1956)

Many mammals mark areas within their home range or territory. Zebras and African buffaloes may rub a termite nest until the ground around the nest is bare. European bison may select a particular tree and remove portions of the bark by rubbing it with their horns. Dung heaps and other signposts usually have a territorial function.

28-E   Field exercise: Choose an area with at least two distinct "habitats" (such as an open field and an adjacent wooded area). Mark off two parallel lines 100 meters long and 2 meters apart. These transects should extend into both "habitats." Carefully examine the area between the lines for mammal sign. Plot the location and nature of all sign on a map or piece of graph paper. Using Murie (1954) or a similar reference, identify the sign as precisely as you can. How many kinds of mammals can you now say inhabit the area? Does there seem to be a higher concentration of mammals in any one part of the transect? Where would you set traps to catch the mammals that produced the sign that you have identified?

Habitat Analysis

The **habitat** of a mammal, i.e., where it lives, is defined by geographic, physical, chemical, and biotic characteristics (Brower et al. 1990). Brower and Zar considered the **macrohabitat** to represent the overall habitat of a community of organisms and the **microhabitat** to represent the specific habitat occupied by a population of a given species. In ecological studies, we are most interested in techniques for characterizing microhabitats, but attributes of macrohabitat may also be important in helping to identify areas suitable for reserves, refuges, and parks.

Selection of Samples

The investigator should carefully consider how many samples are needed to characterize the habitat of a particular species. Replicated samples or observations taken at a site allow one to conduct statistical analyses of data. Your instructor or statistical consultant will advise you on the number of samples needed for characterizing a particular habitat. As a rough guideline, plant ecologists recommend that samples include about 10% of a study area to get a reasonable estimate of the vegetation in the area. In the following sections, we describe techniques for analyzing soil samples and for measuring vegetation using line transects and quadrats.

Analysis of Soils

Fossorial species of mammals are specialized for living most of their lives underground. One would expect that different species of fossorial mammals might exhibit preferences for different soil types. Different species of pocket gophers (Geomyidae), for example, are known to live in different microhabitats characterized by soils with extremely small particle sizes (clays) to large particle sizes (sands and small pebbles).

Soil samples can be taken to analyze soils for particle size distribution, organic matter, or nutrients. Soil scientists of the U.S. Department of Agriculture use particle size as one of the factors in determining soil type. Table 28.1 shows the USDA System for classifying soil particle sizes. Soil laboratories most often use hydrometer techniques to estimate soil particle sizes (see procedures described in Cox 1996 and Brower et al. 1990). From these analyses, the USDA System groups soils into eleven basic types: *sand, loamy sand, sandy loam, sandy clay loam, loam, sandy clay, clay loam, silt loam, silt, silty clay loam,* and *clay.*

TABLE 28.1 Size Categories of Soil Particles Based on U.S. Department of Agriculture System

Category	Particle Diameter in mm	μm
Clay	< 0.002	< 2
Silt	0.002–0.05	2–50
Very fine sand	0.05–0.10	50–100
Fine sand	0.10–0.25	100–250
Medium sand	0.25–0.50	250–500
Coarse sand	0.5–1.0	500–1,000
Very coarse sand	1.0–2.0	1,000–2,000

Adapted from Brower et al. (1990:45).

A qualitative technique can also be used to characterize soils based on how well a squeezed lump of soil remains intact in the hand after the grip on the sample is released. Soils with a high sand content (e.g., *sand, loamy sand,* and *sandy loam soils*) will quickly fall apart and not remain as a clod. In contrast, a lump of soil with a high clay content (e.g., *clay* and *clay loam soils*) will generally remain as clods after the hand pressure is released.

28-F   Examine several soil samples on display and make a clod in your hand of each type. What is the feel of these soils? Do the soils break apart when your grip on the clod is released?

Vegetation Line Transects

Line transects can be used to get quantitative estimates of plant density, of cover area, and frequency and can produce results comparable to those obtained by quadrat sampling techniques (Cox 1996). Line transects are often used in studies of mammals, in which quantitive data are needed to characterize plant resources for diet or resource analyses. In these studies, separate line transects are used for each vegetation stratum (Fig. 28.9) in the community because different species of mammals may occupy different strata in the plant community.

In line-transect sampling, a 30- to 50-meter tape is extended along a line that is determined by a random procedure (e.g., use of a table of random numbers to obtain the compass bearing to orient the line). The tape is secured at one end (using a spike or screwdriver) and then threaded through the habitat in a *straight line*. With a metric tape, the centimeter (cm) marks provide a convenient way to subdivide the tape for recording purposes and make calculations easy. For example, if a plant touches the line, lies above the line, or lies below the line, then that interval is part of the **line intercept** for that individual plant or clump of plants. You should also record the intercept lengths for intervals that have no plant cover, in order to get an estimate of the extent of bare ground.

In Figure 28.10A, four species of plants are depicted. In intervals "b" and "d," the *crown* of the plant is projected to the tape to determine the line-intercept length. In intervals "a" and "c," the *basal* area is shown projected to the transect line. You should indicate if you used the "basal" or the "crown" technique to record line-transect lengths.

In Figure 28.10B, three species of plants (labeled a, b, and c) are depicted. In this example, measurements were taken on crown intercepts, with data recorded to the nearest centimeter (cm). Thus, species "a" has an intercept length of 3 cm (25–27 cm), "b" has an intercept length of 4 cm (20–23 cm), and "c" has an intercept length of 1 cm (plant touches line only in the 24-cm interval). On the data sheet (Table 28.2), record the exact

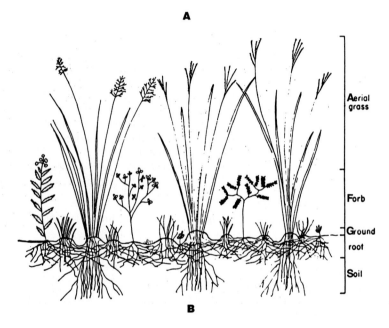

Figure 28.9 Stratification in plant communities. (A) Mixed deciduous forest; (B) prairie plants.
(Brower et al. 1998:37)

interval lengths (e.g., 24–27) for each species [or group of plants (e.g., forbs or grasses)]. With the exact interval lengths recorded, you can look for areas of bare ground. In this example, bare ground between species "a" and "b" would be a 1-cm section in the interval 23–24. The procedure described above can be used to calculate **cover** (or **dominance**) or to calculate **relative dominance** values as a percentage of the ground surface intercepted, using the following formulas:

$$\text{cover} = \frac{\text{total of intercept lengths for a species}}{\text{total transect length}} \times 100$$

$$\text{relative dominance} = \frac{\text{total of intercept lengths for a species}}{\text{total of intercept lengths for all species}} \times 100$$

28-G Use a random procedure to extend a 10-meter tape in a straight line along the ground. Record the line-intercept lengths for all species of forbs, sedges, and grasses along this line. Compute the cover (dominance) and relative dominance values for at least three species of forbs. Which species of forb was dominant?

To record density and relative density of plants, you must also record the maximum width of the plant on a horizontal line perpendicular to the transect. Cox (1996) and Brower et al. (1990) provided additional formulas and instructions for calculating these other vegetation indices for plant communities.

Vegetation Quadrats

Vegetation **quadrats** are plots of ground defined by rectangular or circular frames, or, when large quadrats must be established, by staking out an area on the ground. The larger quadrats are most often used for shrub and tree

Sign and Habitat Analysis   187

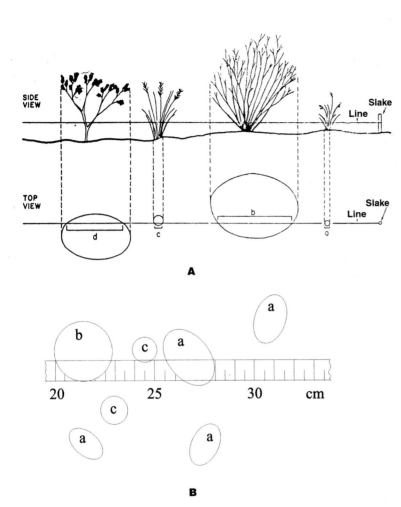

Figure 28.10 Technique for determining vegetation line-intercept lengths. (A) Example of *basal area* (c and a) and *crown area* (d and b) intercepts; (B) example of intercept lengths using the crown area method. Note that in (B) the intercept length for species "a" is the interval 25.5–28 cm for a total of 2.5 cm; the area of bare ground between species "b" and "c" is only the interval 23–24 for a total of 1 cm.
(A, Brower et al. 1998:98; B, R.E. Martin)

TABLE 28.2	Example of Data Sheet for Recording Vegetation Line-Transect Data*

Line-Transect Data

Locality: _____
Date: _____ Recorder(s): _____
Length of Transect: _____ meters   Interval Length: _____   Intercept Method (circle): BASAL CROWN
Stratum (circle): GROUND HERB SHRUB TREE   Station/Grid Point: _____

Plant/Clump Number	Species: Intercept Length	Species: Intercept Length	Species: Intercept Length	Species: Intercept Length	Species: Intercept Length

Page ____ of ____

*Data from each sheet are transferred to a summary sheet to get totals for each species.

TABLE 28.3	Example of Data Sheet for Recording Quadrat Vegetation Data Data from Each Sheet are Transferred to a Summary Sheet to Get Totals for each Species.

Quadrat Data

Locality: _____

Date: _____ Recorder(s): _____

Area of Quadrat: _____ Units of Area: _____ Area Method (circle): BASAL CROWN

Stratum (circle):   GROUND   HERB   SHRUB   TREE   Station/Grid Point: _____

Plant/Clump Number	Species: Area	Species: Area	Species: Area	Species: Area	Species: Area

Page _____ of _____

---

species, whereas the frame technique is used when it is possible to position the frame over the plants to be measured (Cox 1996). This technique can be more time-consuming than the line-transect technique because it requires that the area of each individual plant be measured within a quadrat (Table 28.3). For grasses, the measurement of the individuals of each species is often impractical with this technique. Thus, a measurement of the area covered by a class of plants (e.g., all of the forbs, or all of the grasses) might be determined. Alternatively, Cox (1996) described a semiquantitive technique for using cover classes (Table 28.4) to estimate the percent ground cover.

TABLE 28.4	Method for Ranking Vegetation Using Cover Classes	
Cover Class	Range of Percent Cover	Midpoint of Percent Cover
1	0–5	0.5
2	1–5	3.0
3	5–25	15.0
4	25–50	37.5
5	50–75	62.5
6	75–100	87.5

After Cox (1996:90).

You can calculate several density and dominance values of plants using the following formulas:

density = number of individuals/area sampled

relative density = (density for all species/total density for a species) × 100

dominance = total area covered by plant/area sampled

relative dominance = (dominance of a species/total dominance for all species) × 100

28-H Position a 1-meter square frame over a plot of ground that has been located using a random technique. Count the number of individual forbs in the quadrat and determine the area occupied by each. Use these data to calculate the density and dominance for each species of forb. What are the densities and relative densities of the forbs? What are the dominance and relative dominance values for the different species of forbs?

CHAPTER 29

RECORDING DATA

Field observations yield valuable data on mammal distribution, abundance, habitat tolerances, feeding habits, reproductive cycles, behavior, etc. In order to record these observations for later reference, it is essential to develop good note-taking practices. Accurate and complete field notes will be of value to the observer and to other scientists at later dates. With the passage of time, environmental conditions change, and old field notes can be valuable sources for documenting and analyzing changes.

Primary data refer to original notes, observations, or measurements made on a specimen, live animal, or habitat. These data can be recorded in field notes, tally sheets, specimen tags, tape recordings, or computer media. **Secondary data** refer to any of these observations after they have been subsequently transcribed (e.g., rewritten) or transferred to another medium (e.g., entered into a computerized database). When secondary data are utilized, the researcher must *proofread* the secondary data file against the primary data to insure that transcription of the primary data has been done accurately.

The format of the field notes should be flexible and permit changes to fit a given situation. A suggested method for recording data follows, with various examples of notes and labels. Additional suggestions can be found in Hall (1962) and Mosby (1969a:61–72; 1980:45–54).

EQUIPMENT

Paper and Notebook

Good quality white paper (of neutral or slightly alkaline pH) should be used for notes. A bond paper having a rag content of 50% to 100% is preferable. This paper is durable and will resist water damage and deterioration due to age. Sulfide papers (e.g., computer printout and most lightweight notepaper) should *not* be used because they are highly susceptible to deterioration and water damage. Paper with dimensions of 6 1/4" × 8 1/2" is of a convenient size to work with.

We recommend that the paper be placed in a sturdy three-ring binder or notebook. Index dividers may be used to separate different sections of the field notes. The use of a ring binder has advantages and disadvantages. With the binder, it is easy to lose a few sheets of paper and thus render your field notes incomplete. This possibility could be avoided by using a prebound volume of blank pages, as is often done on extended field expeditions. A danger to the field notebook is loss or damage in the field. Thus, many field-workers use two notebooks. One contains past field notes and is stored in a safe place in the office, whereas the other, containing only current notes, is taken into the field. At the end of each portion of fieldwork, the notes are removed from the field notebook and inserted into the permanent one. This system, impossible with a prebound volume, frees the bulk of the data from the chance of loss or destruction in the field.

Inks and Pens

A *black waterproof ink* should be used for writing field notes and data on specimen tags. The ink must also be resistant to alcohol and formalin, to grease and other animal fluids, and to the ammonia or detergent frequently used in cleaning skeletal material. Refer to Williams and Hawks (1986) for information on suitable inks. Ballpoint, nylon- or felt-tipped pens (or washable fountain pen inks), should *never* be used because these pens rarely use permanent ink or may have broad tips that bleed through the paper. If permanent ink is not available, a No. 2 or 2 1/2 pencil can be substituted until a permanent-ink pen is available.

Permanent inks will quickly clog most fountain pens. The Rapidograph made by Koh-I-Noor and similar pens made by Pelikan, Castell, and other companies are among the few pens that will work with these heavy inks. Although less convenient (but more economical), a staff pen and point may be used. Some disposable pens (e.g., Uniball Deluxe) have fine tips and permanent ink and are suitable for writing on specimen tags. Some fine-tip permanent-ink felt pens are suitable for making marks on tubes to be placed in liquid nitrogen containers. *You should always verify that the ink in any pen used for writing data and specimen labels is permanent by independent testing under "real world" conditions.*

Specimen Labels

Labels for study skins, for tanned skins, skulls, and entire skeletons should be made of 100% rag stock white paper and should be of sufficient thickness to allow string to be attached without danger of tearing the paper. Labels for fluid preparations (entire animals, organs, etc.) must be made of heavy weight, 100% rag stock.

FIELD NOTES

Some mammalogists and other vertebrate biologists organize their field notes into three sections: (1) **journal,** (2) **catalog,** and (3) **species accounts.** The catalog, a numbered listing of all specimens preserved, is essential if any capture and preservation of mammals is done. The journal, a field diary of all activities and observations, is highly recommended for all types of fieldwork. The species account section includes detailed observations on particular species. Special data-recording forms (see: "Special Data Forms") may be filed separately or included with the journal and species accounts sections.

Journal

The field journal is a complete, chronological record of the activities and observations of an investigator. The what, when, where, who, and how of all fieldwork should be recorded here. Results of the work should be described, along with supplemental information on habitat, general impressions of mammal populations, conversations with residents of the area, and any additional information that is potentially helpful.

The name of the investigator and the year should be recorded at the top of each page. Number the pages consecutively beginning with 1. Record the exact locality and date of each observation or account. The locality should list country, state, county, and miles (or kilometers) and direction from a permanent map feature (e.g., town, city, mountain). The date should be written out fully (e.g., 14 July 2000). Figure 29.1 is an example of a journal page. Methods of recording location, date, and other data follow.

K. G. MaTocha
1968
Journal

December 23. Texas: Kleberg County, 1 1/2 miles West Loyola Beach on F.M. 628. -- Set 19 Sherman traps and 1 live can trap. Omaline & lab. chow as bait. Set about 5:00 p.m. Along ditch with some water. Short grass along edges. -- With Rick McDaniel
10:00 p.m. Check traps -- Took:
1- Peromyscus leucopus (escaped)
3- Oryzomys sp.
1- Sigmodon hispidus
Reset traps that contained animals.

December 24. Picked up traps at 10:30 a.m. -- Took:
1- Oryzomys sp.
3- Mus musculus (2♂, 1♀)
2- Peromyscus leucopus (2♂)
2- Peromyscus maniculatus (1♂, 1♀)

December 27. Texas: Kleberg County, -- Two trap sites:
#1 -- 0.9 mi. W. Loyola Beach on F.M. 628
Set along ditch (with water), edge of road.
5:30 p.m. Total of 23 traps as follows:
10 - Sherman -- Omaline as bait
13 - Museum Specials -- Peanut butter as bait.
#2 -- 1.5 miles North of F.M. 628 on F.M. 772.
Much Chloris and Cenchrus in the area. Set at 5:45 p.m.

Figure 29.1 Sample field journal page. Original size 6 1/4 × 9 inches.
(K. G. Matocha)

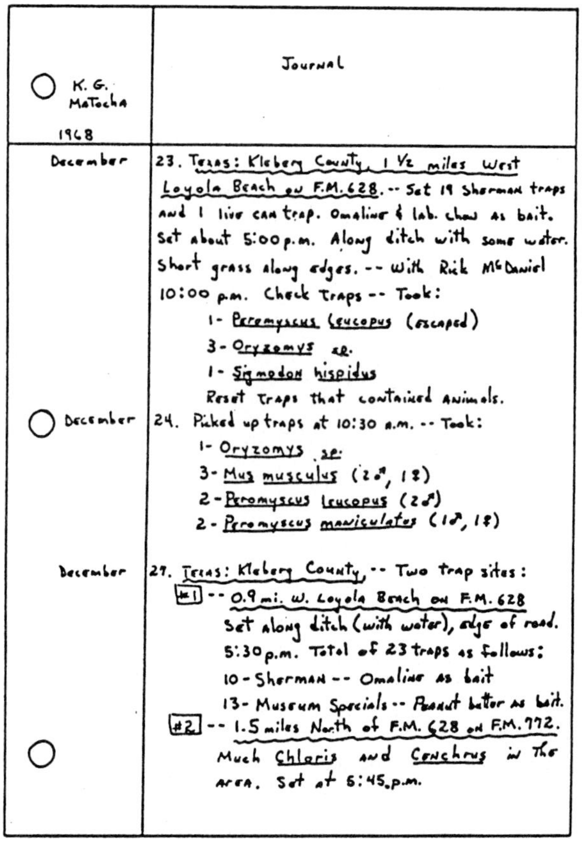

Catalog

The catalog (Fig. 29.2) is a record of all specimens that are preserved in any manner. Each specimen is assigned a number that is associated with the name of the collector. If you have never captured and preserved animals previously, your first entry will be designated number 1. Throughout your life, *never repeat a number* once it has been used for an animal.

Species Accounts

This section of the field notes can be very useful when many observations are being recorded on a single species. Information on a particular species is called a species account. List the scientific name and common name of the animal at the top of the page. Give the date and location of the observation. As time permits, record

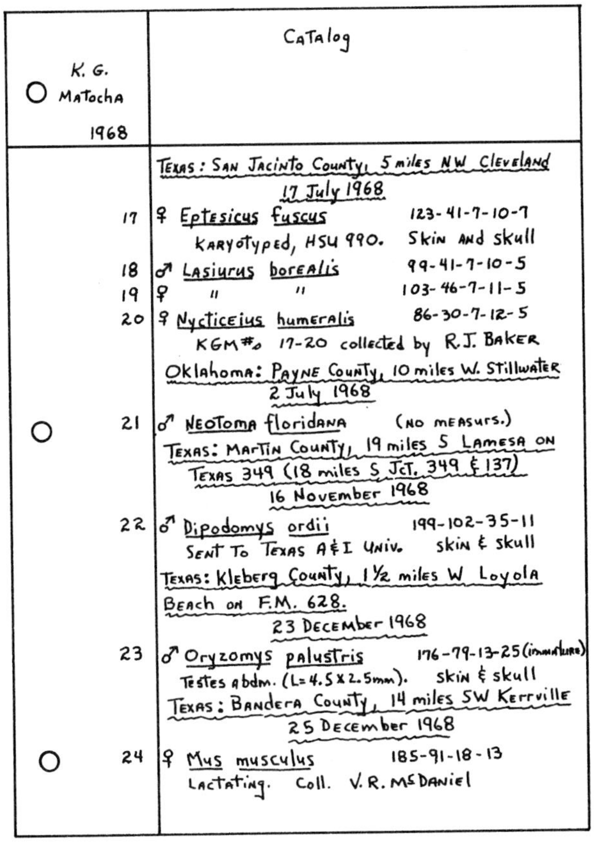

K. G.
MaTocha
1968

CaTalog

Texas: San Jacinto County, 5 miles NW Cleveland
17 July 1968
17 ♀ Eptesicus fuscus 123-41-7-10-7
Karyotyped, HSU 990. Skin and skull
18 ♂ Lasiurus borealis 99-41-7-10-5
19 ♀ " " 103-46-7-11-5
20 ♀ Nycticeius humeralis 86-30-7-12-5
KGM#s 17-20 collected by R.J. Baker
Oklahoma: Payne County, 10 miles W. Stillwater
2 July 1968
21 ♂ Neotoma floridana (no measurs.)
Texas: Martin County, 19 miles S Lamesa on
Texas 349 (18 miles S Jct. 349 & 137)
16 November 1968
22 ♂ Dipodomys ordii 199-102-35-11
Sent To Texas A&I Univ. skin & skull
Texas: Kleberg County, 1½ miles W Loyola
Beach on F.M. 628.
23 December 1968
23 ♂ Oryzomys palustris 176-79-13-25 (immature)
Testes abdm. (L=4.5×2.5mm). skin & skull
Texas: Bandera County, 14 miles SW Kerrville
25 December 1968
24 ♀ Mus musculus 185-91-18-13
Lactating. Coll. V.R. McDaniel

Figure 29.2 Sample field catalog page. Original size 6 1/4 × 9 inches.
(K. G. Matocha)

observations made on the species, even those that may seem insignificant at the time. These additional observations may prove valuable in later analyses of data. Reference can be made to pages in the journal for additional information. Figure 29.3 is an example of a species account.

Locality

Geographic Information Systems, Maps, and Map Scales

August (1993) provided a very useful and detailed discussion of how to create accurate locality data for use in geographic information systems (GIS). His account is must reading for anyone interested in creating accurate geographic data and using those data in GIS applications. Manufacturers of GIS software, global positioning system (GPS) receivers, state agencies, and universities also have vast quantities of information available. Using search terms such as *GIS, GPS,* or *mapping,* will help you find this material. The sections that follow and the paper by August (1993) will also help guide you in the search for information and data.

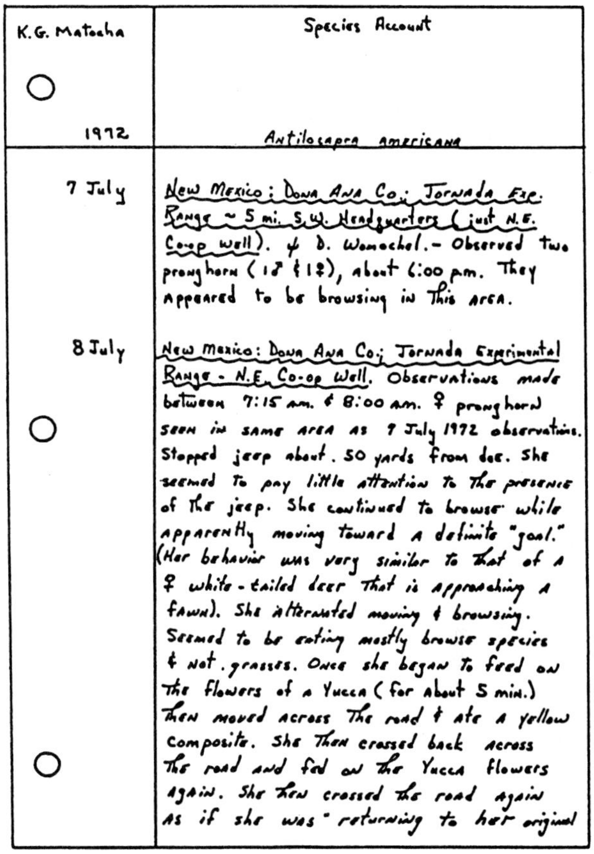

K.G. Matocha
1972

Species Account

Antilocapra americana

7 July New Mexico: Dona Ana Co.; Jornada Exp. Range ~ 5 mi. S.W. Headquarters (just N.E. Co-op well). & D. Womochel.- Observed two pronghorn (1♂ & 1♀), about 6:00 p.m. They appeared to be browsing in this area.

8 July New Mexico: Dona Ana Co.; Jornada Experimental Range - N.E. Co-op Well. Observations made between 7:15 a.m. & 8:00 a.m. ♀ pronghorn seen in same area as 7 July 1972 observations. Stopped jeep about 50 yards from doe. She seemed to pay little attention to the presence of the jeep. She continued to browse while apparently moving toward a definite "goal." (Her behavior was very similar to that of a ♀ white-tailed deer that is approaching a fawn). She alternated moving & browsing. Seemed to be eating mostly browse species & not grasses. Once she began to feed on the flowers of a Yucca (for about 5 min.) then moved across the road & ate a yellow composite. She then crossed back across the road and fed on the Yucca flowers again. She then crossed the road again as if she was "returning to her original

Figure 29.3 Sample species account page. Original size 6 1/4 × 9 inches.
(K. G. Matocha)

Accurate determination of locality depends on an understanding of maps and map scale. Typically, the scale of a map is determined by the representative fraction. The **representative fraction** of a map is the ratio between the map distance and the ground distance between equivalent points (Campbell 1998). The representative fraction is independent of units, although typically maps are designed to be multiples of either metric or English measuring units. For example, the standard quadrangle map (quad) in the United States is the 7 1/2 minute series that has a representative fraction of 1:24,000. For this **large-scale map,** 1 inch on the map is equal to 24,000 inches on the ground or 2,000 feet; in a 1:50,000 scale map, 1 mm on the map is equivalent to 50,000 mm (50 meters) on the ground. The larger the denominator of the representative fraction, the less detail will be present on the map. On a typical 7 1/2 minute quadrangle map, one can theoretically plot a location to about ± 31.25 feet (calculated by dividing 2,000 feet by 64, where the 64 represents the denominator of a 1/64 inch fine pencil line). On a 1:250,000 **small-scale map,** that same1/64 inch line could mark a position accurate to only about ± 325.52

feet (calculated by dividing 20,833.33 feet by 64). United States Geological Survey map standards state that a given point has only a positional accuracy of ± 40 feet on a 1:24,000 scale map, and a positional accuracy of ± 443 feet on a 1:250,000 scale map (Ian Martin, *in litt.*). Scale on published maps is often indicated by a graphic scale drawn on the map. A graphic scale on a map has an advantage when the map is enlarged or reduced because the scale will always be correct no matter what the ultimate size of the published map.

Large-scale (e.g., 7 1/2 minute quads) and small-scale (e.g., 1:100,000 or 1:250,000) maps are available in the United States from the U.S. Geological Survey, special depository libraries, or private vendors. All of these maps are available in digital form (usually distributed on CDs) from private vendors (e.g., TopoDepot, SureMaps, or LandInfo) or, in the small-scale versions, from federal and state agencies. You can find these federal and state agencies and the private companies by Internet searches or by checking the appendices in references such as Campbell (1998).

Universal Transverse Mercator, State Plane, and Latitude/Longitude

In the United States, standard quadrangle topographic maps (7 1/2 minute) have important reference marks (called **tics**) positioned along the neat line of the map. The **neat line** of the map is the line that forms the boundary of the map. This neat line is inset from the paper boundary of the map, and at each corner of the neat line you will see the latitude and longitude values for that corner. The 7 1/2 minute quads in the United States are referenced to the **1927 North American Datum (NAD 27)**. The quad maps updated in recent years also show position marks for the **1983 North American Datum (NAD 83)**, which is a geographically more accurate datum (August 1993; Campbell 1998). Position data derived from the NAD 27 maps can be corrected to the 1983 datum by using routines in GIS software *or* by setting a GPS receiver to receive data in the updated datum.

As you move away from the corners of the map, you will see three different tic marks that are positioned at right angles to the neat line of the map. These tic marks will specify the **latitude/longitude** positions, the **state plane coordinates,** and the **Universal Transverse Mercator (UTM) coordinates**. (Some older quad maps may not have UTM coordinates.) These marks are easy to recognize once you understand the labeling system, as follows:

State plane coordinate values: Black lines that cross the neat line on either side. *These lines are 10,000 feet apart* and *show all digits* of the state plane values. These numbers may or may not have feet associated with the printed number.

UTM coordinate values: Blue lines that also cross the neat line. *These lines are 1,000 meters apart.* The numbers for the UTM values *will not show the last three zeros* in the number and will have an "m" for meters at the end of the written number. Newer 7 1/2 minute quads and 1:100,000 scale maps will have the 1,000 × 1,000 UTM grid superimposed onto the topographic maps, which makes alignment and measurement much more precise.

Latitude and longitude coordinate values: These values are indicated by the corners of the map and by *black lines at 2.5´ intervals* that end at the neat line and do not go into the white space beyond the neat line. Because the distance between successive minutes of longitude depends on one's location on earth, there is no set distance between a minute of longitude for all maps. For example, at the equator (0 latitude), a degree of longitude is 111.319 kilometers (km), while at a latitude of 35°, a degree of longitude would be only 91.288 km (Campbell 1998). There is also a very slight difference between successive minutes of latitude due to the flattening of the earth at the polar regions (Campbell 1998).

Latitude and longitude values are often converted to decimal degrees to avoid the complexity of dealing with degrees, minutes, and seconds and to facilitate processing of data by GIS software. For example, in decimal degrees, a latitude of 35 degrees, 30 minutes, and 15 seconds would be 35.504167 and would be calculated as follows:

decimal degrees	35.000000
decimal minutes	0.500000 [30/60]
decimal seconds	0.004167 [15/3600]
Sum of values =	35.504167 decimal degrees

Once you have found your location on a map, the appropriate tic marks on the neat line of the map can be used to interpolate your position with respect to a given coordinate system. Use a straight edge, triangle, and a finely sharpened pencil to make the north/south line (the X-values or "eastings") and east/west lines (the Y-values or "northings") on the map. Then, measure from these map lines to the locality and either subtract or add the appropriate distances to get the coordinate values that are closest to the position you are trying to describe. (Fig. 29.3 shows the procedure for UTM coordinates.) Be sure to align the ruler carefully because even a slight movement of the ruler can introduce hundreds of feet of error into the final result. In the map legend, you will see information about the correct zone for either the state plane coordinates or the UTM coordinates. These zones must be recorded along with the appropriate coordinates to correctly locate the position on the face of the earth. The zone information is also required in GIS software when digital map positions are converted to a different coordinate system (e.g., projecting latitude and longitude values to state plane coordinates). Also make note of the datum (e.g., NAD 27 or NAD 83) if coordinate conversions are anticipated.

Recording Data 193

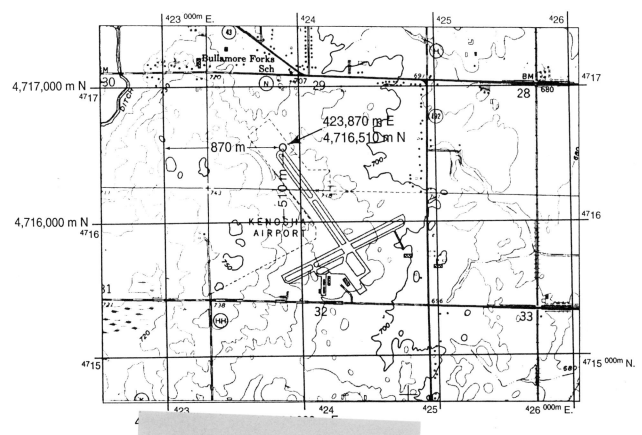

Figure 29.4 Use of a U[TM grid to locate a feature on a t]opographic quadrangle map, Pleasant Prairie, Wisconsin, 1:24,000 scale[. ...] shown drawn on the map, and the located feature (the Kenosha Airport) [... po]sition would be 4,716,510 m N, 423,870 m E, Northern Hemisphere, Zo[ne ...]
(Campbell 1998:57)

In Figure 29.4, th[e ...] Hockley County; 3 m N Ropesville or end of the Kenosha (W[... ...] Hardeman County; 2 mi. N, 3 mi. E Quanah the arrow. The coordin[ates ...] ions in countries other than the United m N, and 423,870 m E [...] untry, political subdivision (e.g., department This information, alon[g ...]ce), and kilometers (or miles) from a also be part of the loc[ality ...] ninent map feature. or specimen tag. The [...] be sufficient for use b[y ...] ars Province; 15 km N Shiraz

[...pr]ecise locality data are valuable and should [... ...] when possible (see "Other Coordinate Systems" section that follows). Geographic coordinates (latitude and longitude, accurate to minutes and seconds; UTM coordinates), legal land descriptions, and elevations provide the most accurate and lasting data.

NEW MEXICO: Doña Ana County; south edge of Red Lake, T 18S, R 1E, SW 1/4 of Section 27

or

NEW MEXICO: Doña Ana County; south edge of Red Lake, 32°42´ N, 107°10´ W

In decimal degrees, the latitude and longitude values for the New Mexico example would be 32.7000000 N,

Typical Locality Descriptions

As described by Williams et al. (1977), there are two principal systems for written locality descriptions: *specific to general* and *general to specific*. Various institutions have adopted standard ways for locality descriptions to be written on specimen labels. In this manual, we give examples of the general to specific scheme because, among other things, it clearly has a distinct advantage for retrieval using data-processing techniques. Data for localities in the United States should include *state, county* (or *parish*), and *miles* (or *kilometers*) and *direction* from a recorded permanent map feature such as a courthouse (or a principal intersection in smaller towns).

107.1666667 W, although this implies more precision than is actually present in the original data.

Sometimes the reference landmark may be in one county (or other political subdivision), while the specimen was actually taken in an adjacent county. Always list the county in which the specimen was taken.

OKLAHOMA: Harper County; 3 miles N Fort Supply (Note: Fort Supply is a town in Woodward County, Oklahoma)

We recommend that only cardinal compass directions (N, S, E, W) be utilized and that N or S precede E or W, where appropriate. A direction given as NE is imprecise and difficult to pinpoint (i.e., is the specified NE direction *exactly* 45 degrees from the N-S axis?). In addition, one should avoid use of road junctions or railroad intersections as locality descriptors for specimens. If used, they should be listed as additional comments in the journal.

CHILE: Santiago Province; 1 km N, 0.5 km E Cerro Manquehue

Other Coordinate Systems

Crawford (1983) proposed using a grid system for more accurately specifying specimen localities. Under this system, a specimen taken in North America can be pinpointed to about 111 meters by using decimal degrees (latitude and longitude with minutes accurate to 0.001). Hamaker and Koeppl (1984) described a method for transferring points on a map to geographic coordinates. These coordinates could then be used to prepare an updated map on the same or another projection. Geographic Information Systems (GIS) is a multidisciplinary field that links geographic coordinate data with an information database of attributes. Mammalogists are advised to collect geographic data in a manner that is suitable for use in GIS programs. See McLaren and Braun (1993) for additional information on the use of GIS in mammal studies. Also, refer to the "Coordinate Data from Global Positioning Receivers" and the "Universal Transverse Mercator, State Plane, and Latitude/Longitude" sections in this chapter for methods to generate the most accurate locality data and make these data suitable for use in a GIS.

Coordinate Data from Global Positioning Receivers

Several systems are available worldwide to receive coordinate data from global positioning satellites. In the United States, the Department of Defense has a system of 24 or more satellites that send position data by radio signals. On the ground or in the air, civilian **global positioning system (GPS)** receivers provide very accurate coordinate data *if care is taken to correct the data for the effects of selective availability.* **Selective availability (SA)** is an intentional random error introduced into the GPS signal by the U.S. Department of Defense so that the position data on civilian GPS receivers may be off by ± 100 meters (± ca. 300 feet). At some locations, this error in the position can be corrected by use of **differential correction,** which uses data from a known location, contemporaneous in time with the GPS receiver (the **rover**) data, to correct the coordinate data collected by the rover.

Through purchase of a subscription, differential data can be received in *real time* by the GPS rover through a radio signal received from satellites (e.g., *Omnistar*) or from radio signals received from certain FM stations (Differential Corrections, Inc.). Close to the coast and along navigable rivers, the U.S. Coast Guard provides radio signals with differential data that can be picked up at no cost by radios and transmitted to the GPS receiver for real-time correction. Alternatively, one can *postprocess* the rover data using differential data from a known location. Often, manufacturers of GPS equipment have links to GPS reference station data that can be accessed on the Internet and downloaded to your computer for postprocessing. Universities and state agencies (e.g., Texas Department of Transportation) also provide differential data on the Internet to be used for postprocessing.

Low-cost GPS receivers generally do not have the capability for postprocessing of coordinate position data. Thus, unless you are able to receive a radio link to have these rover data processed in real time by a radio link, then you must be satisfied with coordinate data with an error of ±100 meters. Even these position data can be better than locality coordinates determined using odometers on a car (typically ±528 feet when read to 0.1 mile) or localities determined using typical small-scale highway maps or other small-scale maps with scales of 1:100,000 or 1:250,000.

DATE

The date of collection should be written out completely. Do not write "4/1/00" because this may be interpreted as either "1 April 2000" or "4 January 2000." Likewise, never use only the last two digits (e.g., "00") for the year. Because collections have lasted and should last for hundreds of years, the above date could in the future be interpreted as "1 April 1800" or "1 April 1900." The recommended method for writing the date is "1 June 2000" or "15 February 2000" because placing the numeral before the month eliminates the necessity of a comma. If the specimen is taken alive and dies or is killed at a later date, the date of death and the date when captured should both be recorded.

MEASUREMENTS, WEIGHT, AND SEX

Mammal specimens should be measured prior to preparation. The standard measurements for a mammal are always listed in the following order: (1) *total length,* (2) *tail length,* (3) *hindfoot length,* (4) *ear length,* and in bats,

(5) *tragus height*, and (6) *forearm length*. The measurements are taken in millimeters, and each measurement is separated by hyphens because this punctuation mark is unlikely to be interpreted as a numeral. The weight is recorded in metric units: grams or kilograms.

The sex should be recorded using the symbols ♂ for male and ♀ for female. Write "sex?" if the sex cannot be determined. Indicate immature, juvenile, or subadult if one of these terms is appropriate.

Refer to Chapter 31 for the techniques of measuring, weighing, and sexing mammals.

REPRODUCTIVE CONDITION

If the specimen is a male, measure and record the length (exclusive of epididymis) and breadth of a testis. For example, if the testis measured 15 mm in length and 7 mm in breadth, the notation on the tag might read **Testis 15 × 7 mm.** If the species is one in which the testes descend seasonally, record their position, such as **testis descended** or **scrotal,** or **testis abdominal** or **nonscrotal.** If the specimen is a female, check the uterus for embryos, and record your observations. If embryos are present, record their number, their locations in the uterus, and their crown-to-rump length (CR) in millimeters. For example, if a female specimen had six 15-mm embryos, of which two were in the right uterine horn and four in the left uterine horn, then the abbreviated notation on the label might read: 6 Embs. = 15 mm CR, 2R, 4L. If the female is lactating, make note of this. If embryos or any portion of the reproductive tract are preserved for later examination, attach a label with the collector's initials and catalog number of the female and note the type of preservation (see Table 29.1) in the catalog. Refer to Chapter 31 or to Taber (1969), Larson and Taber (1980), or Brown and Stoddart (1977) for additional information on ascertaining the reproductive condition of a specimen and for precautions to be observed when doing the necropsy.

PARASITES

In epidemiological surveys, extensive records are kept on individual hosts. An ectoparasite survey data sheet utilized by personnel of the Division of Mammals, Smithsonian Institution, is shown in Figure 29.5. One data sheet was completed for each host captured and sampled for ectoparasites. See Chapter 32 for additional information on collecting and preserving ectoparasites.

PORTIONS PRESERVED (TYPE OF PRESERVATION)

The usual mammal specimen consists of a study skin and skull. If the postcranial skeleton, baculum, embryos, stom-

TABLE 29.1 Types of Preservation Often Found in Collections of Mammals*

Code	Definition
AL	Alcoholic
SS	Skin and skull
SB	Skin, skull, and body skeleton
SN	Complete skeleton
SK	Skull only
SO	Skin only
SA	Alcoholic with skull removed
KB	Skin and body skeleton (no skull)
AN	Anatomical
PS	Partial skeleton
CO	Cranium only
HM	Head mount
BM	Body mount
SC	Skin, skull, and alcoholic carcass
BS	Body skeleton
OT	Other, with explanation in comments
HO	Horn(s) only
AO	Antler(s) only
BO	Baculum only
MO	Mandible only
TH	Tooth(teeth) only
TK	Tusk(s) only
SM	Skin, skull, and baculum

*Refer to American Society of Mammalogists Committee on Information Retrieval (1996) for additional information.

ach contents, or any other portions of the specimen are preserved, these should be noted in the catalog and on the skin tag (if any). If ecto- and/or endoparasites are preserved, these should be mentioned. If the specimen is preserved in liquid or in any manner other than a standard study skin and skull, note this fact. If either the skin or skull is badly damaged, or if, for some reason, it is not preserved, then record exactly what is included as a specimen.

Standard types of preservation (Table 29.1) recognized by the Committee for Information Retrieval of the American Society of Mammalogists can be found in Williams et al. (1979) and American Society of Mammalogists Committee on Information Retrieval (1996). Explain the nature of the specimen if it does not fit one of the standard categories.

METHODS OF COLLECTION

A brief mention should be made of the method used to secure the specimen. "Snap-trapped" or "shot" are examples of adequate catalog or label descriptions, but more detailed notes should be made in the journal. The abbreviation **DOR** is frequently used to designate an animal found dead on the road. If the animal is caught in a baited trap, it is useful to name the bait used.

(1-10) ECTOPARASITES
(1)
(2) Mites
(3) Ticks
(4) Siphonaptera
(5) Diptera
(6) Anoplura
(7) Mallophaga
(8) Hemiptera
(9) Coleoptera
(10) Other

PARASITE LOCATION (Notes)

(11) TIME
0 Day
1 Night
2 Morning
3 Afternoon
4 1800
5 1900
6 2000
7 2100
8 2200
9 2300

(12) TIME
0 2400
1 0100
2 0200
3 0300
4 0400
5 0500
6 0600
7
8
9

(13) SKY
0 Clear
1 Partly Cloudy
2 Overcast
3 Fog
4 Mist
5 Showers
6 Steady Rain
7 Heavy Rain
8
9

(14) MOON
0 New
1 1st Qr
2 Full
3 3d Qr
4 None

(17) WIND
0 None
1 Light Breeze
2 Gusts
3 Strong Wind
4 Gale

(15-16) SPECIFIC LOCALITY

(18-19) Temp, Max

(20-21) Temp, Min

(22-25) Latitude

(26-29) Longitude

(30-33) Elevation (Meters)

(34-36) No. Trap Nights or Net Hours

(37) TOPOGRAPHY
0 Plain
1 Valley
2 Ridge Side
3 Ridge Top
4 Undulating

(38) COVER
0 Deciduous Forest
1 Evergreen Forest
2 Cloud Forest
3 Nam Forest
4 Swamp or Marsh
5 Savannah or Pasture
6 Paramo
7 Cropland
8 Orchard
9 Yard

(39) FOREST SUCCESSION
0 Mature
1 Scrub

(40) SITE MOISTURE
0 Dry
1 Moist
2 Near Stream
3 Beside Stream
4 In Stream

(41) CAPTURE SITE
0 Ground
1 Log
2 On Log
3 Rock
4 Base of Tree
5 In Tree
6 Cavern
7 House
8 Fence or Wall
9

(42) COLLECTING DEVICE
0 Mist Net
1 Fish Net
2 Live Trap
3 Snap Trap
4 Steel Trap
5 Gun
6 Hand Caught
7 Found Dead
8 Purchased
9 Other

(43) BAIT
0 Rolled Oats
1 Banana
2 Corn
3 Coconut
4 Peanut Butter
5 Fruit
6 Vegetable
7 Meat
8 Nut
9 Other

(44-47) HOST CONDITION
(44)
0 Well Fed
1 Poor
(45)
0 Molting
1 Not Molting
(46)
0 Wet
1 Not Wet
(47)
0 Damaged
1 Not Damaged

HOST MEASUREMENTS:
Total Length
Tail Vertebrae
Hind Foot
Ear from Notch
Weight
Testes

(48-58) HOST IDENTIFICATION
(48-49) Order
(50-51) Family
(52-54) Genus
(55-57) Species
(58) Verified

(59) AGE
0 Nestling
1 Juvenile
2 Subadult
3 Adult
4 Old Adult

(60) SEX
0 Male
1 Male, Testes Enlarged
2 Female
3 Female, Lactating

(61-64) EMBRYO CR LENGTH

(65) EMBRYOS
0 None
1 One
2 Two
3 Three
4 Four
5 Five
6 Six
7 Seven
8 Eight
9 More

(66) STOMACH CONTENTS
0 Insects
1 Fruit
2 Nectar
3 Meat
4 Nuts
5 Vegetation
6 Empty
7 Liquid
8 Mud
9 Other

(67) NATURE OF SPECIMEN
0 Skin & Skull
1 Skin & Skull & Skel
2 Skull Only
3 Skull & Skeleton
4 Skin Only
5 Formalin
6 Discarded
7
8
9

(68) OTHER PARTS SAVED
0 Yes
1 No
List

NOTES:

LAMC FORM 2-193
ECTOPARASITE SURVEY DATA SHEET
(69-74) DATE
(75-80) COLLECTION NO.

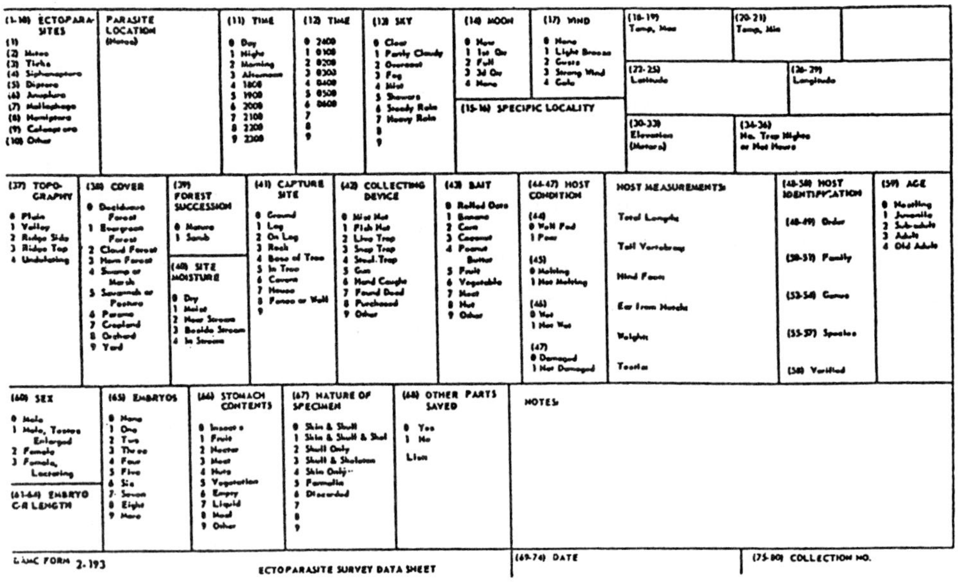

Figure 29.5 Ectoparasite survey data sheet. Original size 5 1/2 × 8 1/2 inches.
(Division of Mammals, Smithsonian Institution)

The six items or elements of data listed earlier (total length, tail length, hindfoot length, ear length, tragus height, and forearm length) are usually included with each catalog entry and on skin tags. The locality, date, sex (if ascertainable), and type of preservation are essential elements that must be included with each entry. At times, it may not be possible to record the other data. For example, it is not possible to record tail length, sex, or reproductive data for a weathered skull found in the field.

SPECIMEN LABELS

For each specimen recorded in the field catalog, a corresponding data tag (or tags) should be attached to the specimen (and to each of its separate parts). Depending upon the specimen preparation technique, several different kinds of tags may be required for each mammal. If you are collecting for a museum or university collection, the institution will usually provide you with the necessary specimen tags. If these are not provided, use the types of paper recommended earlier.

STUDY SKINS

Labels or tags for study skins are usually about 3" × 3/4". All of the data from the catalog entry should be recorded on the tags in permanent ink. Include your field catalog number, first initials, and entire surname (Fig. 29.6). The exact arrangement of data on the tag

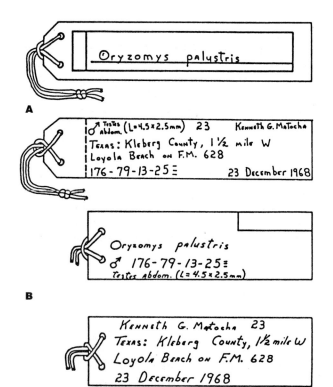

Figure 29.6 Both sides of two specimen tags. (A) The style used by the University of Kansas Museum of Natural History. (B) The style used by the Oklahoma State University Museum. Note that a space is always left blank for the museum's catalog number.
(K. G. Matocha)

varies from collection to collection. If you are collecting for an institution, use the preferred format. All data (except, perhaps, ecological notes) should generally go on one side of the tag. This will make your specimens easier to work with. Although most of the data are entered in permanent ink, the identification (scientific or common name) will be entered in pencil once identification is made at the museum collection.

Thread the tag through two holes punched at least 1/8 inch from one end and tie an overhand knot in the string about 1 inch from the tag. See Figure 29.6A for the correct way to string a tag. Use a square knot to securely attach the tag just *above the right ankle* of a study skin. With bat skins, the label is sometimes attached above the knee joint to allow for greater visibility of the calcar and to protect the feet (see Fig. 31.21).

Use the same type of label for a skull or any other bones (e.g., a weathered skull or skeleton) that will not require cleaning.

SKULLS AND SKELETONS TO BE CLEANED

The type of label described earlier will usually not survive the cleaning processes used for skeletal material. Thus, the skull tag is usually a small piece of resistant paper that has only the collector's name and field catalog number and the sex of the specimen. After the skull or skeleton has been cleaned, a permanent label, including data from the museum catalog, will be placed with the specimen.

Figure 29.7 illustrates a skull tag. Again, the string should be knotted about 1 inch from the tag. Attach the tag loosely around a mandibular ramus of an uncleaned skull, and secure with an overhand knot incorporating both strands (see Fig. 31.22). Attach the tag to the pelvis of a complete skeleton or to secure locations on each portion of a partly disarticulated skeleton.

SKINS TO BE TANNED

Because few tags can survive the tanning process, either of the two types described earlier may be attached to a skin to be tanned. A secure point of attachment for the tag is through an eye hole. Prior to tanning, special code marks are punched into the skins to insure correct reference to the catalog number of the specimen. After the specimen has returned from the tanner, the permanent specimen tag bearing complete data will be reattached to the skin.

FLUID PRESERVATION

The type of specimen labels described earlier are usually not substantial enough to survive long immersion in alcohol or formalin. Thus, special, heavy-duty tags (usually parchment) are required (Fig. 29.8). Complete data from the field catalog are entered on these tags with permanent ink. If a whole animal is preserved, tie the tag securely *above the ankle* of the *right hindfoot*. Several specimens may then be placed in one container.

If embryos, stomach contents, parasites, or some detached portion or portions of the specimen are preserved in fluid, a separate container must be used for each, *or* each portion may be securely tied in cheesecloth and a tag attached to each cheesecloth package. Several of these packages can then be placed in one container (*Note!* This recommendation holds only if mixing of fluids from several specimens would not alter the results of subsequent studies; mixing might be a problem with certain biochemical studies.) A label with complete data should be inserted *into* the container. Labels attached to the outsides of the containers all too frequently come off and are lost.

Figure 29.7 A skull tag. Note that a knot is tied in the string about 1 inch from the tag.
(K. G. Matocha)

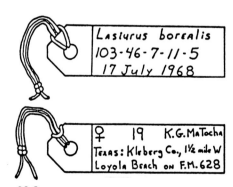

Figure 29.8 The two sides of a tag for a specimen preserved in fluid.
(K. G. Matocha)

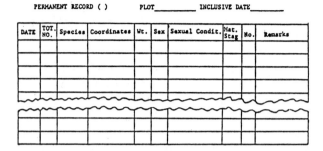

Figure 29.9 Data recording sheet for small mammal censusing. Original size 8 1/2 × 11 inches.
(P. L. Meserve)

Special Data Forms

In many situations, it is appropriate and convenient to utilize special forms for recording data in the field and laboratory. In live-trapping studies, data on spatial location, condition of animal, and other information must be maintained. For these purposes, a tally sheet (Fig. 29.9) or a more elaborate form is utilized to facilitate subsequent transcription of the records onto computer disks or other input media.

Recording Measurements

For quantitative analyses, extensive measurements are generally recorded on coding sheets that permit accurate transcription to computer databases. Semiautomatic systems for making linear measurements are in use (Anderson 1972; Calaprice and Ford 1969; Sneath and Sokal 1973:452–53). Consult advertisements in the journals *Science* and the *Journal of Mammalogy,* among others, for details on some of the available systems. Voice-activated tape recorders and voice-activated computers also offer the capability for data entry.

Recording Behavior

Because behavioral events can occur in quick succession, coding sheets are designed for rapid recording. Even so, instrumentation and keyboards are usually necessary for recording sequences of behavior. Instrumentation is available for direct recording and transcription of behavioral events under field conditions (Butler and Rowe 1976; Lehner 1996; Stephenson et al. 1975). With the advent of notebook and palm computers, there is increasing use of these devices to gather primary data under field and laboratory conditions. Techniques for recording behavior were described by Altmann (1974) and Lehner (1996) and were briefly summarized by Feldhamer et al. (1999:29).

Computers

Computers provide fast and relatively easy means for getting handwritten or typed primary **data** into a **database.** These devices also serve as a primary method for entering data in the laboratory or in the field. *Notebook computers* and *palm computers* are miniature versions of these microcomputers. Voice-activated data entry is also available but not yet very reliable. Most database programs on the market today offer the capability to transfer information from one computer system or database to another kind. To effect this transfer, data can be transferred by *modem,* or the two computer systems can be directly connected through a network, a cable, or infrared beam. In all cases, the user is encouraged to proofread data after transfer to a database or after transfer to another system, to insure that no errors in characters or formatting have been introduced as a result of the transfers. Texas Tech University developed a relational database (Wildcat III) to capture primary data (field notes, personal preparation catalog, and specimen tags) in the field for later transfer, following verification, into the museum's collection database (King et al. 2000).

Documentation Standards for Museum Specimens

Williams et al. (1979) drafted a set of "Documentation Standards for Automatic Data Processing in Mammalogy," under the auspices of the Committee on Information Retrieval in Mammalogy of the American Society of Mammalogists. In 1996, version 2.0 of these standards was published under the auspices of the Carnegie Museum of Natural History and the American Society of Mammalogists Committee on Information Retrieval (American Society of Mammalogists Committee on Information Retrieval 1996). In the updated version 2.0 standards, the data fields were divided into *essential, preferred,* and *optional* categories of data (Table 29.2). The essential data fields would be those data elements that should be transmitted between institutions that have networked connections. The preferred fields would be those categories of data that are useful for research or collection management. The optional fields would be data categories that might be useful to some but not all collections.

Mammalogists and other investigators working with mammals should strive to record specimen data in accord with these standards. This will not only promote consistency in data recording but will help insure more accurate and useful specimen data for researchers. In Table 29.2, we have listed the essential, preferred, and optional categories that are used for specimen data. Refer to the earlier publication by Williams et al. (1979) and the 1996 publication (American Society of Mammalogists Committee on Information Retrieval 1996) for additional comments about these data categories.

TABLE 29.2 Essential, Preferred, and Optional Data Categories Used in Collections of Mammals

Data Category

Essential Data Categories	Preferred Data Categories	Optional Data Categories
Institutional acronym (for data transfer only)	Availability status	Divisional acronym
Collection catalog number	Accession number	Special number
Genus	Family	Donor
Species	Subspecies	Date cataloged
Type of preservation	Specific locality or reference point/reference point modifier	Published records
Sex		Type description
Date collected	Elevation	Order
Collector or preparator or both	Latitude and longitude (for old specimens)	External measurements
Collector's number or preparator's number or both	UTM (for old specimens)	Weight
Country		Age
State, province (first-level political subdivision)		Reproductive data
County, district (second-level political subdivision)		Ecological notes
Ocean		Continent
Sea		Township and range
Bay, inlet, strait, gulf, channel, major island group		Coordinate precision index
		Remarks or comments
Latitude and longitude (for new specimens)		Ancillary collections
UTM (for new specimens)		Specimen condition reporting

Refer to American Society of Mammalogists Committee on Information Retrieval (1996) for additional information.

CHAPTER 30

COLLECTING

The study of mammals frequently requires that they be captured and/or prepared as specimens. Many detailed investigations can be carried out only when the mammal is in captivity or has been prepared as a museum specimen. Taxonomists require specimens to document the presence of species in particular localities, to assess geographic and other types of variation, and to prepare reviews and keys on various groups of mammals. Scientists interested in life history studies of mammals collect specimens to assess reproductive patterns or food habits. Population biologists capture mammals alive, mark them for permanent identification, and later try to recapture the marked animals, to obtain data on population sizes and population structures. All behavioral and many cytological studies require live mammals for field or laboratory investigations. These are only a few of the many types of research that require the collection and/or capture of mammals. Additional information on the use of mammals in research and educational settings can be found in publications of the American Society of Mammalogists (*ad hoc* Committee for Acceptable Field Methods in Mammalogy 1987; *ad hoc* Committee for Animal Care Guidelines 1985; Animal Care and Use Committee 1998) and the Association for the Study of Animal Behavior (Association for the Study of Animal Behavior 1996).

HEALTH AND SAFETY

Some large mammals (e.g., proboscideans, artiodactyls, perissodactyls, and others) pose threats to researchers attempting to capture, handle, and mark these animals. For such mammals, specialized handling facilities or drug immobilization from a distance may be required to minimize danger to the investigator. The researcher should always be aware of the behavior and habits of the animals under study to minimize risk and to lessen the chance for potentially dangerous encounters.

Some mammals may harbor potentially dangerous pathogens that can be life-threatening (Kunz and Gurri-Glass 1996). The site where the work is to be conducted may also present risk to the researcher due to the prevalence of particular human diseases or toxins in the area of study. The investigator should always consult with a physician prior to undertaking any field work in a foreign country or with particular groups of animals that may pose a health risk (Animal Care and Use Committee, 1998; Kunz and Gurri-Glass 1996). Investigators planning to work in tropical areas of the world should consult with appropriate health agencies (e.g., in the United States, the Centers for Disease Control; and the Public Health Service), medical professionals, and papers such as Kunz and Gurri-Glass (1996) for specific guidelines. In the sections that follow, we have provided information on some of the most serious diseases that may be transmitted from mammals to humans or from habitats frequented by mammals.

Rabies

Rabies is an almost invariably fatal disease caused by *Lissavirus* and transmitted most often in the saliva of an infected animal to a wound caused by the infected animal or to broken skin. In humans, the incubation period is typically two to 12 weeks with a range of 10 days to 15 months (Kunz and Gurri-Glass 1996). The World Health Organization (WHO) and the U.S. Centers for Disease Control recommend that all individuals who have regular contact with bats or other high risk mammals (e.g., carnivores) should be immunized against rabies (Kunz and Gurri-Glass 1996). Persons bitten by animals should wash the wound thoroughly with soap and water followed by a disinfectant (e.g., 70% alcohol). Postexposure vaccination against rabies following such a bite is always advisable (Kunz and Gurri-Glass 1996).

Hantavirus

At least two variants of hantavirus are known to infect humans, and certain species of rodents are the primary vectors (Mills et al. 1995). In the United States, a particularly virulent form of hantavirus that causes hantavirus pulmonary syndrome (HPS) was recognized in 1993. HPS symptoms appear within 45 days after exposure and the initial symptoms including fever, headache, fatigue, vomiting, shortness of breath, and a dry cough; later symptoms include rapid heartbeat, breathing difficulties, and abrupt respiratory failure (Kunz and Gurri-Glass 1996).

The principal vector of this disease in the United States is the deermouse, *Peromyscus maniculatus,* although other species of *Peromyscus* have been implicated. Mills et al. (1995) provided specific precautions for working with rodents that may be infected, along with recommendations for decontamination of work surfaces, instruments, cages, and traps. Workers should avoid feces and urine of infected animals unless appropriate protective gear is utilized.

Histoplasmosis

Histoplasmosis is a disease of mammals, caused by the soil fungus *Histoplasma capsulatum,* and is often associated with high concentrations of animal feces such as guano deposits in caves (Animal Care and Use Committee 1998; Kunz and Gurri-Glass 1996). Workers visiting caves where this organism is potentially present should wear a respirator equipped with filters that can exclude particles as small as 2 microns (µm) or use a self-contained air supply (Kunz and Gurri-Glass 1996).

Lyme Disease

Lyme disease is caused by the bacterium *Borrelia burgdorferi,* which is carried by ticks in the *Ixodes ricinus* complex (Kunz and Gurri-Glass 1996). This disease was first noticed in the United States in 1975 near the town of Lyme, Connecticut, but is now known to occur in virtually every part of the United States. In the northeastern United States, the principal reservoirs of the disease are the deer tick (*Ixodes dammini*) and the white-footed mouse (*Peromyscus leucopus*), but humans get the disease by receiving a bite from the tick (Kunz and Gurri-Glass 1996). Symptoms of the illness include a red, ring-shaped rash that appears about two to five days after the tick bite; weakness, dizziness, muscle aches, sore throat, and swollen lymph glands are also noted (Kunz and Gurri-Glass 1996). Prior to field work, use tick repellents (particularly those that contain DEET) and, following field work, search the body for any crawling or attached ticks (Kunz and Gurri-Glass 1996).

Other Diseases

Refer to Kunz and Gurri-Glass (1996) and Animal Care and Use Committee (1998) for information on *tularemia* and *plague.*

COLLECTING, CONSERVATION, AND THE LAW

Ethics and the Law

Investigators should collect only the number of specimens needed for a particular project and in accordance with guidelines of professional societies (*ad hoc* Committee for Acceptable Field Methods in Mammalogy 1987; Animal Care and Use Committee 1998). Further information on the ethics of collecting and care of mammals can be found in Rudran and Kunz (1996) and Friend et al. (1996).

In recent years, increasing concern on the part of many individuals for the conservation of flora and fauna and for the protection of threatened and endangered species prompted the passage of many federal and state laws. These and earlier laws regulate the salvage, capture, possession, transport, and sale of certain species of animals and affect, in many instances, the manner in which scientific and educational activities involving these animals can be conducted.

Federal and International Regulations Pertaining to Collecting

In the United States, the Endangered Species Act (Code Federal Register, CFR 50.17) and the Marine Mammal Protection Acts (CFR 50.18, 50.216) provide important protections through listing and regulation of endangered species. In addition, the Lacey Act, in part, regulates the importation and exportation of nonendangered species and injurious wildlife (CFR 50.10, 50.13, 50.14, 50.16).

The Convention on International Trade in Endangered Species (CITES) of Wild Fauna and Flora, TIAS 8249, to which the United States is one of 146 member nations, bans commercial international trade of listed endangered species by member nations. The Convention defined three categories of species (listed in Appendices I, II, and III, as listed below):

Appendix I: All species or other taxa threatened with extinction.

Appendix II: All species or other taxa not presently threatened with extinction, but those in which trade must be restricted to ensure their survival.

Appendix III: All species or other taxa identified by a particular country as subject to conservation regulations within that country and requiring the cooperation of other parties to the convention.

For importation (or exportation) of animals and animal products into (or out of) the United States, the provisions of CITES, the Endangered Species Act, the Marine Mammal Protection Acts, and applicable regulations of the Department of Agriculture (e.g., those related to animal care and use) must be adhered to. Refer to Genoways and Choate (1976), Berger and Phillips (1977), and McDiarmid et al. (1996), and recent updates of the pertinent sections of the U.S. Federal Code, for specific guidelines and the provisions of these regulations.

Federal Laws and Regulations Pertaining to Care, Handling, and Transport of Mammals

Mammals maintained in captivity for more than a few days are generally subject to the provisions of the Animal Welfare Act of 1985. This act, enforced by the U.S. Department of Agriculture, is designed to ensure the humane handling, care, treatment, and transportation of regulated animals used for research or exhibition, sold as pets, or transported in commerce (Friend et al. 1996). Most institutions that have animal-holding facilities are generally subject to provisions of this act, as are the recipients of many federal grants from agencies such as the National Science Foundation and the National Institutes of Health.

State Regulations

The complexity, number, and lack of uniformity in state laws regulating scientific collecting have prompted some scientists to urge the adoption of uniform regulations among states for this activity (see McGaugh and Genoways 1976). Until realistic and uniform regulations exist for all states, it is necessary for scientists and students to follow the appropriate nonuniform state and local regulations pertaining to specimen acquisition. Refer to McGaugh and Genoways (1976) and to Berger and Neuner (1981) for a summary of state laws regarding scientific collecting permits and for addresses where copies of the latest regulations for each state can be obtained. This is a rapidly changing area, and you are advised to consult with the appropriate state and wildlife agency for the latest laws and regulations.

CONSERVATION

Populations of many species (e.g., certain cave bats) not considered threatened or endangered at present can nonetheless be permanently harmed by indiscriminate collecting. All collecting should be carried out for a valid purpose (*ad hoc* Committee for Acceptable Field Methods in Mammalogy 1987; Animal Care and Use Committee 1998), and the purpose should dictate the amount and type of collecting. For instance, a project to document the occurrence or distribution of a species in a given area would normally require collecting only a few individuals at a variety of locations, whereas a study of individual variation within a population might require larger numbers from fewer locations.

Many bat colonies are particularly susceptible to disturbance. The slightest provocation will cause many species to desert roost sites, and such provocation can be extremely harmful if it disturbs a nursery colony or a hibernaculum. Whenever possible, get an expert opinion before disturbing bat colonies, and when collecting in them, create as little disturbance as possible.

LOCATING A COLLECTING AREA

Obtain permission from landowners or lease holders before collecting in an area. Failure to do so may subject you to arrest for trespassing. Often, the landowners will be able to direct you to favorable collecting sites when you observe this courtesy. A large percentage of the federal land in the United States is available for collection, with permission, but state game laws and applicable federal laws must be followed (see McGaugh and Genoways 1976).

After you reach the collecting area, make a visual survey to determine where to concentrate your efforts. Some species may be restricted to particular types of habitat, but others may occupy several habitats. After the appropriate habitat area is located, look for "sign" before placing traps (see Chapter 28).

After you finish collecting in an area, always remove all trap markers, wire, plastic, cloth, paper, and any other debris from the site.

RECORDS

Keep a careful record of all collecting activities. It is very important to know when, where, and how the specimens were collected. Always keep the data closely associated with live animals housed in cages and with dead animals stored in a freezer. Refer to Chapter 29 for record-keeping techniques.

Methods of Trapping Mammals

Kill Traps

The traps listed below are valuable for most forms of collecting when it is not necessary to acquire live animals. They are frequently more effective and less expensive than live traps.

Snap Traps

Snap traps are by far the most important trap type for general collecting. Snap traps come in three basic sizes. The largest, the **rat trap** (Fig. 30.1A), and the smallest, the **mousetrap,** are generally available and familiar to most people. The rat trap is used to catch rats, *Rattus;* woodrats, *Neotoma;* chipmunks, *Tamias;* the smaller ground squirrels, *Spermophilus;* and similarly sized mammals. Smaller mammals, (e.g., most *Peromyscus, Microtus,* and *Blarina*) can be taken in rattraps, but the strength of the springs on these traps may cause damage to the specimens. Because of the small distance between the sprung wire bail and the bait pan of mousetraps, the skulls of mammals collected with them are usually crushed. Because the skull is an important part of any museum specimen, mousetraps are very seldom used for scientific collecting. **The Museum Special** (made by Woodstream Corp., Lititz, Pennsylvania) is especially designed for collecting small mammals for scientific purposes (Fig. 30.1B). It is larger than a mousetrap but less powerful than a rattrap. Its design greatly reduces the frequency of skull damage. West (1985) reported that the new design of the Museum Special trap, compared with

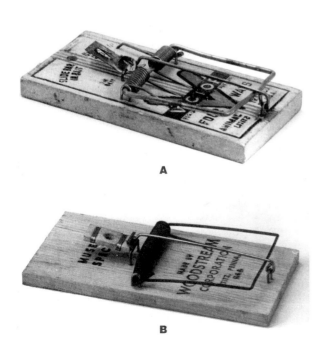

Figure 30.1 Examples of snap traps. (A) Rat trap; (B) Museum Special.
(R. E. Martin)

Figure 30.2 Conibear trap, shown set.
(Stains, H.J. 1962. Game biology and game management; a laboratory manual. Burgess Publishing Co., Minneapolis. 143 pp., fig 7e. With permission of the publisher.)

the old style of this trap, was less adept at catching chipmunk-sized mammals but better able to secure some smaller species of mammals such as shrews.

To set any of the snap traps, place a small quantity of bait on the pan and adjust the trigger mechanism so that it will be set off with a light touch. Because woodrats are adept at collecting these traps and adding them to their den debris, it is wise to wire the traps to a rock or branch when trapping near woodrat dens.

Steel Traps

Steel leghold traps are used primarily by fur trappers and infrequently for scientific collecting (*subject to state regulations*). These traps usually catch mammals by one leg and generally do not kill. However, they are included under the "kill trap" classification because they usually severely damage the animal and necessitate killing it. These traps cannot be considered humane and should not be used unless they are padded and checked frequently (Animal Care and Use Committee 1998).

If steel leghold traps are used, set the traps under water so that the caught animal quickly drowns. The **Conibear Trap** (Fig. 30.2) is a humane alternative to the steel leghold trap. This trap comes in a variety of sizes and often can be substituted for the smaller-sized leg traps. The Conibear trap generally kills animals very quickly.

Gopher Traps

Several traps designed specifically to catch gophers are manufactured in the United States. A style that is widely available and easily set is shown in Figure 30.3. The traps are set in underground runways, singly (Fig. 30.4A) or in pairs (Fig. 30.4B), and are tied to a stake to prevent them from being dragged deep into the burrow system by the

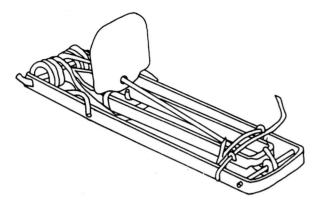

Figure 30.3 Macabee-type gopher trap, shown set. Suitable for capturing all but the largest pocket gophers.
(R. E. Martin)

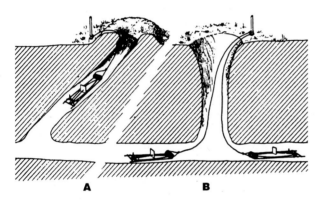

Figure 30.4 Single- and dual-trap sets for pocket gophers. (A) Single set in lateral tunnel; (B) dual set, back to back, in main tunnel.
(Crouch 1933:17)

captured animal. The tunnel is left open so that the gopher will push dirt over the trap and in so doing set off the trigger mechanism. These traps must be checked every 15 to 30 minutes to get maximum benefit and minimize discomfort to captured animals.

Mole Traps

The **harpoon mole trap** (Fig. 30.5) is the most commonly used mole trap in the United States. It is set on the surface. The mole tunnel is pressed down and the trap pushed into the ground so that it straddles the collapsed tunnel. The trigger mechanism is cocked, and the height of the trap is adjusted so that the mechanism will be raised and thus set off by the mole as it travels just under the surface of the ground.

Live Traps

Live traps offer several advantages over kill devices. Animals that are needed alive can be taken back to the laboratory for further investigation or studied or marked at the trap site and then released. In areas where endangered or threatened species may occur, the use of live

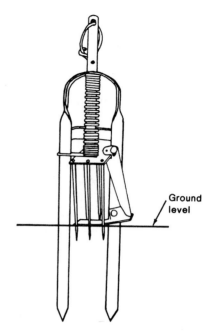

Figure 30.5 Harpoon mole trap, shown set.
(R. E. Martin)

traps will minimize the chance of harming these forms and may be required by federal or state regulations governing the issuance of a permit to the investigator.

Arrange the traps so that the animals will be protected from extremes of cold and heat, and check them frequently for any captures. In cold weather, provide cotton (but *not* polyester because animals may be entangled and injured in this material) or other suitable insulating material.

Commercial Traps

Several manufacturers produce and market live traps in a range of sizes to capture everything from mice to coyotes. Some firms make types large enough to capture small deer. In the United States, welded wire traps (Fig. 30.6) are made by several companies including Allcock Manufacturing Co. (Havahart® brand traps) and Tomahawk Live Trap Co. These traps (made in collapsible or noncollapsible varieties) have either one or two doors and are activated by a small bait pan in the center of the

Figure 30.6 Havahart® trap, one of several varieties of welded wire live traps.
(Allcock Manufacturing Company)

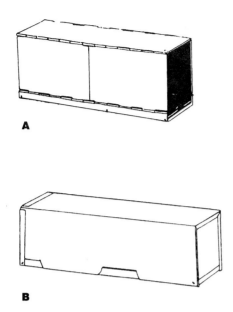

Figure 30.7 Folding (A), shown set, and nonfolding (B). with door closed, models of Sherman live traps.
(G. A. Moore)

trap. The **Sherman trap** (H. B. Sherman Traps) is made of galvanized sheet iron or aluminum in folding (Fig. 30.7A) or nonfolding (Fig. 30.7B) models. The **Longworth trap** (Longworth Scientific Instrument Co., Ltd.) is made of aluminum and is composed of two sections: a trap and trigger mechanism, and a detachable nest box This added feature provides greater protection of the captured animal from the elements and makes for easier handling of the live mammal in the field. Several investigators have reported potential bias with the use of this trap. Grant (1970) reported that lighter-weight animals are not caught as readily in Longworth traps as the heavier animals. Boonstra and Rodd (1982) found that some heavier species of voles (e.g., *Microtus pennsylvanicus*) sometimes sprung the traps without being caught; they suggested replacing the standard treadle in the Longworth trap with one that was about 1 inch (2.54 cm) longer.

Homemade Traps

Many types of live traps can be made relatively easily and inexpensively. If a particular trap design does not suit your needs, the plans may be modified so that the trap will capture the species that are desired. The **Gen trap** (Fig. 30.8) was originally designed by Shemanchuk and Bergen (1968) for use in ground squirrel burrows. However, this trap is adaptable to other species by modifying the diameter of the pipe and the method of setting. The Gen trap is constructed of economical materials and is easily made and transported. The **Fitch trap** is dependable and widely used for many species of rodents. This trap is constructed of hardware cloth, galvanized metal, wire, and a metal can (see Fitch 1950, for details on original design, and Rose 1973, for modified version). Pocket gophers, Geomyidae, are somewhat difficult to live-trap, but Baker and Williams (1972) reported good results with their design (Fig. 30.9). Earlier, Howard (1952) designed a live trap for pocket gophers that was the standard for many years.

According to Moore (1940), moles can be captured alive using modified kill traps (Fig. 30.10). Arboreal mammals can be captured in a variety of homemade traps including one designed to capture sugar gliders, *Petaurus breviceps* (Mawbey 1989).

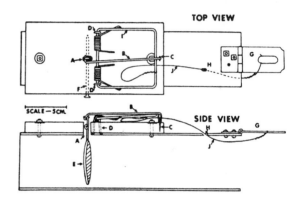

Figure 30.9 Live trap for pocket gophers.
(Baker and Williams 1972)

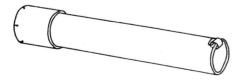

Figure 30.8 Gen trap, with removable rear closure.
(R. E. Martin)

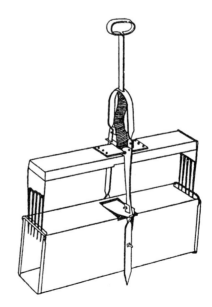

Figure 30.10 Live trap for moles.
(Moore 1940:224)

Additional designs for live traps can be found in Burt (1927), Garlough et al. (1942), Gilmore (1943), Howard (1953), LoBue and Darnell (1958), Taber and Cowan (1969), and Jones et al. (1996).

Pitfalls

Tall cans or jars, placed in the ground with their rims even with the surface, are useful in catching small mammals, particularly shrews. Provide an adequate quantity of food in the container and check frequently for captures. Baffles (drift fences) pressed vertically into the ground increase the effectiveness of pitfalls by guiding mammals toward them. Williams and Braun (1983) compared the effectiveness of pitfall and conventional traps for securing small mammals.

Runways, Corrals, and Box Traps

Larger mammals (even rabbits and hares) may be herded along runways into corrals. Care must be taken to avoid injuring the animals during the chase because most will be released. Most states restrict this activity, and usually only qualified professional biologists or game wardens engaged in management or research programs may utilize this technique. Check with game officials before attempting this procedure, and secure the necessary permits. Foreyt and Glazener (1979) provided plans for a box trap suitable for capture of deer or feral hogs.

Special-Purpose Traps

Beavers may be taken in either Bailey or Hancock traps, and muskrats in Snead traps (see Couch 1942; Taber and Cowan 1969). Because these animals are furbearers, they may generally be captured during furbearer season only or under the provisions of a scientific collecting permit.

METHODS OF TRAP PLACEMENT

Snap traps and most types of commercial and homemade live traps can be distributed by any of the following three methods. The other types of traps are almost always placed according to the sign method. Additional information on methods for placing traps can be found in Jones et al. (1996).

Sign Method

With the **sign method** of trap placement, appropriate fresh mammal sign is located, and a trap is positioned to catch the animal. Although this method generally produces the best results, it is more time-consuming than the other methods that follow. To facilitate relocating traps placed in this manner, either each trap must be marked with a conspicuous trap marker or else detailed notes must be kept. The sign method could result in a biased sample of the species present because those small mammals that leave little or no conspicuous sign are less likely to be collected.

Paceline Method

The **paceline method** places traps at regular intervals. One technique is to select a starting point and mark it with a strip of conspicuously colored cloth or plastic surveyor's tape. Sight toward a distant landmark and place traps at regular intervals as you walk toward that point. Trap markers are usually placed at the beginning and end of each line. If the line is particularly long or if the habitat has dense ground cover, it is prudent to also mark every fifth or tenth trap location. The paceline method generally allows for the placement of a maximum number of traps in a minimum amount of time and is also free of the bias mentioned earlier. However, it will usually result in a lower percentage of captures than the sign method.

Grid Method

A trap grid is a series of parallel trap lines. Grid trapping is most frequently utilized in life history studies where mammals are marked and recaptured (see Chapter 36). The size of the grid is dependent on the species to be captured and the type of study. In small rodent studies, for example, the grids are often arranged in 12 rows and 12 columns (144 traps total), with each trap station 15 to 20 meters from an adjacent station. The traps are usually placed without regard to sign.

TIMING OF TRAPPING ACTIVITY

Because most small mammals are crepuscular or nocturnal, traps are usually set just before dusk and checked or collected in early morning. Ants and other small invertebrates will eat the bait if the traps are set too early in the day, and they can ruin trapped specimens if the traps are not checked soon after dawn. It is frequently productive to check traps during the night, remove trapped mammals, and reset the traps. Fleas will leave a specimen as soon as the body cools, so frequent checks of kill traps are necessary if you are interested in collecting these ectoparasites.

To catch diurnal mammals such as ground squirrels, traps can be set any time during the day. But they should be checked at frequent intervals, particularly in hot weather, to minimize decomposition in killed animals or heat stress or death in individuals that are live-trapped.

Weather conditions will frequently influence trapping success. A heavy rain may wash the bait off of traps or may set them off. A heavy snowfall can bury traps and make them very difficult to relocate. The activity patterns of some mammals may be influenced by the weather or by the phase of the moon. Kangaroo rats, *Dipodomys,* for example, are more active and therefore more readily collected when there is no moonlight.

Calculating Trapping Success

Trapping success is given as the percentage of the traps that produce specimens. Thus, if 100 traps were set and 10 mammals were collected, the trapping success would be 10%. The number of traps used multiplied by the number of nights they were in position is referred to as the number of **trap nights.** Thus, 100 traps set for one night equals 100 trap nights, 20 traps set for five nights equals 100 trap nights, and 20, then 40, then 30, and then 10 traps set on four successive nights still equals 100 trap nights. Trapping success is also frequently given as a percentage of trap nights.

The trapping success to be expected varies considerably with the habitat, type of trap, trapping method, population density, and many other variables. In the United States, a paceline of snap traps in moderately good habitat will usually produce about 10% success. In the American tropics, trapping success is generally much lower.

See Chapter 36 for additional references and for techniques that can be used to determine relative abundance or densities.

Baits and Scents

Baits

Bait may be preferred seasonal food or a substance entirely new to the animal. For rodents, the most commonly used bait in temperate North America is a mixture of peanut butter and oatmeal (mixed bird seed is sometimes added). To catch insectivores, it is helpful to add a small quantity of meat or fish to the mixture. Most carnivores will be attracted by canned dog food or fish products, whereas rabbits and similar herbivores require apples, carrots, or lettuce. Dead animals may also be used to attract mammals to traps.

In many areas, ants can kill mammals in live traps, remove bait, or otherwise be a big problem. A change of bait may be effective in minimizing its loss to insects. Getz and Prather (1975) mixed shredded cotton into heated peanut butter to retard the removal of the bait by insects.

Scents

Scent baits are very effective with many mammal species because the odor may arouse sexual interest, antagonism, or curiosity. Scents may be superior to food baits for mammals that communicate by odor. In general, valerian (prepared from *Valeriana* sp.) is attractive to carnivores of the family Canidae, and catnip (*Nepeta* sp.) is attractive to the black bear, *Ursus americanus.* Beaver castor or castoreum (prepared from musk sacs of both sexes of *Castor*) elicits a positive response from beavers. Refer to Taber and Cowan (1969) for instructions on how to prepare and use various scents.

Minimizing Damage to Trapped Mammals

At times, larger mammals (particularly carnivores) may disturb or injure smaller mammals caught in live traps. Getz and Batzli (1974) designed a cage that is placed over a live trap to minimize this possibility.

In areas where ants may cause damage to specimens (e.g., tropics, deserts), traps must be checked frequently (including nighttime hours) to remove captured mammals. Frequent checking of kill traps may also minimize damage resulting from the activities of shrews, carnivorous rodents (e.g., *Onychomys*), and small carnivores.

Traps must be checked frequently during warm weather. Mammals held in live traps may overheat and die, and specimens in kill traps may decompose quickly under such conditions.

Adjusting Trapping Methods for Regional and Climatic Differences

Most of the advice contained in this chapter applies to North American temperate conditions. You may have opportunities to collect in other countries and other climatic areas. Often, you can learn important techniques by talking with persons experienced in trapping in these areas. Papers in the *Journal of Mammalogy* and similar journals are also helpful, as are the introductory sections of many regional faunal works.

Other Methods of Capturing or Collecting Mammals

Hunting

Larger species of mammals are generally collected using a rifle or shotgun. Small species, if not trapped or collected by some other means, can be collected using a shotgun. Specimens collected for scientific purposes must be killed in a manner that causes the least damage to the specimen or to the body parts required for study. Remember that firearms impose an obligation upon the user to handle them safely and in compliance with local and state laws.

Shotguns of 12, 16, 20, or .410 gauge are generally used to collect smaller mammals. Very small species are best collected with .22 caliber rifles or pistols (preferably smoothbore) loaded with bird shot, or a .410 shotgun with half-loads of No. 12 shot. Rifles loaded with ball ammunition are generally used on the larger species. A combination rifle-shotgun, e.g., a ".22–.410 overunder" is a very versatile firearm for collecting. A .22 cartridge will handle most species up to the size of a raccoon or badger. Larger calibers are necessary for larger carnivores and large herbivores. If you anticipate collecting some of the larger mammals, it would be well to seek the advice of a person experienced in firearms and hunting and comply with all laws and regulations.

Drug Immobilization

Large mammals that must be captured alive are usually immobilized with drugs to prevent harm to the animal and the captor. The drugs most commonly used are nicotine salicylate and succinylcholine chloride. To achieve proper dosages, it is necessary to have a fairly accurate method of estimating weights of the target animals. Taber and Cowan (1969) listed approximate dosages for a list of selected mammals. Harthoorn (1965, 1975) provided a comprehensive treatment of this specialized field, and Twigg (1975b:89–91) gave a particularly succinct account.

The drugs are administered by means of specially designed syringes attached to arrows or firearm projectiles. The Cap-chur gun is a commercial form of such a device. In most states, drug immobilization can be conducted only by qualified scientists, game management personnel, or zoological garden staffs. Consult local game officials before attempting this procedure because the procedure is regulated, as are the drugs used in the delivery devices.

Corral Traps and Nets

Taber and Cowan (1969:286–90) described and illustrated several types of herd and drift traps for the capture of large mammals. A radio-controlled drop net was used by Ramsey (1968) to capture white-tailed deer (*Odocoileus virginianus*) and axis deer (*Axis axis*). All of these methods are subject to state regulations and are generally not used by personnel not employed by state or federal game agencies.

Den Investigation

Woodrats, *Neotoma*, some species of kangaroo rats, *Dipodomys*, and several other species can frequently be dug out of their dens and then captured by hand, with a net, or shot with firearms. In most states, it is illegal to collect game and furbearing species and many other species in this manner. Consult applicable regulations and game officials before using this procedure.

Hand Capture

Many nocturnal mammals, particularly certain rodents, can be captured by hand or with a net. A light is shone in the animal's eyes, causing the animal to remain motionless while approached and captured. You will feel safer wearing gloves, but many species make no attempt to bite. Hand capture can be conducted by walking through fields with a lantern or by driving slowly along infrequently traveled roads. Spot-lighting for game species is illegal in most states. Consult state and federal wildlife laws before attempting this procedure.

Bat Collecting

Prior to collecting bats, consult papers such as Kunz et al. (1996) to learn about the ethics involved and potential harm that might be done to populations. In general, bats may be collected by netting, with the Tuttle trap, with harp traps, or by hand, all of which allow you to ascertain the species involved prior to removal from the population.

Bats may be caught in **mistnets** (Fig. 30.11), bat traps (Fig. 30.12), or by hand. The nets are strung over water holes (especially in desert regions), across streams, or across trails or narrow clearings in timbered areas. Bats can also be collected by hand from bridge expansion joints, old mine tunnels, caves, attics, belfries, or other sites (see Greenhall and Paradiso 1968). Francis (1989) compared the success of mistnets with two different styles of harp traps for securing bats.

Be especially careful about collecting bats from suspected breeding or nursery colonies. In the United States, the gray bat, *Myotis grisescens,* the Indiana bat, *Myotis sodalis,* and several others are endangered species and must not be disturbed or captured without permits from the U.S. Fish and Wildlife Service and appropriate state agencies. Determine the specific identity of any bats in a newly found colony before collecting any. Some nursery colonies are extremely sensitive to disturbance (e.g., movement, light, noise) and may abandon the location, with adverse results to the popula-

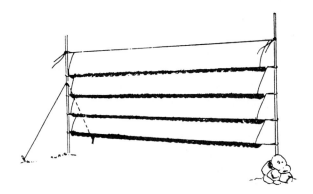

Figure 30.11 Mistnet used in capturing bats.
(Greenhall and Paradiso 1968)

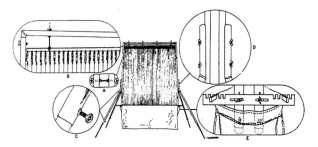

Figure 30.12 Double-frame Tuttle trap showing details of construction.
(Tuttle 1974:476)

tion. Disturbing bats during hibernation can result in stored energy loss, which may keep them from surviving the winter.

Several specialized techniques have been devised for capturing bats in conjunction with population, movement, and other life history studies. Many of these techniques were discussed and illustrated by Tuttle (1974), Greenhall and Paradiso (1968), Barbour and Davis (1969), Constantine (1958, 1969), and Kunz (1988). Kunz et al. (1996) and Kunz (1996) provided detailed information, with excellent illustrations, on how to capture, handle, and mark bats for study. Also, consult issues of *Bat Research News* (1960 to present).

Salvage

Many good records result from collecting mammals that are found dead on the road (abbreviated **DOR**). The specimen may be damaged, and often the skull is crushed, (making it less useful to neontologists but more useful to paleontologists), but most tears can be sewn up to make a suitable preparation. Often the specimen will be bloated and have a foul odor, but a good preparation can still be made if the hair has not started to **slip** (that is, if large patches cannot be easily pulled free from the underlying dermis).

Raptor pellets (both hawk and owl) frequently yield valuable mammal specimens. These are a particularly good source of shrews, which are not often caught in traps. Only skeletal material and matted hair will remain, but specific identity of the prey species is usually possible.

Occasionally, weathered skulls, leg bones, vertebrae, and other skeletal elements are found in the field. Such remains of animals that died of natural causes or predation can be identified and often provide useful locality records.

Frequently, it is possible to obtain hides, skulls, and/or other material from animals killed by hunters and trappers. With appropriate state permits, such specimens can be sources of important records.

Whenever possible, salvage should be used to increase the size of series, document distribution, record variation, provide internal organs for analysis, etc. Salvage is the one type of collecting that does not require that the scientist disturb a population.

Permits must be obtained from federal and state agencies to salvage specimens of endangered species, marine mammals, or any specimen found in a national monument or a national park in the United States. Otherwise, you may be subject to arrest and stiff fines for possession of these animals (Berger and Phillips 1977; McGaugh and Genoways 1976:80–81). Collecting of game animals and forbearers, and many nongame species, is generally subject to state regulations, and you must have the necessary license or permit to secure or handle these mammals.

MARKING MAMMALS

It is frequently necessary to mark individual mammals for studies of population density, home range, homing, migration, or other behavior. Detailed information on how to mark mammals is found in Kunz (1996) and Rudran (1996).

Tagging

Tagging involves the attachment of a metal or plastic tag to an animal. The tags are frequently numbered for individual recognition and may be attached to the ear, in blubber (whales), or around the neck or leg. Small rodents are frequently marked with fingerling tags attached to the ear. Microchiropteran bats are tagged (banded) with lip-end bands secured loosely around the distal portion of the forearm (Kunz 1996). Greenhall and Paradiso (1968) and Bonaccorso et al. (1976) described banding techniques for use with bats and improvements in procedures. **PIT transponders** have been used in recent years to provide permanent individual identification for animals (Fagerstone and Johns 1987). A cylinder containing a transmitter with a unique frequency is injected into the dermis of the mammal, and the identification is read with a special receiver. Twigg (1975c), Kunz (1996), and Rudran (1996) provided useful information on how to mark mammals.

Collar tags are frequently marked with reflective tape or plastic streamers so that individuals of large species (usually carnivores or ungulates) can be identified at a distance. Taber and Cowan (1969) and Knowlton et al. (1964) discussed several of these methods in detail. Buchler (1976) designed a chemiluminescent tag for tracking bats and other small nocturnal mammals; Heidt et al. (1967) utilized an implanted magnet to track the movements of voles (*Microtus*) in an enclosure.

Coloring and Freeze Marking

Dyes are useful in marking mammals for behavioral investigations. The black dye, Nyazol D, can be applied to any species that has pale pelage. Several individuals can be marked with the same color by applying the dye to different regions of the body. Refer to New (1958), Haresign (1960), Taber and Cowan (1969), and Taylor and Quy (1973) for additional dyes and methods.

A pressurized refrigerant (dichlorodifluoromethane, CCl_2F_2) was used by Lazarus and Rowe (1975) to apply permanent markings (by changing the natural pigmentation pattern of the hair) on individuals of several species of rodents. More recent techniques use a mixture of dry ice and alcohol to cool the branding iron for freeze marking (Rice and Kalk 1991). In contrast, marks made by pigment dyes such as Nyazol D last only until the dyed hair is shed in a molt.

Radio-Location Telemetry

Collars with radio transmitters and battery packs (or solar cells) can be attached to the necks or backs of animals to provide information on movements, activity, and physiological states (e.g., resting, moving, etc.). Individuals can be distinguished by differences in frequencies or pulse rates (clicks). Although the initial cost of the equipment (transmitters, receivers, antennas) is costly, the system offers the advantage of providing many location coordinates at low cost per location, compared with mark-recapture methods (see Chapter 36).

Radio-location telemetry techniques were used by Banks et al. (1975) to study the activity and home range of the brown lemming (*Lemmus sibiricus*) and by Mineau and Madison (1977) to study the movements of white-footed mice (*Peromyscus leucopus*). Brander and Cochran (1969:95–108) provided additional references on the use of radio-location telemetry with a variety of mammals, and a summary of procedures and equipment.

CARE OF COLLECTED MAMMALS

Care of Live Mammals

Live traps should be checked frequently, and any captured mammals should be removed to a suitable holding cage. Mammals held in captivity impose an obligation upon the captor to provide adequate housing, water, and food. Check the animals frequently, especially during the first days after capture, when they are adjusting to captive conditions. In the United States, mammals in captivity are generally subject to the provisions of the Animal Welfare Act, administered by the U.S. Department of Agriculture, which mandates that institutions establish an Animal Care & Use Committee to monitor the treatment of animals in captivity.

Most captive mammals require a constantly available source of water or succulent vegetation. Some mammals need a complex diet for good health, but others will thrive on one that is relatively simple. Most rodents live well on a diet of commercial rat or mouse chow, although a seed diet is required for some species. Green food such as lettuce should be provided occasionally. Most carnivores can be fed canned and/or dry dog food. Rabbits will thrive on commercial food pellets, supplemented with fresh lettuce or carrots. Consult Crandall (1964) or Lane-Petter et al. (1967) for more information on care of live mammals. Greenhall (1976) provided comprehensive treatment on the care of bats in captivity.

Killing Mammals Humanely (Euthanasia)

It is sometimes necessary to kill a mammal that has been in captivity, is severely wounded, or caught in a live trap. This can be done humanely by a variety of methods. Small mammals can be asphyxiated almost instantaneously by tightly compressing the thoracic cavity until the heartbeat ceases.

Small or large mammals can be killed by placing them in a closed container and introducing a cotton wad saturated with chloroform or ether. (**Caution: these chemicals are highly flammable.**) Carbon monoxide fumes, as from automobile exhaust, can also be piped into the container. (**Danger: Carbon monoxide** is an **extremely** toxic, colorless, and odorless gas that must be used only where there is adequate ventilation.)

Care of Dead Mammals

Mammals that are to be prepared as museum specimens should be processed quickly to prevent spoilage. Large mammals (e.g., deer) should be eviscerated if skinning will be delayed for several hours. Small mammals must be skinned quickly, be preserved in chemical solutions, or frozen for subsequent processing. If frozen or refrigerated, always enclose a full set of data within the bag that contains the specimen. The freezer bag should be airtight, fastened securely, and have little air in it so as to minimize specimen dehydration and conserve space. Refer to Chapter 31 for detailed specimen preparation techniques.

CHAPTER 31

Specimen Preparation and Preservation

A museum research specimen of a mammal usually consists of a study skin (filled or tanned) and a cleaned skull. Frequently, the postcranial skeleton is also saved and cleaned. Sometimes the specimen consists of a complete cleaned skeleton or the entire animal or parts of it preserved in fluid. The following instructions should enable you, with practice, to prepare specimens of high quality. A demonstration of the preparation of a study skin by an experienced preparator will be a valuable addition to these instructions.

Supplies and Equipment

Supplies needed for preparation include the following:

Cotton. Long staple, unabsorbent. Cotton batts used in making quilts usually serve well.

Wire. For reinforcing the legs and tail. Use monel or other wire that does not corrode.

Labels. For skin, skull, and each additional portion preserved (see Chapter 29).

Permanent ink. See Chapter 29 for discussion of suitable inks.

White cotton thread. Sizes 40 and 8 (button and carpet).

Corn meal (or fine hardwood sawdust). For absorbing fat, blood, and other body fluids. If tissues will be removed for chemical analyses, do the removals *before* use of the corn meal.

Borax. *Optional.* For drying and preserving specimens; salt is frequently used for very large specimens. Borax may discolor skins. Remove tissues needed for chemical analyses prior to use of this chemical.

Equipment preceded by an asterisk (*) are essential. The others are helpful to have.

*Scissors. Good quality, surgical or dissecting, advisable to have a small pair with fine, sharp points and a larger pair with one sharp and one blunt point.

Scalpel. Use one with *sharp* disposable blade. For larger specimens (raccoons and larger) a sharp-bladed **knife** is probably more useful.

*Forceps. Straight or bent, fine or medium points.

Hemostat. Used to hold cotton body while inserting into skin of smaller specimen.

*Pliers with wire cutter. Or separate pliers and wire cutter.

*Toothbrush. For brushing the fur.

*Pens. For use with permanent inks. One type of pen may be needed for paper tags and special permanent-ink pens needed for writing labels or identification numbers onto cold-storage containers. (See Chapter 29 for details.)

*Millimeter rule. 6" and/or 12". Transparent plastic is helpful but not essential. For larger animals, a steel tape marked in millimeters is needed.

*Needles. Straight, assorted sizes. Large eyes are advisable.

*Pins. Glass-headed preferred, insect, or common straight pins acceptable. A long, slender, corrosion-resistant pin with a large head is best.

Prior to Preparation

Specimens should ideally be prepared or processed soon after death to prevent deterioration of tissues that may be needed for genetic or chemical analyses. If the specimen is frozen, thaw it in a warm place. The time required for thawing will vary. Do not place specimens directly under a hot lightbulb. This frequently causes differential thawing, with some areas starting to "slip" (the hair and epidermis separate easily from the dermis, resulting in bald spots on the finished skin), while others are still frozen. (Shrews tend to slip especially soon after death under all circumstances and therefore usually have skinning priority over other specimens.)

Tissue Samples

Read the section on "Chromosome and Tissue Preservation" if bone marrow or tissue samples will be removed for chromosome, genetic, or chemical analyses. Follow the guidelines for this removal or consult appropriate literature or experts about how to proceed.

Ectoparasites

Remove the fresh or freshly thawed specimen from its bag and examine it and the bag for fleas, lice, ticks, mites, and other ectoparasites. See Chapter 32, "Collecting Ectoparasites of Mammals" for techniques for collecting and preserving ectoparasites.

Catalog Entry

The preparator's catalog or field catalog discussed in Chapter 29 is a numbered list of all specimens preserved. Enter locality, date, and other pertinent data in the catalog and assign the specimen a number.

Measurements

In the process of removing the skin from a carcass, it is inevitable that the skin will be stretched, and while drying, it will shrink. Thus, a finished specimen will rarely be exactly the same size as the original animal. Because size and proportion of a mammal are important in identification and in many other ways, a set of standard measurements is taken of the specimen prior to skinning. Four measurements (total length, tail length, length of hindfoot, and length of ear) are recorded for all species, where possible. Two additional measurements, length of tragus and length of forearm, are recorded for bats. These measurements are always recorded in millimeters and always in this order. Instructions for taking these measurements follow.

Total length (TL). From the tip of the nose to the distal end of the tail vertebrae. Place the animal on its back so that the backbone is straightened but not stretched. Position the head with the rostrum extending straight forward in line with the backbone. Measure the distance between the tip of the nose and the end of the last tail vertebra; exclude hairs extending beyond the tip of the tail (Fig. 31.1). If your millimeter rule is not long enough, place pins at the tip of the nose and tip of the tail vertebrae, remove the specimen, and measure the distance between pins.

Tail length (T). Bend the tail up at a right angle to the body, straighten it out, and measure the distance from the vertex of the angle to the distal end of the last tail vertebra; exclude protruding hairs (Fig. 31.2).

Hindfoot length (HF). Measure from the back edge of the heel to the tip of the longest toe plus claw. Include the claw in the measurement (Fig. 31.3).

Ear length (E). Measure from the *notch* at or near the base of the ear to the furthermost point on the edge of the pinna (Fig. 31.4). European mammalogists typically measure the entire length of the ear (see Fig. 31.7 for comparison).

Tragus length (Tr). (Chiroptera only) The tragus is a leaflike structure projecting up from inside the

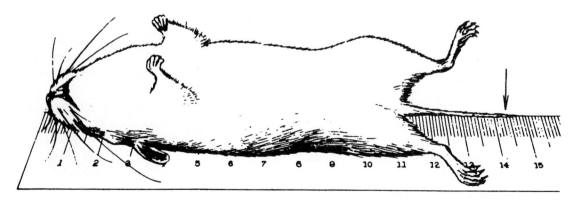

Figure 31.1 Measuring total length (TL).
(Setzer 1963)

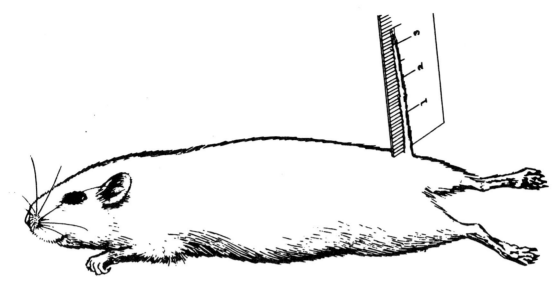

Figure 31.2 Measuring tail length (T).
(Setzer 1963)

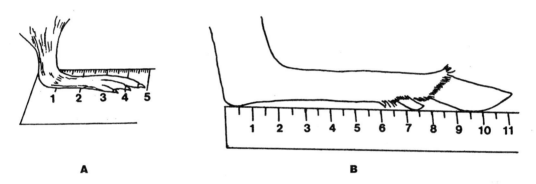

Figure 31.3 Measuring the hindfoot (HF) of a rodent (A) and a hoofed mammal (B).
(A, R. E. Martin; B, after Hershkovitz 1954)

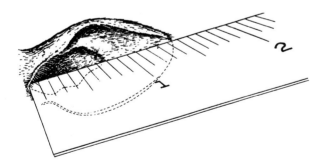

Figure 31.4 Measuring the ear length from the notch. (E).
(Setzer 1963)

base of the ear pinna in most bats. Measure from the base to tip (Fig. 31.5).

Forearm length (FA). (Chiroptera only) Fold the wing and measure from the tip of the elbow to the furthermost point on the wrist (Fig. 31.6).

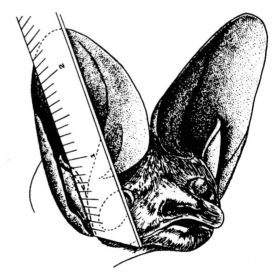

Figure 31.5 Measuring the tragus (Tr) from the notch.
(Setzer 1963)

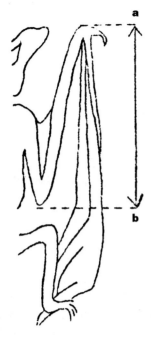

Figure 31.6 Measuring the forearm of a bat.
(After Hershkovitz 1954)

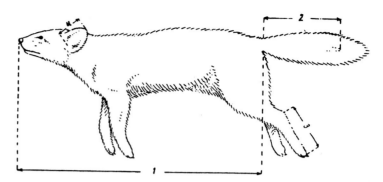

Figure 31.7 The four standard measurements taken by European mammalogists. 1, Head and body length (HB); 2, tail length (T); 3, hindfoot length excluding claws (HF s.u.); and 4, ear length (E). Note that European mammologists measure the *total length of the ear* rather than the *ear length from the notch*.
(Novikov 1956)

Record these measurements in the catalog. Be sure that they are in the proper order. The abbreviations may be used, but because the order is constant, no abbreviations are generally necessary, and the measurements may be given in sequence, separated by dashes. Thus, a rodent may be 310-150-40-12, and a bat may be 110-45-7-15-TR6-FA50.

If a portion of a mammal is missing, or if, for any other reason, the exactness of a measurement is in doubt, we recommend that the measurement be set off in brackets (or circled). Thus, if the tip of a mouse's tail was missing, the measurement would be recorded as [270]-[110]-40-12. Note that the damaged tail also affects the total length measurement. If the ears are torn or mutilated, the measurement might be 270-124-40-[10]. Head and body length should be recorded and labeled as such if part of the tail is missing. Alternatively, the tail measurement should be placed in brackets (or circled).

The measurements presented above are in standard use in North America. Collectors in other countries may take slightly different measurements. They record head plus body length rather than total length. Thus, the head plus body length must be added to the tail length to get the total length, or the tail length must be subtracted from the total length to get head plus body length. Regardless of which system is used, all three of these measurements are obtainable by simple arithmetic.

Europeans (and others) and North American collectors also differ in the method used for recording hind foot length. North Americans include the claw in this measurement, whereas Europeans omit the claw. The initials "c.u." for the Latin meaning "with claw" included or "s.u." for the Latin meaning "without claw" frequently are included after the hindfoot measurement. Thus, measurements of the same specimen by North American and European collectors might be

North American: TL 140, T 38, HF 14 (c.u.), E 12

European: HB 102, T 38, HF 12 (s.u.), E 14

Figure 31.7 illustrates the four standard measurements taken by Europeans (and some others).

Measurements of cetaceans and sirenians differ from those discussed above, because there are no hindfeet or external ears to measure. Figure 31.8 illustrates some of the measurements most commonly taken on the smaller fully aquatic mammals.

Weight

Weight (actually **mass**) should be recorded in grams (kilograms for a very large mammal) while the specimen is fresh. If the specimen is not weighed before freezing, weigh it before skinning and indicate in the catalog how long the specimen had been frozen before weighing. The weight figure is frequently placed after the measurements and preceded by three horizontal lines (e.g., ≡ 310 g).

In the field, the weight can be determined by using high-quality spring-loaded scales (e.g., *Pesola*®) that come in a number of weight ranges (5 grams to 2,000 grams). In the laboratory, a triple-beam balance is often used to obtain the mass of the specimen.

Sex and Reproductive Condition

The sex of most mammals is easily determined by examination of the external genitalia (Fig. 31.9). In some species, particularly shrews, an internal examination for testes or ovaries may be necessary. Record sex in the catalog using the symbol ♂ for a male and ♀ for a female. Use "sex?" if for any reason the sex cannot be determined.

Reproductive conditions such as lactation or descended testes should be noted in the catalog. Refer to

Specimen Preparation and Preservation **215**

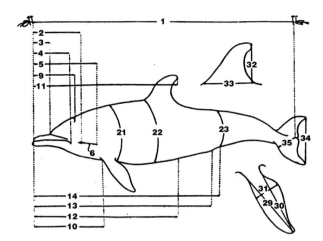

Figure 31.8 External measurements to be recorded for smaller cetaceans, as recommended by the American Society of Mammalogists' Committee on Marine Mammals. LENGTH: 1, total; 2, tip of upper jaw to center of eye; 3, tip of upper jaw to apex of melon boss; 4, gape; 5, tip of upper jaw to external auditory meatus; 6, center of eye to external auditory meatus; 9, tip of upper jaw to blowhole along midline or to midlength of two blowholes; 10, tip of upper jaw to anterior insertion of flipper; 11, tip of upper jaw to tip of dorsal fin; 12, tip of upper jaw to midpoint of umbilicus; 13, tip of upper jaw to midpoint of genital aperture; 14, tip of upper jaw to center of anus; 29, anterior insertion of flipper to tip; 30, axilla to tip of flipper; 33, dorsal fin base; 35, distance from nearest point on anterior border of flukes to notch. WIDTH: 31, flipper (maximum); 34, flukes (tip to tip). HEIGHT: 32, dorsal fin (fin tip to base). GIRTH: 21, on a transverse plane intersecting axilla; 22, maximum; 23, on a transverse plane intersecting the anus. Refer to Norris (1961) for further details.
(Norris 1961:475)

Dimmick and Pelton (1996) for further information on determination of sex and ascertaining the reproductive condition of a specimen.

The Standard Study Skin

Most mammals the size of the striped skunk, *Mephitis mephitis,* or smaller are preserved as cotton-filled study skins. Larger animals are usually preserved as tanned skins; directions for preparing skins for tanning will be presented later in this chapter. Rabbits, hares, porcupines, some armadillos, and certain other mammals that may be larger than a skunk are, for various reasons, preserved as filled study skins. Special instructions for each of these will be presented.

1. Once ectoparasites have been removed and measurements and other data are recorded in the catalog and on the tag, place the specimen on its back in a shallow pan or on a newspaper-covered surface. If tissues will be removed from the specimen for chemical or genetic analyses, be sure to have the necessary containers and storage facilities nearby to accept these tissues.

2. To make the first incision, pinch the skin of the lower abdomen between the fingers of one hand and lift it so that it is separated from the muscular body wall underneath. Use scissors to make a longitudinal cut through the skin (Fig. 31.10). Be careful to cut only through the skin and not through the muscles of the abdominal wall. Use the scissors to extend the cut backward to a point just in front of the anus and forward to the posterior edge of the

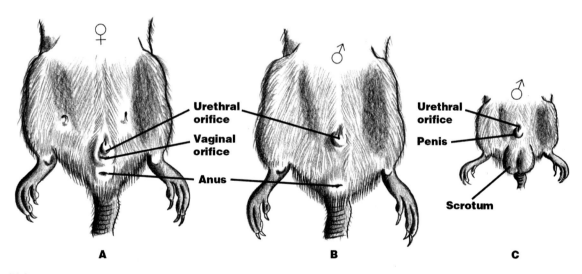

Figure 31.9 Appearance of external genitalia in an adult female (A), an immature male (B), and a mature male (C) cotton rat, *Sigmodon hispidus*. In the female (A), regardless of reproductive condition, the distance between the *anus* and the open *urethral orifice* is much shorter, and the *vaginal orifice* (may be closed in immature or nonbreeding females) will be immediately caudal to the urethral orifice. In males (B & C), note the greater distance between the anus and the base of the *penis.* An immature male (B) will not show a pronounced *scrotum* (due to smaller size of the testes) and is said to have *nonscrotal testes* or *testes abdominal* (TA). Adult males of wild species of rodents, in the nonreproductive season, will often lack a pronounced scrotum because the testes are smaller and often confined to the abdominal cavity. In contrast, a mature male in reproductive condition (C) will have *scrotal testes* or *testes descended* (TD).
(Terry Maxwell)

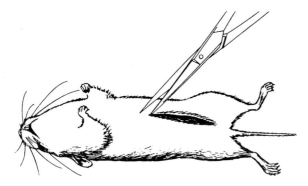

Figure 31.10 Making the midventral incision.
(Setzer 1963)

Figure 31.11 Severing the hindleg at the knee joint.
(Setzer 1963)

breastbone. If the abdominal wall is cut, use liberal amounts of corn meal or hardwood sawdust to absorb the body fluids that will leak out.

3. In males, evert the penis and sever this organ internally (inside the skin) as close to the body wall as possible, taking care not to cut through the baculum (if present). In small mammals, leave the penis containing the baculum attached to the skin. In larger carnivores and other large mammals possessing a baculum, this bone should be removed, tagged, and saved.

4. Using fingers or forceps, loosen the skin along the sides of the incision until the knee joints are exposed. Use corn meal or hardwood sawdust liberally to absorb blood or other body fluids.

5. Push the knee toward the incision and use fingers or forceps to free the skin from upper and lower legs. If the postcranial skeleton is not to be saved, cut through the knee joint (Fig. 31.11) and strip the muscle from the lower leg bones. Repeat for the other hindleg. If the postcranial skeleton is to be saved, separate the lower leg bones (tibia, fibula) from the foot bones (tarsals) at the ankle joint (*do not cut* the leg bones).

6. When both legs are free, cut through the rectum and urogenital duct(s) to free the skin from the abdomen. Use corn meal to absorb fluids from the severed tracts. Then loosen the skin around the base of the tail.

7. With the fingers and fingernails of one hand, hold the proximal end of the tail skin tight against the underlying vertebrae as you pull the tail out of the skin with the other hand. The tail skin should not be allowed to turn inside out; it should fold up accordion-fashion behind the restraining fingers as the tail is pulled from it. If the tail is difficult to pull, it sometimes helps to roll it between the tabletop and the handle of a scalpel. (*Note!* The tails of armadillos, large scaly-tailed mammals such as beavers and certain opossums, and very bushy-tailed mammals such as foxes and skunks require special techniques. Refer to the "Special Specimen Preparation Techniques" section.)

8. After the tail is free, work anteriorly, using fingers to separate the skin from the body. Try to minimize stretching of the skin during this procedure.

9. When the forelegs are reached, treat them in the same manner as the hindlegs. (*Note!* With bats, to retain the forearm measurement in the specimen and to provide support for the wing, sever the humerus near the shoulder joint or disarticulate it from the shoulder joint.)

10. Work forward until the bases of the ears are exposed. Fat and glandular tissue may have to be picked away so you can see where the ear cartilage enters the skull. In small mammals, the base of the ear can be grasped between thumb and forefinger as close to the skull as possible and pulled free from the skull. With larger specimens, it is necessary to cut the auditory canal (Fig. 31.12). The canal should be cut as close to its entrance into the skull as possible.

11. When both ears are free, work forward cautiously until the posterior edges of the eyes can just be seen through a layer of transparent tissue. Hold the skin slightly away from the head and cut *through this membrane* just over the eye (Fig. 31.13). Be careful not to cut into the eyeball or through the eyelids. When the membrane has been cut, the skin will be attached only at the front corner of the eye. Carefully sever this anterior attachment with a scalpel or fine-pointed scissors. If you are successful in this operation, you will see a complete "eye ring" of darker skin tissue that surrounded the eye.

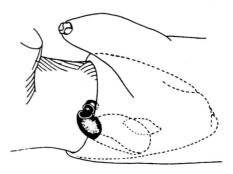

Figure 31.12 Freeing the ear cartilage.
(Setzer 1963)

Specimen Preparation and Preservation 217

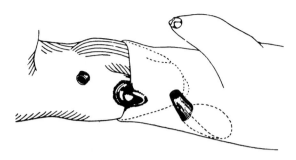

Figure 31.13 Freeing the eyelids.
(Setzer 1963)

If, by chance, the eye ring is accidently damaged in the process, it can easily be repaired at the cut or tear by using a fine needle and thread.

12. Using the fingernails or scalpel (depending on size of animal), work the skin forward until the lips are reached. As you remove the skin from the underlying bone and tissue, you will see the roots of the vibrissae on the underside of the skin. Using scalpel or scissors, free the lips from their attachments to bone and connective tissue but be careful not to cut through the bases of the vibrissae. Work the skin forward until the skin is attached to the body only at the tip of the rostrum. Use scalpel or scissors to sever the nasal cartilage, and be careful not to damage the nasal bones located immediately posterior (Fig. 31.14).

13. Lay the carcass to one side. Remove all fat and any large pieces of other tissues from the inside of the skin (rubbing the inside of the skin with the corn meal or sawdust makes the skin easier to handle and removes fats) and then turn it right side out. Check the fur for dirt or blood. Apply corn meal (or sawdust) to dirty spots and brush with an old toothbrush. If the grease or stains remain, treat them in the following manner. To remove blood stains, sponge the area with cold water. It may be impossible to remove all traces of blood stains from white hair. To remove grease from a small area of the skin, sponge the area with soapy water or with a solvent such as white gasoline or benzene. (**Caution:** These two solvents are highly flammable and poisonous.) If the skin is excessively greasy,

wash it in soapy water or one of the above solvents. Take particular care not to stretch the skin during washing. If the skin is washed, note this fact in the catalog and name the solvent used.

To dry a small area that has been washed, work dry corn meal into the fur and brush it out. Repeat until the area is dry. If a major portion of the skin needs drying, place it in a large container with a quantity of corn meal and shake. Remove the skin and brush out the wet corn meal. Repeat this procedure until the hair is dry. Compressed air may also be used for drying instead of or in addition to corn meal.

14. Sew the mouth shut using a three-cornered stitch (Fig. 31.15). If conditions are humid or the skin is excessively greasy, dust the leg bones and inside of the skin with borax. (*Note!* Borax is *not recommended* for species with red pelage.)

15. Select an appropriate diameter of monel wire for use in the specimen (e.g., 020 for pocket gophers and wood rats; 022 for large mice; 024 for smaller mice and all bats; 026 for the smallest species). To straighten monel wire, secure the end of a length of wire (a few meters) and pull on it until you have stretched it sufficiently. Then cut the wire to convenient lengths (25 to 50 cm) and store until needed.

16. Cut straight wires long enough to extend from the sole of the foot well into the body skin (Fig. 31.16). Insert a wire along the bone into the tip of the longest toe. Except for mouse-sized animals, wrap cotton around the wire and leg bones (if these were left in place) or, with forceps, insert an appropriate-sized cone of cotton into the leg skin alongside the wire. (*Note!* If the specimen is of a rabbit or hare, refer to the "Special Specimen Preparation Techniques" section of the chapter before proceeding.) For small mammals (mouse-sized or smaller), it is generally unnecessary to wrap the leg bones or wire with cotton.

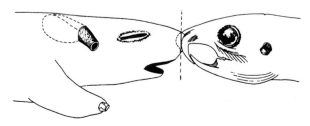

Figure 31.14 Severing the nasal cartilage.
(Setzer 1963)

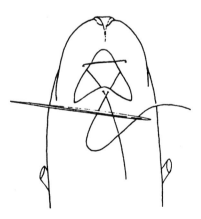

Figure 31.15 Sewing the mouth. After stitches are made, draw the thread tight and knot.
(Setzer 1963)

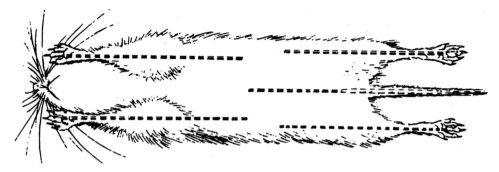

Figure 31.16 Position of the leg and tail wires in the completed study skin prior to pinning. Refer to Figures 31.19 and 31.20 for correct placement of feet.
(Setzer 1963)

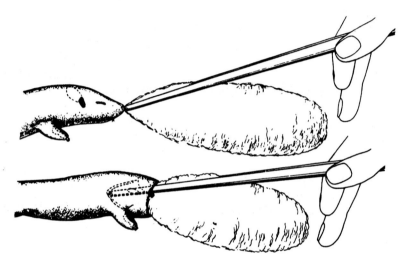

Figure 31.17 Beginning the insertion of the cotton into the skin.
(Setzer 1963)

17. For small mammals, prepare the cotton for the body by loosely folding a thin sheet into a flattened cylinder that is longer and slightly larger in diameter than the skinned carcass. Use forceps or a hemostat to fold one end of the cylinder into a pointed cone.
18. Retaining a grip on the cotton point with the forceps or hemostats, insert the cotton body into the skin, making sure the cotton *completely fills out the head and nose* (Fig. 31.17). Tear excess cotton from the posterior end of the roll and tuck the end into the rump area of the skin. Check to make sure that leg wires are correctly aligned (see Fig. 31.16).
19. Cut a length of monel wire sufficient to reach from the tip of the tail skin to just beyond the anterior end of the ventral incision (see Fig. 31.16). Use *small* wisps of cotton to tightly wrap the wire to a taper approximating that of the skinned tail but of a slightly smaller diameter (Fig. 31.18). (***Do not** wrap tail wire for bats.*) Insert the wrapped wire into the tail skin so that it will be on the ventral surface of the cotton body. (Note the position of the tail in Fig. 31.19.) Moistening the cotton with water may ease entry.

20. Close the incision using a baseball stitch (Fig. 31.20). Be careful not to catch tufts of hair under the stitches.
21. If the specimen is a male, measure the greatest length and width of each testis, exclusive of the epididymis, and record these in the catalog and on the specimen tag. If the specimen is a female,

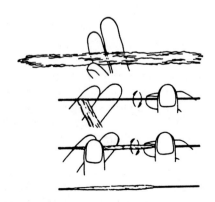

Figure 31.18 Wrapping a tail wire.
(Setzer 1963)

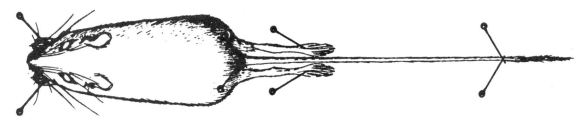

Figure 31.19 A mouse correctly pinned and positioned.
(Modified from Setzer 1963)

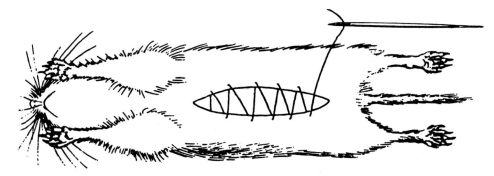

Figure 31.20 Sewing the incision.
(Setzer 1963)

check the uterus for embryos. If embryos are present, they should be counted and crown-to-rump measurements taken. These data are recorded in the catalog and on the specimen tag. Fill out the remainder of the specimen label and use a square knot to attach it *tightly just above the ankle* of the right hindfoot. With bats, the specimen label may be attached to the tibia (Fig. 31.21) to prevent distortion of the calcar.

22. Place the specimen belly-side down on a piece of cardboard, fiberboard, Styrofoam, or other substrate that will hold pins and not hamper drying. Use your fingers to manipulate the specimen into a symmetrical shape with appropriate tapering of the head. Place the front feet under each side of the head and pin them in place. Be sure that the two feet are placed evenly and that the pins do not crease the sides of the face. Draw the hindfeet posteriorly and pin them sole down, parallel and adjacent to the tail. Cross a pair of pins over the base of the tail and another pair near its tip. Be sure that the tail is in a straight line with the body and that the tail skin is right side up throughout its length (see Fig. 31.19). Dry the skin where ants, flies, and mice cannot get at it. In some cases, it may be necessary to inject the feet or other fleshy portions with 10% buffered formalin (Table 31.1) to protect against or eliminate maggots.

With bats, the pinning technique is different. The wings are folded alongside the body, and the wrist is pinned level with the nose. The phalanges and metacarpals are spread just enough so that each

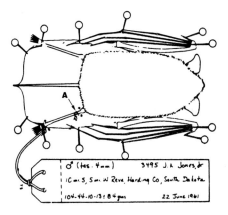

Figure 31.21 A bat correctly pinned and positioned. Attachment of the string at point A is more secure and prevents distortion of the calcar (if present). See Hall (1962:22) for elaboration.
(Hall 1962)

bone is accessible for measuring, and the wing is pinned in position. The thumb is pinned alongside the second metacarpal. The feet are pulled back and out but not spread farther than the pinned position of the wings. A pin is used behind each of the calcars (spurs supporting the tail membrane) to draw the membrane tight. A properly pinned bat is illustrated in Figure 31.21.

If the head of a pinned specimen has a tendency to rise up off the pinning surface, pin the head down with a pin at an angle through the lower lip. Be careful not to crease the side of the face.

TABLE 31.1	Some Commonly Used Fixatives and Preservatives for Mammal Tissue		
Material	Ingredients	Quantity	Nature and Use
Alcohol, ethyl, 70%	95% ethanol	70 ml	Preservative; fix and harden organs; storage
	Water	25 ml	
Alcohol-acetic acid-formaldehyde (AFA)	95% ethanol	50 ml	Preservative and fixative, especially organs; storage time not critical
	Acetic acid, glacial	2 ml	
	Formalin (40% formaldehyde solution)	10 ml	
Bouin's fluid	Picric acid, saturated aqueous solution	750 ml	Fixative; organs, tissues; several weeks storage
	Formalin (40% formaldehyde solution)	250 ml	
	Glacial acetic acid	50 ml	
Embalming fluid	Formalin (40% formaldehyde solution)	5 parts	Preservative; especially whole mammals used for anatomical dissection
	Glycerin	5 parts	
	Phenol	5 parts	
	Water	85 parts	
Formalin, 10%, neutral buffered	Formalin (40% formaldehyde solution)	100 ml	Preservative and fixative; organs, tissues, whole mammals; storage time indefinite
	Distilled water	900 ml	
	Sodium acid phosphate ($NaH_2PO_4 \cdot H_2O$)	4.0 g	
	Anhydrous disodium phosphate (Na_2HPO_4)	6.5 g	

Based on information in Guyer (1953), Humason (1967), Mosby and Cowan (1969), and Wobeser et al. (1980).

23. Embryos may be preserved by placing them in a container of 10% buffered formalin (Table 31.1) along with a labeled tag. Alternatively, each set of embryos from a specimen can be wrapped in cheesecloth, tagged with the collector's number, and placed in a container of preservative.
24. If stomach contents (see Chapter 34) or any portion of the internal anatomy is to be checked or saved, it should be done at this time.
25. If the postcranial skeleton is not to be saved, detach the skull by severing the neck. Take care not to damage the occipital condyles. Remove the eyeballs and cut away major muscle masses. Attach a labeled skull tag as shown in Figure 31.22 and see step 27.
26. If the postcranial skeleton is to be saved, do not detach the skull of small- to medium-sized animals. Remove viscera, fat, and major muscle masses. The baculum should be prepared separately, although labeled with the same preparator's number, from the rest of the skeleton. Be careful not to remove the muscles that in some mammals (e.g., felids) contain the clavicles. Attach tags to the mandible and pelvis. Fold legs and tail along the body, and if the neck is long, fold the head back as well. Use string to wrap the skeleton into a compact bundle with no protruding pieces that might be broken off (Fig. 31.23). Be sure that a tag is visible outside the wrapping.
27. Place any detached skull in a container of cold water and allow it to soak for 12 hours (in hot

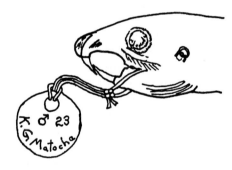

Figure 31.22 Filled-out skull tag and proper method for attachment of tag to the skull. Note that the tag is secured to the mandible but some slack is left between the knot and the bone.
(R. E. Martin)

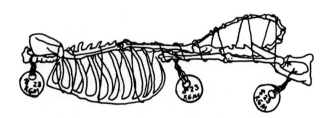

Figure 31.23 Roughed out and labeled skeleton properly arranged for drying or shipment.
(Modified from Hershkovitz 1954)

weather, change the water during this period to prevent excessive decomposition of the flesh and strong odors). Then, using an atomizer bulb and pipette or a hypodermic syringe fitted with a blunt needle, remove the brain by flushing the cranial cavity with water. In larger animals, the brain may also be removed with a brain spoon or hook that can be fashioned from a piece of heavy wire, one end of which has been flattened and curved to form a hook. Anderson (1965) suggests a wooden spoon whittled out of a stick. Then flush the brain cavity with water.

28. Place the skull, skeleton, and baculum in a safe place to dry. Uncleaned skull or skeletal materials should *never* be enclosed in an airtight container, because such treatment may cause the flesh to mold or macerate and make subsequent cleaning by dermestid beetles difficult. Do not allow skulls or skeletons to become infested with fly larvae. Also remember that various wild and domestic animals will eat uncleaned skeletal material if they get a chance. On field trips, a sturdy screened cage may be necessary to prevent loss of skeletal material to animals and to prevent flies from laying their eggs on the carcasses. For techniques to use in cleaning skulls and skeletons, refer to the "Cleaning Skeletal Material" section.

SPECIAL SPECIMEN PREPARATION TECHNIQUES

Armored Mammals

Armadillos, pangolins, porcupines, and other "armored" mammals are generally prepared as standard study skins. Because the skins of these animals cannot be turned inside out, it is necessary to extend the primary incision from the neck to the tip of the tail and work the skin free progressing from the midventral to the middorsal line. In some armadillos, the tail is encased in complete bony rings that must be cut with tin shears or a hammer and chisel to remove the tail. If it is impossible to remove the distal portion of an armadillo's tail (and with some species, it always will be), inject the area with formalin and carefully pin it so it will dry in a straight position.

Splitting Tails

In addition to the armored mammals just mentioned, many medium- to large-sized species having scaly tails (e.g., beavers, muskrat, *Didelphis* opossums) or very bushy tails (e.g., skunks, foxes) must have the tail skin cut from the tail vertebrae. Slit the skin of round or laterally flattened tails along the midventral line, being careful not to cut the hair. Then use a knife or scalpel to free the skin from the tail. A beaver tail is split along a lateral

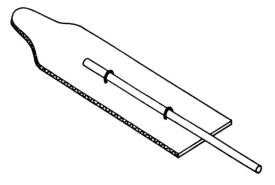

Figure 31.24 Cardboard form and wooden leg support for rabbit or hare skin. Wrap a thin layer of cotton around the form before inserting it into the skin.
(R. E. Martin)

edge. If the specimen is to be prepared as a standard study skin, the tail wire is formed, wrapped, and inserted and the tail sewn shut around it. If the skin is to be tanned, it should be treated as described in the section on "Preparation of Skins to Be Tanned."

Rabbits

Rabbits and hares require special treatment because the hindfeet are large and heavy, and the skin is thin and tears easily. Proceed as described through step 16, then spread the skin out on a table, belly side down. Cut a piece of cardboard to fit the skin. The head and neck will be narrower than the body and should be roundly tapered in front (Fig. 31.24). Cut a piece of 1/4" dowel (or a similarly sized straight twig) to a length sufficient to extend from the tips of the hind toes to the midpoint of the body. Wire this stick in position along the midline of the cardboard body (Fig. 31.24). Then wrap the cardboard with a thin sheet of cotton. Cut and wrap a tail wire (see step 19) and insert into the tail. Insert the cardboard body into the skin, leaving the dowel extending out between the hindlegs. Sew the incision carefully (lagomorph skins are thin and tear easily). Tie the hindlegs to the dowel at two points. Orient the front feet anteriorly, and with a stitch or two attach them to the underside of the neck. Lay the ears back and use a single stitch to loosely attach each to the skin of the back. Attach a specimen tag securely above the ankle of the right hindfoot. Return to step 21, then to step 23, and continue through to step 28. An alternative technique for preparing rabbits and hares was presented by Anderson (1961).

Skin-Plus-Skeleton Preparation

Hafner et al. (1984) recommended the use of a single specimen to obtain a specimen consisting of a skeleton and of a skin. Under this system, the specimen is prepared as a study skin in the normal fashion with the following exception: The entire manus and pes are removed on the left side and kept with the rest of the skeleton, and the manus and pes on the right side are kept with the study skin. Because mammals are bilaterally symmetrical, it offers a way to obtain both "complete" skeletons and "complete" skins. We recommend this method for any species where the number of specimens available for preparation is limited due to difficulty in capturing the animals or where permits from regulatory agencies may not permit larger samples to be taken.

Chromosome and Tissue Preservation

Increasingly, mammalogists utilize **tissue biopsy** and blood samples to conduct cytological, genetic, and toxicological analyses. For these studies to be meaningful, the researcher must utilize proper protocols to ensure that the samples are preserved in a manner that is appropriate for the particular study. In this specialized field, the student is advised to consult the relevant literature and to work with researchers who are familiar with proper preservation techniques. Sherwin (1991) reviewed methods for collecting mammal tissues and data for genetic studies. Baker and Haiduk (1985) provided important guidelines for collecting, preparing, and storing biopsy samples for tissue cultures.

In general, if you plan to collect tissue or blood samples from a specimen, you should do so soon after the animal dies. For tissue-culture, the tissues most often removed from dead mammals include skin, lung, heart, and connective tissue (Baker and Haiduk 1985). When animals cannot be killed, tissue samples can also be obtained from blood, skin biopsies, ear punches, toe clippings, or tail tips (Baker and Haiduk 1985; Sherwin 1991). These samples should be collected under clean conditions, placed in special cold-storage vials (e.g., Nunc tubes), and placed on dry ice or in storage containers (Dewars) of liquid nitrogen. The cold-storage vials must be properly labeled using appropriate permanent ink pens (see Chapter 29) with an appropriate reference number that contains the complete data on the specimen. At Texas Tech University, home of one of the largest repositories of frozen tissues of wild mammals, the specimen container is assigned a TK number. The TK number is used to link information about that tissue to a computerized database that contains pertinent data on the specimen. Refer to Baker and Haiduk (1985) for further details on how to process and store tissues of mammals.

Fluid Preservation of Specimens

Frequently, it is advisable to preserve all or a portion of a mammal in a liquid preservative. Whole mammal specimens should first be hardened or **fixed** in 10% *buffered formalin* (see Table 31.1). The fixative is injected into the specimen or allowed to enter through an incision in the abdominal and thoracic body walls. Body fluids will dilute the formalin so the volume of fixative should exceed the volume of the animal by at least six times (animal:fixative, 1:6). If many specimens will occupy a container, the strength of the formalin must be increased to prevent deterioration of tissue.

Unbuffered formalin is acidic and will degrade bone if specimens are left in it for long periods. Thus, formalin-hardened specimens are soaked in running water for 24 hours and then transferred to 10% neutral buffered formalin (Table 31.1), to 70% ethanol, or 50% isopropyl alcohol for permanent storage. The containers should be as airtight as possible, and tags of high-rag content or parchment tags should be securely attached just above the ankle of the right hindleg of each specimen. A label should never be attached to the outside of a container but always placed inside with the specimen. In museum collections, a stainless steel or monel metal tag (with catalog number) may be used instead of a paper or parchment tag, to minimize the chance that the number will become illegible with time.

Standard Study Skin with Cardboard Body

A study skin can be prepared using a cardboard body rather than a cotton body. This technique is widely utilized in Europe and, to some extent, in the United States. A slight amount of time may be saved in preparing skins using this method, although space is not often saved in storage cabinets because the specimens have wide lateral dimensions and tend to occupy more tray space in comparison with round skins. Some institutions (University of Kansas, Natural History Museum, London) store these skins in file cabinets. Refer to Brown and Stoddart (1977) for a description of this technique. For a variety of reasons, we do not recommend it.

Cased Skins

The smaller mammals for which tanned skins are desired (foxes, raccoons, etc.) may be prepared as modified **cased skins,** with the feet still attached. To prepare an acceptable cased skin for tanning, make the initial incision as in step 2. Extend this cut posteriorly to the tip of the tail but do not extend it anteriorly. From this incision, make cuts along the insides of the hindlegs to the feet. Remove the skin from the carcass, turning it inside out over the body. When the forefeet are reached, make an incision from the underside of the upper leg to the foot and remove the skin. Proceed as above. The only difference between a flat skin and a cased skin is that the latter is not slit from the abdomen to throat. This results in a cylindrical skin on which the ventral pelage may be more easily studied. Leave the skin turned inside out and coat both sides *liberally* with salt.

After the skin has been removed and salted, treat the carcass and skeletal material as described in step 21 plus steps 23 through 28.

Preparation of Skins to Be Tanned

by Keith A. Carson[1]

The skins of mammals larger than raccoons or foxes are usually preserved by tanning. This process changes the skin into leather.

Skinning Large Mammals

29. Make a ventral incision from the tip of the tail to the base of the head as shown in Figure 31.25. Extend the cuts up each of the legs from a point at the base of the toes (on the palm side) to the ventral incision (Fig. 31.25). To ensure symmetry, it is important to make all cuts before beginning to free the skin from the carcass.
30. With a knife or scalpel, free the skin from the Achilles tendon and the adjacent portion of the hindleg. Then pull the skin away from the leg. A slight flexing of the leg will produce enough slack in the skin to permit the inversion of the skin over the foot. Insert a scalpel (bladeless) handle between the skin and fleshy part of each digit so that the toes can be pried free of the adherent skin up to the distal phalangeal joint. This joint is then cut through, leaving the terminal phalanx in the skin. See step 41 for special instructions for skinning out the feet.
31. Skin carefully around the tail vertebrae and work the skin over the rump. Pull the skin tight and cut through the rectum.
32. Skin the forelegs in the same manner as the hindlegs and then work the skin free along each side of the ventral incision.

[1]This section written, in part, by Keith A. Carson, former tanner, Field Museum

33. If the mammal has no horns or antlers, proceed to step 34. If it has horns or antlers, make a cut beginning about midway on the back of the neck and extending straight to a point directly above the occipital condyles. Then extend the incision to the base of each horn or antler (Fig. 31.26). From the flesh side, cut the ears loose from the skull as close to the bone as possible and work the skin forward. To attempt to skin the remainder of the head at this point would be very difficult because the skin is still attached to the body.
34. With most specimens, especially large ones, it is helpful to suspend the carcass. Insert a hook

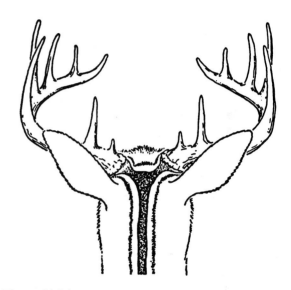

Figure 31.26 Y-shaped incision through which horns or antlers are removed.

(Anderson, R. M. 1965. Methods of collecting and preserving vertebrate animals, 4th ed., revised. Bull. Nat. Mus. Canada, 69:1–199, fig. 24. With permission of National Museums of Canada.)

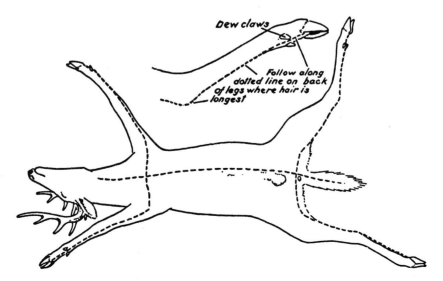

Figure 31.25 Incisions used when preparing a large skin for tanning.

(Anderson, R. M. 1965. Methods of collecting and preserving vertebrate animals, 4th ed., revised. Bull. Nat. Mus. Canada, 69:1–199, fig. 23. With permission of National Museums of Canada.)

between the tough Achilles tendon and the rest of the leg and hang the carcass from a hoist or convenient tree. Grasp the skin at the level of the tail (use an old towel or other cloth to provide a good grip) and pull the skin away (or cut free with knife) from the body all the way to the base of the skull.

35. Proceed with the skinning of the head. On mammals with horns or antlers, pry or cut the skin free at the base of these structures. Insert the index finger into the eye socket from the hair side of the skin and pull the skin gently away from the skull, stretching the tissues that hold the eyelids to the skull. These tissues can then be cut close to the bone with no danger of damaging the lids.

36. In some artiodactyls, there is a deep depression (the antorbital pit) located immediately in front of each eye. Keep tension on the skin and cut close to the bone to free the skin.

37. Cut through the lower lip at the midline all the way to the bone. Then cut posteriorly along each dentary (retaining as much of the inner surface of the lip as possible on the skin) until the skin is completely freed from the lower jaw. Continue skinning the head. Keep an even tension on the skin and work toward the rostrum. If care is taken, the skin can even be separated from much of the nasal cartilage. At this point, cut straight down through the remaining nasal cartilage leaving about 1/4 inch on the skin. Continue to free the upper lip until the entire skin is cut loose. The lips should be "pocketed" to prevent slippage (loss of hair) due to decomposition. This is done by splitting the lips all the way to the bottom of the fold that is formed by the skin and the inner mucous membrane. If the lips are thick, remove as much muscle as possible without damaging the lips.

38. The ears must be skinned all the way to their tips (on the back or hair side) or else the ear will shrivel up in the tanning process. Much of this work is done with the fingers (although a round-tipped butter knife or similar tool works well) by inserting them between skin and cartilage and separating the connective tissue. Invert the ear as the work proceeds until the ear is inside out (Fig. 31.27). It is usually not necessary to remove the cartilage from the front of the ear unless the skin of the ear is fatty or densely haired. Turn each ear right-side out.

39. The thick tissue on the inside of eyelids must be "slit" or opened to prevent the loss of the eyelashes due to decomposition. Make a number of small cuts (extend parallel to the edge of the eyelid) through the tissue back of the eyelid but not through the skin itself. Be careful not to cut into the papillae that contain the roots of the eyelashes. These appear as a line of small yellow bumps near the rim of the eyelid.

40. Split the 1/4 inch of nasal cartilage left on the inside of the skin down the middle almost to the

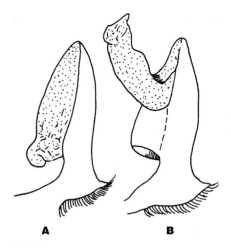

Figure 31.27 Inverted ear of larger mammal showing adherent cartilage (A, stipple = cartilage) and the process of the cartilage being separated from the skin of the pinna (B). (Redrawn from Hershkovitz 1954)

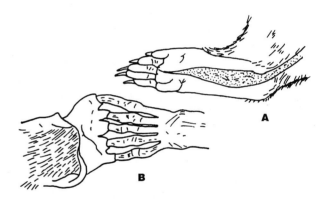

Figure 31.28 Foot of carnivore showing incision on ventral side (A) and skinned digits (B). (Redrawn from Hershkovitz 1954)

skin of the nose. This cut will help to maintain a "natural" shape to the nose. On many mammals, there are large pads of tissue adhering to the inside of the nasal skin and covering the base of the whiskers. Remove this tissue but be careful not to damage the roots of the whiskers.

41. On nonungulates, make an incision on the ventral side of each foot (Fig. 31.28A). Then skin out the foot to the base of each distal phalanx (Fig. 31.28B). The distal phalanges are left attached to the skin. In ungulates, skin the digits out completely and remove the distal phalanges from the skin. (Occasionally, the distal phalanges adhere so tightly to the hoof that their removal is not possible.) Leave the skin of the feet inverted for degreasing, defleshing, and drying. The skin is now ready for tanning.

42. After the skin has been removed, treat the carcass and skeletal materials as described in step 21 plus steps 23 to 28. The bag formed by the skin of each ear should be filled with salt, and the entire skin should be liberally coated with salt on both sides.

Tanning Skins

Most institutions ship their salted and/or dried hides to commercial tanneries specializing in hair-on tanning. (See Dowler and Genoways, 1976, for a list of commercial firms.)

If tanning is not to proceed immediately (such as under field conditions), allow the skin to dry partially and then fold it into a loose bundle (feet inside) with the flesh side on the outside.

Each specimen sent to the tanner should have the museum's catalog number punched into the skin from the flesh side to avoid confusion at the tanners (paper labels generally do not survive the tanning process). A three-cornered file that has been ground to a point is ideal for this purpose because the triangular hole it makes will not close up during the tanning process.

Cleaning Skeletal Material

by Laurie Wilkins[2]

Methods for cleaning skeletal material include boiling, maceration, the use of chemicals, and the use of various arthropods. The procedure for cleaning a mammal skeleton depends upon the size, age, and condition of the specimen; the number of specimens to be cleaned; the ultimate use of the specimen; and the facilities available for processing. There is variation in the procedures used, and some experimentation and judgment will be required to develop the most suitable methods for your needs.

Cleaning with Dermestid Beetles

Although mealworms, ants, and various crustaceans have been used for skeletal preparation, the use of beetles of the genus *Dermestes* (family Dermestidae) has generally proved to be the most favorable method. Beetles are easily acquired, and colonies are easily maintained. The size of the colony can be controlled depending upon the amount of material to be processed. In addition, beetles will satisfactorily clean all sizes of skeletons. A carefully maintained colony will produce meticulously cleaned and articulated skeletons. There is generally no loss of teeth, and very little effort is required on the part of the preparator.

Establishing and Maintaining a Beetle Colony

A starter population of dermestid beetles can usually be collected from a road-killed carcass during the warmer months in temperate North America (or obtained from an existing colony). *Dermestes maculatus* (= *D. vulpinus*) is frequently used in established colonies, although species of *Anthrenus* have also been reported (Voorhies 1948). In general, the adults of *Dermestes* are 5 to 12 mm in length,

oval-shaped, with shiny reddish-brown to black dorsal surfaces and predominantly white ventral surfaces (Fig. 31.29A). Larvae (Fig. 31.29B) are clearly segmented, elongated, bristly, and vary in length from 2 to 12 mm, depending upon the number of molts each has undergone. To establish the colony, collect a minimum of 25 or 30 adults and larvae. Adults are needed for reproductive purposes, but it is the larvae that do most of the cleaning. For further information on Dermestidae, including their identification, see Hinton (1945) and Russell (1947).

The container for the colony should be of glass, metal, or Plexiglas, with tightly sealed seams and smooth vertical walls (Voorhies 1948). Wood, Styrofoam, and cardboard are not suitable because the larvae will excavate sites in these materials to pupate and will escape. For a small colony, a coffee can or wide-mouth gallon jar will suffice. Aquaria of various sizes with screened lids are excellent containers. Large stackable trays with adjustable compartments have also been utilized (Williams et al. 1977). Replace portions of the lid or sides of the container with fine wire screening for ventilation and to prevent the escape of the insects. The container should be well-ventilated and escape-proof. Several layers of sheet cotton placed at the bottom of the container will provide pupation sites for the larvae and an egg-laying medium for the

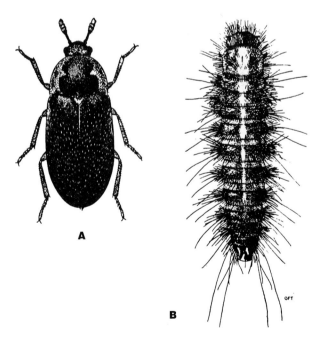

Figure 31.29 Adult (A) and larva (B) of dermestid, *Dermestes maculatus,* a species frequently used for cleaning skeletal material in natural history museums. The adults of this and other species are generally dark with some sort of pattern and are covered with scales or "hairs." The antennae are generally short and club-shaped. Larvae are covered with many long "hairs."

(Hinton, H. E. 1945. A monograph of the beetles associated with stored products. British Museum (Natural History), figs. 341 & 391. With permission of the Trustees of the British Museum (Natural History).)

---

[2]Formerly, Division of Mammals, Florida Museum of Natural History, Gainesville, Florida 32611.

adults. A metal screen with 1/4 inch openings placed inside the container and supported slightly above the cotton minimizes the accumulation of residue (excreta and shed exoskeletons) from immediately around specimens (see Sommer and Anderson 1974 for details). It will also permit air flow around specimens. The screen can be raised from time to time as the debris accumulates.

Place containers in a darkened, well-sealed room away from other biological material. Every precaution must be taken to prevent the beetles from escaping because they can cause serious damage in a collection of specimens. A well-ventilated room will keep the air relatively free of the odors and dust that result from the colony. High humidity in the room may create conditions that cause the formation of beetle-repellent molds on the specimens or mite infestations. A room temperature of 70° to 80° F is satisfactory. Fluctuations of humidity, temperature, and light intensity, within reasonable limits, do not appear to hamper the colonies to any significant degree.

Initially, any fresh material can be used to build the colony, although a sheep or pig head that may be obtained from a butcher works very well. Smaller specimens (e.g., raccoon or opossum skulls) can be used for small colonies. Introduce the starter colony directly onto the specimen after it has been properly prepared (see the "Preparing Specimens for Cleaning by Dermestids" section). It may take a month or two, depending upon the desired size of the colony, before the population has increased to the point where the beetles will be ready to clean specimens at a regular rate. Thereafter, it is necessary to introduce fresh material frequently in order to maintain the colony. How often will depend upon the colony size or level of activity desired.

Place each specimen in a separate container (cardboard box lids, metal trays, or mesh wire baskets) to ensure against loss of elements or mixing of specimens. Mesh baskets work best (cardboard boxes are quickly destroyed by the larvae as they excavate sites for pupation) but be sure to select a mesh that is fine enough to retain any elements that may become disarticulated (phalanges, claws, bacula, etc.).

The time required to clean a skull or skeleton will depend on the size and activity level of the colony. In an active colony, a small skull or skeleton will be cleaned in a day. Larger skeletons the size of a raccoon's may take two or three days. Check the condition of the skeletal material often.

Preparing Specimens for Cleaning by Dermestids

Each specimen must be properly prepared before it is placed in the colony. The procedure differs slightly depending upon the condition of the specimen (that is, whether it is fresh, was dried in the field, or previously preserved in fluids).

The first steps in preparing fresh specimens are similar to those outlined in steps 25 to 28. When these are completed, soak specimens in several changes of cold water for 6 to 24 hours, depending on size, to remove excess blood and to produce a whiter skeleton. Never use warm water because this hastens the decomposition process and will set blood in the bone.

Allow specimens to dry but do not allow them to become completely dehydrated. Proper drying of specimens improves the quality of the preparation and hastens the cleaning. Proper amounts of moisture in the specimens will attract the beetles without creating conditions favoring mold or decomposition. Dry specimens by placing them in indirect sunlight, in a hood, in front of a fan, or by applying heat in the form of a lamp. Protect the material from house pets or wild animals (e.g., raccoons). In the field, hang carcasses from a tree (in the shade) out of reach of climbing animals. As an added precaution, specimens may be placed in ventilated (screened) containers. In the summer months, if not kept free of flies, specimens may become infested with maggots. Should this unfortunate situation arise, freeze the carcass, or douse it with household bleach and rinse immediately. In either case, thoroughly dry the specimen before placing it back in the bug colony.

Dehydrated and mummified specimens that have been stored for long periods of time must be softened to make them more acceptable to the dermestids. Soak them in several changes of cold water for 6 to 24 hours. Soak fragile skeletons (e.g., of bats or immature animals) only a short time and dry them well. Larger specimens require much longer soaking periods and may still be damp when given to the beetles. In this case, tissue that is barely sticky to the touch is perfect. In extreme cases, simmer mummified specimens in water for several hours and dry before introducing them to the colony. Case (1959) suggested that presoaking specimens in concentrated ammonium hydroxide (**Caution:** Fumes are toxic) prior to the water treatment gives good results.

Occasionally, it is desirable to clean material that has been preserved in alcohol or formalin. Remove the specimen from the preservative, immerse it in running water for two days, and then dry (de la Torre 1951). If the dermestids fail to clean the skeleton adequately (especially with formalin-preserved material), several replications of the above treatment may be necessary. In extremely difficult cases, it may become necessary to resort to one of the methods mentioned in the "Maceration" or "Enzyme Digestion" sections.

Special Considerations

Immature animals create special problems. They will be quickly and completely disarticulated (including the epiphyses of long bones and all the elements of the skull in which the sutures have not fused) by a densely populated, active colony. It may take longer, but you will get better results if the specimens are put into a less active colony consisting mostly of smaller larvae.

It is sometimes necessary or desirable to limit or prevent cleaning by beetles in certain areas of the skeleton where disarticulation of delicate bones seems likely. A strong formalin solution (one part water to one part stock formalin) applied to the specific area by dipping or painting with a brush will discourage the dermestids from cleaning that portion of the skeleton (Sommer and Anderson 1974).

Never put a horned animal into a bug colony unless the horn sheaths have been removed. In addition to the damage the beetles will do to the horn sheaths, skulls in this state are difficult to fumigate, and you will increase the risk of introducing beetles into a collection.

CAUTION: Long-term exposure to beetles can cause allergies. Take precautions by wearing disposable gloves and a dust mask at all times.

Precautions with Dermestids

Long-term exposure to dermestid beetles can lead to health problems (Timm 1982). Increased sensitivity over years of exposure may result in allergies to the microscopic remains (feces, fine hairs on larval exoskeletons) that remain in the air or on the skeletons as fine dust. The allergic response includes itching, hives, irritation of the eyes and respiratory passages, cold sweat, weakness, and in the extreme case, anaphylactic shock. Take precautions by always wearing disposable gloves and a dust mask when working in a bug chamber.

It is also essential to properly clean and soak skeletons (see following sections) before installing them into your collections. Lack of attention to these precautions can result in dermestid infestations in the collection or in skeletal material with adherent remains from the bug colony.

Fumigation

When material is cleaned to your satisfaction, transfer it to a bug-free tray or muslin bag, after checking to be certain all elements are present. Quickly move specimens to a fumigation container to kill any remaining live insects. The container is best housed in the bug room or in an adjoining room sealed off from your lab or collection area. Fumigate specimens in a well-sealed glass jar or an airtight metal box or specimen case. The fumigant presently being used to protect your collection may be used. Use only the most effective approved fumigant available. Fumigate small specimens for three days and large specimens for a week. Carefully inspect specimens removed from the fumigation chamber for signs of live beetles—an indication that the fumigant must be replenished and that you must refumigate.

Subsequent Cleaning

The final stage of cleaning involves a treatment with ammonium hydroxide and water. Remove any pins or wire that might have been used in the field to attach labels, and replace these with string. If you are uncertain about the durability of the label or ink, augment the original tag with a permanent label before proceeding. (*Important:* Keep the original tag associated with the specimen—to prevent the original tag from being damaged by the solution, attach it to the outside of the container during the treatment.) Rinse skeletal material and flush out the foramen magnum with water to remove any "bug" debris. All rinsing should be done over wire screening. Then place the skeleton in a solution of one part ammonium hydroxide (using 28% to 30% stock solution) to three parts water. Use vials, jars, or stainless steel trays (one skull or skeleton per container). Cover the container and soak for 8 to 12 hours (4 hours for small skulls or skeletons). This treatment softens the remaining tissue for subsequent cleaning. The ammonium hydroxide is then drained off (the same solution may be used several times), the container refilled with water, and the specimen allowed to soak for an additional 4 hours. Large specimens may require a longer soaking period with several changes of water.

CAUTION: Work in a well-ventilated area or under a fume hood. Ammonium hydroxide fumes are toxic. Quickly and thoroughly wash areas of the skin that come in direct contact with the solution. Use rubber gloves and eye protection.

Remove any tissue remaining on the skeleton by carefully scraping with a scalpel or a stiff-bristled brush. A fine pair of scissors and forceps are useful tools for removing resistant tissue. A fine stream of water with a cross stream of compressed air may be preferable to the use of tools, especially for small or fragile skulls (Hall and Russell 1933). Once cleaned, each specimen is placed in an individual tray to prevent mixing of elements. The specimen is then allowed to dry slowly at room temperature.

Degreasing

The treatment in ammonium hydroxide may adequately degrease small skeletons, but medium-sized to large material often require additional attention.

Agents that have been used for degreasing are acetone, chloroform, benzene, carbon tetrachloride, ethylene trichloride, methylene chloride, and trichloroethylene (see Sommer and Anderson 1974 for degreasing procedure). Hildebrand (1968) has given the subject of degreasing a thorough treatment, and his account is recommended reading before undertaking this phase of the operation, should it be necessary. Toxicity, flammability, and expense are the main deterrents to a practical degreasing solution.

CAUTION: Degreasing agents are hazardous, and extreme care should be exercised. They should be used only in well-ventilated areas or under fume hoods with appropriate respirators and filter. Check current Environmental Protection Agency (EPA) or similar agencies for use of these chemicals.

Bleaching

Bleaching is *not used* for study materials because it loosens teeth, disarticulates skeletons, and degrades bone. If it is desirable to whiten bones for exhibit purposes, hydrogen peroxide (5% to 15% solution) is recommended.

Cleaning by Boiling, Maceration, and Enzyme Digestion

If conditions do not favor the use of beetles, or if there is only an occasional specimen to be cleaned, other methods can be used for cleaning skeletons. These include boiling and maceration. For all three methods, prepare the specimen in the same way as for cleaning by dermestid beetles.

Boiling

Boiling is a method for cleaning fresh, dried, or fluid-preserved specimens of medium to large size. This method usually results in a completely disarticulated skeleton. Although this may be an asset for convenient storage of large skeletons, a completely disarticulated skeleton is often not desirable. There are other disadvantages to the process. Cartilaginous elements are destroyed, teeth may fall out, bones may crack, and the thin bones of immature animals may warp. In spite of these drawbacks, boiling can be used successfully if the preparator is careful. It is *not recommended* for general cleaning of skeletal material.

Presoak prepared specimens in cold water for 6 to 24 hours. This step is not essential, but it will remove blood from the fresh carcass and soften the tissue of dried or fluid-preserved specimens. Dismember large animals into conveniently sized pieces for handling. Choose a container that will allow the specimen to be completely immersed. This will keep grease (which rises to the top) from accumulating on the bones. Wrapping the feet tightly in cheesecloth will keep the elements of the feet together if they become disarticulated. For a skull with horns, position the head so that the skull (but not the horns) is immersed. Simmer in water, to which a little detergent has been added, for up to 2 hours. Large, dried, and fluid-preserved specimens take longer than small, fresh specimens. When the tissue takes on a gelatinous appearance, remove the container from the fire, cool slowly to prevent the teeth from cracking, and wash under running tap water. The loosened flesh should easily fall away from the bone, but it may be necessary to use a brush, scraper, scissors, or forceps to remove tougher ligaments or tendons. Be careful not to cut into the bone. Always work over a screened drain, and be sure to retrieve any separated bony elements or teeth. It may be necessary to repeat the treatment. Allow the bones to air-dry at room temperature. Long bones dried too quickly may crack. Hoffmeister and Lee (1963) described a method of cleaning difficult specimens by boiling in ammonium hydroxide. This method may damage small fragile skulls and is *not recommended*.

Maceration

There are several maceration procedures for cleaning skeletons, but only bacterial maceration will be discussed here. Chemical maceration is faster, but it requires the addition of chemicals—a procedure that makes it impractical for many students. If allowed to proceed too far, chemical maceration may damage the skeleton, and it is therefore *not recommended*. Nevertheless, one such technique, using antiformin solution, has been successfully used for the preparation of skeletons from preserved materials (Green 1934; Harris 1959). For more information on chemical maceration methods, refer to Hildebrand (1968) and Mahoney (1966).

For bacterial maceration, soak the carefully roughed-out skeleton in water. Use a glass, enamel, plastic, or earthenware container with a tight-fitting cover. Allow to stand at room temperature for one to four weeks, agitating frequently and changing the water occasionally. The length of time required for soaking depends on the type of preparation desired.

It is possible to obtain an articulated skeleton using this process if the maceration is controlled by removing the skeleton before the ligaments are destroyed. This may take up to a week, but check the specimen daily. To obtain a disarticulated skeleton, maceration can be complete and may take from two to four weeks. Wrap the feet in cheesecloth to keep the elements of each foot together. Always use a screen or strainer when pouring off the water. Rinsing the skeleton in a dilute solution of household bleach and water will eliminate odors. Thoroughly rinse in water to stop the action of the bleach and dry slowly.

CAUTION: **Bacterial maceration produces strong odors that may escape from the cleaning containers and be offensive to coworkers. Also, maceration often results in specimens that are more disarticulated than is desired in most research collections when workers do not check the progress of the operation often enough.**

Enzyme Digestion

The use of enzymes for cleaning is much faster than bacterial maceration. Trypsin and papaine lead to complete disarticulation, and these are not recommended for immature animals. Pancreatin produces articulated skeletons and has been successfully used in the preparation of small and fragile skeletons. It has been noted (Williams et al. 1977) that the use of certain enzyme-activated detergents may continue to work even after the cleaning has been completed, sometimes leading to deterioration and even destruction of a specimen. For information on these methods, refer to Hildebrand (1968), Harris (1959), and to Mahoney (1966) before proceeding.

Histological Preparations

Although formalin is an excellent fixative, it may be desirable to use other fluids if special histological studies are contemplated. Standard fixatives, their suggested uses, and preparation and storage times are given in Table 31.1.

A 100% rag stock tag should be attached to the specimen, which is then placed in a container along with the fixative. In addition to the standard specimen data (see Chapter 33), record the type of fixative and the date it was added.

CHAPTER 32

Collecting Ectoparasites of Mammals[1]

Key to Arthropod Ectoparasites

When a mouse is trapped in the field, the collector frequently thinks that just one specimen has been secured. But, in reality, by trapping a single mouse, numerous zoological specimens have been captured, because a diverse community of insect and arachnid external parasites may be found on the mouse's body, not to mention the *endo*parasites. For the purpose of this chapter, we are using the term *ectoparasite* to mean all arthropods that are either found externally on the host or that are at least partially visible upon examination of the external surface of the host (e.g., bot fly larvae).

The collection of ectoparasites ("**ectos**" for short) is important for many reasons. Ectos may provide data that improve our understanding of the ecology of the host. Many parasites are important vectors for diseases affecting humans and domestic animals. Because of this economic importance, some of the organizations that have funded the study of vertebrates are more interested in the parasites than in the hosts themselves. Ectoparasites are thus well worth the little extra time it takes to collect and properly preserve them.

Segregation of Hosts

In Chapter 30, we recommended placing each mammal specimen collected in a separate container. This practice is essential if maximum data are to be gathered on an ectoparasite fauna. Specimens should never be placed together in a common container until they have been thoroughly examined for ectoparasites. The parasites collected from one host must be kept in a separate vial and never mixed with those from other hosts.

As soon as possible after death, place a small mammal killed by a trap or other method in a clean bag. Paper bags are best because they are inexpensive and may be used once and then discarded. Although this may appear wasteful, it insures that a parasite will not be transferred via the bag from one host to another, and thus contaminate a later sample. Cloth bags may be used, but they must be very carefully searched, thoroughly washed, and dried between each use.

Live mammals may be placed in paper or cloth bags. Obviously, paper bags cannot be used for specimens that will quickly gnaw their way out. To kill such specimens, place the bags, along with a piece of cotton saturated with ether or chloroform, into a container with a tight-fitting lid.

Large mammals that cannot be placed in a bag must be searched for ectoparasites in the field immediately upon being collected.

Kinds of Ectoparasites

Numerous kinds of insect and arachnid ectoparasites occur on mammals. Some groups are quite host-restricted, being found only on certain groups of mammals, whereas other parasite groups are found on a wide variety of hosts. Some parasites occupy very restricted parts of the host's body, whereas others may occur almost anywhere. Below is a list of the major groups of ectoparasites found on mammals.

---

[1]Prepared by Dr. Eric H. Smith, formerly with the Division of Insects, Field Museum of Natural History, Chicago; presently with Dodson Bros. Exterminating Co. Lynchburg, Virginia.

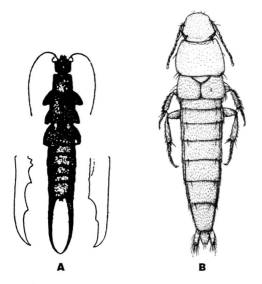

Figure 32.1 (A) Earwig: *Chelisoches morio,* (Dermaptera), forcepslike cerci variable in form; (B) rove beetle *Amblyopinus* sp. (Coleoptera: Staphylinidae).
(A, Helfer 1953:14 ; B, Machado-Allison and Barrea 1964:188)

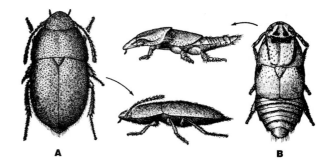

Figure 32.2 (A) mouse-nest beetle, *Leptinus testaceous* (Coleoptera: Leptinidae); (B) beaver parasite beetle, *Platypsyllus castoris* (Coleoptera: Leptinidae).
(White 1983:125)

You may wish to identify the kinds of ectoparasites that you have collected. A key to the major groups of ectoparasites found on mammals has been prepared for this purpose and follows the list. All illustrations are of mammal ectoparasites with the following exceptions: Figures 32.1A, 32.3A, 32.5A, B, D; 32.6; and 32.12A. For these exceptions, each species illustrated resembles members, of the same group, which are ectoparasites.

Insecta

Fleas (Siphonaptera, Fig. 32.7) are laterally flattened, wingless insects found on almost all mammals. They may be present anywhere in the fur, but some kinds are found embedded in certain areas of the skin. Nonembedded fleas are among the first parasites to leave a dead mammal, so the host specimens must be captured alive or taken from traps soon after death if fleas are to be collected. The nests of mammals are also good sources of fleas because that is where the slender, whitish, and legless larvae usually live, and adults may spend extended periods of time there when off the host.

Chewing/biting lice (Mallophaga, see Fig. 32.10) and sucking lice (Anoplura, see Fig. 32.11) are dorsoventrally flattened wingless insects. Lice of one order or the other are found on most mammals (except bats), but no mammalian species is known to harbor members of both orders of lice. Their presence is often indicated by eggs or egg cases glued to hairs. The entire life cycle is spent on the host, and they do not readily leave the host when it dies. Sometimes dead lice can be found on old skins in museum collections.

Bat bugs and **bed bugs** (Heteroptera: Cimicidae, see Fig. 32.9) are relatively large, wingless, dorsoventrally flattened insects. They can be found in the nests of certain rats and mice, in roosts of bats, homes of humans, and occasionally on the mammals themselves. Another heteropteran family, Polyctenidae, is rare and is found only on bats and in bat guano.

Louse flies (Diptera: Hippoboscidae, see Figs. 32.4, 32.8A) are winged or wingless insects; some species are winged when they first emerge but shed their wings when a suitable host is found. They are fairly rare on most wild mammals in most areas and are difficult to capture because most are agile and readily leave the host when it dies or is disturbed.

Bat flies (Diptera: Streblidae, Figs. 32.4D, 32.8C, D, and Nycteribiidae, Fig. 32.8B) are insects found only on bats and in bat roosts. Most emerging streblids are winged and streblids, like hippoboscids, are very agile. Because bats can be frequently plucked from a roost and placed in a bag with a minimum of disturbance (see precautions, Chapter 30), these flies are more easily collected than are louse flies. Some streblids are on their host for only a short time, usually when feeding, whereas others are on the host most of the time. Nycteribiid flies are all wingless and are spider-like in appearance. They are found in the fur and on the patagia of the host.

Bot flies (Diptera, see Figs. 32.4A, B, 32.5C: Oestridae: Cuterebrinae, Gasterophilinae, Hypodermatinae, and Oestrinae) have larvae that are parasitic on rodents, artiodactyls, and certain other mammals. Warbles, one type of bot fly larvae, are located subcutaneously and breathe through a hole in the host's skin. Their presence can usually be detected by a lump in the skin or by a denuded patch around the breathing hole.

Earwigs (Dermaptera, see Fig. 32.1A) that are ectoparasitic are rarely encountered. The suborder Arixenia contains Malayan species that are ectoparasites of bats, and the suborder Diploglossata contains South African species that are ectoparasites of rodents.

Fur moths (Lepidoptera: Pyralidae, see Fig. 32.5B) include several species that are found in the fur of sloths. The immature stages (larvae or caterpillars) feed on the algae that grow on the hair of these mammals.

Beetles (Coleoptera) include five families that contain at least some species that are found on mammals, as follows:

- **Silken fungus beetles** (Cryptophagidae) are found in the fur of mice; questionably parasitic.
- **Mammal-nest beetles** (Leptinidae, Figs. 32.3A, B, 32.5E) include a number of beetles that occur in the fur of shrews, moles, and other insectivores, as well as the fur of beavers and mice. These beetles are often found in mammal nests. Those on beavers are parasitic, and other species are questionably parasitic.
- **Dung beetles** (Scarabaeidae, Fig. 32.3B) occur on wallabies, "rat"-kangaroos, and sloths. They are found in the fur, especially near the anus, and drop off when the mammal defecates, and they then deposit eggs in the excrement or droppings. These beetles are not parasitic on the mammal.
- **Rove beetles** (Staphylinidae, see Figs. 32.1B, 32.5D) are found in the fur of rodents and marsupials. Their most common place of attachment is behind the ears, but they will detach and scurry through the fur if the animal is distressed. They leave the dead host when it begins to cool but are then sluggish and do not wander far from the host. Some of these beetles are questionably parasitic, while others feed on the ectoparasites present on the mammal.
- **Small carrion beetle** (Leptodiridae; Fig. 32.3A) include some species that live in the nests of small mammals and some that are found on living mammals and may be parasitic.

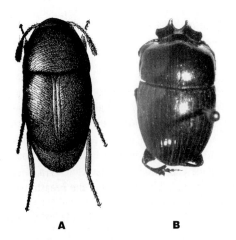

Figure 32.3 (A) Small carrion beetle, *Ptomaphagus* sp. (Coleoptera: Leptodiridae); (B) dung beetle, *Uroxys gorgou* (Coleoptera: Scarabaeidae).
(A, White 1983:123; B, H.F. Howden photograph, with permission)

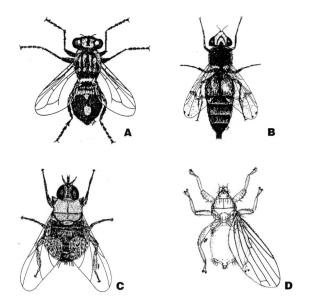

Figure 32.4 Representatives of winged Diptera: (A) common cattle grub, *Hypoderma lineatum* (Oestridae: Oestrinae); (B) horse bot fly, *Gasterophilus intestinalis* (Oestridae: Gasterophilinae); (C) louse fly, *Icosta americana* (Hippoboscidae); (D) bat fly, *Trichobius sphaeronotus* (Streblidae), left wing removed.
(A, B, & C, Bland and Jaques 1978:350, 354; D, Jobling, B. 1939:495. On some American genera of the Streblidae and their species, with the description of a new species of *Trichobius*. Parasitology, 31:486–497, fig. 4A. With permission of Cambridge University Press.)

Arachnida

Hard ticks (Acari: Metastigmata: Ixodidae, see Figs. 32.14, 32.16A, B, C) are usually firmly attached to the host, with the small head embedded in the skin and the large abdomen protruding above the surface. They are common on mammals and may be found anywhere on the body, especially in the region of the ears. They do not leave the host until their feeding is completed, and they are sometimes difficult to detach without damage to either tick and/or host.

Soft ticks (Acari: Metastigmata: Argasidae, see Figs. 32.15, 32.16D, E, F) are sometimes found on bats and on certain burrow-nesting rodents in arid regions. They more commonly occur in crevices in caves, in hollow trees, and in burrows occupied by their hosts.

Chiggers and **follicle mites** (Acari: Prostigmata, see Figs. 32.22, 32.23). Chiggers are parasites as larvae and are common on many mammals, forming clumps in, on, or around the ears, inside the nasal passages, around the lips, eyes, and urogenital area, or in other constricted regions. They do not readily leave the host. Follicle mites occur in the apocrine sweat glands of humans or the Meibomian glands of certain bats and rodents.

Body mites (Acari: Mesostigmata, see Figs. 32.13, 32.17) are commonly found on many marsupials, bats, and rodents. They occur in the fur and on naked areas of skin such as the bat's patagium. Many species occur only in very specific body locations such as the ears, nasal cavities, and anal opening. They leave the host fairly soon after death.

Mange and **fur mites** (Acari: Astigmata, see Figs. 32.19, 32.20, 32.21) are extremely minute and frequently overlooked because of their small size. They burrow under the skin or attach to hairs. They are slow to leave the host after death. Species in several families produce nonfeeding forms called hypopi, which are adapted to cling to mammals, etc. These phoretic hypopi are not parasitic as are the larvae and adults, and their function is presumed to be dispersal.

KEY TO ARTHROPOD ECTOPARASITES OF MAMMALS

1 With one pair of antennae, may be short and/or concealed in grooves on head (Fig. 32.2, 32.8); at most and usually with three pairs of jointed legs (rarely with additional tubular leglike appendages on venter of abdominal segments) .... (INSECTA, in part) **3**

1' Without antennae; leg number variable, usually not three pairs **2**

2 (1') Legless, without jointed legs or prolegs (tubelike abdominal legs) .... (INSECTA, in part; larvae) **13**

2' Usually four pairs of jointed legs (some immatures with three pairs; rarely with only two pairs) .... (ARACHNIDA: Acari) **15**

3 (1) One or two pairs of well-developed wings, may be membranous or horny/leathery; if front wings horny/leathery, then hindwings may be absent **4**

3' Wingless, or if with vestigial or rudimentary wings, then front wings not horny/leathery **8**

4 (3) Wings membranous (like cellophane), but may also be covered with scales **7**

4' Front wings horny/leathery **5**

5 (4) Front wings short, leaving most of abdomen exposed (Figs. 32.1, 32.2B) **6**

5' Front wings long, covering most of abdomen (Fig. 32.2A), Leptinidae in part (Fig. 32.3A), Leptodiridae (Fig. 32.3B), Scarabaeidae; Cryptophagidae .... **Coleoptera**
beetles

6 (5) Abdomen with relatively long, heavily sclerotized cerci (Fig. 32.1A), cerci either straight (on rodents) or forcepslike (on bats) .... **Dermaptera**
earwigs

6' Abdomen without such cerci (Figs. 32.1B, 32.2B); six abdominal segments exposed and 11-segmented antennae (Staphylinidae, Fig. 32.1B) or 4 abdominal segments exposed and antennae short and indistinctly segmented (Leptinidae in part; on beavers, Fig. 32.2A) .... **Coleoptera**
beetles

7 (4) With only one pair of wings, these not covered with scales (Fig. 32.4) .... **Diptera**
flies

7' With two pairs of wings, largely or entirely covered with scales (on sloths) .... **Lepidoptera: Pyralidae**

8 (3') Body appearing insectlike, with distinctive head, thorax, and abdomen, and with jointed legs suitable for locomotion **9**

8' Body larviform, appearing caterpillarlike (Fig. 32.5A, B) or maggotlike (Fig. 32.5C) with no apparent head and legs, or like an elongate grub (Fig. 32.5D, E) **13**

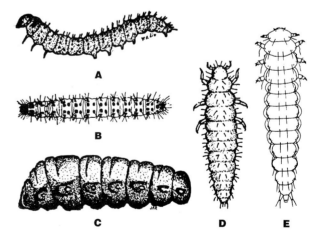

Figure 32.5 Representatives of larvae: (A) caterpillar (Lepidoptera) showing prolegs on abdominal segments 3–6 and 10; (B) garden webworm, *Loxostege similaris* (Lepidoptera: Pyralidae); (C) bot fly larva (Diptera: Oestridae); (D) rove beetle larva, *Oligota oviformis* (Coleoptera: Staphylinidae); (E) mammal-nest beetle larva, *Platypsyllus castoris* (Coleoptera: Leptinidae).
(A, B & D, Chu 1949:164, 154, & 83; C, Bland and Jaques 1978:354; E, Wood 1965:53)

9 (8)   Tarsi five-segmented (Fig. 32.6); antennae short and usually concealed in grooves on head; mouthparts piercing-sucking, forming a beak **10**

9'   Tarsi with fewer than 5 segments; antennae and mouthparts various **11**

10 (9) Body flattened laterally (at least thorax flattened, abdomen of embedded gravid female swollen and not flattened); usually jumping insects with large coxae and relatively long legs (Fig. 32.7) .... **Siphonaptera**
        fleas

10'   Body flattened dorsoventrally; not jumping insects (Fig. 32.8) .... **Diptera**
        bat flies and louse flies

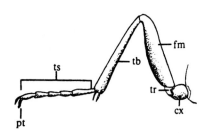

Figure 32.6 Generalized insect leg; cx, coxa; tr, trochanter; fm, femur; tb, tibia; ts, tarsus; pt, pretarsus (claws).
(Bland and Jaques 1978:32)

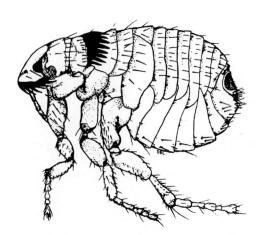

Figure 32.7 Cat flea, *Ctenocephalides felis* (Siphonaptera).
(Bland and Jaques 1978:357)

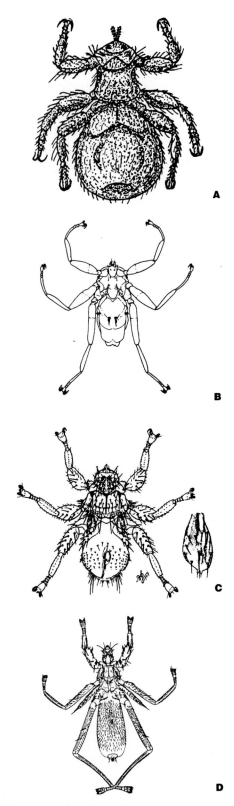

Figure 32.8 Representatives of Diptera with reduced or no wings: (A) sheep ked, *Melophagus ovinus* (Hippoboscidae); (B) nycteribiid bat fly (Nycteribiidae); (C and D) streblid bat flies (Streblidae): *Mastoptera guimaraesi* (enlargement of wing on right) and *Neotrichobius stenopterus*.

(A, Bland and Jaques 1978:350; B, From Manual of Medical Entomology, Third Edition, by Deane P. Furman and Elmer P. Catts, by permission of Mayfield Publishing Company. Copyright © 1961 and 1970, Deane P. Furman; C & D, Wenzel and Tipton 1966:513, 537)

11 (9')   Antennae distinctly longer than head; tarsi three-segmented (Fig. 32.9) ...... **Heteroptera**
bed bugs and bat bugs

11'   Antennae not longer than head; tarsi one-segmented **12**

12 (11')   Mouthparts mandibulate; head as wide as or wider than prothorax (Fig. 32.10); or if head narrower, then mandibles borne at end of long proboscis and found on elephants and warthogs .... **Mallophaga**
chewing lice

12'   Mouthparts piercing-sucking; head usually narrower than prothorax (Fig. 32.11) ...... **Anoplura**
sucking lice

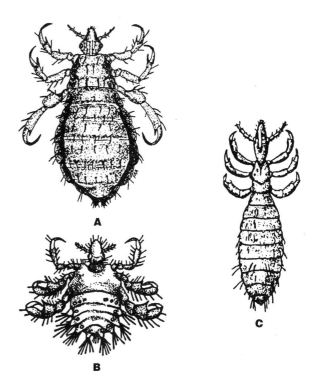

Figure 32.11 Representatives of sucking lice (Anoplura): (A) body louse, *Pediculus humanus humanus;* (B) crab louse, *Pthirus pubis;* (C) longnosed cattle louse, *Linognathus vituli.*
(Bland and Jaques 1978:127, 128)

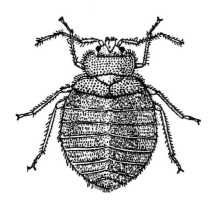

Figure 32.9 Bed bug, *Cimex lectularius* (Heteroptera).
(Bland and Jaques 1978:42)

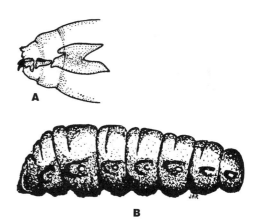

Figure 32.12 Fly larva (Diptera) without head capsule and legs: (A) enlargement of anterior end where head is represented by two sclerotized hooks; (B) bot fly larva (Oestridae: Oestrinae).
(A, Chu 1949:29; B, Bland and Jaques 1978:354)

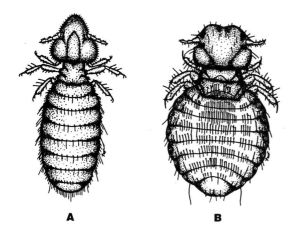

Figure 32.10 Representatives of chewing lice (Mallophaga): (A) cattle chewing louse, *Bovicola bovis;* (B) dog chewing louse, *Trichodectes canis.*
(Bland and Jaques 1978:125)

13(2/8')   Body without head capsule, head represented by one or two median sclerotized hooks that move vertically (Fig. 32.12A), legless (Fig. 32.12B) .... **Diptera**
botflies

13'   Body with distinct head capsule (Fig. 32.5A, B, D, E), mouth parts mandibulate; with jointed legs **14**

14(13') With prolegs (tubelike unjointed legs, Fig. 32.5A) on venter of some abdominal segments (on sloths) .... **Lepidoptera**
Pyralidae

14' Without prolegs but may have terminal appendages on abdomen (Leptinidae, on beavers, Fig. 32.6E; Staphylinidae, on rodents and marsupials, Fig. 32.6D) .... **Coleoptera**
beetles

15 (2') With body stigmata (spiracles/breathing pores) located behind coxae IV or laterad between coxae III-IV, each surrounded by a stigmal plate (Figs. 32.13B, 32.14B, F, 32.15B) **16**

15' Without such stigmata/spiracles (Figs. 32.19, 32.20, 32.21, 32.23) **17**

16 (15) With body stigmata/spiracles located behind coxa IV (Fig. 32.14B, F); palpal tarsus without apotele; hypostome well-developed, with recurved teeth ventrally (Fig. 32.14C); with Haller's organ on tarsus of first pair of legs (Fig. 32.14D); Fig. 32.16 .... **Metastigmata**
ticks

16' With body stigmata/spiracles located laterad between coxae III–IV (Fig. 32.13B); with a terminal, subterminal or basal apotele on palpal tarsus (Fig. 32.13C); hypostome without recurved teeth; without Haller's organ; Fig. 32.17 .... **Mesostigmata**
body mites

17 (15') Pretarsus without paired lateral claws, empodium either clawlike, or usually, suckerlike (Fig. 32.18A); chelicerae typically chelate-dentate; genital papillae usually present; genital field/pore located ventrally, anterior to coxae IV; palpi small, two-segmented; dorsum of body never covered by overlapping sclerites and never vermiform (Fig. 32.19); (Figs 32.20, 32.21) .... **Astigmata**[2]
mange and fur mites

17' Pretarsus with paired lateral claws (Fig. 32.18B), empodium rarely clawlike or suckerlike (Fig. 32.18A); chelicerae typically stylettiform or hooklike; genital papillae usually absent; genital field/pore terminal; palpi usually three- to five-segmented and conspicuous, if palpi small and with fewer segments, then body dorsum either vermiform or with overlapping sclerites (Fig. 32.22); chiggers (Fig. 32.23A, B), and follicle mites (Fig. 32.23C, D, E, F) .... **Prostigmata**
follicle mites and chiggers

Supplies and Equipment Needed

In addition to the supply of bags discussed earlier, a few other items are needed for collecting ectoparasites.

Cotton, chloroform or *ether,* and a *tightly closed chamber* are needed to anesthetize live hosts and the ectoparasites.

A *white enameled tray* approximately 12" × 18" provides a good background against which to search for ectos.

An old *toothbrush* or a similar implement is useful for brushing the fur to remove concealed parasites.

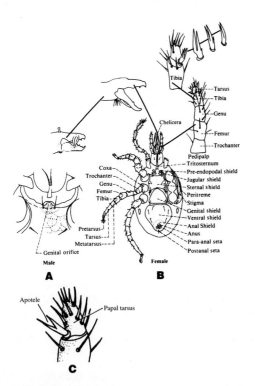

Figure 32.13 Body mites (Mesostigmata). (A & B) generalized mite with structures labeled: (A) male venter showing genital area; (B) female venter; (C) palpal tarsus of mite showing an apotele, *Haemogamasus sp.*

(A, B, McDaniel 1979:11; C, from *Manual of Medical Entomology*, Third Edition, by Deane P. Furman and Elmer P. Catts, by permission of Mayfield Publishing Company. Copyright © 1961 and 1970, Deane P. Furman)

[2]Some with a specialized non-feeding form called a hypopus, which looks completely different from all other developmental stages; see Krantz (1970) or McDaniel (1979).

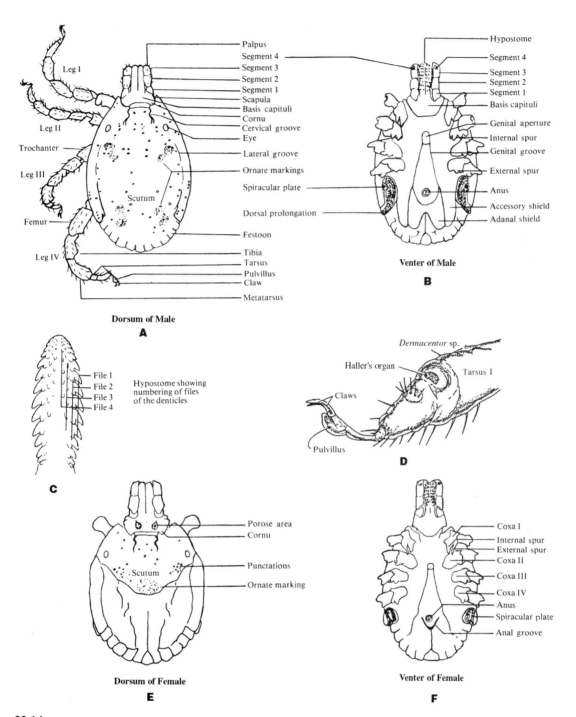

Figure 32.14 Hard ticks (Metastigmata: Ixodidae). Generalized tick with structures labeled: (A) male dorsum; (B) male venter; (C) hypostome with recurved teeth; (D) tarsus of first pair of legs showing Haller's organ, *Dermacentor* sp.; (E) female dorsum; (F) female venter.
(A,B,C,E, & F, McDaniel 1979:18; D, from Manual of Medical Entomology, Third Edition, by Deane P. Furman and Elmer P. Catts, by permission of Mayfield Publishing Company. Copyright © 1961 and 1970, Deane P. Furman.)

Very fine-pointed *jeweler's forceps,* a *camel's hair brush,* a *dissecting needle,* an *aspirator,* or a combination of these is needed to pick up the ectos and transfer them to alcohol vials.

Numerous *small vials* of 70% *ethanol* are needed for storage of ectoparasites. The vials should be equipped with tight-fitting caps or stoppers that will not allow the alcohol to leak or evaporate. One vial is needed for each host specimen examined.

Small *tags* of good-quality paper (see Chapter 29) are needed for inserting data into the vials with the parasites.

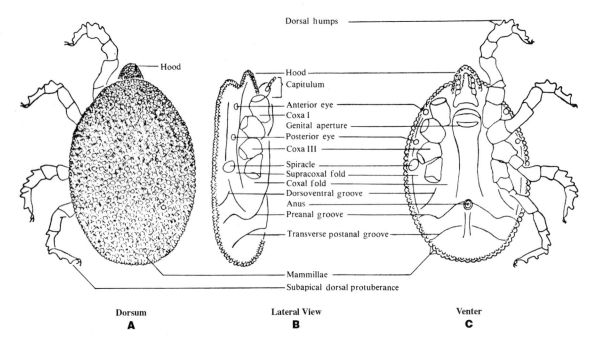

Figure 32.15 Soft ticks (Metastigmata: Argasidae), generalized tick with structures labeled: (A) dorsum; (B) lateral view; (C) venter.
(McDaniel 1979:19)

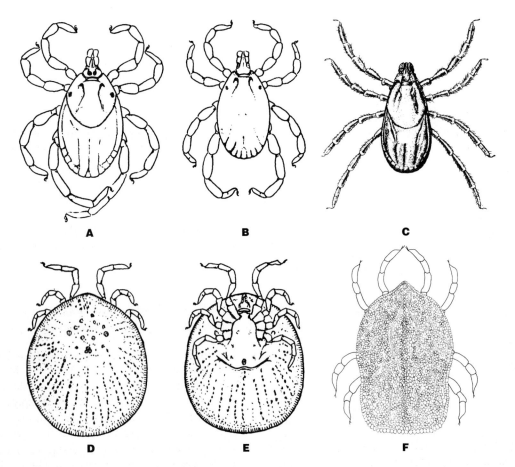

Figure 32.16 Ticks (Metastigmata): hard ticks (Ixodidae): (A) female, (B) male, and (C) brown dog tick, *Rhipicephalus sanguineus*, commonly on dogs; soft ticks (Argasidae): (D) dorsum and (E) venter of female, and (F) *Ornithodoros talaje*, found on wild rodents and humans.
(A,B,D, & E, Baker et al. 1958:81; C, F, McDaniel 1979:101,107)

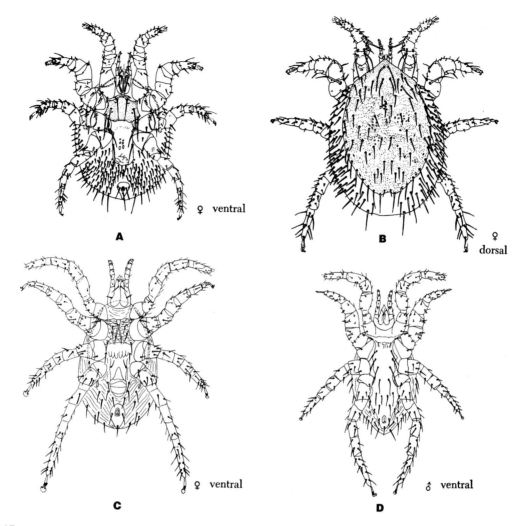

Figure 32.17 Body mites (Mesostigmata: Laelapidae): (A & B) *Gigantolaelaps mattogrossensis,* commonly on the rice rat, *Oryzomys palustris*. (A) Female venter, (B) female dorsum. (C & D) *Haemolaelaps geomys,* commonly on pocket gophers, (C) female venter, (D) male venter.
(McDaniel 1979: 90, 91)

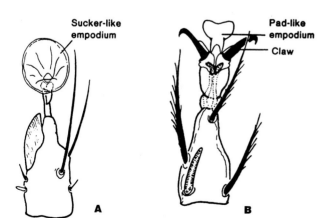

Figure 32.18 Tarsi of mites showing empodium: (A) tarsus II showing sucker-like empodium; (B) tarsus II showing pad-like empodium.

(Krantz, G. W. *A manual of acarology.* Oregon State University Book Stores, Inc., Corvallis. 335 pp., figs. 5-3 & 5-7. Copyright 1970 G. W. Krantz, with permission of the author.)

A solution of *detergent and water* may be needed to thoroughly wash dead hosts and remove adherent ectoparasites.

SEARCHING FOR AND REMOVING ECTOPARASITES FROM THE HOST

Carefully remove the dead host from the bag while holding them over a clean white tray. Reseal the bag and set it to one side. If the host has not been placed in ether or chloroform, place it in the center of the tray and immediately begin to search for moving parasites such as fleas, mites, and the various dipterans. Pick these up with jeweler's forceps, a dissecting needle, or a camel's hair brush, which have been moistened with alcohol or with an aspirator.

Some collectors do not etherize or chloroform their hosts feeling that they can more easily locate moving

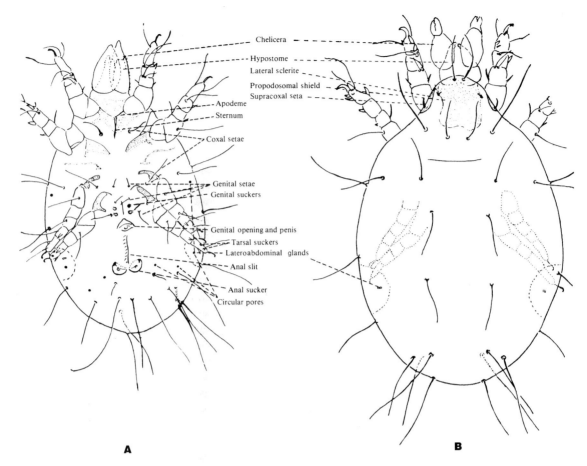

Figure 32.19 Mange and fur mites (Astigmata), generalized mite with structures labeled: (A) venter, (B) dorsum. (After McDaniel 1979:21)

parasites than dead ones. However, because winged insects, fleas, and other active parasites may escape during the search, other collectors prefer to kill all parasites before looking for them.

Once all visible ectos have been gathered, search the areas of the ears, eyes, lips, axillae, and urogenital openings for ticks, chiggers, and other embedded parasites. These embedded creatures cannot be removed without damaging either the parasite or the mammal's skin. Because the damage to the mammal is usually only a small hole in the skin, and the damage to the parasite could render it impossible to identify, it is best to remove embedded ectos by cutting free the small piece of skin to which they are attached.

Rub through the mammal's fur with your finger, forceps, or dissecting needle to find concealed organisms. Then, hold the host above the tray and brush the hair vigorously in all directions with a toothbrush or similar utensil. Carefully pick through the debris in the tray and do not neglect to examine the brush itself.

When you are certain that you have removed all ectoparasites from the host, examine the bag that it was in. Pay particular attention to seams. If the bag is paper, tear it open for close examination. If it is cloth, turn it inside out and shake or brush it over the tray.

Ectoparasites may also be removed by placing a dead host in a closed container (e.g., glass jar) partially filled with a solution of liquid detergent (one or two drops) and water. By vigorously shaking the container, the parasites may be dislodged from the host. The host is then removed from the container and the solution allowed to stand for a few minutes. (It may be necessary to add a few drops of 95% alcohol to disperse bubbles in the solution.) The supernatant fluid is then decanted several times, and the remaining fluid is examined for ectoparasites with the aid of a binocular dissecting microscope. Hosts that must be kept alive can be placed on a wire screen above a detergent solution for 24 to 48 hours. Detaching or mobile ectoparasites will often fall into the solution and can then be recovered.

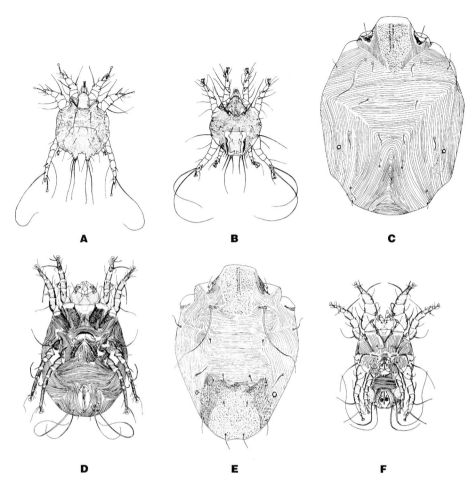

Figure 32.20 Mange and fur mites (Astigmata): (A) *Psoroptes oris,* causes mange to horses, cattle, sheep, goats, elk, rabbits, and mountain sheep; (B) *Otodectes cynotis,* the ear mite of cats and dogs; (C–F) *Dermatophagoides farinae,* causes scalp dermatitis of humans and is also found on the Norway rat, *Rattus norvegicus* and mice of the genus *Peromyscus* sp., (C) Female dorsum, (D) female venter, (E) male dorsum, (F) male venter.
(McDaniel 1979: 228, 231, 259, 260)

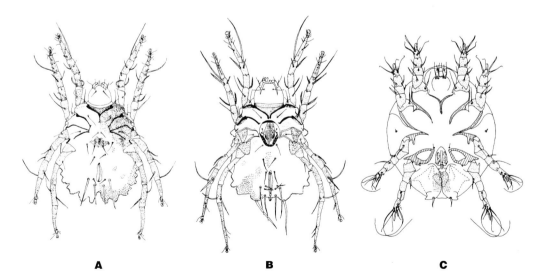

Figure 32.21 Hypopial form and comparison to male and female in the Glycyphagidae from North American mammals: *Labidophorus talpae* from the hairy-tailed mole, *Parascalops breweri*. (A) Male venter, (B) female venter, (C) hypopus venter.
(McDaniel 1979:234, 235)

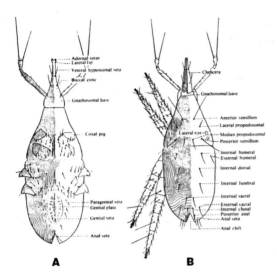

Figure 32.22 Follicle mites and chiggers (Prostigmata), generalized bdellid mite (Bdellidae) with structures labeled: (A) venter, (B) dorsum.
(McDaniel 1979:11)

When you have located all ectoparasites and transferred them to a vial of 70% ethanol, proceed with the preparation of the mammal specimen as outlined in Chapter 35. If warbles are found, these should be left in place until the animal is skinned, then they can be removed easily from the inside of the skin and added to the vial with the other ectoparasites.

Note in the field catalog and on the tag that ectoparasites were collected. If you can identify these to order or family, do so in the catalog and give at least approximate numbers. If certain types of ectos were found only in certain regions of the body (e.g., mites in the anus), note this fact in the catalog. If pieces of skin-bearing embedded parasites have been cut away, note this in the catalog and give the location on the host from which they were removed. On a small durable tag, write the catalog number of the host (including collector's name). If space permits, add date and place collected. Insert this tag *into* the vial with the parasites and seal the vial.

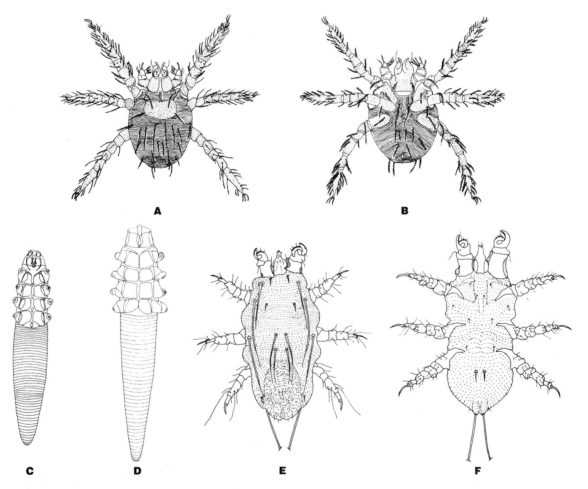

Figure 32.23 Follicle mites and chiggers (Prostigmata): (A & B) chigger *Trombicula alfreddugesi* (Trombiculidae), (A) larval dorsum, (B) larval venter; (C & D) follicle mite *Demodex canis* (Demodicidae), commonly on dogs in the United States, (C) male venter, (D) female venter; (E & F) myobiid mite, *Blarinobia cryptotis,* (Myobiidae) in the United States found on mammal orders Insectivora, Chiroptera, Rodentia, and Marsupialia, (E) male dorsum, (F) male venter.
(McDaniel 1979:128, 174, 193)

Before proceeding to the next specimen, be sure that no parasites or debris containing parasites remain on the tray, on the brushes, or on any of the other tools.

IDENTIFICATION OF ECTOPARASITES

Identification of ectoparasites to order or family is relatively easy, but beyond the family level it is difficult or impossible for the novice. There are, however, experts in the United States and elsewhere who are willing to receive collections and provide identifications (Arnett 1993). The entomology departments of most large universities and major natural history museums usually are willing to help you locate an expert or group of experts on various taxa. The specimens themselves, however, should never be sent until the expert has expressed an interest in them and willingness to work with them.

CHAPTER 33

AGE DETERMINATION

Being able to ascertain or estimate age is important in studies of mammals. Knowledge of age structures of populations is necessary for their management (Morris 1972; Taber 1969) and for understanding the life history strategies of mammals (Caughley 1966; Cole 1954; Deevey 1947; Wilson 1975). Methods of determining the ages of individuals in a population are reviewed by Gandal (1954), Morris (1972), Spinage (1973), Pucek and Lowe (1975), and Harris (1978); also see the bibliography by Madson (1967) and articles in journals such as *Acta Theriologica,* the *Journal of Mammalogy,* and the *Journal of Wildlife Management.*

TYPES OF AGING CRITERIA

When samples of mammals are taken from a population, it is usually impossible to assign a **known age** to any specimen unless birth was observed and the individual was uniquely marked for later identification. An **absolute age** can often be determined by counting incremental growth lines in various structures of wild-caught individuals. More commonly, a **relative age** may be assigned to an individual based on comparison with other individuals in the sample. The procedures for establishing absolute and relative ages should be standardized by study of individuals of known age (Morris 1972). However, if the individuals of known age were captive animals, the morphological changes that occurred during development may have been different than the changes experienced by individuals living in the wild, and thus the results obtained from captives may not be accurate for purposes of aging.

A rough estimate of relative age is used by many taxonomists to segregate adult from younger individuals of a sample. In mammalian population biology, the category **adult** generally refers to the larger and potentially reproductive members of a population. A **subadult** individual is generally a young of the year that may or may not be in reproductive condition. This individual is typically smaller than an adult but is otherwise similar. A **juvenile** individual is smaller than a subadult and often (but not always) has a pelage coloration that is different from that of subadults and adults. A **nestling** is a recently born individual that is still confined to a nest. In precocial species (e.g., artiodactyls, hares), the nestling stage is virtually nonexistent, and only a juvenile stage is present.

33-A Assemble a sample of skulls of one species from a single locality and, preferably, representing a small interval of dates of collection. Arrange the skulls in the order of their presumed age by examining tooth wear, tooth eruption, degree of ossification of bones, etc. (see Chapter 2). Can you separate these skulls into adult, subadult, and juvenile categories?

33-B Examine skins from the same sample of specimens assembled in Exercise 33-A. Does information from pelage morphology and coloration aid in assigning these specimens to age groups? Do these procedures tell you the absolute ages of these specimens?

Use of Statistics and Known-Age Samples

It is important to have a reference sample of known-age individuals so that the efficacy of an aging procedure can be evaluated. Some of the aging techniques discussed will be more accurate than others, although they may involve more time and care in processing the samples. The nature of the investigation will determine what level of accuracy you should strive for and whether you can accept a relatively larger error rate to save time for other procedures.

In studies of relative age, the technique of regression analysis is often used (Sokal and Rohlf 1969). With known-age individuals, age is a nonrandom variable (i.e., it is measured without error) and thus the quantity measured (Y, dependent variable) must be regressed on age (X, the independent variable). With samples of unknown age, age can be estimated by the method of inverse prediction (Dapson 1973; Dapson and Irland 1972; Sokal and Rohlf 1969:446–448) using values obtained from the measured variable (e.g., weight). In such cases, the computation of confidence limits for these regression lines is different from that of a Model I regression (Sokal and Rohlf 1969).

A representative sample of relative and absolute age-determination methods will be described below under the type of structure involved.

Growth of Skull, Skeleton, and Body

Dimensions and Weight

In early life, growth in most kinds of mammals is continuous and thus provides a means for separating the youngest individuals from adult members of a population. Some structures cease growth sooner than others. Not all morphological features are useful for aging individuals in a population. Increases in linear dimensions and weight may be useful indicators of age during the earliest portions of a mammal's life (Fig. 33.1) but rapidly lose their usefulness once adult dimensions are reached (Hoffmeister and Zimmerman 1967; Kirtpatrick and Hoffman 1960). These measurements are frequently utilized in live-trapping studies because of their simplicity and because of the frequent lack of other suitable criteria obtainable from live mammals.

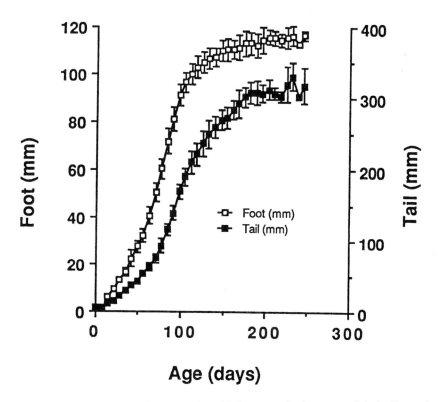

Figure 33.1 Increase in mean foot and tail lengths (± SD) from birth to maturity in a marsupial, the Tasmanian bettong, *Bettongia gaimardi*, Diprotodontia, Potoroidae.
(Rose 1989: 255)

Degree of Fusion of Epiphyseal Cartilage

The degree of fusion (determined by X-ray analysis or analysis of skeletons) of the distal epiphyseal cartilages of the radius and humerus has been used in many management studies to age specimens of bats, rabbits, and foxes (Sullivan and Haugen 1956). The technique is less accurate than most other methods (Wight and Conaway 1962).

33-C Assemble a sample of known-age skulls (perhaps obtainable from laboratory colonies, a mink ranch, or a research project). Select several cranial measurements (see Chapter 2) and measure the skulls. Using age as the independent variable (X) and a measurement variable as the dependent variable (Y), perform a regression analysis on these data (see Sokal and Rohlf 1969 or similar reference).

33-D Examine skeletal elements (pelves, distal elements of limbs) of a sample of known-age material (see Exercise 33-C). Is the degree of fusion of the epiphyses a useful indicator of age? At what age (in the species studied) do the epiphyses completely fuse with the diaphysis (i.e., shaft of longbone)?

Baculum

For male mammals, the weight, length, and volume of the baculum often provide a means for separating a juvenile from an adult (Fig. 33.2) individual (Elder 1951; Friley

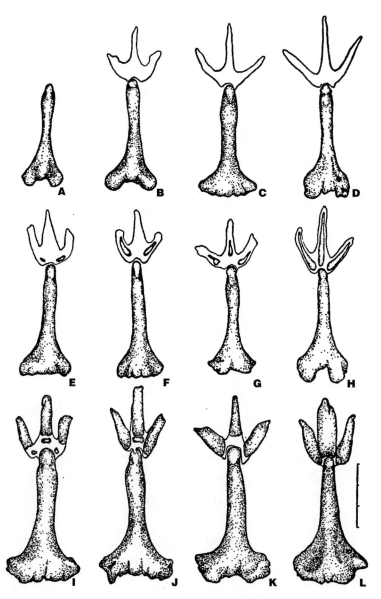

Figure 33.2 Age changes in bacula of muskrats, *Ondatra zibethicus,* Rodentia, Muridae, Arvicolinae. The top (A–D) and middle (E–H) rows are from juvenile individuals five to eight months of age. The bottom row (I–L) shows patterns in adults 15 months or more of age. The scale is in millimeters.
(Elder and Shanks 1962:146)

1949). More precise age determinations are generally not possible using this bone (Harris 1978; Morris 1972).

RELATIVE GROWTH AND MORPHOLOGY OF TEETH

Tooth Wear

Relative wear on teeth has been widely utilized to separate mammals into age groups for taxonomic studies (Fig. 33.3) and for population and management studies (Figs. 33.4 and 33.5). Spinage (1973) noted that the pattern of wear on teeth generally follows a negative exponential curve. Thus, younger animals tend to be classified as older than their true age. Harris (1978) found that 65.5% of the red foxes (*Vulpes vulpes*) he examined could be aged correctly with an error of 1 year or less. Aging was accurate in 93.3% of the cases with foxes up to 4 years of age. The accuracy of this technique also

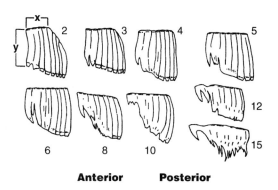

Figure 33.4 Lateral views of upper molars (M^3) in warthogs *Phacochoerus* sp., Artiodactyla, Suidae, of 2 to 15 years of estimated age. The X (occlusal surface) and Y refer to measurements used to derive ratios (X/Y) useful for aging this species.
(Spinage, C.A., and G.M. Jolly. 1974. Age estimation of warthog. J. Wildlife Manag. 38(2):229–233, fig. 1. Copyright 1974 The Wildlife Society. Reprinted with permission.)

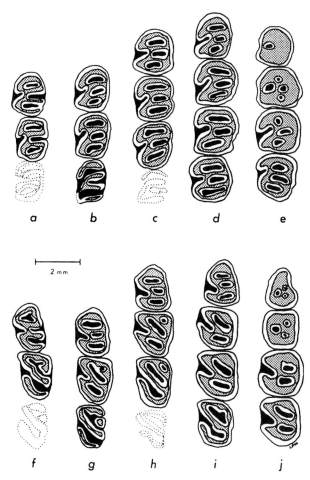

Figure 33.3 Relative age classes based on tooth wear in museum specimens of spiny rats, *Proechimys* sp., Rodentia, Echimyidae. Tooth rows (a–e, left upper; f–j, left lower) arranged from left to right in order of increasing tooth wear and presumed age. Dotted lines indicate teeth that are not at occlusal level. True ages unknown.
(Martin 1970:4)

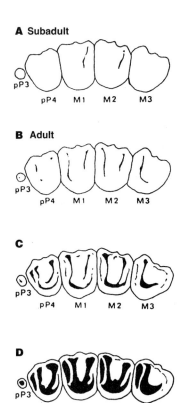

Figure 33.5 Occlusal views of the left upper molariform teeth of four known-age gray squirrels, *Sciurus carolinensis*, Rodentia, Sciuridae, from Wake County, North Carolina, to show the sequende of dental wear through two age classes. (A) Subadult, 338 days old; (B) adult, 494 days; (C) adult, 1,122 days; (D) adult, 2,413 days. Dark areas represent exposed dentine on the occlusal surface. Dental notation: pP3 and pP4 = permanent third and fourth premolars; M1, M2, and M3 = molars.
(Hench et al. 1984. Age classification for the gray squirrel based on eruption, replacement, and wear on molariform teeth. J. Wildlife Manag. 48(4):1409–1414. Copyright 1984 The Wildlife Society. Reprinted with permission.)

depends on the type of tooth examined. The incisors are useful for aging many species of canids (Harris 1978), whereas molars are useful for aging cervids (Taber 1969). Generally, this technique is less accurate for aging than methods based on discrete data (e.g., tooth eruption, annulations in tooth or cementum; Morris 1972).

Tooth Eruption

In many rodents, the pattern of tooth eruption is a useful indicator of age during the first few months following birth. In artiodactyls, relative degrees of tooth eruption may be used to group individuals into year or seasonal classes during the first 2 to 3 years of life (Dimmick and Pelton 1996; Larson and Taber 1980; Taber 1969).

Pulp Cavities

Radiographs are used to measure the extent of pulp cavities in selected teeth in order to group animals into age categories. Tumlison and McDaniel (1984) found that the pulp cavities of juvenile gray foxes, *Urocyon cinereoargenteus,* were significantly more open than those in adult foxes.

33-E Examine teeth in the sample of skulls assembled for Exercise 33-C. Can you group these specimens into age categories (or an age sequence) based upon tooth eruption and tooth wear without relying upon the known ages? How do your tooth wear age categories compare with the true ages of the specimens?

Growth of Eye Lenses

Weight of Eye Lens

The weight of the eye lens increases with age (Fig. 33.6) and thus many studies of age structures of populations have relied on weighing the lenses removed from mammal specimens (see Lord 1959 and review in Friend 1968; Morris 1972). Several investigators (Adamczewska-Andrzejewska 1973; Myers et al. 1977) found eye lens weight to be accurate for estimating ages of several species of myomorph rodents. Morris (1972) reported that the technique has been most successfully applied with animals of medium size (e.g., rabbits, hares) during the period of rapid growth prior to the attainment of adult size. Harris (1978), studying red foxes (*Vulpes vulpes*), did not find lens weights to be useful for separating year classes of animals that had attained adult size.

Refer to Friend (1968), Morris (1972), and Harris (1978) for precautions that must be followed to insure accuracy and precision in the measurement of lens weights.

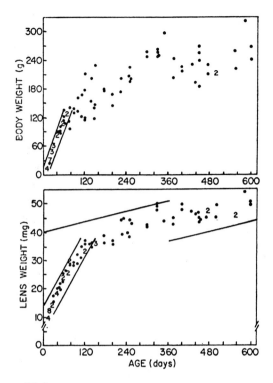

Figure 33.6 Changes with age in body weight and eye lens weight in cotton rats, *Sigmodon hispidus,* Rodentia, Muridae, Sigmodontinae. Heavy lines indicate 95% confidence intervals.

(Birney, E.C., R. Jenness, and D.D. Baird. 1975. Eye lens proteins as criteria of age in cotton rats. J. Wildlife Manag. 39(4):718–728, fig. 1. Copyright 1975 The Wildlife Society. Reprinted with permission.)

Measurements of Lens Protein

Dapson and Irland's (1972) technique found that the amount of the soluble tyrosine (lens protein) increased linearly during approximately the first year of life in old-field mice (*Peromyscus polionotus*). They also found that the increase in the insoluble tyrosine fraction was a curvilinear function of age during the first 750 days of life in these rodents. Birney et al. (1975) demonstrated the accuracy of this technique (Fig. 33.7) for aging cotton rats (*Sigmodon hispidus*) but cautioned that the most accurate technique is not necessarily the one that should be used if time is limited or if the needed specialized equipment is lacking.

33-F Compare the changes in body weight and lens weight in Figure 33.6 with the changes in soluble and insoluble lens protein in Figure 33.7. How do the results differ? At what ages do all of the techniques produce the best results?

Growth Lines

The growth of teeth and bones in mammals is not uniform throughout the year. Thus, narrow and wide layers

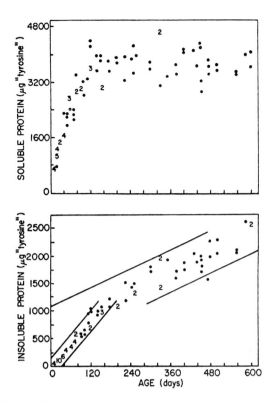

Figure 33.7 Changes with age in soluble and insoluble lens protein in cotton rats, *Sigmodon hispidus,* Rodentia, Muridae, Sigmodontinae. Heavy lines indicate 95% confidence intervals.

(Birney, E.C., R. Jenness, and D.D. Baird. 1975. Eye lens proteins as criteria of age in cotton rats. J. Wildlife Manag. 39(4):718–728, fig. 2. Copyright 1975 The Wildlife Society. Reprinted with permission.)

of dentin, cementum, and bone may be laid down in different seasons of the year or in annual increments (Klevezal and Kleinenberg 1967; Morris 1972). Generally, these **growth lines** or **incremental lines** are the only useful criteria for distinguishing year classes of adult individuals in or from a population, other than marking groups of the same age (i.e., cohorts) when they are young or recently born. These growth lines also provide an absolute age for an individual because the units are discrete and not subject to the continuous variation inherent in the criteria for relative growth described previously (but see cautions in Harris 1978). Phillips et al. (1982) stated that considerable confusion exists in the literature in the use of these terms, along with confusion about the types of dentin layers added to a tooth. They defined *primary dentin* as that formed in the initial phase of tooth development and *secondary dentin* as that formed after the tooth erupts. They considered incremental lines to be those appearing in the secondary dentin.

Teeth: Dentin and Cementum

Growth lines in teeth (Fig. 33.8) have been utilized widely for the aging teeth of marine mammals since publication of the paper by Laws (1952). Subsequently, the

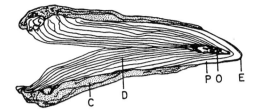

Figure 33.8 Outline drawing of tooth section from a bottlenose whale, *Hyperoodon ampullatus,* Cetacea, Ziphiidae showing growth lines in the dentin. C, cementum; D, dentin; E, enamel; O, osteodentin; P, prenatal dentin.
(Christensen 1973:333)

technique has been applied successfully to other long-lived species such as moose, bears (Stoneberg and Jonkel 1966), caribou (Miller 1974), and others. Harris (1978), in studies of red foxes, found that sections of only the tooth cementum gave good resolution of growth lines that could be demonstrated consistently. Growth lines of dentin were less reliable for aging, and he also cautioned that the determined age should be based on sections of at least two teeth from the same individual.

Erickson and Seliger (1969) and Morris (1972) summarized the various methods used to prepare sections of teeth for study. For larger teeth (whales, pinnipeds, artiodactyls), a section is made of a tooth and it is then ground down to a thin layer using carborundum powder or a grindstone. Smaller teeth can be treated similarly if the teeth are mounted on a cork block (or slide) for easier handling. A second method for making sections involves decalcification and then sectioning of the teeth using a microtome. Formic acid is the most commonly used decalcifying agent although, with caution, a dilute solution of nitric acid is also permissible (Morris 1972). Tumlison and McDaniel (1983) used a decalcifying solution prepared as follows:

0.7 g. ethylenediaminetetraacetic acid (EDTA)

0.2 sodium potassium tartrate

1.0 liter 10% HCl,

which decalcified canines of bobcats (*Lynx rufus*) and river otters (*Lontra canadensis*) within two to three days. Hematoxylin stain generally gives good resolution of the growth lines (Harris 1978), but Gridley's silver impregnation, Masson's trichrome, or the periodic acid-Schiff reaction stains have also been used (Phillips et al.1982).

33-G Select a carnivore tooth or artiodactyl tooth (incisors are frequently utilized because the permanent set of these teeth appears early in development) with which you can practice the preparation of thin sections. Cut thin slices of the tooth and mount in resin on a slide. Grind the tooth to a thin section using various abrasive corborundum powders. Examine the section microscopically under transmitted light (students should select teeth of different ages for later comparison). Can growth lines be detected in the sections? (It may be necessary to stain the teeth lightly with hematoxylin; Morris 1972.) If known age material is available, how well do the numbers of growth lines correspond with the true ages? Is there always only one growth line per year?

Periosteal Lines in Bone

Millar and Zwickel (1972) and Franson et al. (1975) successfully utilized sections of mandibles to age, respectively, pikas (*Ochotona princeps*) and mink (*Mustela vison*). Usually, sections of mandibles are prepared by decalcification and are sectioned using histological procedures. Morris (1972) recommended that formalin-fixed material be utilized because the process of cleaning the bones may result in the loss of some structural detail. Minor accessory growth lines may be seen among the true annual growth lines in bones (Morris 1972). Periosteal growth lines are generally thicker than growth lines in cementum and are thought by some workers to be easier to interpret for this reason.

Growth Lines in Horns

Seasonal changes in forage quality cause the deposition of the keratinized epithelial layers in horns to be unequal (Taber 1969). Thus, these growth lines (Fig. 33.9), with adjustments, can be used to estimate year classes of members of the family Bovidae (Artiodactyla) that possess horns (Morris 1972). Murie (1944) successfully utilized this technique in a classic study of Dall sheep (*Ovis dalli*) mortality. Caughley (1965, 1966) summarized the use of this technique in the preparation of a life table for the Himalayan tahr (*Hemitragus jemlahicus*) introduced into New Zealand in 1904. Geist (1966), working with known-age bighorn sheep (*Ovis canadensis*), found that counts of growth lines were satisfactory for aging males but gave erratic results with females.

33-H Examine a number of bovid horns (preferably from animals of known age) of a species. Can growth lines be detected on the horns? Can you differentiate between annual and minor growth lines? If possible, compare horns of species living in temperate climates with those of species living in tropical climates. What differences, if any, are apparent in the growth lines?

Growth Ridges in Baleen

Growth ridges are formed in the baleen plates of mysticete whales as new material is added at the base of these keratin structures. Seasonal variation in growth rate results in ridges that can be related to annual increments of age, although Morris (1972) stated that it is difficult to obtain precise results. Further details and a review of the literature can be found in Jonsgård (1969).

Epithelial Earplugs of Whales

Each auditory meatus or canal of a mysticete whale is closed. Consequently, the epithelial lining that sloughs off the walls of this canal cannot exit from the body of the animal and instead forms a layered earplug (Morris 1972). Alternate light (higher fat content) and dark layers apparently represent seasonal feeding patterns. Roe (1967) provided a review of the literature in this field, along with data concerning the fin whale (*Balaenoptera physalus*).

Counts of Antler Tines Inaccurate for Aging Individuals

There is a common misconception among hunters and laymen in general that the number of tines or "points" possessed by a deer is an accurate index of the age of the animal. Cahalane (1932) reviewed the evidence for this claim and found no convincing evidence for it.

AGE DETERMINATION IN LIVE MAMMALS

Certain population and behavior studies and those involving critically endangered species require that animals not be killed when one is attempting to get age estimates under field conditions. Several methods have been developed to place animals into age classes.

External and Pelage Characteristics

Most mammals exhibit very rapid growth following birth, after which body size (e.g., mass and dimensions)

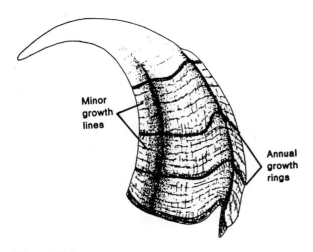

Figure 33.9 Horn of Himalayan tahr (*Hemitragus jemlahicus*, Artiodactyla, Bovidae) showing minor growth lines and true annual growth lines.
(After Caughley 1965)

is not a good indicator of age (see Fig. 33.1). Thus, for species maturing within the first year of life, adult body size may be reached in a few months after birth. This pattern is especially pronounced in rodents in which estimation of age by body mass becomes problematic in as little as three to 12 months after they are born.

For similar reasons, characteristics of the pelage may be sufficient to separate juveniles from subadults or adults. In murid rodents, individuals of the juvenile age class often have a distinctly gray color in contrast to the the nongray pelages of subadults and adults. In gray squirrels, *Sciurus carolinensis*, characteristics of the ventral surface of the tail can be used to separate individuals into juvenile and subadult categories (Sharp 1958). Methods for determining age in many game and furbearing species can be found in Taber (1969), Larson and Taber (1980), and Dimmick and Pelton (1996).

Molar Wear (Attrition)

Wear on molariform teeth has often been used to age dead mammals but only rarely to age live mammals. Hoogland and Hutter (1987) were able to measure molar wear on live, restrained black-tailed prairie dogs, *Cynomys ludovicianus*. From the degree of wear, they were able to establish four age categories corresponding to 0.5, 1.5, 2.5, and ≥ 3.5 years. Cox and Franklin (1990) found that premolar gap width (Fig. 33.10), the distance between the protoconid and paraconid, increased with age in black-tailed prairie dogs, enabling them to establish five age categories (Table 33.1) verified by known-age animals.

TABLE 33.1 Ages of black-tailed prairie dogs, *Cynomys ludovicianus*, Rodentia, Sciuridae, estimated by the premolar gap technique on live, restrained individuals[1]

Age Class (year)	Age (weeks)	Premolar Gap Mean (mm)
0[2]	20	1.80
1	72	2.47
2	124	3.14
3	176	3.82
4	228	4.49

Table adapted from Cox and Franklin 1990:145.
[1] Refer to Cox and Franklin (1990) for details on the procedure and methods for calculating confidence limits for the age estimates.
[2] Age class 0 = juvenile

Strength of Collagen

The strength of collagen fibers in the tails of Belding's ground squirrels, *Spermophilus beldingi*, was used by Sherman et al. (1985) to estimate the ages of individuals in a wild population. The collagen fiber was extracted from the dorsal side of the tail after a small area was shaved and a transverse cut was made through the skin to expose the fibers. The incision was treated with an antiseptic, and the animals did not experience any long-term adverse effects from the procedure. The equation describing the strength of the collagen in relation to age was the following (Sherman et al. 1985:876):

$$\log_{10} (\text{collagen breaking time}) = 0.42 \ln (\text{age in years}) + 5.20$$

$$(r^2 = 0.79, N = 143)$$

Size of Footprints and Shoulder Height

Because of the size and logistical difficulties associated with capture of very large mammals, some investigators have developed ways to estimate ages of individuals indirectly. An interesting technique was developed by Western et al. (1983) to estimate the ages of African elephants, *Loxodonta africana*, Proboscidea, by measuring the lengths of footprints. With this technique, 12 age categories could be recognized, corresponding to chronological ages 0–1 years up to > 15.1 years. The footprint length method was correlated strongly ($r^2 = 0.983$, $N = 13$, $P < 0.001$) with shoulder-height data that had been used previously (Laws 1966) to estimate the ages of elephants.

LIMITATIONS OF AGE-DETERMINATION METHODS

Few studies of age determination of mammals produce essentially equal age estimates when different methods

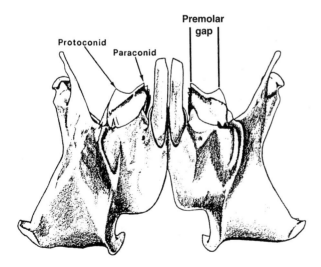

Figure 33.10 Anterior view of mandible of black-tailed prairie dog, *Cynomys ludovicianus,* Rodentia, Sciuridae, showing location of premolar gap. The width of this gap (see Table 33.1) was used to estimate age in live, restrained individuals.

(Cox, M.K., and W.L. Franklin. 1990. Premolar gap technique for aging live black-tailed prairie dogs. J. Wildlife Manag. 54(1):143–146, fig. 1. Copyright 1990 The Wildlife Society. Reprinted with permission.)

are compared (Morris 1972). Body weights, body dimensions, degrees of epiphyseal fusion, and tooth-wear indices generally produce very different results. For aging mammals prior to adulthood, the weight of the eye lens and the lens protein content generally produce the most accurate results (Dapson and Irland 1972). For animals that live in strongly seasonal environments and for more than 1 year, the development of annuli in teeth (dentin or cementum) or in bones has proved to be a very successful and accurate method for aging mammals (Harris 1978; Morris 1972). Phillips et al. (1982) found that the use of incremental lines in secondary dentin and cementum, to ascertain age in bats, was of questionable accuracy. Additional comparative studies of age determination can be found in the *Journal of Mammalogy* and the *Journal of Wildlife Management,* among others (see, e.g., Fiero and Verts 1986; Root and Payne 1984).

For any species studied, tests should be made with known-age animals, using a variety of methods, so as to evaluate the accuracy of the various procedures and to determine whether measurements can be made precisely so as to minimize measuring error.

CHAPTER 34

DIET ANALYSIS

Studies of mammalian diets are important for understanding niche relationships, competitive processes, predation, and the influences that mammals exert on natural and cultivated ecosystems. Today, mammalian diet studies involve quantitative determinations of foods consumed (Fig. 34.1; Bennett and Baxter 1989), quantitative analyses of trophic structure (Fig. 34.2; Meserve et al. 1988), and quantitative estimates of food resource abundance so that resource partitioning (Fig. 34.3; Hickey et al. 1996) and dietary preferences (Ellis et al. 1998) can be examined.

For brevity, coverage in this chapter is limited principally to discussions of some techniques for analyzing diets by macroscopic and microscopic means. References are also provided on how to determine resource levels and how to calculate preference indices. The chapter does not include instructions on how to calculate niche breadth because this topic is generally covered in laboratory textbooks on ecology such as Brower et al. (1990) or Cox (1996).

Further information on the topics covered in this chapter can be found in Baumgartner and Martin (1939), Scott (1943), Dusi (1949), Adams (1957), Storr (1961), Gebczynska and Myrcha (1966), Sparks and Malechek (1967), Williams (1962), Korschgen (1969, 1980), Hansson (1970), Westoby et al. (1976), and Litvaitis et al. (1996).

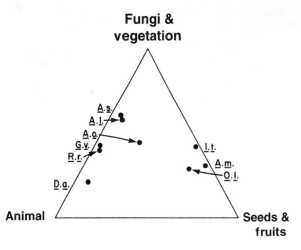

Figure 34.2 Average overall diets for nine species of small mammals at La Picada, Chile, as depicted by DeFinetti diagram. Proximity of points to triangle apexes is proportional to the percentage composition of the corresponding food category in the diet. Abbreviations: A.o. = *Akodon olivaceus;* A.s.= *Akodon sanborni;* A.l. = *Akodon longipilis;* O.l. = *Oligoryzomys longicaudatus;* I.t. = *Irenomys tarsalis;* A.m. = *Auliscomys micropus;* G.v. = *Geoxus valdivianus;* R.r. = *Rhyncholestes raphanurus;* and D.a. = *Dromiciops australis.*
(Meserve et al. 1988:727. Journal of Mammalogy, 69(4):721–730, fig. 1. Copyright 1988 American Society of Mammalogists. Reprinted with permission.)

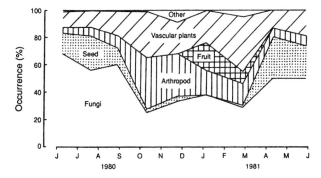

Figure 34.1 Seasonal variation in the occurrence of main food types in the diet of the long-nosed potoroo, *Potorous tridactylus,* Diprotodontia, Potoroidae, in southwestern Victoria, Australia.
(Bennett and Baxter 1989:265, with the permission of CSIRO, Australia)

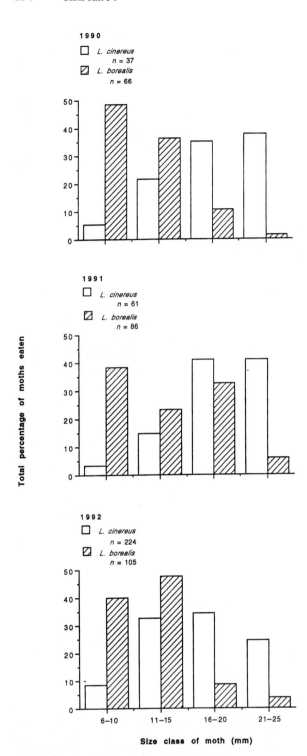

Figure 34.3 Size classes of moths of the five most commonly eaten families as represented by the wings of moths dropped by foraging bats in 1990, 1991, and 1992. Numbers of moths caught in each size class by *Lasiurus cinereus* and *L. borealis*, Chiroptera, Vespertilionidae, are expressed as the total percentage of moths eaten by each species. Sample sizes indicate the total number of wings (= moths) collected per year.
(Hickey et al. 1996:331. Journal of Mammalogy, 77(2):325–334, fig. 4. Copyright 1996 American Society of Mammalogists. Reprinted with permission.)

COLLECTION OF SAMPLES

Types of Samples

Feces and Stomach Contents

Most studies of diet in mammals rely on fecal samples obtained from free-living or live-trapped mammals, or stomach or intestinal samples from dead mammals. A tube may also be used to collect stomach contents from live mammals (e.g., in rodents, see Kronfeld and Dayan 1998). In studies of domestic ruminants, a fistula operation may be performed to allow the sampling of rumen contents of live animals at periodic intervals. In collecting fecal samples from free-living mammals, an investigator must be confident that a sample to be analyzed is derived from a single species. Fecal samples from live-trapped individuals or stomach and/or intestinal content samples from dead specimens are preferred because these generally allow accurate determination of the specific identity of the individual the diet of which is being analyzed. Samples obtained from stomach contents generally present the least biased picture of the food items eaten by the individual (Hansson 1970; Kronfeld and Dayan 1998). Some investigators have found fecal analyses to produce results similar to those obtained from analyses of stomach contents, in species as diverse as prairie dogs, *Cynomys ludovicianus*, (Wydeven and Dahlgren 1982) and insectivorous mammals (Dickiman and Huang 1988) in the orders Insectivora and Dasyuromorphia.

Live traps provide a way to obtain fecal samples without killing the animals. Generally, the baits (e.g., peanut butter, seeds, rolled oats) used in the traps are readily distinguishable from the foods that the animals are feeding on under wild conditions. This technique, most often used with rodents (Meserve 1976), is potentially adaptable to other species and offers a way to monitor individual and population dietary patterns throughout the seasons of the year. The live traps must be thoroughly cleaned before they are reset. Feces from bats captured in mist nets or harp traps can be used to analyze diets. Moth wings dropped by foraging bats have also been used to determine diets in two species of bats of the genus *Lasiurus* (Hickey et al. 1996).

Cheek Pouch Contents

Contents of the cheek pouches of geomyid and heteromyid rodents may provide information on the diet of these mammals, although the results of Reichman (1975) suggest that the diet determined by these samples underestimates the variety of total diet.

Food Caches

Analyses of food caches of mammals can provide information on dietary preferences of a species (Korschgen 1969), but does not yield information on the diet of a single individual or how the diet might change seasonally.

Raptor Pellets

Analysis of hawk and owl **pellets** (the undigested prey remains that are regurgitated by these birds) do not provide information on the diets of mammals but can be helpful in determining the species of mammals that occur in an area. For analysis, these pellets are generally treated with a weak solution (about 2M) of NaOH to dissolve the hair and make removal of the bone fragments easier. Process these samples in a manner similar to the way you would work with carnivore scats (see "Preservation and Labeling of Samples").

Number of Samples

The number of samples collected for analyses is determined partly by the nature of the investigation (e.g., local or regional applicability) and the amount of time available for collecting the samples. Hanson and Graybill (1956) utilized a statistical procedure to determine the number of samples required to estimate the food habits of a population. Korschgen (1969) provided additional guidelines for determining adequate sample sizes.

Preservation and Labeling of Samples

Korschgen (1969) and Hansson (1970) discussed methods used to preserve samples for dietary analyses. Freezing of samples produces the least alteration of material but is frequently impractical for field situations. Drying of scats or pellets is a convenient method for most mammals, especially carnivores, if precautions are taken to prevent mold (oven dry at 80° to 85° C for several hours) and, later, insect infestations (store in tight container with moth flakes, naphthalene).

Stomach or intestinal contents may be frozen or preserved in *5% formalin* (one part stock formalin [37% to 40% formaldehyde] in 19 parts of water) for small- to medium-sized species and in *10% formalin* (one part stock formalin in nine parts of water) for larger species. Preservation in 70% alcohol can also be utilized, although this preservative will extract chlorophyll and is thus less satisfactory for studies on herbivorous species. Hansson (1970) found that formalin-preserved material was superior to alcohol-preserved material, for quantitative determinations, although formalin caused slight overstaining in final slide preparations. To achieve the best preservation, the stomachs or intestines of larger species should be injected with the fixative and then immersed in it. For the largest species (e.g., ruminants), a quart or liter sample may be taken from the stomach or intestine and then preserved in a 5% formalin solution (Korschgen 1969, 1980).

The sample for dietary analysis should be properly tagged (labeled) with full information as to species, sex (if known), date of collection, locality, and the collector's number (see Chapter 29). For samples from cataloged animal specimens, the collector's number is adequate if the remainder of the field data are available in the catalog (see Chapter 29). In live-trapping studies, the number of the individual, sex, date of trapping session, and grid position should be stated on the label or package that holds the sample. For fluid-preserved material, always place the label *inside the container* that holds the sample, so as to prevent its accidental loss. Stomachs or intestines may be individually wrapped in cheesecloth, labeled, and then placed in a single container of preservative to save space.

Identification of Food Items

Reference Samples

Properly identified reference materials are essential for analyzing dietary samples of mammals (Hansson 1970; Korschgen 1969). For larger mammals that consume other vertebrates as prey, these reference materials should include skulls and skeletons of the prey species, and their epidermal elements such as scales (fish), feathers (birds), and hair (mammals). Mammalian prey items can be identified most easily by using skull and skeletal elements. When skeletal elements are fragmentary or inconclusive, it may be necessary to have a reference collection of guard hairs or impressions of the cuticular scales of the guard hairs. To observe pigmentation patterns and medullary characteristics, hair samples can be cleaned in ether (**Caution: Extremely flammable and volatile**), dried, and covered with Permount® (or similar mounting medium), and then sealed with a coverslip. Cross sections of hairs can be prepared using the technique of Mathiak (1938).

Cuticular scale patterns provide some of the most useful characters for identifying hair samples (see Fig. 4.4). The techniques of Williamson (1951) and of Korschgen (1969, 1980), or that described in Exercise 4.1, will enable you to obtain impressions of these cuticular scale patterns. This technique has proved useful in analyses of the diets of ermines and other weasels (*Mustela,* Day 1966) and wolverines (*Gulo gulo,* Myhre and Myrberget 1975).

34-A   Select samples of guard hairs from several species of prey mammals that occur in a local area. (Collect hairs from comparable areas of the body because hair shape, scale pattern, and hair color may differ between regions of the body.) Prepare reference slides using the technique described in Exercise 4-I. Do you see differences in cuticular scale patterns? Would these patterns be useful for identifying species? If not, to what taxonomic level could the prey items be identified?

For invertebrate prey items, appropriate dry and wet specimens of identified taxa should be available for reference. Invertebrate prey items are generally partially digested and broken up in the intestinal tract of the predator. Consequently, Hansson (1970) utilized the following technique to prepare similar-sized fragments for reference slides of these prey animals.

1. Place reference animals in a 1% solution of pepsin at pH 1.0–2.0 and incubate at 40° C for about four hours.
2. Wings, legs, and other sclerotized structures should be cut into small pieces and placed in a vial of clearing solution.
3. Krantz (1978) discussed several agents that can be used for clearing sclerotized tissues. **Lactophenol** is one of the most commonly utilized clearing agents for invertebrates. This agent should be prepared with the following ingredients added *in this order:*

Lactic acid	50 parts
Phenol crystals	25 parts
Distilled water	25 parts

   Immersion of material in lactophenol for 24 to 72 hours should be sufficient for clearing most specimens. Rinse in three to four changes of water until cloudiness disappears.
4. Following clearing, the fragments should be spread in a thin layer on a microscope slide. Place a drop or two of an aqueous mounting medium (Krantz 1978), followed by the animal fragments, onto a microscope slide. Then, using a pair of forceps, pick up a coverslip at its rim, apply the opposite edge to the droplet of mounting medium, and then allow the coverslip to fall into place.

   Hoyer's medium is a widely utilized aqueous mounting medium and can be prepared by mixing the following ingredients *in the following order* (make sure each solid ingredient is completely dissolved before adding additional reagents; *do not heat* the mixture):

Distilled water	50 ml
Gum arabic (crystalline)	30 g
Chloral hydrate[1]	200 g
Glycerine	20 ml

5. Mark the slide with an identifying number and collector's name until a permanent label can be prepared. Allow the newly prepared slide to dry in an oven at 45° C for 48 hours to one week.
6. Remove the slide from the oven, allow it to come to room temperature, and then place a ring of Glyptal®[2] (waterproof paint for electrical circuits; General Electric Company) around the perimeter of the coverslip, to prevent movement of water into or out of the mounting medium. (Infiltration of atmospheric moisture causes a breakdown of the mounting medium and deterioration of the slide preparation.)
7. Prepare a permanent label for the slide, stating the name of the reference taxon, the components preserved, the date and locality, and the names of the clearing and mounting media.

34-B   Collect several species of invertebrates (insects, other arthropods, etc.), preserve in 70% alcohol, rinse in decreasing concentrations of alcohol, followed by a final rinse in water. Then, separate the specimens by taxa and place in labeled vials. Freshly dead or dried materials and those already in an aqueous medium can be passed directly to the water rinse. Use a pepsin solution to dissociate the elements (see above or Hansson 1970). Prepare slides using an aqueous mounting medium. What types of structures appear to be diagnostic for each taxon? To what taxonomic level can the organism be identified?

Reference materials of plants should include seeds and intact seed pods (generally, items > 0.4 mm in greatest dimension), and various tissues taken from the plants, for microscopic analysis (Hansson 1970; Korschgen 1969). Often seeds can be identified by using references such as Martin (1946), USDA (1948), Martin and Barkley (1961), and Musil (1963), Johnson et al. (1983), and Green et al. (1985).

Microscope reference slides (or photomicrographs) are necessary for the identification of finely masticated plant materials found in the stomachs, intestines, or feces of mammals. These slides are generally prepared using the techniques pioneered by Baumgartner and Martin (1939) and by Dusi (1949), and discussed by Hansson (1970) and Johnson et al. (1983). Ideally, plant material for reference slides can be prepared by feeding plant samples to a target species and then making slide preparations from samples (feces, stomach or intestinal contents) obtained from the individuals involved (Hansson 1970). However, this method is tedious (Hansson 1970) and not often utilized. Another alternative is to grind dried plant material to uniform size in a **Wiley Mill** and to then prepare reference slides (Reichman 1975). Lastly, epidermal tissues can be stripped from plant samples, treated with **Hertwig's solution** (Baumgartner and Martin 1939), and then mounted on microscope slides for use in identification.

Epidermal tissues of plants are frequently utilized for identification purposes because they possess useful diagnostic features (Figs. 34.4, 34.5) such as siliceous cells, stomata, spines, and hairs (Dusi 1949; Green et al. 1985; Johnson et al. 1983; Williams 1962). But reference slides should include other material (e.g., portions of flowers with pollen, seed husks and seed endosperms) that may be encountered in the samples to be analyzed. The following technique, adapted, in part, from Baumgartner and Martin

---

[1] Controlled substance; available only by permit for research and professional purposes.

[2] Clear nail polish or Permount® (or similar synthetic mounting medium) can also be used but is not as permanent.

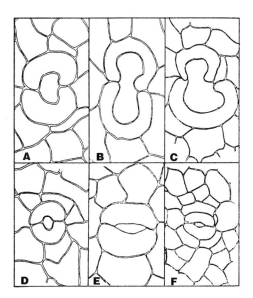

Figure 34.4 Examples of types of stomata found in different food species of fox squirrels (*Sciurus niger*). (A–C) *Julans nigra;* (D) *Carya alba;* (E) *Prunus serotina;* (F) *Cornus alternifolia.*
(Baumgartner and Martin 1939)

(1939), Dusi (1949), Hansson (1970), Reichman (1975), and Meserve (1976), can be utilized to prepare reference slides of a variety of materials:

1. From the reference sample, strip pieces of epidermis from leaves (both sides) and stems, section portions of other plant material (flowers, seeds), and place on a microscope slide (the material may need to be macerated in Hertwig's solution before it can be stripped; Dusi 1949). Alternatively, the plant material can be dried and a sample ground in a Wiley Mill to uniform particle size (this helps to minimize bias because, in examining slides, larger particles tend to be noticed more readily than smaller particles; Westoby et al. 1976).

2. Place a few drops of Hertwig's solution (Baumgartner and Martin 1939) on the slide containing the reference material. Prepare Hertwig's solution (add reagents *in this order*) as follows:

Distilled water	150 ml
HCl	19 ml
Glycerin	60 ml
Chloral hydrate[1]	270 g

[1]Controlled substance; available only by permit for research and professional purposes.

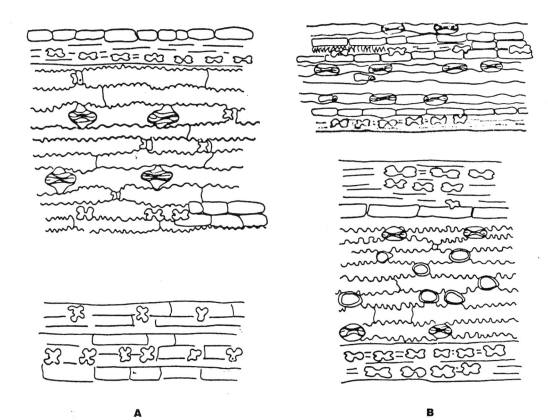

Figure 34.5 Examples of stomata and silica bodies in epidermal layers in two species of grasses: (A) common carpet grass, *Axonopus affinis;* (B) yellow indian grass, *Sorghastrum nutans.* In *Sorghastrum,* the silica bodies are deeply notched, and the stomata are oval. In *Axonopus,* the stomata are diamond-shaped.
(Johnson et al. 1983:20)

Allow the solution to boil for a few seconds on a hot plate (100° to 105° C).

3. Allow the slide to cool quickly in air and rinse *very gently* with tap water to remove most of the Hertwig's solution (be careful not to lose the plant material).
4. The sample can now be stained, utilizing any aqueous stain, unless it is determined that staining is not necessary for identification of the material. Hansson (1970) recommended hematoxylin as a general-purpose stain because it does not stain suberin (a constituent of some cell walls) and thus provides good contrast with tissues that are stained by this dye. For a hematoxylin stain, 10 to 15 minutes should be sufficient to produce the desired color (pale blue). Overstaining can be corrected with acid alcohol, using techniques described in Guyer (1953) or Humason (1967).
5. Rinse the slide *gently* with tap water to remove excess stain and then touch a piece of blotting paper to the slide to draw off excess water.
6. Add one or two drops of Hoyer's medium (see section on preparing slides of invertebrates), or any commercially available water-soluble mounting medium, to the slide preparation, arrange the reference material in the desired positions, and then drop a coverslip on top. Mark the slide with a temporary label (e.g., I.D. number and initials) and then place it in an oven (45° C) to dry for 48 hours to one week. Remove the slide from the oven, allow it to come to room temperature, and then seal the edges of the coverslip with Glyptal® to make a permanent preparation (Krantz 1978).
7. Prepare a permanent label for the slide, listing the collector, identity of the plant, the portions of the plant preserved, the date collected, and the techniques utilized.
8. A photomicrographic "key" can then be prepared using these slides. The identification key is generally a set of photographs or drawings that the observer can refer to as needed (see, for example, the drawings in Johnson et al. 1983 or the photographs in Green et al. 1985).

34-C Collect samples of several species of woody and herbaceous plants and of grasses from a riparian area. Each student should prepare a reference slide using the technique just described. Exchange and compare slides. What structures appear to be most useful for identification of the taxa? Are species easily identified as shrubs, herbs, or grasses? Are genera easier to separate than species? What structural types of plants are most easily identified? Repeat the procedure using plants from a grassland area.

34-D Examine reference slides and prepare a diagnostic key (see Chapter 8 for hints on writing keys) for identifying at least 10 taxa. What characters seem to be most useful for identification purposes? Would it be easier to learn to identify the taxa on sight rather than to key them out?

ANALYSIS OF DIETARY SAMPLES

Preparing Samples for Study

The contents of cheek pouches and the dried scats of large mammals (other than herbivores) can be examined with little or no special preparation. If only bones or chitin fragments are to be examined, the hair in the scats can be separated with an 8% (2M) solution of NaOH (Green et al. 1986). The dietary items in these samples can be compared with a reference collection in order to identify them. Weight, volumetric, and/or frequency of occurrence determinations can then be made (Hansson 1970; Korschgen 1969, 1980).

Microscope slides must generally be prepared for the dietary samples (stomach contents, feces) of herbivorous species (Baumgartner and Martin 1939; Dusi 1949; Hansson 1970) because characteristics may be apparent only under microscopic examination. Generally, fresh, dried, or fluid-preserved material can be prepared, using the technique described for making reference slides. Additional recommendations for analytical procedures can be found in Hansson (1970).

Quantitative Determinations

Ecologically, one of the most meaningful statistics in dietary analyses is a measure of the *weight* of food consumed of each species (or higher-level taxon). This measure can then be directly related to energy flow and analyzed by parametric statistical procedures. A volumetric determination is also meaningful and is often utilized with species that do not masticate their food into fine particles. Measurements of volumes are made by using a graduated cylinder or by displacement of a known quantity of water in a burette (Inglis and Barstow 1960).

Volumetric determinations are very difficult to accomplish with finely masticated food items, and investigators utilize two approaches to obtain estimates of dietary intake. One method estimates the *percent coverage* of a species on a slide (generally estimated to nearest 10% to 20%) to obtain an index of volume (Gebczynska and Myrcha 1966; Keith et al. 1959; Meserve 1976; Myers and Vaughan 1964). Usually, 10 randomly selected fields are examined under a microscope (35–125X), and the percent coverage of fragments is noted for each field.

Another widely used method (Free et al. 1970; Hansen and Ueckert 1970; Reichman 1975; Sparks and

Malechek 1967) is based on determining the *relative frequency of occurrence* of a food item:

Number of occurrences of a species

Total number of occurrences of all species

and then converting these values to *relative densities,* using the table in Fracker and Brischle (1944).

As detailed by Hansson (1970) and by Westoby et al. (1976), neither of these techniques yields results that are satisfactory for all types of plants and plant communities. Plant taxa that are very rare may be missed completely, and those that are abundant in the plant community may be overestimated in the analyses. However, these techniques are the best that are presently available and, as Hansson (1970) noted, may represent the practical limits for further quantitative refinements. Standardization using uniform diets (see next section) can be an important component of dietary studies so as to identify sources of error in the analyses. Holechek et al. (1982) found that observer experience was the most important factor in accurate determination of diets by microscopic analyses.

Standardization Using Uniform Diets

It is useful to conduct feeding trials (Meserve 1976; Westoby et al. 1976) using plant and/or animal mixtures of known composition in order to minimize bias in analyses of dietary samples. These mixtures can be fed to animals, and samples can then be collected from the feces or stomachs. Then, after preparing microscope slides of the samples (or examining the samples macroscopically in the case of larger carnivores or other large nonherbivorous species), the dietary composition can be estimated by using the volumetric and relative frequency methods described earlier. Experimenter bias can be minimized by randomizing the samples so that the estimator does not know the true dietary composition until the completion of the estimates. Westoby et al. (1976) standardized the procedure further by preparing diet mixtures and then grinding them to uniform particle sizes (to eliminate bias in degree of mastication in different species and individuals). Then, slides were made from these artificial samples, analyses made, and the estimates compared with the known values.

34-E Obtain a series of microscope slides of fecal samples or stomach contents. Examine 100 fields at 100× magnification. Obtain frequency data by recording the presence or absence of a taxon in each field. Obtain volumetric data by determining the proportion of the field occupied by each taxon. Convert the frequency data to relative frequency by dividing by the total number of fields that contain fragments of any taxon. How do these two methods compare in ease of application and in results? Are the results comparable? If not, what could be done to test the accuracy of these techniques?

34-F Examinine a series of owl pellets provided by your instructor. Disassemble these pellets with the aid of forceps or teasing needles. Alternatively, use a weak solution (2M NaOH) to loosen the hair around the bones and teeth, followed by a rinse in water over a fine screen. Allow the bones and teeth to dry on paper towels and, prior to examination, make sure that the contents of each pellet are placed in a labeled container (a small cardboard tray is ideal). Use reference material to identify the hard fragments to genus and/or species. What genera or species are represented in each pellet? What is the minimum number of individuals represented in each pellet?

DETERMINATION OF RESOURCE LEVELS

Before preferences (and, conversely, aversion or disregard) of animals for food items can be determined, the resource levels of the plant and animal community must be measured. For plants, these measurements are generally straightforward (see Chapter 28), and techniques can be found in most plant ecology and general ecology texts (Brower et al. 1990; Cox 1996; Oosting 1956). The determination of mammal level numbers (and those of many vertebrates) may involve trapping or line censuses (see Chapter 36 or references in Wilson et al. 1996). Population levels of nonmammalian vertebrates can be estimated using techniques described in Pettingill (1956) for birds and Tinkle (1977) for reptiles and amphibians.

The measurement of arthropod population levels may be more difficult due to the tremendous variety of habitats that these animals occupy and the seasonal and diurnal variations in their activity. Southwood (1966) and Bram (1978) give methods for estimating population levels in many species of arthropods. Hickey et al. (1996) measured insect resource levels with light traps, as part of their study of resource partitioning in two species of *Lasiurus* (Chiroptera, Vespertilionidae).

Seed resource levels can be determined by removing seed pods or heads directly from the plants along transects or from quadrats. Seeds can also be extracted from soil samples using a flotation technique (Franz et al. 1973; Reichman 1975).

CALCULATING INDICES OF PREFERENCE

Mammals can, and often do, exhibit dietary preferences; thus, it is important to be able to measure such preferences in the analyses of samples. Food preferences can be determined once resource levels have been quantified and analyses completed on the food samples (Ellis et al.

1998). Indices of preference are based on a comparison of the frequency of an item in the dietary sample with that of the resource. In the method detailed by Reichman (1975), values greater than 1.0 indicated preference, and those less than 1.0 suggested avoidance or disregard of the resource:

$$\text{Preference} = \frac{\text{Relative frequency in dietary sample}}{\text{Relative frequency of resource}}$$

The **electivity index** of Ivlev (from Alcoze and Zimmerman 1973) is another method for calculating dietary preference. This index (E) is computed using the formula

$$E = \frac{(r_i - p_i)}{(r_i + p_i)}$$

where r_i = percentage volume of a food item in the diet and p_i = percentage density of the same food item in the sample area. A value of zero indicates randomness in resource selection, a positive index value indicates presumed preference for a dietary item, and a negative value indicates presumed avoidance of or indifference to a dietary item.

34-G Compute dietary preference indices using data in Reichman (1975) or Meserve (1976). Compare results from using the two techniques just described. Do the methods produce similar results? Recompute the indices after changing the proportions of the food items in the resource and/or samples. What differences did these changes produce in the values of the indices?

CHAPTER 35

Analysis of Spatial Distribution

An understanding of the spatial distribution of mammals and of the movements that they make within an area is important for interpreting many ecological and evolutionary processes. It also aids in formulating management plans for various species. In this chapter, we will examine the types of spatial organization and movements in populations of mammals and how these phenomena can be studied and measured. General reviews of this subject can be found in Brown (1966), Jewell (1966), Sanderson (1966), Fisler (1969), Brown and Orians (1970), and Flowerdew (1976).

Spatial Organization

The **dispersion** or distribution of animals in an area can be categorized as uniform, clumped, or random (Brown and Orians 1970). In a **uniform dispersion** (Fig. 35.1A), the points in space occupied by individuals are approximately equidistant from one another. In a **clumped,** or patchy, **dispersion** (Fig. 35.1B), individuals are concentrated in some areas and absent from others. In **random dispersion** (Fig. 35.1C), there is equal probability that an individual will occupy any given point in space, and the presence of another individual nearby will not affect this probability. One can test for the type of dispersion pattern that one is dealing with by using various statistical tests. Morisita (1962) provided an **index of dispersion (I_δ)** that can be used to test these patterns:

$$I_\delta = N \frac{\Sigma n_i (n_i - 1)}{\Sigma x (\Sigma x - 1)},$$

where N equals the total number of observations, n_i equals the number of animals observed in the ith observation, and Σx equals the total number of animals found in all observations. Randomness in dispersion is indicated by a value of 1 while values less than or greater than 1 indicate, respectively, that the distribution is uniform or clumped. The value obtained can be checked for significance using the following F-statistic:

$$F = \frac{I(\Sigma x - 1) + N - \Sigma x}{N - 1}$$

The F_{cal}-value is then compared with an F_{tab}-value (Rohlf and Sokal 1969:168–95) using the following **degrees of freedom (df)**:

numerator $df = N - 1$

denominator $df = \infty$ (infinity)

Refer to Morisita (1962) and Flowerdew (1976) for more details on this procedure.

Home Range

The **home range** of a mammal is the area that it occupies during the course of its life, exclusive of migration, emigration, or unusual erratic wanderings (Brown and

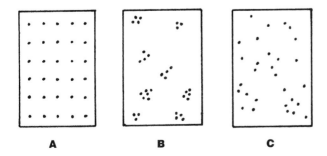

Figure 35.1 Uniform (A), clumped (B), and random (C) dispersion patterns.
(Modified from Brower and Zar 1974:118)

Orians 1970). Burt (1943), in differentiating the concept of home range from the related concept of territory, defined home range as the area traversed by an individual in its normal activities of food gathering, mating, and caring for young, and specifically excluded areas traveled in " . . . occasional sallies outside the area."

The sizes of mammalian home ranges are related to the energy demands of the species, with "croppers" having smaller home ranges than "hunters" (Harestad and Bunnell 1979; McNab 1963). Home ranges may also involve a third dimension (e.g., vertically into shrubs or trees or down into burrows [Koeppl et al. 1977a; Meserve 1977]); thus, measuring area alone may be insufficient in calculating the size of the home-range of certain species.

The home ranges of individuals may overlap partially (Fig. 35.2) or may be entirely separate from one another. The nonoverlap of home ranges may be due to territoriality, mutual avoidance, preferences for particular food sources or habitats, or physical barriers (Brown and Orians 1970). Jorgensen (1968) provided formulae for estimating the probability that occupants of two home ranges will meet by chance, and Adams and Davis (1967) gave a method for estimating the amount of overlap of home ranges.

The area of a home range can be estimated using a variety of techniques (see "Methods for Studying Movements" section in this chapter). Stickel (1954), Sanderson (1966), and Jennrich and Turner (1969) reviewed the procedures for estimating home ranges, using data from grid trapping, and Harris et al. (1990) reviewed home-range estimation methods that involved radio-tracking data. Boulanger and White (1990) used Monte Carlo simulation to test several home-range estimators. They concluded that the harmonic mean method was the least biased but also the least precise. Their overall conclusion was that the available home-range estimators provided only rough measures of the activity of animals.

In the sections that follow, we briefly describe some of the techniques that have been used to estimate home ranges. There is a wealth of literature on this subject, and this is a rapidly evolving field. Consult the latest issues of such journals as the *Journal of Mammalogy,* the *Journal of Wildlife Management,* or *Wildlife Review* for recent developments in techniques for estimating home range.

Polygon and Range Length Methods

1. **Minimum area** (Fig. 35.3A). Capture points are connected to enclose a polygon. Jennrich and Turner (1969) suggested that points be connected in a counterclockwise fashion.
2. **Minimum convex polygon** (Fig. 35.3A). Capture points are connected to form the smallest convex polygon (Southwood 1966). This is a widely used method, but the results, like many of the home-range estimators, are greatly affected by sample size (Boulanger and White 1990).
3. **Boundary strip.** A boundary strip equal in width to half the distance between traps is drawn around the minimum area. In the *inclusive boundary strip* method (Fig. 35.3B), the peripheral points of capture are considered centers of rectangles (each side of which equals the distance between traps), and the home-range area is delineated by connecting the exterior corners of these rectangles to form a maximum estimate of the space utilized. In the *exclusive boundary strip* method (Fig. 35.3C), the boundary strip rectangles are drawn in a manner to minimize the area enclosed.
4. **Range length** (Fig. 35.3D). This measure is the distance between the most widely separated capture points.
5. **Adjusted range length** (Fig. 35.3E). The range length *plus* one-half the distance to the next trap is added onto each end.

Stickel (1954), in empirical experiments, found that the exclusive boundary strip method was the most accurate measure of the actual area of the home range, whereas the adjusted range length most closely approximated the true range length. Jennrich and Turner (1969) pointed out that methods based on connecting points are biased by the size of the sample, (i.e., the area and range length increased with an increase in the number of unique capture points). Hayne (1950) found that the distance between traps also affected the estimated size of the home range.

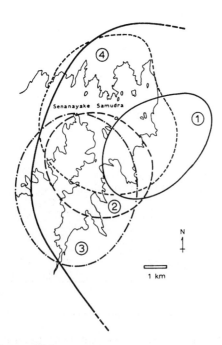

Figure 35.2 Individual home ranges of four adult male Asiatic elephants (*Elephas maximus*) superimposed upon a portion (solid line dashed at ends) of the group home range. (McKay 1973:82)

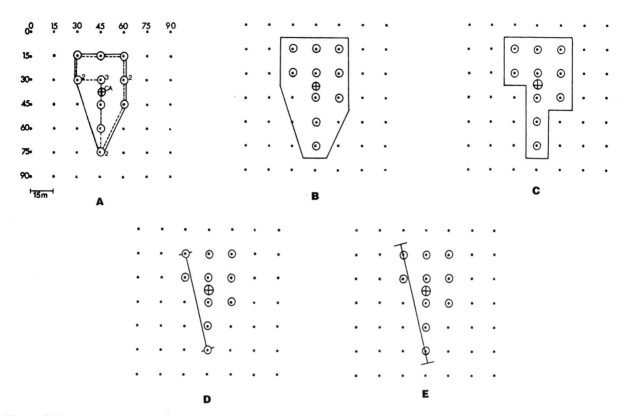

Figure 35.3 Polygon (A–C) and range length (D–E) methods for estimating the two-dimensional home ranges and movements of animals. (A) Minimum area method (dotted lines) and convex polygon method (solid lines); (B) inclusive boundary strip method; (C) exclusive boundary strip method; (D) range length; (E) adjusted range length. The cross within a circle denotes the center of activity.
(Mary Ann Cramer)

Methods Based on Recapture Radii

6. **Center of activity.** Because the number of captures (N) affects the estimated size of the home range, various workers (Calhoun and Casby 1958; Hayne 1949a; Jennrich and Turner 1969) have sought to minimize this bias. The center of activity (Fig. 35.4) is the mean of a set of capture coordinates (Hayne 1949a). It is computed by summing each coordinate (X, Y) of the capture points to arrive at an average or geometric center. Smith et al. (1973) found that the burrows of old-field mice (*Peromyscus polionotus*) were outside the estimated home range of the animals and were peripheral to the aboveground center of activity as revealed by trapping.

7. **Standard deviation of capture radii.** This measure was used by Calhoun and Casby (1958) and other workers to estimate the dimensions of the home range of an animal. One must assume that the home-range shape is circular in order to utilize this technique. Calhoun and Casby (1958) and Maza et al. (1973) provided tests and examples of the use of this method.

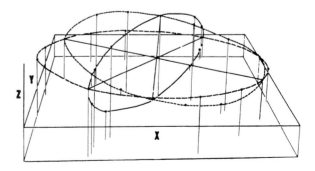

Figure 35.4 Graphic illustration of the three-dimensional home range of a gray squirrel, *Sciurus carolinensis,* showing 95% confidence ellipses for each dimension of the home range.
(Koeppl et al. 1977a:214, 215)

8. **Determinant of the covariance matrix.** This determinate estimate was used by Jennrich and Turner (1969) to eliminate biases due to lack of circular symmetry in home-range shape and to variations due to differing sample sizes. Refer to Jennrich and Turner (1969) for details and computational procedures.
9. **Bivariate home range.** This method was developed by Koeppl et al. (1975) and extended the concepts of Calhoun and Casby (1958) and Jennrich and Turner (1969) into a general model. The method of Koeppl et al. (1975) enables calculation of standardized distances from activity centers and calculates probability ellipses around these centers.
10. **Three-dimensional home range.** Because many species utilize a vertical dimension in their movements, Koeppl et al. (1977a) extended the methodology of the bivariate home-range model (Koeppl et al. 1975) to include this extra dimension. Location data obtained by trapping or observation (Fig. 35.4A) can then be projected into three-dimensional space, and confidence ellipses can be plotted (Fig. 35.4B) for each dimension.
11. **Distance between observations indices.** Koeppl et al. (1977b) developed another method to measure the average size of the home range when the assumptions of the bivariate and three-dimensional models could not be met (e.g., lack of adequate sample sizes).
12. **Harmonic mean.** Dixon and Chapman (1980) developed the harmonic mean method that is not based on the arithmetic center of the animal's activity. With this method, the activity center will be inside the animal's home range, and an animal's movements within the home range have little effect on the value, in contrast to methods that calculate an arithmetic center of activity. Boulanger and White (1990) found the harmonic mean method to have the least bias of several methods investigated, although it also tended to be the least precise in estimating the home range. A computer program to calculate the harmonic mean is available in the MCPAAL software package, version 1.2 (M. Stuwe, Conservation and Research Center, National Zoological Park, Smithsonian Institution, according to Boellstorff and Owings 1995).

35-A Examine several papers (including French et al. 1975; McNab 1963) that report sizes of home ranges of mammals. What trends are apparent from these data?

35-B Use McNab (1963) or other references to obtain values of home-range size. Prepare circular and oval transparent templates scaled to the sizes of these home ranges. Using the same scale, prepare a map of a trapping (observation) grid that is several times larger than the area of each home range. The grid should be drawn to permit the testing of various trap spacings. *Alternatively,* the instructor may wish to prepare templates and the grid in advance, using the suggestions in Stickel (1954). For each toss of the template onto the grid, record in Table 35.1 the observed range length and adjusted range length (range length plus one-half the distance of the spacing between stations added to each end of the range length). Continue for 20 trials. Which method for computing range length came closest to the actual value (population parameter) of range length? Plot observed range length (y-axis) as a function of trial number (x-axis). What effect does the number of trials ("captures") have on the shape of this curve? How could this result be used to plan a sampling program?

35-C Verify for the following set of capture coordinates that the arithmetic center of activity is 9.4, 7.2 (X, Y). Compute the standard deviation (s) for these data by using the following formula (Calhoun and Casby (1958):

$$S = \sqrt{\frac{\Sigma(\bar{x} - x_i)^2 + \Sigma(\bar{y} - y_i)^2}{2(N - n)}}$$

Capture No.		Capture No.		Standard Deviation
X	Y	X	Y	s
1	10,8	11	10,6	
2	10,7	12	9,8	
3	11,5	13	9,9	
4	9,8	14	9,7	
5	10,5	15	10,8	
6	9,9	16	10,7	
7	9,7	17	10,7	
8	9,7	18	10,8	
9	11,6	19	10,7	
10	10,7	20	9,8	

35-D Using the capture coordinate data in Exercise 35-C or another set of data, plot two-dimensional home ranges, using the minimum area, minimum convex polygon, and inclusive and exclusive boundary strip methods. Which method produces the smallest estimate of home range? The largest? What is the basis for adding a boundary strip? What effect would adding a boundary strip have on calculating population density (see Chapter 36)?

The program CALHOME, described by Kie et al. (1996), can be used to calculate several estimates of home range. This program was used by Ribble and Stanley (1998) to get minimum convex polygon estimates of the home ranges of *Peromyscus boylii* and *P. truei*.

TABLE 35.1 — Data Sheet for Results of Exercise 35.5

Trial No.	Actual Range Length	Circular Template		Actual	Oval Template	
		Observed Range Length	Adjusted Range Length		Observed Range Length	Adjusted Range Length
1						
2						
3						
4						
5						
6						
7						
8						
9						
10						
11						
12						
13						
14						
15						
16						
17						
18						
19						
20						
	Σ			Σ		
	X			X		

35-E Use the program CALHOME, or a comparable program, to calculate home-range estimates, using different techniques. How do the estimates compare with one another? Which methods seem to be very sample-size dependent?

Most of the discussion of home range has concerned itself with the *individual* animal. However, chimpanzees, gorillas, giraffes, female red deer (*Cervus elaphus*), and other mammals often possess **group home ranges** (Fisler 1969). **Core areas** of intensive use in the group home range are found in some species of baboons and in coatimundis, *Nasua narica* (Fisler 1969; Kaufmann 1962).

Territory

By the most generally accepted definition, a **territory** is an area defended by an individual or a group. According to Brown and Orians (1970), a territory should possess the following characteristics: (1) a fixed area that may change its boundary slightly over time (Fisler 1969); (2) the possessor of the territory exhibits overt and directly interactive acts of territorial defense (attacks, threat vocalizations, or threat displays) or indirect acts of defense (e.g., scent marking—very characteristic of mammals); and (3) these acts are effective in keeping out rivals. Pitelka (1959) argued that a territory should be defined solely as an exclusive area, without necessarily involving overt defense. A species may exhibit **territoriality** during only a portion of the year or defend an area only near a particular resource (females, food, nest site). The papers in Stokes (1974) presented a balanced view of the concept of territory and, together with Fisler (1969) and Brown and Orians (1970), will give much insight into this subject.

35-F How does the concept of home range differ from that of a territory? What sort of observations would be necessary to determine if an animal is territorial?

Types of Territories

Fisler (1969) discussed the dynamic nature of territorial systems in mammals. An **individual spatial territory** involves a fixed, and sometimes changing, area (Fig. 35.5). One variant of this type of territory is the **nidic territory,** which includes only the immediate area of and around a den or nest site and is typically possessed by females. Another is the **arena territory** that is found in species with polygynous mating systems, and in which the males exhibit overt territorial aggressive behavior at traditional mating grounds. A **lek** is a special type of arena territory where males aggregate for the purpose of attracting females. Leks were described by Buechner (1961) and Leuthold (1966) in classic papers on the Uganda kob (*Kobus kob thomasi*). Species with leks include brindled gnus (*Connochaetes taurinus*), gazelles (*Gazella thomsonii* and *G. granti*), the gray seal (*Halichoerus grypus*), and hammer-headed bats

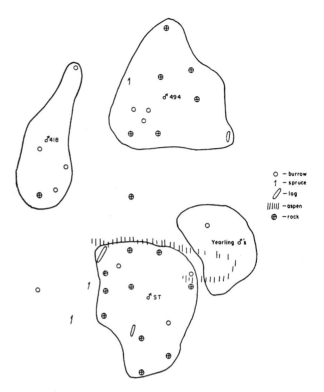

Figure 35.5 Individual spatial territories of three adult males and the home range of four yearling males of the yellow-bellied marmot, *Marmota flaviventris*. (Armitage 1974:250)

(*Hypsignathus monstrosus*). Most pinniped species with polygynous mating systems have a *harem type* of *arena territory* (Fig. 35.6). Males vigorously defend these territories from intrusions by other males and try to maintain an assemblage of females inside the territory.

35-G Would you expect a female mammal to exhibit a nidic territory year-round? Why are arena territories associated with polygynous mammals? (Refer to Feldhamer et al. [1999], or reference books on behavior, for possible explanations).

Group spatial territories (Fig. 35.7) are found in many mammals such as gibbons (*Hylobates* spp.), prairie dogs (*Cynomys* spp.), and several other species (see review in Fisler 1969). In prairie dogs, the **coterie** (a "family" group consisting of a male, several females, and their young) typically occupies a group territory (Fig. 35.7) that is defended by the group against other coteries.

35-H Examine Figure 35.5 showing spatial territories of yellow-bellied marmots, *Marmota flaviventris*, Sciuridae, as described by Armitage (1974). Why do you think the territories of the males are separate? How do you think King (1955), who studied black-tailed prairie dogs, *Cynomys ludovicianus,* defined the boundaries of the group territories shown in Figure 35.7?

Habitat Utilization and Preference

As noted by M'Closkey and Lajoie (1975), studies of the distribution of animals may allow us to make definitive statements about the evolution of habitat selection, predict colonization or range extensions, and permit examination of species-packing and diversity.

The utilization of particular areas and habitats within the range of a species may depend on the distribution of available resources, on climatic conditions, and on the presence of other species (Krebs 1972).

Due to space limitations, this subject will not be covered further in this manual. However, the publications by Harris (1952), Rosenzweig and Winakur (1969), Brown et al. (1972), Grant (1972), Rosenzweig (1973), Colwell and Fuentes (1975), Feldhamer et al. (1999), and textbooks of ecology will give an introduction to this field.

Movements

Migration

Mammals may migrate seasonally to reach feeding and/or breeding grounds and/or more favorable climatic conditions (Feldhamer et al. 1999; Sanderson 1966). **Latitudinal migration** occurs in several species, including the gray whale (*Eschrichtius robustus*), that travels about 9,000 km from summer feeding grounds in arctic waters to winter calving grounds off the coast of Baja California (Rice and Wolman 1971). Many northern fur seals (*Callorhinus ursinus*) migrate about 5,000 km from the Pribilof Islands to more southerly Pacific coastal areas. Some species of bats also make extensive latitudinal migrations (Griffin 1970).

Elevational migration occurs in some populations of elk (*Cervus elephus*) that move from winter ranges at lower elevations to summer feeding ranges at higher elevations.

Seasonal migration also occurs in a variety of other terrestrial mammals, particularly ungulates. For example, caribou (*Rangifer tarandus*) move in large herds from winter ranges in central Manitoba and northwestern Ontario to summer ranges some 500 to 600 miles to the north (Harper 1955). In the Serengeti Plains of Africa, brindled gnus (*Connochaetes taurinus*) move in large aggregations (tens of thousands) over a wide area, whereas in the Ngorongoro area, they are sedentary (Leuthold 1977).

35-I What factors seem to be important in the migrations of baleen whales? Why would ungulates in North America and Africa make long-distance migrations? What methods of study could be used to determine if some species of bat migrates?

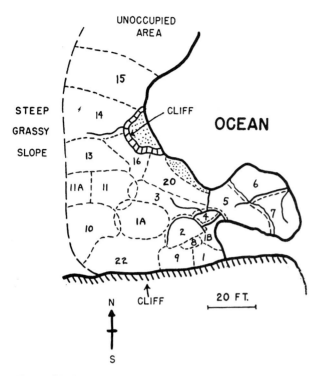

Figure 35.6 Distribution of arena territories (harem type) of the northern fur seal, *Callorhinus ursinus*.
(Bartholomew and Hoel 1953:430)

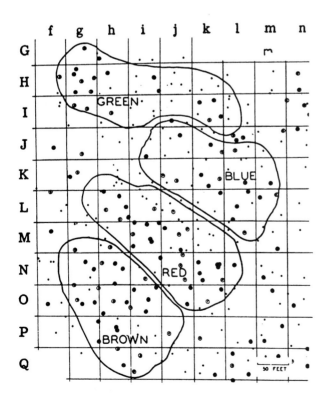

Figure 35.7 Group territories of four coteries of black-tailed prairie dogs, *Cynomys ludovicianus*.
(King 1955:55)

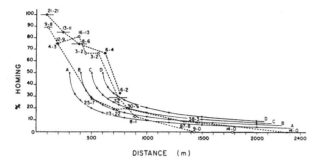

Figure 35.8 Relationship between distance to percent homing success in the deer mouse, *Peromyscus maniculatus gambelii.* (Furrer 1973:471)

Immigration, Emigration, and Dispersal

The composition of a population (see Chapter 36) changes due to the influx or **immigration** of individuals from outside the population and the loss or **emigration** of individuals from the population. When individuals leave a population during a relatively short period of time, the emigration is termed **dispersal.**

Homing, Nomadism, and Exploration

Mammals experimentally removed to areas well away from their home ranges may return to them within a short time (Fig. 35.8). This phenomenon is known as **homing** (Anderson et al. 1977; Henshaw and Stephenson 1974).

Nomadism has been defined as the tendency for individuals to make erratic or wandering types of movements outside of their "normal" territories or home ranges. Brown (1966) suggested that understanding these movements may aid in interpreting the population dynamics of a species. All mammals engage in **exploration** when placed in a novel environment or when moving about their home range.

METHODS FOR STUDYING MOVEMENTS

Tagging

Accurate estimation of individual or group movements and of space utilization often requires that individuals be marked in a distinctive fashion (see Chapter 30). The individual characteristics of flukes (e.g., porpoises and whales) or ears (e.g., elephants) may enable individual recognition without capture and marking of the animals (Hrdy 1977:78–81; Schaller 1963:24).

Record Keeping and Mapping

Records of capture locations of mammals can generally be recorded on the same forms (see Fig. 29.9) as those used for analyses of populations. A detailed map of the study area is also useful. This map can be made from an aerial photograph onto which coordinates or special features have been added. Then copies of the map can be used to plot daily or hourly movements of an individual, a group, or a population. Mosby (1969b, 1980) gives techniques and recommendations for preparing these maps.

Trapping

Many estimates of home-range size are based on data obtained from grid trapping. This method is time-consuming but can produce reasonable quantities of location data. One disadvantage, if traps are checked only once or twice per day, is that many days must elapse before sufficient data are available for accurate estimates.

Blinking Light Transmitters

Light-emitting diodes (LED) can be attached to animals to provide a visual means of detecting their movements at night. These battery-operated devices can be programmed to produce a unique flashing pattern for each recognized individual. Braun (1985) used LEDs to monitor the home ranges and activities of giant kangaroo rats, *Dipodomys ingens,* Heteromyidae. Braun also equipped her study grids with posts that were illuminated by Cyalume brand lights. More information about this technique and the use of radioactive *Betalights* can be found in Nietfeld et al. (1996).

Radioisotopes

Radioisotopes can be implanted in the skin of mammals to follow the movements of an individual at frequent intervals (Nietfeld et al. 1996; Peterle 1969). The main disadvantages of this technique are the limited range for detection of the radioactivity, and the inability to mark many individuals (due to possible confusion of the signal) in a given area (Sanderson 1966).

Radiotelemetry

This technique offers the capability of obtaining large quantities of location data (e.g., one individual for each channel of a receiver). Some of these transmitters, with battery, weigh little (0.8 to 1.2 grams) and will work on small rodents (as small as *Microtus*, 30–40 grams). These transmitters generally are powered by batteries, which add a significant amount of the weight to the entire package (Samuel and Fuller 1996). The major disadvantage of this technique is the cost of the transmitters and receivers and the relatively short life (one to two weeks) of the power packs for some transmitters used on small mammals. Brander and Cochran (1969) and Samuel and Fuller (1996) provided additional information on this technique and precautions to be observed to prevent potential harm to the animal that is equipped with the transmitter and battery.

Fluorescent Pigments

Pigments that can be dusted onto mammals and that fluoresce when exposed to ultraviolet (UV) light can also be used to estimate the area that an animal utilizes. In one technique, the animal is briefly enclosed in a plastic bag that contains a small quantity of the fluorescent powder. The animal is briefly shaken in the bag and then immediately released to move about while the researcher waits in another area. The animal drops particles on the ground and vegetation as it moves. The investigator then returns with the UV light and makes note of where the animal went. One can use flagging, thread, or other markers to make note of the locations visited by the animal, and then return during daylight hours to map the locations. This technique is particularly helpful to determine movements and habitat use of small rodents. Often, one can detect that the animals are utilizing the vertical dimension of the home range by this technique. Refer to Lemen and Freeman (1985), Mullican (1988), and Nietfeld et al. (1996) for additional details on the procedure. Liu et al. (1988) compared the effectiveness of "powder-tracking" with radio-tracking, and found that use of the pigments gave results comparable to those of radio-tracking.

Spool-and-Line Tracking

Spool-and-line tracking has been used to follow the movements of animals in dense forests. With this method, a spool of thread is attached inside a case that is then attached to the animal. One end of the thread is secured, and the animal is released. As the animal walks about, the thread is played out of the holder and deposited behind it. The distribution of the played-out line can then be mapped to get an estimate of how the animal uses its environment. Anderson et al. (1988) used this technique to study the movements of a New Guinean bandicoot, the common echimypera, *Echimypera kalubu,* Peramelidae, during its nocturnal movements.

35-J   In a natural area, under the supervision of your instructor, place a small quantity of fluorescent powder into a plastic bag into which a small nocturnal rodent is then placed. Shake the animal in the powder for three to five seconds and then immediately release the animal onto the ground. Leave the area and allow the animal to move about on its own. Return several hours later to the site of the release and use a portable UV light to follow the "trail" of the animal. Mark the locations of the pigment with flagging or thread. Return the next day to observe in the daylight where the animal moved and to plot the locations of those movements. What was the smallest bit of fluorescent powder that could be detected under the UV light? Was there any evidence of arboreal activity by the rodent? What types of vegetation did the animal touch? How would you mark different individuals on the same night in the same area so that you could distinguish their trails?

Photographs

Photographic techniques have also been used to study the movements of small mammals (Pearson 1960; Wiley 1971). Commercial weatherproof cameras equipped with sensors can be used to detect animals that would otherwise be difficult to observe (nocturnal and rare species). The cameras are placed in selected locations and can be used to measure intensity of utilization of an area, although this requires that several units be operating simultaneously to prevent bias in the data gathering. Refer to the *Journal of Mammalogy* and suppliers of field equipment for examples of these cameras.

Individual Observation

For diurnal mammals living in groups, direct observation is a valuable means of determining space utilization. However, this method is time-consuming and may be misleading if proper precautions are not made to reduce bias in the observation procedure. Altmann (1974) and Feldhamer et al. (1999:28–29) provided valuable suggestions for making observations in a manner that minimizes bias. Armitage (1974) used this technique to study the territorial behavior of the yellow-bellied marmot (*Marmota flaviventris*), and it is widely used to study the movements of many primates (Hrdy 1977; Schaller 1963).

Other Techniques

Sanderson (1966) provided examples of the use of dyes in urine and feces in studies of movements. Justice (1961) and Metzgar (1973) used smoked paper to study the movements of rodents, and found that trap- and track-revealed home-range estimates differed.

CHAPTER 36

ESTIMATION OF ABUNDANCE AND DENSITY

Accurate estimates of population densities are important for answering many questions in ecology and population biology. Excellent comprehensive reviews of methods used in this field include Cormack (1968), Southwood (1966), Seber (1973), Delany (1974), Smith et al. (1975), Flowerdew (1976), Caughley (1977), Otis et al. (1978), White et al. (1982), Pollack et al. (1990), and Lancia et al. (1996). Comparisons of some of the methods for estimating abundance and density can be found in Mares et al. (1981), Nichols and Pollack (1983), Lancia and Bishir (1996), Nichols (1986, 1992), and Nichols and Dickman (1996).

BASIC POPULATION CONCEPTS

A **population** may be defined as a group of all interbreeding individuals of the same species and that occur in an area at a given time. These local populations can grow or decline in a number of ways. **Natality** is the increment of young into a population through births, whereas **mortality** is a decrement to the population through deaths of individuals. Animals may also be added to the population by **immigration**, as individuals move into the population. When animals leave the population, the process is termed **emigration**. Details on age-based population concepts can be found in most textbooks of ecology, and in Caughley (1977), Johnson (1996), and Feldhamer et al. (1999). Sauer and Slade (1987) provided arguments for basing animal demography on body size rather than age, because age determination under field settings is often difficult.

TYPES OF POPULATION ESTIMATES

The terms **abundance** and **population size** pertain to the number of individual animals, based on actual counts, on estimates of their numbers, or on a relative ranking of their numbers (Lancia et al. 1996). Unlike density measures, abundance does not necessarily indicate the number of animals in an area of given size. Lancia et al. (1996) stated that **absolute abundance** is an actual count or an estimate of the number of individuals present, whereas **relative abundance** refers to a ranking of population size (e.g., area A has more than area B). In most studies, a count is made or an estimate made of the number of individuals in a study area to get absolute measures of abundance. Less frequently, a **census**, or **complete enumeration**, is made of all individuals in an area. Such a census might include a count of all bats in a roost or a count of all members of a family group of monkeys that visit a waterhole. **Population density** is the number of individuals per unit area. In field studies, we make an estimate of this number, based on the size of the area sampled and the number of individuals found in that area. For mammals, this quantity is generally reported as the number of individuals per hectare, or, for large mammals, as the number per kilometer^2. The term **relative density** is sometimes used when one is comparing numbers of animals caught in different trap lines, even though the area sampled by the trap line may not be known.

USE OF COMPUTERS FOR ESTIMATION OF ANIMAL ABUNDANCE

Those interested in using computers to estimate abundance and density might want to use the program CAPTURE (White et al. 1982) for closed models (where emigration and immigration effects are not considered important) or the programs capable of calculating Jolly-Seber estimates in an open model (where such movement is apparent). See pertinent sections below for additional references.

SAMPLING CONFIGURATIONS USED IN ESTIMATING DENSITY

In many studies, particularly those involving small mammals, a grid (Fig. 36.1A) of traps is established in an area of uniform habitat. Generally, a minimum of two grids should be established so that information on the variability of the estimate can be obtained (Hayne 1978). In practice, this suggestion is sometimes difficult to follow if the grids are large, and help is limited. The **standard minimum grid** (16 rows by 16 columns) with 256 trap stations (each with two traps per station = 512 traps total) or a 12 × 12 grid are widely used in studies of small mammal populations. The trap spacing in the standard minimum grid is 15 meters. Distance between traps is known to affect estimates of home-range size (Hayne 1950) and estimates of density (Gurnell and Gipps 1989).

A **transect** (Fig. 36.1B) is a line of traps spaced at regular intervals. If an **assessment line** (Fig. 36.1B) of traps is placed at an angle to the transect, then some measure of area and resultant measure of density can be obtained (Smith et al., 1975). In general, a **trap line** is any line of traps placed at regular or irregular intervals in order to secure specimens for identification, for study skins, or for necropsy (see Chapter 30).

Many mammal populations cannot adequately be censused using grids or transects. Thus, population numbers of bats and many colonial species are often estimated by photographic means (Humphrey 1971) or by direct enumeration (Mills et al. 1975; Peterson and Bartholomew 1967). Large terrestrial, large semiaquatic, and large aquatic species are generally counted by aerial census techniques (Caughley 1977; Firchow et al. 1990; Martinka 1976). Additional information on specialized census methods can be found in Overton (1969), Caughley (1977), or Lancia et al. (1996).

ESTIMATING SIZE OF SAMPLING AREA

In grid studies, the simplest way of defining the size of the area for density estimation is to use the perimeter of the grid as the limit of the sampling area. Generally, this practice overestimates population densities because mammals are not sedentary and may have movements that are only partly within the grid area (although they may be trapped on the grid). The discrepancy between the population estimates of animals living on the grid and those living partly on the edges of the grid has been termed the **edge effect.**

Because the area of the grid is not the area sampled (see reviews by Stenseth et al. 1974; Smith et al. 1975), various approaches have been used to estimate the actual sampling area. In density-estimation studies, an **effective trapping area** is the area actually utilized by individuals in a local population. Movement data may be used to calculate a **mean distance traveled** by individuals, and one-half of that number used to add a boundary strip to the study grid to estimate this effective trapping area. Another method to calculate the effective trapping area is to compare the number of captures at outer grid stations with those of a "smaller grid" inside the larger grid (Hansson 1969; Pelikan 1970; Smith et al. 1969). In this system, the **probability of capture** (number of animals caught during the study divided by the number of traps) is plotted as a function of the distance of the traps from the outer edge of the grid (Fig. 36.2). Then, the point in meters where the probability of capture changes abruptly to the horizontal represents the **boundary strip** that should be added to the size of the overall grid. This method assumes random dispersion of individuals, no migration, circular home ranges, circular trappable area for each trap, and traps

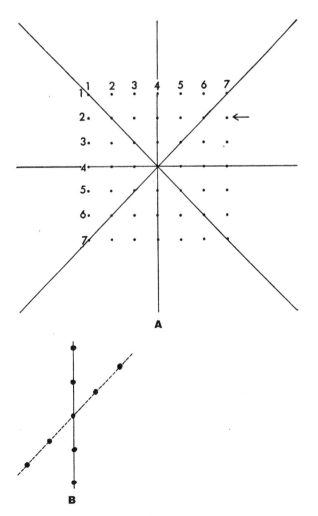

Figure 36.1 (A) Example of a 7 × 7 grid with eight assessment lines (solid); (B) transect (solid line) with assessment line (dashed) intersecting at an angle. The arrow marks the location of trap station 27 (i.e., 2nd row, 7th column).
(Mary Ann Cramer)

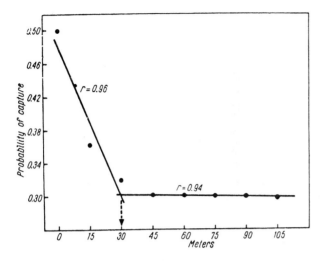

Figure 36.2 Probability of capture (number of animals caught during the study divided by the number of traps) as a function of the distance of the traps from the outer edge of the grid. The width of the boundary strip is indicated by the arrow; r is the regression coefficient.
(After Smith et al. 1969:28)

spaced such that each home range includes more than one trap (Smith et al. 1969). In practice, few of these assumptions can be met, and thus the method is not ideal.

Identification of resident animals on a grid by means of colored bait (which later appears as colored feces), placed prior to the time of trapping (i.e., **prebaiting**) is another technique used to determine the size of the trapping or sampling area (see Smith et al. 1975 and Flowerdew 1976 for review).

The use of one or more assessment lines (see Fig. 36.1A) has been recommended by a number of investigators (Smith et al. 1975; Stenseth et al. 1974) in order to detect movement of animals onto (immigration) or away from (emigration) the trapping grid. These assessment lines are trapped immediately after the regular census.

Accuracy of Estimates

The intensity of trapping (increases with number of days, number of traps), the number of individuals marked, and many other factors affect the accuracy of population density estimates. Roff (1973a) recommended that reliable estimates of population density should have a **coefficient of variation (CV)** less than 0.05 (5%) and confidence limits of $N \pm 0.1N$.

$CV = s (100/X)$, where s = standard deviation and X = mean of the population estimates

The sampling intensities required to obtain these ranges (Overton 1969; Roff 1973a; Smith et al. 1975) limit the usefulness of mark-and-recapture approaches for estimating population densities.

Recording Data

General information on recording data of various types can be found in Chapter 29. For all types of population-estimation techniques, accurate records must be kept on the number of animals observed or handled. Special forms (see Fig. 29.9) are often used to record the species, sex, identifying number (if the estimation technique requires a unique number for each individual), location of capture, and condition of the animal captured. For mark-and-recapture studies, it is generally necessary to record whether an animal is a **new capture** (i.e., first time captured on grid) or a **recapture** (i.e., marked or tagged individual). Techniques for marking mammals with individual numbers or marks are described in Chapter 30.

The location of a capture is frequently recorded using a grid coordinate system (see Fig. 36.1A) or matrix notation. For example, the location of an animal captured at the third station (i.e., row 3) on the second line (i.e., column 2) would be recorded as location 32. Similarly, an animal captured on row 2 and column 3 would be recorded as location 23. This procedure will facilitate data processing by computers. Alternatively, the cumulative distance between the traps may be used to indicate the grid coordinate (Jennrich and Turner 1969). With this system, grid coordinate 23 might be represented as 3045 (i.e., 15 meters × 2 = 30; 15 meters × 3 = 45).

Estimation by Census or Direct Counts

It is sometimes feasible to count or census the total number of animals inhabiting an area. This procedure has been utilized for African ungulates (Talbot and Stewart 1964) and for pinnipeds (Peterson and Bartholomew 1967). Making **total counts** is often not practical for non-colonial, non-herded species because (1) it requires too much effort, (2) some animals may be counted twice, and (3) it disturbs the entire population (Caughley 1977). In addition, a "total count" in almost all cases is actually still an estimate, and thus confidence limits should be calculated on the value obtained, by replicating the count several times (Caughley 1977).

Sampled counts eliminate most of the practical problems associated with total counts. Sample areas are located at random (or, if not truly at random, in a manner that minimizes bias), and then counts are made on the sample areas. Caughley (1977) recommended that sampling intensity in high-density areas be greater than in low-density areas. An error of estimation in the low-density area will not affect the error rate greatly, but undersampling in the high-density area would cause the error rate to increase significantly.

Aerial censusing of larger species of mammals is an important technique for managing wildlife. According to

Caughley (1977), random sampling of areas is more difficult for the navigator and takes more time per unit of area covered. As a result, systematic sampling of areas is more commonly utilized for aerial censusing.

36-A The instructor will set up a mammal community that has taxa representing several species. Each species is represented by a different symbol. For sampling, use a transparent overlay with an area of 1/50 or 1/100 of the total area occupied by the community. In the real world, mammals move about, so this simulation is unrealistic in that sense. Prepare two sampling schemes: (1) *random,* in which 15 sampling areas are located by means of a table of random numbers and (2) *stratified,* in which the 15 sampling areas are placed at a predetermined set of locations according to the scheme provided by your instructor. Select a "species" (e.g., one for each student in the group) and count the number of individuals sampled by the overlay. The instructor will let you compare your estimates with the true population density (N) for each species. Record you results in Table 36.1.

RELATIVE ESTIMATES OF DENSITY

Trap Lines and Trap Nights

If the area of the trapped area is known, then data obtained from trap lines or transects can give estimates of density. Generally, it is not possible to obtain actual density estimates because the size of the trapping area is not known. An assessment line placed at an acute angle to the transect (see Fig. 36.1B) can be used to obtain an estimate of the area sampled (Calhoun and Casby 1958 and review in Smith et al. 1975). A removal procedure seems best for estimating numbers using this method (see "Removal Trapping and Catch per Effort Methods"). A very crude index is the number of animals captured per **trap night** (trap night = one trap set for one night). An estimate of the relative number of individuals is less useful than an estimate of absolute density because no estimate is available of the area sampled, and differences in the behavior of the animals, the habitat, and the weather may make inter-area comparisons meaningless.

Other Relative Estimates

If the area sampled is known, density may also be estimated by counting the number of animals that pass a single point (**point-area count**), those that are flushed along a transect (**flush census**), those that are counted by the roadside (**roadside counts**), or by counting the maximum number of individuals observed along a transect at a given time (**bounded counts**). The signs of mammals may also give an indication of relative numbers. For example, counts of feces, runways, feeding residues, mounds, and other signs may serve as indices to relative numbers (Lord, et al. 1970; Neff 1968). Wauters and Dhondt (1988) found that counting the exposed nests (dreys) of Eurasian red squirrels (*Sciurus vulgaris*) provides a useful population index of potential use for managing this species. However, Carey and Witt (1991) found that track counts were not useful indicators of the abundance of Northern flying squirrels (*Glaucomys sabrinus*) and Townsend's chipmunks (*Tamias townsendii*). Additional techniques of importance for wildlife managers were discussed by Overton (1969), Caughley (1977), and Lancia et al. (1996).

DENSITY ESTIMATES BASED ON MARK AND RECAPTURE

When animals are captured on more than one occasion, a mark-and-recapture technique can be used. In this technique, animals are captured, marked with a unique mark, and released. A certain portion of these marked animals may be recaptured. With a measure of the area sampled, this technique is widely used for estimating density of small mammals, although the estimates obtained may not be very accurate (Caughley 1977; Roff 1973a). General reviews of these techniques can be found in Cormack (1968), Southwood (1968), Eberhardt (1969), Overton (1969), Roff (1973a, 1973b), Seber (1973), Delany (1974), Smith et al. (1975), Flowerdew (1976), Caughley (1977), Lancia et al. (1996), and Nichols and Dickman (1996).

Mark-and-recapture techniques can provide estimates of population densities and provide an opportunity to gather other information about the animals. With this technique, one can gather data on movement patterns, growth rates, age-specific fecundity and mortality, combined rates of birth and immigration, combined rates of death and emigration, and rate of increase (Caughley 1977). Generally, these methods require a considerable amount of effort and expense. Thus, one should determine if these techniques are required to answer the questions posed or whether other less time-consuming and costly methods could be applied.

Assumptions for Use of Mark-and-Recapture Methods

All mark-and-recapture methods assume the following (Delany 1974; Flowerdew 1976; Smith et al. 1975):

1. That the animals do not lose their marks.
2. That the animals are correctly recorded as marked or unmarked individuals.
3. That marking does not affect the probability of survival of the marked individuals as compared with that of unmarked individuals.

TABLE 36.1 | Data sheet for recording results of Exercise 36-A sampling.

Taxon/Replicates:	Random Sampling Locations	Stratified Sampling Locations
1		
2		
3		
4		
5		
6		
7		
8		
9		
10		
11		
12		
13		
14		
15		
Area sampled by overlays		
Total area of study site		
Total individuals = $\sum_{1-15}$		
Estimated density = total individuals/area		
True density (value supplied by instructor)		
Your estimate as percentage of true population density		

How similar are these estimates to one another (an estimate of *precision*)?

How similar are the density estimates to the true density (an estimate of *accuracy*)?

In addition, most methods may require that some of the following assumptions be made:

4a. That the population is *closed* and no gain or loss of members occurs due to natality, mortality, emigration, or immigration, *or*

4b. That the population is *open,* with natality and immigration occurring but with mortality and emigration affecting marked and unmarked animals equally.

5a. That marked animals disperse randomly into the population and that every animal, both marked and unmarked, has the same probability of capture, *or,*

5b. That if different probabilities of capture exist, they are proportionally distributed among all marked and unmarked animals in the population.

The assumptions of 4a or 4b can be checked by experimental techniques (Smith et al. 1975). Refer to Otis et al. (1978), Pollock et al. (1990), and Nichols (1992) for additional guidelines. Pollock (1982) provided a robust model for estimation, that relies on both closed and open models for populations that have unequal probabilities of capture. Generally, the assumptions of 5a and 5b are the ones most frequently violated (see reviews in Smith et al. 1969, 1975) and often cannot be statistically tested (Caughley 1977; Roff 1973b). Seber (1973) described the available statistics and computational procedures for testing these assumptions. See also the review in Lancia et al. (1996).

Petersen (Lincoln) Index (Single Marking)

The **Petersen Index** or **Lincoln Index** is one of the simplest methods for estimating abundance and density. Unfortunately, this mark-and-recapture technique also requires that a stringent set of assumptions (1, 2, 3, 4a, 5a) be met. In the Petersen Index method, the estimated number ($\hat{N}$) of individuals in a sampled population is determined by capturing and marking a sample of animals (M) on one occasion and then capturing a second sample (n) on a subsequent occasion and checking for the number of marked animals that are recaptured (m). Then the number of individuals is estimated using the following formula:

$$\hat{N} = \frac{Mn}{m}$$

Because $\hat{N}$ overestimates N by $1/m$, Bailey (1952) and Roff (1973a) recommended the use of the following formula:

$$\hat{N} = \frac{M(n+1)}{m+1}$$

For example, suppose that 50 animals were captured, marked, and released back into a population. Then, a short time later (to reduce bias due to immigration), the population is resampled and 100 unmarked and 25 marked individuals captured. Then, using the formula $\hat{N} = M(n+1)/(m+1)$, the population is estimated:

$$\hat{N} = \frac{50(125+1)}{25+1} = 242$$

The standard error of this estimate can be approximated by using the formula

$$S_{\bar{x}} = \sqrt{\frac{M^2(n+1)(n-m)}{(m+1)^2(m+2)}}$$

Thus, for this example,

$$S_{\bar{x}} = \sqrt{\frac{2500(126)(100)}{(676)(27)}} = \sqrt{1725.84} = 41.5$$

and $\hat{N} = 242 \pm 42$

How many animals should be marked and recaptured to provide an estimate of the abundance and/or density ± a given standard error? This varies, and formulae for estimating adequate sample sizes may be found in Overton (1969), Smith et al. (1975), and in other sources. Caughley (1977:144) provided a graph that can be used to determine the number of recaptures required to have a 10% standard error of the Petersen estimate of the population.

36-B Obtain a bag containing at least 200 "individuals," as represented by beans. Then, using a table of random numbers, select a two-digit number. Draw this number of beans out of the bag, mark them with a distinctive dot (with pen, crayon, paint, etc.). Return the marked beans to the bag. Select another two-digit number from the table of random numbers, mix the marked and unmarked beans well, and then draw out (without looking!) a number of beans equal to the second two-digit number. Then, compute an estimate of the number of beans in the population by using the modified formula for the Petersen Index.

$$\hat{N} = M(n+1)/(m+1)$$

$$\hat{N} = \underline{\hspace{3cm}}$$

What value did you obtain for the number of beans in the bag? How close was this estimated value to the population parameter (N)? Is this an open or a closed population? Why or why not? Remove the marked beans, replace them with unmarked ones, and repeat the exercise *without randomly mixing* the marked beans in with the unmarked (just drop them in). Did this affect the result (replicate several times for each method)?

Estimates Based on Multiple Marking Occasions (Deterministic)

The Petersen Index method requires that animals be marked on a single occasion. Often, this requirement results in too few marked animals for accurate estimates. To overcome this difficulty, Schnabel (1938) devised a method that is essentially a series of Petersen estimates. For ease in computation, the formula of Schumacher and Eschmeyer (1943) provides an explicit solution:

$$\hat{N} = \frac{\sum M_i^2 n_i}{\sum M_i m_i}$$

This procedure, sometimes termed the **Hayne method**, was developed independently by Hayne (1949b) and requires the same set of assumptions as the Petersen Index method. Refer to Overton (1969) and Caughley (1977) for additional information and for formulae to use in calculating standard errors for these estimates.

An increase in the number of individuals (N) in the sampled population, due to immigration and/or natality during the sampling period, violates the assumptions of the Petersen-type estimates. **Bailey's triple-catch method** detects and counteracts the effect of immigration and provides estimates of birth rates and death rates in addition to estimates of abundance and/or density population size (Caughley 1977). Roff (1973a) recommended against its use because a better estimator, the Jolly-Seber method, is available.

The Jolly-Seber Stochastic Method

All of the models for estimating abundance or for estimating density are considered to be *deterministic* methods because they assume constant rates of death, birth, immigration, and emigration. In contrast, *stochastic* methods are more realistic because the rates of death (along with emigration) and birth (along with immigration) are estimated independently on each sampling occasion.

The **Jolly-Seber method,** developed independently by Jolly (1965) and Seber (1965), is the most well-known of the stochastic population models. This method requires that assumptions 1, 2, 3, 4b, and 5a be met. Again, the requirement for equal probability of capture is the most serious limitation for the use of this and all other mark-and-recapture methods. The arithmetic computations required for this method are somewhat tedious, but computer programs can be used to calculate the Jolly-Seber estimates (White, 1971). Good reviews of the Jolly-Seber technique and associated computation procedures can be found in Southwood (1966), Overton (1969), Seber (1973), and Caughley (1977).

Removal Trapping and Catch per Effort Methods

Assuming no births, no deaths, no immigration, or emigration, and an equal probability of capture of individuals, the number of animals trapped and removed in an area (assuming constant trapping effort) on succeeding occasions should decrease (Grodzinski et al. 1966; Hayne 1949b; Zippin 1956). In these removal trapping procedures, animals may be removed by kill traps or the animals may be "effectively removed" by livetrapping and marking the animals to be removed. In the latter case, the animals are removed "on paper" for the purpose of the calculations.

The **Zippin method,** (Zippin 1956), based on two trapping occasions, is one of the simplest removal trapping (catch per effort) methods. In this procedure, the number of individuals (N) is estimated using the following formula:

$$\hat{N} = \frac{n_1^2}{n_1 - n_2}$$

where n_1 = number of animals caught on the first day and n_2 = number of animals caught on the second day. For example, 100 traps were set for two days. On the first day, the catch was 50 (n_1) and on the second day, 20 (n_2). Substitution into the above equation yielded the following estimate of abundance ($\hat{N}$):

$$\hat{N} = \frac{50^2}{50 - 20} \cong 83$$

Seber (1973) provided additional discussion of this procedure.

The Hayne method (Hayne 1949b) uses data obtained over a number of trapping occasions. Smith et al. (1975) cautioned that for the use of this technique, or any other regression technique, a significant r-value must be obtained. Hansson (1969) discussed another technique based on removal catches that also provides a method for calculating the effective trapping area. The method of Kaufman et al. (1971), using assessment lines, compensates for unequal probabilities of capture.

Enumeration Method

Equal probability of capture is a severe practical limitation for the use of most mark-and-recapture methods of estimating the number of individuals in a population. Consequently several investigators, including Petrusewicz and Andrzejewski (1962), Krebs (1966), and Krebs et al. (1969), sought instead to conduct a complete census of resident animals on their trapping grids. Roff (1973b) sharply criticized such an approach because "The problems involved in a complete enumeration are vastly greater than even in the mark-and-recapture method and the complete inability to measure the accuracy of the estimates makes it a technique that cannot be accepted." This method is generally referred to as the **Calendar of Captures method** or the **Minimum Number Known Alive.** Flowerdew (1976) provides a clear description of the technique. Nichols and Pollack (1983) and Nichols (1986) stated that population estimates based on enumeration techniques were inferior to those based on the Jolly-Seber method. Pollock (1982) provided suggestions for how to effectively use the Jolly-Seber technique even when unequal trappability is assumed.

Change-in-Ratio Methods

Changes in the relative abundance of two categories of animals (e.g., adults versus subadults; males versus females) can be used to estimate population numbers (Seber 1973). Hanson (1963) and Rupp (1966) presented a review of these **Change-in-Ratio methods** but pointed out the general lack of procedures for determining confidence limits for these estimates. Paulik and Robson (1969) remedied this problem by providing formulae to estimate standard errors. These methods are widely utilized to study populations of game animals. Consult the above papers and Overton (1969), Seber (1973), and Lancia et al. (1996) for details.

CHAPTER 37

LITERATURE RESEARCH

The literature pertaining to wild mammals is voluminous. In 1976, Jones, Anderson, and Hoffman reported that it included 115,000 titles and was increasing at a rate of 5,000 to 6,000 papers per year. Obviously, it is impossible for a mammalogist to be familiar with all that has been published, but when a scientist is engaged in a particular study, it is necessary for him or her to know all that others have written in that field. Thus, the mammalogist frequently finds it necessary to conduct a literature search.

THE LITERATURE OF MAMMALOGY

Many full-length books have been published about mammals, but most of the literature of mammalogy is in technical journals. Jones and Anderson (1970) reported that 50% of the existing literature on mammals is contained in about 40 journals. The remaining 50% is widely scattered, with at least 150 different journals needed to encompass only 70% of the literature.

Several journals are devoted exclusively to publications in the field of mammalogy. Some of these are highly specialized and cover only one aspect of mammalogy or one taxonomic group of mammals. Others publish articles on all aspects of mammalogy. The primarily mammalogical journals include the following:

Acta Theriologica. 1954–present. Polish Academy of Sciences, Mammals Research Institute, Warsaw, Poland. General.
Australian Bat Research News. 1964–present. Lyneham, Australia.
Australian Mammalogy. 1972–present. Australian Mammal Society, Inc., Lyneham, Australia. General.
Batchat. 1983–present. Nature Conservancy Council Publications, Peterborough, United Kingdom. Bats.
Bat Research News. 1960–present. Potsdam, New York, USA. Bats.
Bibliotheca Primatologica. 1962–present. Basel, Switzerland. Primates.
Cetology. 1971–present. St. Augustine, Florida, USA. Cetaceans.
Current Mammalogy. 1987–present. Plenum Publishing Co., New York, USA. General reviews.
Folia Primatologica. 1963–present. Basel, Switzerland. Primates.
Journal of Mammalogy. 1919–present. American Society of Mammalogists, Lawrence, Kansas, USA. General.
Laboratory Primate Newsletter. 1962–present. Providence, Rhode Island, USA. Primates.
Lutra. 1959–present. Leiden, Netherlands. General (in Dutch, with English, French, or German summaries.)
Lynx. 1962–present. Prague, Czech Republic. General (in Czech, English, German, Russian, and Slovak).
Mammalia. 1937–present. Paris, France. General (in French and English).
Mammalia Depicta. A supplement to *Zeitschrift für Säugetierkunde.*
Mammalian Species. 1969–present. American Society of Mammalogists, Lawrence, Kansas, USA. Species accounts.
Mammal Review. 1970–present. London, United Kingdom. General.
Marine Mammal News. 1975–present. Cetacea and Pinnipedia.

Monkey. 1957–present. Primates (in Japanese).
Myotis. 1953–present. Bats (in English and German).
Nyctalus. 1969–present. Halle, Germany. Bats.
Primate News. 1957–present. Beaverton, Oregon. Primates.
Primates. 1957–present. Aichi, Japan. Primates.
RE'EM. Israel Mammal Information Center, Society for the Protection of Nature, Tel Aviv, Israel. Mammals of Israel (in Hebrew).
Säugetierkundliche Mitteilungen. 1953–present. Munich, Germany. General.
Teriologia. 1972–present. Novosibirsk, Russia. General.
Whales Research Institute. 1948–present. Tokyo, Japan. Cetaceans.
Zeitschrift für Säugetierkunde. 1926–present. Hannover, Germany. General.

Kosin (1972) reported that of the animal science papers covered by the 1969 *Biological Abstracts,* 63.1% were published in English, 14.8% in Russian, 6.5% in German, 4.1% in French, 2.1% in Japanese, 2% in Spanish, and the remaining 7.4% in other languages. Unfortunately, similar data for the literature in mammalogy are not available, but English is the primary language, with Russian, German, and French all being very important. Japanese and Spanish constitute significant percentages, with the remaining languages of the world only poorly represented. Many of the journals in the list above publish articles in English, German, or French. Some of the journals published in non-English speaking nations include a high percentage of papers in English (e.g., *Acta Theriologica*), but usually the native language of the country predominates. English summaries are frequently included with articles published in other languages.

Many journals published in all parts of the world include papers on mammals, as well as on other groups. In the United States alone, the number of such journals is large. Most states and many cities have academies of science, in the publications of which mammal papers may often appear (e.g., *Transactions, Kansas Academy of Science; Proceedings, California Academy of Science; Proceedings, Biological Society of Washington,* [D.C.]). Similarly, many major universities and museums have one or more publication series that may include mammal papers (e.g., *Occasional Papers, Museum of Zoology, University of Michigan; University of Kansas Publications, Museum of Natural History* [now replaced by *Occasional Papers* and *Miscellaneous Publications, University of Kansas Museum of Natural History*]; *Occasional Papers* and *Special Publications, The Museum, Texas Tech University; Bulletin* and *Novitates, American Museum of Natural History; Annals of Carnegie Museum* and *Bulletin of Carnegie Museum of Natural History; Fieldiana: Zoology,* Field Museum; *Proceedings* and *Bulletin, U.S. National Museum* [now replaced in part by *Smithsonian Contributions to Zoology*]; *American Midland Naturalist,* University of Notre Dame). *Ecological Monographs, Ecology, Evolution,* *The Journal of Wildlife Management, Murrelet, Southwestern Naturalist,* and *Systematic Zoology* are a sample of U.S. journals that are published by various scientific and professional societies and that frequently include papers of direct interest to mammalogists.

37-A   Which of the journals devoted exclusively to mammals are present in your library? Examine the two most recent volumes of each. What languages predominate in each? Which include English summaries with articles published in languages other than English? Do any of the "general mammalogy" journals seem to have a predominance of articles on one particular aspect of mammalogy (e.g., systematics, ecology)?

37-B   Examine the most recent complete volume of each of several other biological journals in your library. List those that include articles that are within the scope of mammalogy.

37-C   List the 10 journals that you think are most important in regard to the aspect of mammalogy in which you are primarily interested.

BIBLIOGRAPHIES AND ABSTRACTS

Several types of bibliographies are published that aid the mammalogist in keeping up with the literature of his or her field. Some of these are also useful in performing systematic literature searches.

Biological Abstracts is one of the most complete and up-to-date bibliographies in the biological sciences. As the title implies, abstracts of papers are included along with the bibliographic references. Although *Biological Abstracts* (*BioAbstracts,* for short) is extremely useful for some fields, as Jones and Anderson (1970) point out, it is "quite incomplete for some branches of mammalogy." In addition, a literature search using only *BioAbstracts* can be cumbersome and time-consuming.

Current Contents® is a weekly publication of the Institute for Scientific Information (Philadelphia). Its specialty is timely reproduction of the tables of contents of recent issues of scientific journals. The version *CC*® *Agriculture, Biology & Environmental Sciences* provides a good (though not complete) selection of journals that are of interest to mammalogists.

Wildlife Review is a publication of the U.S. Department of Interior, that includes abstracts along with most of the bibliographic entries. Entries are arranged alphabetically by author, under several categories and subcategories. *Wildlife Review* covers many American and foreign journals, and although it emphasizes game species, nongame species are also included. This source is particularly good for Russian and other East European journals.

Wildlife Abstracts is issued at about 10-year intervals by the publishers of *Wildlife Review*. The title is a misnomer because this publication contains no abstracts; it is a subject cross-index to *Wildlife Review*.

Probably the most complete and useful bibliography of publications on mammals is *Zoological Record, Section 19, Mammalia*. Each section of *Zoological Record* is divided into three parts. The first is a list of complete bibliographic entries arranged alphabetically by author. The second part is a subject index to the papers listed in the first part, and the third is a systematic (taxonomic) index to those same papers.

Several other bibliographic sources are useful to the student of mammalogy. These include such diverse publications as printed card catalogs of great libraries (e.g., *The General Catalogue of Printed Books* of the Library of the British Museum, London), old bibliographic series now defunct (e.g., *Bibliotheca Historico-Naturalis* covering 1700–1846, *Bibliotheca Zoologica* covering 1846–1860, and *Bibliotheca Zoologica II* covering 1861–1880), and still-active bibliographies of peripheral interest to mammalogists (e.g., *Index Medicus*). For a detailed discussion of these and other bibliographic sources, see Smith and Reid's (1972) *Guide to the Literature of the Life Sciences*.

37-D Examine *Biological Abstracts*. If you have never used this series, have your instructor or librarian show you how to use the various indexes, including the author, taxonomic, and computer-rearranged subject (B.A.S.I.C.) indexes. Although *BioAbstracts* can be cumbersome, it is an important resource. Every biologist should be able to use it.

37-E Examine *Wildlife Review* and *Wildlife Abstracts*. How do these listings compare with issues of *Current Contents* and *Biological Abstracts*? Compare the numbers of entries for game and for nongame species of mammals.

37-F Examine *Zoological Record, Section 19, Mammalia*. In what year did publication of this bibliography begin? What is the most recent year available? Compare the method of indexing in this resource with those of *Biological Abstracts* and *Wildlife Abstracts*.

37-G Examine other similar resources that may be available in your library (e.g., *Current Contents, Citations Index, Index Medicus*). Become familiar with the use of each of these.

Making a Search

One of the most commonly used methods of performing a literature search is to find a recent, fairly thorough article on the subject in which you are interested and examine the bibliography at the end of the article. Then look up all of the articles listed in the first paper and examine their bibliographies and so on. If this procedure is continued long enough, it is possible to locate most of the pertinent publications on a given topic. However, there may well be important papers that will be missed, and this type of search can be extremely time-consuming and repetitious.

To perform a thorough literature search, we recommend beginning with the most recent issue of *Zoological Record*. Check the appropriate index of indexes and prepare a 3" × 5" index card or word processor entry on each pertinent reference. It is good practice to begin immediately to use a standardized system of bibliographic entry such as one of those outlined in the *CBE Style Manual* (Council of Biology Editors 1972). For examples of bibliographic entries for a variety of kinds of publications, see the Literature Cited of this manual.

After checking the most recent issue of *Zoological Record*, continue backward in time, using the same source. *Zoological Record* has been published for over 100 years. A complete literature search would thus require examination of over 100 volumes, but this is usually neither practical nor necessary. The nature of the topic being researched will dictate the depth of time that must be explored. We recommend examining at least the latest 10 volumes of *Zoological Record*, and often it may be necessary to examine up to 25 volumes. Important papers older than this will usually, but not always, be mentioned in the bibliographies included in more recent papers.

Once you have completed your search in *Zoological Record*, it will be necessary to use another source to fill in the gap between the most recent issue of *Zoological Record* and the present date. Again, depending upon the nature of the topic being researched, a variety of possibilities are available. *Biological Abstracts* may prove to be the most useful, but *Wildlife Review* or sources such as *Citations Index* may be helpful.

When you have completed the search of these bibliographic sources, you will probably have a sizable stack of reference cards. Sort the cards by journal and locate each of the articles. You will probably be able to discard several references that are of only marginal interest. As you read each pertinent article, examine the bibliography included in it to locate important references that you may have missed in your search.

This procedure for performing a literature search is only one way of going about this activity. A truly complete literature search is usually impossible and impractical. Whitehead (1971:216) said that an average but conscientious systematic zoologist with access to a large library will generally retrieve only about 80% of the existing information.

As you will see after performing the exercises that follow, the coverages of *Zoological Record* and *Biological Abstracts* are not identical. Many papers will be included in one but missed by the other. Some papers will be missed by both but will appear in other bibliographic sources. The greater the number of bibliographic sources searched, and the farther back in time it goes, the more complete a search will become. However, the time required to perform a full search utilizing many bibliographic sources is prohibitive. Because of the better coverage by *Zoological Record* of papers of a systematic, anatomical, or ecological emphasis, and because of the relative ease of using this source, we recommend the above procedure for general use. However, if the primary research interest centers on other topics, such as genetics or physiology of mammals, *Biological Abstracts* may be better for the in-depth search than *Zoological Record*. Availability may be another factor dictating which bibliographic sources will be used. *Biological Abstracts* will be present in the libraries of most U.S. institutions where biological research is being conducted, but *Zoological Record* may not be available. Thus, *Biological Abstracts* will be used as the main bibliographic source.

The main thing you must do is to become thoroughly familiar with the literature and the bibliographic tools in your field. The scientific literature exists for the use of all scientists. If your own publications are to have merit, you must examine and utilize the pertinent literature.

37-H Examine the most recent 20 volumes of *Zoological Record, Section 19, Mammalia* and prepare a set of bibliography cards on one of the following topics or on another topic approved by your instructor.
   a. *Bet hedging* as an ecological strategy in ground squirrels.
   b. Natural history of bats from northeastern Mexico.
   c. Chromosomes of the spiny rat, *Proechimys trinitatis*, in Trinidad.
   d. Lack of population cycling in the vole *Microtus pennsylanicus*.
   e. Diets of the dasyurid marsupials *Sminthopsis murina* and *Antechinus stuartii*.
   f. Parasites in beavers (*Castor canadensis*) from central Texas.
   g. Use of pelvic tuberosities for aging squirrels of three genera.
   h. Use of ultrasonic detectors to monitor the cries of bats.

37-I Repeat the literature search in Exercise 37-H for the same 20-year period, using *Biological Abstracts*. How do the references obtained from the two sources differ? How much time was required for each of the two searches? What are the advantages and disadvantages of each source?

37-J Repeat the search conducted in Exercises 37-H and 37-I, using *Wildlife Review*. How did time spent and results obtained compare with those of the previous searches?

37-K Use all of the sources available to perform a relatively complete literature search on one of the topics listed in Exercise 37-H or on another topic approved by your instructor.

COMPUTERIZED LITERATURE DATABASES

Comprehensive Databases

In recent years, many organizations and companies have made literature databases available for searching by libraries or individuals. One of the largest database companies is *Dialog Information Services, Inc.* in Palo Alto, California. This company provides a central clearinghouse for a number of literature databases, including *Zoological Record* (file 185) and *Biosis Preview* (files 5 and 55). The electronic version of *Zoological Record* has literature from 1978 on, and that of *Biosis Previews* (the electronic version of *Biological Abstracts*) has literature from 1969. These databases provide a rapid and efficient means for updating knowledge of the literature. Many libraries now offer this service to users for a fee (the user normally pays online charges when connected to the database), or you can subscribe to the service individually.

Since the late 1980s, a number of companies have offered scientists the opportunity to use a subscriber's computer and packaged software to search floppy disks or CD-ROMs that are mailed to the subscriber on a weekly or periodic basis, for a fee. These services are not inexpensive but can be cost-effective when one considers the time lost and commuting costs involved in ferreting out scarce journals and indices, using conventional research techniques. One entry in the field is *Current Contents on Diskette,* the electronic version of *Current Contents,* published by the Institute for Scientific Information. In 1989, the company introduced *Current Contents on Diskette for Agriculture and Environmental Sciences.* This version is the one that is most useful to mammalogists because of its particular mix of journals covered. An alternative version, *Current Contents for Life Sciences,* is less useful because it is primarily oriented to the biomedical field and covers only a few journals that would be of interest to students of mammals. Other entries in this field include *Cambridge Scientific Abstracts* and *Reference Update.*

Subject-Area Databases

Another approach to using computers for literature searching is provided by *Absearch Inc.* This company has assembled a database of literature for a selected set of journals that can be searched, for a fee, either from a CD-ROM supplied to the customer, or at their website: http://www.absearch.com. Most of the Absearch databases are wildlife-related and include papers on mammals, birds, reptiles, amphibians, fishes, and other topics depending on the particular database purchased by the customer. For mammals, selected literature (*only since 1955*) from the *Journal of Mammalogy, Mammalian Species,* and *Special Publications of the American Society of Mammalogists* is included in their "mammalogy" database. The "wildlife" database includes selected literature from the Wildlife Society; the "NAWNRC" database includes selected literature from the *Transactions of the N. American Wildlife & Natural Resources Conferences*. The number of databases produced by this company changes yearly, and the user would be advised to check with the company for the latest information. All of the Absearch databases use a read-only version of *Procite*® as the search engine for locating literature in the database. This particular version of *Procite*® allows you to export the results of your search to a word processor file in several of the bibliographic styles of professional journals (including those of the American Society of Mammalogists). Should the users wish to add their own keywords to the database, they would have to get, *at additional cost,* a complete version of the *Procite*® software.

Reprint Management Software

Individuals may also wish to use available software to manage a bibliography of literature or the bibliographic references for a collection of papers on particular subjects. Several companies make suitable software packages for this purpose. The list of available software is broad and includes titles such as *Papyrus, Procite*®*, Ref-11, Reference Manager, Refmenu, Refsys Super,* and *Sci-Mate*. Refer to advertisements in comprehensive journals such as *Science* or *Nature* to get addresses and website information for some of these companies. In evaluating these packages, make sure that the software can import or export references by way of an ASCII (i.e., text) file and has sufficient capability to handle all of the keywords that code each citation. To insure that the package will perform as you wish, always request a demonstration copy or see a version demonstrated at a meeting or by a colleague before you purchase the software.

COMPUTERIZED LITERATURE SEARCH STRATEGIES

Literature databases and most reference software packages utilize **Boolean logical operators** to define and limit the references selected (Fig. 37.1). Under this system, the logical connectors *or, and,* and *not* are used to specify which references are to be included in the search results. This is especially important when you are searching a database that may contain millions of records, and you do not want to get a printout with citations for thousands of papers because you failed to limit the search parameters.

Suppose, for example, that you want to find papers on echolocation in bats. The following statement might be used to limit the search:

bats *and* echolocation

This creates a subset of references that includes all papers referring to bats using echolocation, even if echolocation

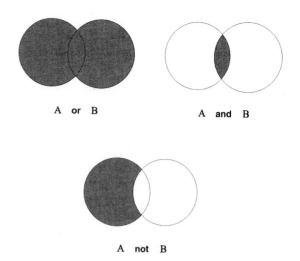

Figure 37.1 Examples of Boolean logical operators (*or, and, not*) used in searches of literature databases. The *or* operator retrieves all records that contain at least one of the search items; the *and* operator retrieves all records that contain all of the search items; and the *not* operator eliminates a search term or group of search terms.
(R. E. Martin)

was mentioned incidentally in connection with studies on echolocation in species other than bats. All papers on bats and all on echolocation would be included under the following search statement:

bats *or* echolocation

This would return a list of all references about bats and all references about echolocation, regardless of whether the papers on bats dealt with echolocation or whether bats were mentioned in the papers about echolocation. Another Boolean operator might involve the following statement:

bats *not* echolocation

This statement would include all papers about bats except those that dealt with echolocation in bats.

Reprints

A **reprint** is a separately printed copy of a paper from a journal or book. Multiple reprints are made available to authors or to the institutions with which these individuals are associated. Normally, an author, or the institution, must pay to receive the desired number of copies (say, 100–200) of the published paper. These are delivered to the author or institution after the article appears in the journal or book. Sometimes the publishers of the paper will provide the author(s) with free copies.

Exchanging Reprints

The **reprint-exchange system** is a procedure in which an interested scientist or other interested party writes to the author (or the institution with which the author is associated) and requests a copy (reprint) of a paper. Normally, the author responds by sending the reprint. (Occasionally, an author may write back to say that the supply has become exhausted and that no further copies are available for distribution.) In return, the person who sends out the reprint may request that he/she be placed on the recipient's mailing list to receive copies of his/her papers as they become available.

A research group in an institution (or an individual scientist) may also prepare and send out numerous copies of an annual list of the group's published works. A fellow scientist who receives such a list can then check off any desired publications and return the list, along with his/her current mailing address or a self-addressed mailing label. Generally, the group that issued the list requests each recipient to send any of his/her published papers to the group as a return courtesy. The advantage of this system is the timely dissemination of scientific findings among interested researchers worldwide. It also provides prospective graduate students and established scientists with an opportunity to see what various research groups are working on and provides a basis for their decision-making in choosing a school for further study or a site to do research.

Cataloging Reprints

Once reprints have been received, they must be arranged in some logical manner so that a particular paper can be found when needed. There are three basic systems for arranging reprints, each with its own advantages and disadvantages.

Reprints may be arranged by *subject* so that all papers on a particular topic are grouped together. One problem with this system is that a given paper may deal with a variety of different subjects, but it can be filed under only one.

Another method is to arrange the reprints alphabetically by the name of the first *author*. This method may work fine as long as you can remember papers by the names of their authors (as most scientists do), but it may not be too helpful to a student who is given access to the reprint collection and who needs to find recent papers on the chromosomes of bats. Also, this system often requires much shifting of reprints to accommodate subsequent papers by the same author or authors.

A third technique is to number the reprints (e.g., within an author-alphabetized group of papers; a group of all reprints received in a given year). This technique requires at a minimum an author index (computerized or card file) but eliminates the constant shifting of reprints to find room for new papers. With the computerized reprint-retrieval systems currently available, it may be best to simply number all reprints consecutively and use the search capabilities of the computer software to locate a particular author's paper or a paper on a given subject. The computerized searches also allow one to prepare a chronological list by year of papers in the reprint file.

37-L   Check with your library to see if it offers a computerized search service. Does the service access references only if they are in the library? Is a national database consulted? What are the charges, if any, for this service?

37-M   Select a topic provided by the instructor. How many references were located? What range of years was covered? Estimate how long the search would have taken with noncomputerized searching.

37-N   Use a computerized literature database to search for references on a topic of your choosing or of your instructor's choosing. How long did it take to enter the program and begin searching? What Boolean operators (i.e., *or, and, not*) were used in the search? What effect did those operators have on the number of references found on the topic? How easy is it to prepare references for entry into the database? Can the program import or export ASCII (text) files?

37-O   Examine a reprint collection made available for your use. What system of organization is used? Is an author index available? A subject or taxonomic index? Are any of these reprints in a computerized database?

GLOSSARY

♀. In biology, the symbol for female; in astrology, the symbol for Venus.

♂. In biology, the symbol for male; in astrology, the symbol for Mars.

1927 North American Datum. See North American Datum (1927).

1983 North American Datum. See North American Datum (1983).

A

abdomen. In insects, the posterior or third major division of the body, consisting of usually at least four segments; with, in the adult stage, no structure functioning as legs. See **head, thorax.**

abdominal cavity. The largest body cavity. Its contents include most of the digestive system and portions of the urinogenital system. In mammals, it is separated from the thoracic cavity by the muscular diaphragm.

absolute age. Age of individual in terms of real time. Compare **relative age.**

abstract. A brief, concise statement or set of statements summarizing the significant content and conclusions of a scientific publication.

Acarina. The order of arachnids that encompasses the chiggers, ticks, and all other mites.

Achilles tendon. Calcaneus tendon. Large tendon connecting the calf muscles (gastrocnemius, soleus, and others) to the calcaneus.

accuracy. The nearness of a measurement to the true value.

acetabulum. Socket in pelvic girdle at area where ilium, ischium, and pubis meet, and with which the head of the femur articulates.

adjusted range length. The range length plus one-half the distance to the next trap added onto each end. See Figure 35.3E.

adult. Generally, a sexually mature individual.

adult pelage. The type(s) of hair covering that is/are characteristic of adults of the species.

aerial. Pertaining to mammals that have the capacity for sustained flight. Bats are the only aerial mammals. See Figure 14.1.

agouti hair. A hair with a specific sequence of alternating pale and dark bands.

alarm pheromone. A pheromone produced by an alarmed individual. See **pheromone.**

albinism. The absence of all external pigmentation, typically inherited as a Mendelian recessive.

albino. An animal lacking all external pigmentation. See **albinism.**

alisphenoid bone. One of the paired bones that form portions of the lateral walls of the braincase, located anterior to the squamosal and in the temporal fossa. See Figure 2.1B, C.

alisphenoid canal. A tubular passageway beneath an arch of bone near the base of the alisphenoid bone and through which a blood vessel passes. Found only in certain groups of mammals. See Figures 2.1B, C and 19.7.

allopatric. Pertaining to disjunct geographic ranges of two or more taxa.

allotype. A single, designated, paratypical specimen of the sex other than that of the holotype.

alveolar sheath. See **incisive alveolar sheath.**

alveolus. A bony socket for the root or roots of a tooth.

ambulatory. Pertaining to walking (e.g., the ambulatory locomotion of a bear).

ammonium hydroxide (NH_4OH). Caustic chemical that in diluted form is used to clean and partially degrease skeletal material.

angora. Pertaining to hair that has continuous growth.

angular process. The posterior ventral projection of the dentary. See Figure 2.1D.

ankle (joint). The joint, in the hindlimb, between the distal ends of the tibia and fibula of the lower leg and the tarsal bones of the foot.

annual molt. A once yearly shedding and replacing of hair in mammals.

antebrachium. Forearm; the portion of the front limb between wrist and elbow joints.

antennae (sing., **antenna**). In Arthropoda, paired, segmented sensory organs located on the head or anterior end of the cephalothorax. In insects, one is attached on each side of the head—commonly called "feelers."

anterior. Of, pertaining to, or toward the front end of an organism.

anterior nares. Anterior common opening of the nasal passages in the skull.

antitragus. Small, fleshy projection on the posteroventral margin of the pinna of some bats. See Figure 14.3D.

antler. A frequently branched, paired, bony cranial projection found in deer. Covered with skin (velvet) during growth. Shed periodically. See Figures 5.5 through 5.9.

apical. At the tip of a structure.

apotele. The last segment of the pedipalps, always reduced, represented by claws or a two- to four-tined structure at the inner basal angle of the tarsus.

appendicular skeleton. That part of the skeleton consisting of the pectoral and pelvic girdles and the bones of the limbs.

aquatic. Pertaining to life in water, particularly fresh water.

arboreal. Pertaining to climbing and/or living in trees.

arena territory. A spatial territory at a traditional breeding ground that is defended by males in polygynous species to maintain a female harem. See Figure 35.6.

arrector pili muscle. A small muscle of smooth muscle fibers, situated in the dermis, and attached at an oblique angle to the hair follicle. Under autonomic control, this muscle contracts to erect the shaft of the hair and, in humans, causes "goosebumps" on the surface of the skin. See Figure 4.1.

articular bone. In vertebrates other than mammals, this bone at the proximal end of the lower jaw bears the surface of articulation of the lower jaw with the rest of the skull. In mammals, the bone becomes incorporated into the middle ear to form the **malleus**. The malleus is the first of three **ossicles** in the chain of small bones leading from the eardrum or tympanum to the oval window of the cochlea. See Figure 1.1.

assessment line. A line of traps placed through and outside a trapping grid (see Fig. 36.1A) or through a trap line (Fig. 36.1B) to measure the extent of movement of animals captured on the grid or on the trap line. The assessment line is used to estimate the trappable area for density estimation.

astragalus. Ankle bone; talus. Most proximal bone of the pes, and articulating with the distal end of the tibia and fibula. See Figure 6.4.

atlas. The first cervical vertebra, which articulates with the skull. See Figure 6.1.

auditory bulla. A more-or-less inflated hollow structure enclosing the middle ear chamber in many mammals.

Australian Region. The faunal region that includes the Australian continent, adjacent islands, including New Guinea and the islands of Indonesia north and west to Wallace's Line. See Figure 1.4.

awns. Guard hairs of relatively uniform length and with slender bases and expanded tips.

axial skeleton. That part of the skeleton consisting of the skull, auditory ossicles, hyoid apparatus, sternum, ribs, and vertebrae.

axis. The second cervical vertebra. See Figure 6.1.

B

baculum. A bone found in the penis of certain kinds of mammals. See Figure 6.2.

Bailey's triple-catch method. A deterministic mark-and-recapture method of population estimation that corrects for immigration, mortality, and natality during the period of sampling.

baleen. So-called whalebone. Material comprising the cornified epithelial plates suspended from the upper jaws of mysticete whales. Used to strain food from the water. See Figure 20.4.

barbs. Small projections on a spine that prevent the spine, once embedded, from being easily removed. See Figure 4.6.

basioccipital. See **occipital bone**.

basisphenoid. The bone forming the floor of the braincase. Anterior to the basioccipital and posterior to the presphenoid and pterygoids. See Figure 2.1B.

beam. The main trunk of an antler. See Figure 5.8.

beast of burden. An animal used to carry loads.

bez tine. The first tine above the brow tine of an antler. See Figure 5.8.

bibliography. A collection of references to publications, usually arranged in alphabetical order by the last name of the first listed or only author and usually covering a stated topic, time period, or both.

bicuspid. 1. A tooth having two major cusps. 2. In human anatomy, a premolar.

binomen. The two-word combination used as the scientific name of a species.

bipedal. Pertaining to locomotion on only two legs.

bivariate home range. Home-range estimate in two dimensions, formulated into general model by Koeppl et al. (1975).

body (of dentary). See **dentary**.

borax. A crystalline, slightly alkaline borate of sodium, $Na_2B_4O_7$, sometimes used for drying and preserving skins. See Chapter 35.

boundary strip. An area that should be added to the basic size of a trapping grid so that an effective trapping area can be determined. The width of the boundary strip can be estimated using assessment lines (see Chapter 36).

boundary strip method. In home-range estimates, this is an area equal in width to half the distance between traps is added around the minimum area. In the *inclusive boundary strip* method (see Figure 35.3B), the peripheral points of capture are considered centers of rectangles (each side of which equals the distance between traps), and the home-range area is delineated by connecting the exterior corners of these rectangles to form a maximum estimate of the space utilized. In the *exclusive boundary strip* method (see Figure 35.3C), the boundary strip rectangles are drawn in a manner to minimize the area enclosed.

bounded count. Census made by counting the maximum number of individuals observed along a transect during a brief time period.

brachiation. Locomotion by means of swinging from one handhold to another, usually through trees. See Figure 7.15.

brachium. Upper arm between shoulder and elbow.

brachyodont. Pertaining to teeth that have low crowns.

bracket key. A key in which the parts of each couplet are in immediate succession. See Figure 8.1.

braincase. 1. That portion of the skull that encloses and protects the brain. 2. The skull posterior to a plane drawn vertically through the anterior margins of the orbits. See **rostrum.**

bristles. Overhairs with angora growth; frequently stiff and wiry.

browsers. Mammals that feed on forbs or on leaves of shrubs and trees.

brow tine. The first tine above the base of an antler. See Figure 5.8.

buccal. 1. Pertaining to the cheeks (e.g., the buccal side of a tooth is that side closest to the cheek). 2. In some usages, pertaining to the mouth cavity as a whole.

bulbourethral glands. Cowper's glands. In male mammals, the small paired glands that secrete mucus into the urethra at the time of sperm discharge.

bunodont teeth. Low-crowned teeth with roughly hemispherical cusps. See Figure 3.6.

burr. Enlarged, rugose area at the base of an antler, where it joins the pedicel. See Figure 5.8.

burrow. A tunnel excavated and inhabited by an animal.

C

cache. A collection of stored items or food provisions accumulated by small mammals such as rodents.

caecotrophic pellets. Soft fecal pellets, rich in vitamin B_1, produced in the caeca of most species of lagomorphs. After the pellets are voided, they are reingested by the animal so that the vitamins can be absorbed. See **coprophagy** and **reingestion.**

calcaneum. Calcaneus. The heel bone, the largest and posteriormost tarsal bone. See Figure 6.4.

calcar. A spur (of cartilage or bone) that projects medially from the ankle in many species of bats and from the wrist of many gliding mammals, and that helps support a patagium. See Figure 14.1.

Calendar of Captures method. See **Minimum Number Known Alive.**

calipers. Instrument for taking precise linear measurements. Available in several styles—those in which the reading is taken from a dial or digital display, those in which it is read from a vernier scale, and others that produce a digital file that can be read by a computer.

callus. Any hard, thickened area of the epidermis.

camouflage coloration. A color pattern that allows a mammal to blend with the background and that renders it less visible to predators and/or prey.

canal. A tubular passage or channel in a bone or other tissue.

canine. One of the four basic kinds of teeth found in mammals. The anteriormost tooth in the maxilla (and its lower counterpart in the dentary). Frequently elongated, unicuspid, and single rooted. Never more than one per quadrant.

caniniform. Pertaining to an incisor or premolar that has the shape and appearance of a canine.

cannon bone. Fused metatarsals or metacarpals. See Figure 27.2.

carcass. The portion of the body of an animal that remains after the removal of the skin.

carnassial dentition. See **secodont dentition.**

carnassial pair. The last upper premolar and first lower molar of adult living mammals with secodont (= carnassial) dentition. (Formed by next to last upper premolar and last lower premolar in deciduous dentition.) The largest pair of bladelike teeth that occlude with a scissorslike action. Possessed by many members of the order Carnivora. See Figures 19.2, 19.3.

carnassial teeth. See **carnassial pair.**

carnivore. A mammal that consumes meat as the primary component of its diet. Also used for any member of the order Carnivora, regardless of diet.

carpal. Any one of the group of small bones in the region of the wrist in the skeleton of the forelimb. Distal to the radius and ulna and proximal to the metacarpals. See Figure 6.3.

carpus. The most proximal portion of the manus. Extends from the distal ends of the radius and ulna to the proximal ends of the metacarpals. Contains the carpal bones.

cased skin. A mammal skin prepared for tanning, by making an incision from the heel of one hindfoot across the urogenital region and to the heel of the leg on the opposite side. For scientific specimens, legs are kept with the skin, which is then removed from the body, stretched on a frame, and dried.

catalog. The portion of the field notes in which full data are entered for each specimen saved. All entries are consecutively numbered. See Figure 29.2.

category. See **taxonomic category.**

caudal vertebrae. The vertebrae of the tail, posterior to the sacral vertebrae. See Figure 6.1.

caviomorphs. Members of a group of rodents of the New World (especially of South America) that have a greatly enlarged infraorbital foramen.

cementum. The layer of bonelike material covering the root of a tooth. Sometimes termed *cement*. See Figure 3.1.

center of activity. The mean of a set of capture coordinates or observation coordinates for an individual animal.

cercus (pl., **cerci**). In arthropods, an appendage, usually paired, of the terminal abdominal segment. Usually slender, filamentous, and segmented, and sometimes heavily sclerotized. See Figure 32.1A.

cervical vertebrae. The vertebrae of the neck. These number seven in almost all living mammals. See Figure 6.1.

Change-in-Ratio method. A procedure for determining population densities by observing changes in the relative abundance of two categories of animals, such as adults and subadults or males and females.

cheek teeth. Collectively, the premolars and molars or any teeth posterior to the position of the canines.

chelate-dentate. Pertaining to mouth parts of Acarina, consisting of a movable digit or chelicera and an immovable digit, one or both digits bearing "teeth."

chelicerae (sing., **chelicera**). The pincerlike first pair of appendages of adult Acarina and of other arachnids.

Chordata. The animal phylum distinguished, in part, by a dorsal hollow nerve cord, a notochord, and pharyngeal pouches, at some stage in the life history. Includes some jawless aquatic forms, fishes, amphibians, reptiles, birds, and mammals.

chorioallantoic placenta. All eutherian mammals and a few marsupials have this type of placenta in which the chorion and allantois of the embryo and fetus make contact with the maternal uterine lining.

cingulid. See **cingulum.**

cingulum. An enamel shelf, frequently with cusps, that borders one or more margins of an upper tooth. A *cingulid* is its counterpart in a lower tooth.

civet. 1. A carnivore in the family Viverridae, order Carnivora, that has specialized and well-developed perianal glands that produce a distinctive musk or odor. 2. A musk produced by the specialized perianal glands of civets, and used in the manufacture of high-quality perfumes.

class. A major subdivision of a phylum, consisting of one or more orders.

clavicle. A ventral bone of the pectoral girdle. Reduced or absent in many mammals. The collarbone in humans.

claw. The most common form of digital keratinization found in mammals. Usually long, curved, and sharply pointed. See Figure 6.5.

cloaca. A chamber into which the digestive, reproductive, and urinary systems empty and from which the products of these systems leave the body.

clumped dispersion. Individuals are concentrated in some areas and less concentrated or absent in other areas. See Figure 35.1B.

coefficient of variation. The standard deviation expressed as a percentage of the mean: $CV = (s)(100/X)$. Used to minimize effects of size when comparing the variability of organisms that differ greatly in size (e.g., the variability of dimensions of a mouse can be compared with that of a horse).

columella. A bone that transmits vibrations from the tympanum to the inner ear in reptiles, birds, and anurans, and is homologous with the hyomandibular of fishes and the stapes of mammals.

commensal. 1. A species that benefits by living in close association with another species that is not directly harmed by the association. 2. A species of mammal, such as *Rattus norvegicus* or *Mus musculus,* that lives in close proximity with human dwellings or structures and benefits from the association. Also and adjective.

commissure. Any sharp, crescent-shaped crista found on a cheek tooth.

concealing coloration. Protective coloration facilitating concealment.

condylarth. One of the primitive ungulate mammals of the order †Condylarthra, from the Paleocene and Eocene epochs, having a slender body, low-crowned teeth, and five-toed feet, each toe ending in a small hoof.

Conibear trap. A steel kill trap used primarily to collect squirrels and smaller furbearers. See Figure 30.2.

conspecific. Set of two nominal taxa that belong to the same species.

continental drift. The movement of the continents relative to one another. Caused by their being carried about on the moving tectonic plates. See **plate tectonics.**

Convention on International Trade in Endangered Species of Wild Fauna and Flora. International convention regulating the exportation and importation of threatened or endangered species of wildlife. See Chapter 30.

convex polygon. Home-range estimate in which the capture points are connected to form the smallest convex polygon. See Figure 35.3A.

coprophagy. Feeding on feces. See **reingestion** and **caecotrophic pellets.**

coracoid. One of the separate elements of the pectoral girdle in nonmammalian vertebrates. Rudimentary and fused to the scapula in most marsupials and placentals. See Figure 6.1.

core area. Area of intensive use by a group within the overall group home range.

cornification. See **keratinization.**

coronoid process. A projection of the posterior portion of the dentary. It is dorsal to the mandibular condyle. See Figure 2.1D.

corpora quadrigemina. Four oval masses that serve as centers of optic and auditory reflexes and form the dorsal part of the mesencephalon in the brain of mammals.

cortex. Any outer layer or "rind," such as the outer layer of the mammalian ovary or the cortex layer of the hair.

costal ribs. Ribs that articulate with the vertebrae and the sternum.

coterie. A "family" group consisting of a male, several females, and their young.

countershading. Concealing coloration of an animal such that parts normally in shadow are relatively pale and parts normally illuminated are dark.

couplet. Each pair of mutually exclusive alternatives that all collectively compose a biological identification key.

cover. See **dominance.**

Cowper's glands. See **bulbourethral glands.**

cranium. See **skull.**

crepuscular. Most active during the twilight periods of dusk and dawn.

crista. A crest or ridge on a tooth, frequently designated by a distinctive name, such as entocrista. A *cristid* is the corresponding structure on a lower tooth.

cristid. See **crista.**

crown. The portion of a tooth (especially a more-or-less flat-topped cheek tooth) extending above the gumline. See Figure 3.1.

cryptic coloration. See **concealing coloration.**

cursorial. Pertaining to running. Cursorial locomotion is running locomotion.

cusp. A point, projection, or bump on the crown of a tooth, usually designated by a particular name, such as hypocone.

cuticle. The thin outer layer of a hair. See Figure 4.1.

cuticular scales. The outer layer of the shaft of a hair, characterized by distinctive shapes and patterns. See Figure 4.4.

D

dactylopatagia. In a bat, thin sheets of skin (flight membranes) that fill the spaces between the digits of the forelimb as follows: *dactylopatagium minus* (between digits 2 and 3), *dactylopatagium longus* (between digits 3 and 4), and *dactylopatagium latus* (between digits 4 and 5). See Figure 14.1.

data (sing., **datum**). Observations or measurements taken on a sampling unit.

deciduous dentition. The juvenal or milk dentition of mammals. In placentals, consists of a complete set of temporary incisors and canines, and a complete or partial set of temporary premolars; in marsupials, the third premolars only. Replaced by permanent or adult dentition.

deciduous teeth. Milk teeth; teeth that appear earliest in the lifetime of a mammal and that generally are replaced by permanent teeth. See **deciduous dentition.**

definitive hair. Hair that grows to a certain length and is then shed and replaced.

degreasing. Process whereby fats and oils are removed from bones by means of various chemical agents. See Chapter 31.

degrees of freedom (*df*). A quantity, generally $N - 1$, that corrects for bias when certain sample statistics are computed.

den. A cave, hollow log, burrow, or other cavity used by a mammal for shelter.

density. See **population density.**

dental formula. A convenient way of expressing the numbers of teeth of different sorts in an individual or species (e.g., I 3/3, C 1/1, P 4/4, M 3/3). The letters designate incisors, canines, premolars, and molars, respectively. The numbers above the line give the numbers of teeth of each sort on one side of the upper jaw; those below the line indicate the number on one side of the lower jaw.

dentary. One of the pair of bones that comprise the lower jaw (mandible) of mammals. The tooth-bearing horizontal portion of this bone is referred to as the **body,** and the vertically projectiing portion is the **ramus.** See Figure 2.1D.

dentine. A hard, generally acellular material between the pulp and enamel portions of a tooth. See Figure 3.1.

dermal bone. Bone that forms without a cartilage precursor (e.g., certain skull bones).

dermal papilla. An extension of the dermis up into one of the hollow, hairlike fibers that compose a rhinoceros horn. See Figure 5.12A.

dermal scale. A bony scale that originates in the dermal layer of the skin. Found primarily in fishes.

dermal skeleton. The portion of the skeleton in which the bones are produced by intramembraneous ossification and without a cartilage precursor (Kardong 1998).

dermestid. A beetle of the family Dermestidae. They feeds on dried muscle and other tissues. These beetles are utilized in many institutions because the larvae are very efficient in cleaning skeletal material. See Figure 31.29.

dermis. The layer of the integument underneath the epidermis. Composed primarily of connective tissue. Also contains vascular, adipose, nervous, and other types of tissue. See Figure 4.1.

determinant of the covariance matrix. Used by Jennrich and Turner (1969) to measure noncircular home ranges of animals.

diaphragm. The septum dividing the abdominal and thoracic cavities, muscular in mammals.

diaphysis. The middle portion of a long bone, between the two epiphyses.

diastema. A pronounced gap between adjacent teeth. For example, the space between incisors and premolars in species lacking canines. See Figure 2.4 between point A and B.

dichotomous. Pertaining to a division into two equivalent or similar branches.

didactylous. Pertaining to a manus or pes that has only two digits, such as the manus in two-toed sloths, genus *Choloepus.*

dietary overlap. The extent to which the diets of two organisms, populations, or species incorporate the same kinds of food items.

differential correction. A method to adjust for the intentional error in a civilian global positioning system (GPS) receiver so that the indicated positional coordinates are close to the true position on earth.

digit. Any finger or toe.

digitigrade. Pertaining to walking on the digits only, with the wrist and heel held off the ground. See Figure 7.1B.

dilambdodont tooth. Type of tooth characterized by a W-shaped ectoloph on the occlusal surface. See Figure 3.5.

diphyodont. Having two sets of teeth: a milk or deciduous set and then a permanent set.

diprotodont. A dental configuration found in the Paucituberculata and Diprotodontia. The lower jaw is shortened, and the first lower incisors are greatly elongated. See Figure 11.2. Compare **polyprotodont.**

dispersal. The permanent emigration of individuals from a population.

dispersion. In statistics, the scatter of values from a central point. In spatial terms, the distribution of animals in an area. See **random dispersion, uniform dispersion, clumped dispersion.**

disruptive coloration. Coloration pattern that interferes with perception of an animal's presence because the pattern does not coincide with the form and outline of the animal's body.

distal. Distad. Situated away from the base or area of attachment.

distance between observations index. Home-range index developed by Koeppl et al. (1977b) for use when data are not normally distributed.

diurnal. Pertaining to activity primarily during the daylight hours. Opposite of nocturnal.

domesticated. Pertaining to a species of animal that has been bred for use by humans. In mammals, domesticated species are generally found in members of the orders Carnivora, Perissodactyla, and Artiodactyla.

dominance. 1. quantitative measure of the total biomass or areal coverage of a species per unit of habitat space. 2. behavioural status that confers priority of access of certain individuals to food, mates, and other needs. (Cox 1996).

DOR. Abbreviation for "dead on road." Used in field records to indicate specimens found in this condition.

dorsad. Toward the dorsum. Dorsal.

dorsal. Pertaining to the back or upper surface.
dorsal fin. In mammals, a middorsal projection of fibrous connective tissue, found only in certain cetaceans. See Figure 20.1.
dorsolateral. Positioned "between" dorsal and lateral—actually encompassing parts of both.
dorsum. The dorsal surface of an animal.

E

eccrine sweat glands. See **sweat glands.**
echolocation. Sonar; the sensing of objects and surfaces by emitting sound pulses and receiving and identifying the echoes reflected by the objects or surfaces. Used by most bats and most cetaceans.
ecological niche. The overall pattern of resource use by a species due to its adaptations for exploiting different types of resources and occupying different habitats (Cox 1996:264).
ecto. Abbreviation for **ectoparasite.**
ectoloph. A crista, generally connecting parastyle, paracone, metacone, and metastyle. See Figure 3.4.
ectoparasite. A parasite on, or in, the integument of an animal (e.g., fleas, lice, ticks).
edentulate. Lacking teeth.
edge effect. The discrepancy between the population estimates of animals living on the trapping grid and those living partly on the edges of the trapping grid.
effective trapping area. In estimation of population density, the area of the trapping grid plus the estimated area outside the grid that is used by animals. See Chapter 36.
elbow (joint). The joint in the forelimb between the humerus and the radius and ulna. See Figure 6.1.
elevational migration. Migration by mammals to different elevations in response to seasonal availability of food or seasonal climatic conditions.
embrasure. A space between two teeth and into which a tooth in the opposing jaw fits when the teeth are occluded.
embrasure shearing. A cutting or scissorslike shearing action produced by two teeth moving past each other in occlusion. Characteristic of the teeth of several early groups of mammals and of the carnassial teeth of carnivores.
emigration. The event of individuals leaving a population of which they have been a part.
empodium. The median structure between the claws of arthropods. It arises from the pretarsus and is padlike, suckerlike, etc. See Figure 32.18.
enamel. Material comprising the extremely hard outer layer of the crown of a tooth, consisting of calcareous compounds and a small amount of organic matrix. Usually white but sometimes brown, red, or yellow in rodents and certain other mammals. See Figure 3.1.
enamel island. An exposed patch of enamel surrounded by a surface of dentine. Typically found in the teeth of some rodents. See Figure 23.16.
Endangered Species Act. A U.S. law, passed in 1973 and subsequently amended, that regulates the capture, possession, and sale of threatened or endangered species of wildlife. See Chapter 30.
entoconid. In lower cheek teeth, a cusp on the anterior, lingual side of the talonid.
entotympanic bone. A bone surrounding the middle ear cavity. Sometimes fused with the tympanic to form a compound auditory bulla.
enucleate. Lacking a nucleus. Mammalian red blood cells (erythrocytes) are enucleate.
enzyme digestion. Use of enzymes, such as trypsin or papaine, to clean bones. See Chapter 31.
epidermal scales. Keratinized scales of the epidermis.
epidermis. The surface layer of the integument, exterior to the dermis. Composed entirely of epithelial tissue. See Figure 4.1.
epiphyseal cartilage. In limbs, the cartilage at the ends of the long bones. In age estimation, the ossification of this cartilage and the adjacent **diaphysis** (portion between the two epiphyses) proceeds along a definite pattern followed by the final ossification of the epiphyseal plate. See Chapter 33.
epiphyseal plate. A narrow band of cartilage between the epiphysis and diaphysis and that ossifies after the epiphysis and diaphysis.
epipubic bones. Paired bones that project anteriorly from the pelvic girdle into the abdominal body wall of most marsupials and all monotremes. See Figure 11.5.
epithelial tissue. A tissue that covers a body or structure or lines a cavity.
equal probability of capture. When the probability of capturing any individual is equal to the probability of capturing any other individual.
erythrocytes. Red blood cells.
ether. Extremely volatile and highly flammable hydrocarbon that is sometimes used to anesthetize mammals and often used to immobilize ectoparasites prior to their removal from a host. See Chapter 32 for precautions on use.
Ethiopian Region. The faunal region that includes all of Africa south of the Sahara Desert, the southern portion of the Arabian Peninsula, and the island of Madagascar. See Figure 1.4.
eumelanin. The pigment that produces various browns and black in mammalian hair and integument.
euthemorphic tooth. A modified tribosphenic tooth, frequently square in outline, but modified in other ways in different groups of mammals. See **dilambdodont tooth** and **zalambdodont tooth.**
Eutheria. One of the three infraclasses of mammals, and the one of the two in the subclass Theria that includes the so-called placental mammals.
excrescences. The outgrowths from the integument such as true horns, pronghors, antlers, rhino "horns," giraffe "horns." The rough bumps on the snouts of whales are sometimes called excrescences. See Chapter 5.
exoccipital. See **occipital bone.**
exploration. Pertaining to movements by animals as they investigate new surroundings.
exploratory movements. Movements by individuals when placed in a novel or strange environment.

external auditory meatus. Canal leading from the surface of the head to the tympanic membrane. See Figure 2.1C.

external nares. The external openings to the nasal passageways found in all vertebrat animals, including mammals.

F

family. The major category, in scientific classification, that is just below order; or a taxon at that level and containing one or more genera.

faunal regions. Major divisions of the earth's land surface, based upon broad faunal differences and similarities. See Figure 1.4.

femur. The single bone of the upper (proximal) portion of each hindlimb. See Figure 6.1.

fibula. The lateralmost of the two bones of each hindlimb that are distal to the femur and proximal to the tarsal bones. See Figure 6.1.

field journal. See **journal**.

field notes. Collectively, the journal, specimen catalog, species accounts, and any other specialized data-collection forms.

Fitch trap. Live trap constructed of hardware cloth, galvanized metal, and a metal can. Easily made, using inexpensive materials, and generally useful for capturing most kinds of small rodents. See Fitch (1950) for details.

fixative. A chemical, such as formalin, that is capable of hardening tissues, organs, or specimens. See Table 31.1 for a list of suitable chemicals.

fixed. A tissue, organ, or specimen that has been hardened by immersion or injection with a fixative such as formalin, embalming fluid, or FAA. See also **fixative**.

flags. Distinctive tail or rump patches in certain mammals, that are thought to have signaling or communication functions.

flippers. Limbs fully adapted for an aquatic life. Digits elongate and connected by intervening tissue. See Figures 19.6 and 19.9.

fluid-preserved. Pertaining to specimens fixed and/or preserved in a fluid such as formalin or an alcohol solution. See Table 31.1.

flukes. Posterolateral flattened projections of a cetacean's or dugong's tail. Supported entirely by fibrous connective tissue (no skeletal support). See Figures 20.1 and 25.5.

flush census. Census of individuals that are flushed along a transect.

follicle. A small cavity or pit (e.g., Graafian follicle, hair follicle).

foramen (pl., **foramina**). Any opening, orifice, or perforation, especially through bone.

foramen magnum. The large opening at the rear of the skull and through which the spinal cord enters the braincase. See Figure 2.1B.

forbs. Usually low, nonwoody, herbaceous plants that are neither grasses nor sedges, and that, in strongly seasonal climates, die back to the ground at the onset of the inclement season.

forefoot. The foot, or hand, on the anterior or pectoral limb.

form. A depression in vegetation or soil and used as a resting place by certain kinds of mammals.

fossa. A pit or depression in a bone. Frequently a site of bone articulation or muscle attachment.

fossorial. Pertaining to life under the surface of the ground. Burrowing. See Figure 7.8.

foxing. A process in which fur, as on study skins, gradually becomes, over the years, more and more yellowish, orangish, or reddish.

frass. Insect feces, such as that accumulating, along with shed larval skins, in a colony of dermestid beetles.

free ribs. Ribs that articulate with the vertebrae but do not extend to the sternum. See Figure 6.1.

friction ridges. Ridges in the epidermis that serve to increase friction (and thus traction) of naked areas, such as the soles of the feet. See Figure 4.2B.

frog. The pad in the central portion of a hoof, especially of equids. See Figure 6.7.

frontal bones. The anteriormost pair of bones in the roof of the braincase. Situated between the orbits, anterior to the parietals, and posterior to the nasals and maxillae. See Figure 2.1A, C.

frugivorous. Feeding on fruit.

fur. More or less soft pelage, especially dense underhair with definitive growth and serving primarily for insulation.

fusiform. Compact, streamlined, tapered at ends. Pertaining to body with shortened projections and no abrupt constrictions. See Figure 7.14.

G

generic name. First word of a binomen or trinomen. The name of a genus.

genital field/pore. The region immediately surrounding and encompassing the genitalia of Acarina.

genital papillae. The small fingerlike/suckerlike protuberances within the region of the genital field of Acarina.

Gen trap. Live trap consisting of a plastic tube and a hinged metal door. See Figure 30.8 and Shemanchuk and Berger (1968) for details.

genus (pl., **genera**). The major category just below family; or a taxon at that level, containing one or more species.

glands of Moll. The wax-producing glands found in the wall of the nonosseus portion of the external auditory meatus.

glenoid fossa. Depression in the scapula, into which the head of the humerus articulates.

global positioning system (GPS). A system of satellites that provide signals to GPS receivers on land, sea, and air to give geographic coordinates. See Chapter 29 for details.

Gloger's Rule. An ecological "rule" that states that races of mammals in arid regions are paler in color than related races in humid regions.

gnathosoma. The anterior body region of Acarina, which bears the food-gathering apparatus.

Gondwanaland. The southern supercontinent that began to separate from Pangaea near the end of the Paleozoic Era, and that ultimately split into South America, Africa, Madagascar, India, Antarctica, and Australia. See **continental drift**.

granivorous. Feeding on seeds.

graviportal. Pertaining to a limb structure adapted for supporting great weight. Found in very heavy mammals (e.g., elephants). See Figure 7.6.

grazers. Mammals that feed on grass.

group home range. The home range of a group of individuals, such as that of a herd or family group.

group spatial territory. A concept of "territory" similar to that of spatial territory, but possessed by members of a group. See Figure 35.7.

growth lines. Absolute, incremental growth lines present in teeth, bones, and other tissues of mammals, and useful for aging purposes.

guard hairs. Coarse, protective hairs forming an outer coat found on most mammals. See **spines, bristles,** and **awns.**

H

habitat. 1. A place with suitable environmental conditions for supporting a population of some species and that is currently supporting such a population. 2. Any place that has the potential to support such a population, whether or not it is doing so at present.

hair. Cylindrical, filamentous outgrowths of the epidermis that consist of numerous cornified epidermal cells. Found only in mammals. See Figure 4.1.

Haller's organ. A structure located on tarsus 1 of Metastigmata (Acarina). See Figure 32.14D.

hallux. The first (medial) digit of the pes. Is frequently opposable in arboreal mammals. See Figure 6.4.

hand. A manus that has prehensile digits and, usually, an opposable thumb.

hard palate. The bony septum between the nasal passages and the oral cavity. Formed by the palatine bones and portions of the maxillae and premaxillae. See Figure 2.1B.

harmonic mean. A method of home-range estimation, in which the the average of the reciprocals of the radial distances to all occurrence locations is calculated. See Chapter 35 and Dixon and Chapman (1980).

harpoon mole trap. A trap designed to catch moles of the family Talipidae, order Insectivora. See Figure 30.5.

Hartner's glands. Modified sebaceous glands located behind the eyeball that lubricate the nictitating membrane of the eye.

haustellate. In Arthropoda, pertains to forming a beak suitable for piercing.

Hayne method. Deterministic method of population estimation, for use when individuals are marked on several occasions. The Schnabel (1938) and Schumacher and Eschmeyer (1943) methods are similar.

head. In insects, the anterior body region on which are found the eyes, mouthparts, and antennae.

head of the femur, or **head of the humerus.** The proximal ball on each of these limb bones that articulates with the appropriate socket to form a ball joint.

herbivore. An animal that consumes plant material as the primary component of its diet.

herbivorous. Feeding primarily or principally on vegetation.

Hertwig's solution. Reagent for clearing plant tissues during preparation of microscope slides for dietary analysis. See Chapter 34.

heterodont. Pertaining to a dentition in which teeth are clearly differentiated into various types such as incisors, canines, premolars, and molars. See Figure 2.1C and D. Opposite of homodont.

hip girdle. The pelvic girdle.

Holarctic Region. Collectively, the Nearctic and Palearctic Faunal Regions combined. See Figure 1.4.

holotype. A single specimen, designated in the original scientific description naming a new species or subspecies as the specimen with which the new name is to always be associated, and that is to serve as the ultimate objective authority as to what kind of organism the name applies.

home range. The area that a mammal occupies during the course of its life, exclusive of migration, emigration, or unusual erratic wanderings (Brown and Orians 1970).

homeotherapy. Pertaining to regulation of a relatively constant body temperature by physiological means regardless of external temperature; endothermy, warm-blooded, as in birds and mammals (Feldhamer 1999).

homing. The tendency of animals to return to their home area when experimentally or otherwise displaced to another area.

homodont. Pertaining to a dentition in which all teeth are very similar in form and function. See Figure 17.4. Opposite of heterodont.

homologous. Said of structures found in different kinds of organisms, if they have a common evolutionary origin from some particular structure in a common ancestor.

hoof (pl., **hoofs** or **hooves**). The digital keratinization in unguligrade mammals, a horny sheath completely encasing the tip of a phalanx and usually providing the animal's only point of contact with the substrate. See Figure 6.7.

horn. 1. One of two or more structures projecting from the head of a mammal and generally used for offense, defense, or social interaction. Cattle, sheep, Old World antelopes, etc. (family Bovidae) have horns formed by permanent, hollow, keratin sheaths growing over bone cores. See Figure 5.2. 2. The keratinized material that forms the sheaths of these horns. See Figure 5.1.

host. The organism parasitized by a parasite.

Hoyer's medium. Aqueous mounting medium used in microscope slide preparations. See Chapter 34.

humerus. The single bone in the upper (proximal) portion of each forelimb. See Figure 6.1.

hunter. According to McNab (1963:136), a mammal species that is principally granivorous, frugivorous, insectivorous, or carnivorous.

hybrid. The offspring resulting from a cross between individuals of two species.

hyoid apparatus. A series of bones and cartilages found in the throat area that provide support for the base of the tongue, and upper portion of the trachea, larynx, and esophagus. Vestigial or absent in many mammals.

hypocone. In upper cheek teeth, a cusp on the posterior lingual side of the crown, sometimes situated on a distinct talon. See Figure 3.6.

hypoconid. In lower cheek teeth, a cusp on the posterior, labial side of the talonid area of the crown.

hypoconulid. In lower cheek teeth, the posteriormost cusp in the talonid area of the crown.

hypodermis. The most interior layer of the integument and consisting primarily of fatty tissue; in this layer are found the base of hair follicles, vascular tissues, parts of sweat glands, and some sensory receptors. This layer is sometimes included as part of the lower layer of the dermis.

hypostome. The anterior region of the ventral surface of the gnathosoma of Acarina. See Figure 32.14B, C.

hypsodont. Pertaining to a high-crowned tooth. See Figures 26.3 and 27.10. Opposite of brachydont.

hysternosoma. In Acarina, the entire region of the body posterior to the insertion of legs III.

hystricognath mandible. Type of mandible found in certain rodents (Bathyergidae, Old and New World hystricomorphs), in which the angular process arises lateral to the outer border of the incisive alveolar sheath. See Figure 23.8.

hystricomorphs. Rodents from the Old and New World in which the infraorbital foramen is greatly enlarged. It includes the caviomorphs (New World) and Old World mammals such as Hystricidae, Thryonomyidae, and Petromyidae.

I

-id. The ending of an uncapitalized "English" word made by dropping the *-ae* from the end of the name of a family—is used to denote any member of that family.

-idae. A suffix added to the stem of the name of the type genus to form the name of a family.

idiosoma. The body region of Acarina posterior of the gnathosoma and including the leg-bearing region and the region posterior to the legs.

ilium (pl., **ilia**). Most dorsal of the three major bones in each half of the pelvic girdle. The pelvic bone that articulates with the sacral vertebrae. See Figure 6.1.

imbricate. Overlapping, as in the shingles of a roof.

immigration. The moving of new individuals into a population.

-inae. A suffix added to the stem of the name of the type genus to form the name of a subfamily.

incisiform. Pertaining to a canine or other tooth that has the shape and appearance of an incisor.

incisive alveolar sheath. The bony socket or alveolus that receives the root of an incisor.

incisive foramina. See **palatal foramina**.

incisor. The anteriormost of the four basic kinds of teeth found in mammals. Usually chisel-shaped. Upper incisors are always rooted in the premaxillae.

incremental lines. See **growth lines**.

Inc. Sed. (Incertae Sedis). In an uncertain position. In classification, signifies that the relationship of a taxon to other taxa is not known.

incus. The middle ear ossicle of mammals, situated between the malleus and stapes. Derived from the quadrate bone of other vertebrates. See Figure 1.1.

indented key. A format of biological key in which the two alternatives of each couplet may be widely separated by intervening couplets. See Figure 8.2.

index of dispersion. One of the statistical measures of the nature of spatial dispersion, such as Morisita's (1962) Index (Id).

index of home-range size. Index of home-range size developed by Metzgar and Sheldon (1974).

individual home range. The home range of an individual animal.

individual spatial territory. Territory defended by a single individual.

inflected. Deflected or bent inward.

infraorbital aperture. See **infraorbital foramen**.

infraorbital canal. A canal through the zygomatic process of the maxilla, from the anterior wall of the orbit to the side of the rostrum. See Figure 2.1A, C.

infraorbital foramen. A foramen through the zygomatic process of the maxilla.

inguinal. Pertaining to the region of the groin.

-ini. A suffix added to the stem of the name of a type genus to form the name of a tribe.

inner fold. An elongated enamel island on the crown of a tooth.

insectivorous. Feeding on insects.

integument. The skin, consisting of two layers, the epidermis and dermis, and the derivatives of these layers, such as scales and hair. See Figure 4.1.

interclavicle. A median bone lying between the ventral ends of the clavicles and the ventral surface of the sternum. In mammals, found only in monotremes.

internal nares. The posterior openings of the nasal passages of the skull.

interparietal. An unpaired bone of the dorsal portion of the braincase and between the parietals and the supraoccipital. Absent in some mammals.

ischium (pl., **ischia**). Most posterior of the three major bones in each half of the pelvic girdle. See Figure 6.1.

J

Jolly-Seber method. A stochastic method of population estimation, that corrects for mortality and natality effects on each sampling occasion.

journal. 1. A serial publication. Usually of a professional or scientific society (e.g., *Journal of Mammalogy*). 2. A portion of the field notes, a chronologically arranged account of all activities. See Figure 29.1.

jugal. The bone that forms the midportion of the zygomatic arch, between the zygomatic process of the maxilla and that of the squamosal. See Figure 2.1.

juvenal. See **juvenile**.

juvenile. An individual which is physiologically (and usually reproductively) immature or undeveloped. In mammals, often with distinctive pelage coloration and texture. Also used as an adjective.

juvenile pelage. The type of pelage characteristic of a juvenile of a species.

K

karyotype. The particular chromosome complement of the cells of a specific individual or species—each karyotype is characterized by number of chromosomes, their shapes, and sizes.

keeled sternum. A breastbone (sternum) with a medial ventral ridge that provides an expanded surface area for muscle attachment. Present in most bats.

keratinization. The process of becoming keratinized.

keratinized. Impregnated with keratin—a tough fibrous protein especially abundant in the epidermis and epidermal derivatives.

key. Tabular, usually dichotomous arrangement of diagnostic characters and used to identify organisms to various taxonomic levels. See Figures 8.1, 8.2, and 8.3.

key characters. Characters used in the construction of a biological key, to distinguish different organisms.

knee (joint). The most proximal joint in the hindlimb, between the femur and the tibia and fibula.

known age. Age established for an individual, based on observation of birth or other equally reliable criteria.

L

labial. Pertaining to the lips (e.g., the labial side of a tooth is the side closest to the lips [or cheeks]).

lacrimal bone. A small bone in the anterior wall of each orbit. See Figure 2.1C.

lacrimal duct. A small duct, perforating the lacrimal bone, extending from the inner corner of each eye to the nasal cavity; serves as drain for tears, the secretions of the lacrimal gland.

lactophenol. Reagent for clearing sclerotized tissues during preparation of microscope slides. See Chapter 34.

lambdoidal crest. See **occipital crest.**

lamina (pl., **laminae**). In some teeth, platelike structures that may (or may not) bear cusps on their exposed edges.

lanugo. Fine, soft hair on the fetus. A type of vellus.

large-scale map. A map in which a small geographic area is shown and which, in turn, provides a lot of detail. For example, city planning maps of 1:200 scale or topographic quads of 1;24,000 scale are large-scale maps.

lateral. Located away from the midline, at or toward the side(s).

latitude. Angular measurement of north-south location, relative to the equator (Campbell 1998). A position north or south of the equator indicated by degrees, minutes, or seconds, or by decimal degrees. In the Northern Hemisphere, the number of degrees is preceeded by the symbol N and is a positive number in geographic information system (GIS) analyses; in the Southern Hemisphere, the value is preceeded by S or a minus (–) sign for GIS analyses.

latitudinal migration. Seasonal, back and forth, migration, by animals, between one latitude and another (e.g., the migration of gray whales).

Laurasia. A supercontinent or compact group of continental landmasses that formerly existed in the Northern Hemisphere.

left aortic arch. The structure through which the blood leaves the left ventricle of the heart and then passes to the body proper in mammals. Both left and right arches are present in reptiles and other vertebrates, except for birds, in which only the right is present. See Figure 1.2.

lek. See **arena territory.**

Lincoln index. See **Petersen index.**

line intercept. The portion of a measuring tape that is intercepted by the canopy of plants, bare ground, or other habitat features during line-transect sampling. See Figure 28.10.

line transect. A method used to estimate plant cover or other habitat features by measuring the line-intercept lengths of the habitat feature along a measuring tape or line. See Figure 28.10.

lingual. Pertaining to the tongue (e.g., the lingual side of a tooth is the side nearest the tongue).

litter. The set of young resulting from a single pregnancy.

longitude. Angular measurement of east-west location, relative to the prime meridian (Campbell 1998). In the Eastern Hemisphere, the number of degrees is preceded by the symbol E and is a positive number in geographic information system (GIS) analyses; in the Western Hemisphere, the value is preceeded by W or a minus (–) sign for GIS analyses.

Longworth trap. A brand of live trap that has a detachable nest box.

loph. A ridge on the occlusal surface of a tooth, formed by the elongation and fusion of cusps. See Figure 3.7.

lophodont. Pertaining to a tooth that has an occlusal surface pattern consisting of lophs. See Figure 3.7.

lordosis. 1. In locomotion, suppleness of the spinal column that allows the hindfeet to be placed well anterior to the forefeet when the animal is running. 2. In behavior, a posture assumed by many female mammals, particularly rodents, during copulation.

lower maxillary process. See **maxillary process.**

lumbar vertebrae. The vertebrae between the thoracic vertebrae and the sacral vertebrae. See Figure 6.1.

M

maceration. Use of bacteria to clean bones. See Chapter 31.

macrohabitat. The overall habitat of a community of organisms (Brower et al. 1990).

malar (= **jugal**). In human anatomy, the cheekbone.

malleus. The outermost of the three middle ear ossicles. Derived from the articular bone. See Figure 1.1.

mammae. The externally visible structures on some adult mammals that contain mammary glands and other tissues. Generally have nipples or teats for delivery of milk to the young. See **mammary glands.**

Mammalia. A class of vertebrates in which the main or sole jaw articulation is between the squamosal and the dentary.

mammary glands. Milk-producing glands unique to mammals, and believed to be modified sebaceous glands. See Figure 4.9.

mandible. The lower jaw. In most present-day mammals, the mandible is composed of a single pair of separate bones, the dentaries. In some mammals (e.g., humans), these bones are completely fused together. See Figure 2.1D.

mandibular condyle. The knob by which each dentary articulates with the rest of the skull. See Figure 2.1D.

mandibular fossa. One of the two concavities with which the mandibular condyles articulate. In living mammals, confined to the squamosal bone. See Figure 2.1B.

mandibular ramus. Ramus. See **dentary** and Figure 2.1D.

mandibular symphysis. The cartilaginous area of attachment between the paired dentaries.

mandibulate. In Arthropoda, with mandibles fitted for chewing.

manual. Pertaining to the manus (= forefoot).

manus. The forefoot or hand. Collectively, the carpus, metacarpus, and manual digits. See Figure 6.3.

Marine Mammal Protection Acts. U.S. laws regulating the salvage, killing, molestation, and capture of marine mammals. See Chapter 30.

marking. Special behavioral means by which a mammal leaves an indication of its presence in its environment. Often accomplished by deposition of scent gland secretions.

marsupial. A mammal in the infraclass Metatheria. Includes the orders Didelphimorphia, Paucituberculata, Microbiotheria, Dasyuromorphia, Peramelemorphia, Diprotodontia, and Notoryctemorphia.

marsupium. External pouch formed by folds of skin on the abdomen wall. Found in many marsupials and in some monotremes (echidnas). Encloses mammary glands and serves as incubation chamber and/or protection for the young.

mass. A measure of the inertia of matter, expressed in kilograms (SI units) or pound mass (English system). **Weight**, a measure of force, is often used incorrectly to refer to this property of matter.

masseteric canal. A perforation through the mandible, at the bottom of the masseteric fossa.

masseteric fossa. A concavity in the outer side of the dentary at its posterior end. See Figure 2.1D.

masseter muscle. Adductor muscle of the mandible of mammals. Generally divided into external, middle, and deep portions. See Figure 23.3.

mastoid process. The exposed portion of the petromastoid bone (most of which is concealed within the auditory bulla) that forms the otic capsule enclosing the inner ear. The mastoid process, if present, is located just posterior to the auditory bulla. See Figure 2.1C.

maxilla (pl., **maxillae**). Either of the pair of relatively large bones that form major portions of the sides of the rostrum, contribute to the hard palate, form the anterior root of each zygomatic arch, and bear all upper teeth except the incisors. See Figures 2.1A, B, C.

maxillary. Pertaining to the maxilla.

maxillary process. A particular projection of the maxilla. It forms the anterior root of the zygomatic arch and is frequently divided into an upper portion (above infraorbital foramen) and a lower portion (below infraorbital foramen).

mean. The arithmetic average of the measured values of a variable in a sample or population.

mean distance traveled. The average distance moved by a mammal in a sample of movement data. One-half of this mean distance is often added to the edge of a trapping grid to estimate the effective trapping area for density estimation. See Chapter 36.

medial. Lying in or near the plane dividing a bilaterally symmetrical anumal into two mirror-image halves.

medulla. The central portion of a structure composed of distinct concentric layers or regions (e.g., the medulla of a hair, an ovary, kidney, or adrenal gland).

Meibomian gland. A specialized sebaceous gland located in the eyelid, that helps to moisten the conjunctiva of the eye (Kent and Miller 1997).

melanism. Unusual darkening owing to the deposition of large amounts of melanins in the integument.

mental foramen. Small opening (sometimes two openings) near the anterior end of each mandibular ramus; carries a nerve and blood vessel. See Figure 2.1D.

mesaxonic foot. Type of foot structure such that the main axis of weight support passes through a single digit. See Figures 7.2 and 26.2.

metacarpal. Any bone of the metacarpus, or any of the bones in the manus between the carpals and phalanges. Only one metacarpal per digit. See Figure 6.3.

metacarpus. The region of the manus, between the distal surfaces of the carpals and the proximal ends of the most proximal manual phalanges. Contains the metacarpal bones.

metacone. In upper cheek teeth, a cusp on the posterior, labial side of the trigon area of the crown.

metaconid. In lower cheek teeth, a cusp on the posterior, lingual side of the trigonid area of the crown.

metaconule. In upper cheek teeth, a cusp on the posterior portion of the crown, between the metacone and hypocone.

metapodial. General term used for either a metacarpal or metatarsal.

metatarsal. Any one of the bones in the pes, between the tarsals and the phalanges. Only one metatarsal per digit. See Figure 6.4.

Metatheria. One of the infraclasses of the subclass Theria. Contains only the marsupials.

microhabitat. The specific habitat occupied by a given population of a species (Brower et al. 1990).

middle ear. The cavity between the tympanic membrane and the otic capsule. Contains the malleus, incus, and stapes. Usually enclosed in an auditory bulla. See Figure 1.1.

middorsal. Along the midline of the back.

migration. See **seasonal migration**.

milk. A liquid secreted by mammary glands and that provides nutrients and some antibodies to nourish and protect the young of mammals.

milk tooth. Any tooth in the deciduous set of teeth of mammals with diphyodont dentition. Replaced by a permanent tooth.

minimum area. Home-range size estimate in which capture points are connected to form a polygon. See Figure 35.3A.

minimum convex polygon. Home-range size estimate in which capture points are connected to form the smallest convex polygon. See Figure 35.3A.

Minimum Number Known Alive. An enumeration of individuals captured on a trapping grid and including those individuals not trapped on one occasion but appearing on a later occasion. The procedure is known as the **calendar of captures method**.

mist net. A net used to capture birds and bats. Usually 3 meters high and ranging from 6 to 30 meters long. See Figure 30.11.

molar. Any cheek tooth situated posterior to the premolars, (if present) and having no deciduous precursor. One of the four kinds of teeth in mammals.

molariform teeth. Teeth that have the shape and appearance of molars, regardless of whether they are true molar teeth.

molt. The process in which hair is shed and replaced.

molt pattern. The pattern of sequential shedding and replacement of the pelage during molt. The pattern is frequently species-specific. See Figure 4.7.

monophyletic. Pertaining to a taxon the members of which are all derived from a single immediate line of descent.

monophyodont. Having a single set of teeth, with none being replaced. In contrast, see **diphyodont**.

monotypic. A taxon containing only a single taxon at the next lower major category level.

mortality. The decrement of individuals to a population through deaths. In life tables, time-specific mortality is indicated by the symbol d_x.

mousetrap. A commercial snap trap designed to kill the house mouse, *Mus musculus*. Owing to its small size, it usually kills by crushing the skull and thus is not useful as a collecting tool to mammalogists.

multiparous. State of female mammals that show evidence of placental scars of different ages and thus produced by two or more pregnancies.

mummified. Pertaining to a specimen that has been preserved by natural dehydration.

muscular diaphragm. The muscular septum between the thoracic and abdominal cavities of mammals.

Museum Special. A brand of commercially produced snap trap specifically designed for collecting mammals for museums. Intermediate in size between the rat- and mouse-traps. See Figure 30.1B.

musk. Scent secretion from any one of a variety of special scent glands in various kinds of animals.

myomorphs. Members of one of the classically recognized three suborders in the order Rodentia; generally, those rodents that have an oval, round, or V-shaped infraorbital foramen. In current classifications, these rodents are included in the suborder Sciuromorpha. See Table 23.1.

N

nail. Flat, keratinized, epidermal, translucent growth protecting the upper portion of the tip of a digit; a nail is a modified claw. See Figure 6.6.

nasal bone. Either of the paired bones on the rostrum that form the roof of the nasal passages. Usually situated between the dorsal margins of the premaxillae and of the maxillae. See Figure 2.1A, C.

nasal branches. See **premaxilla**.

nasal passages. The passages, located dorsal to the secondary (hard) palate, that connect the external and internal nares.

natality. The increment of young into a population through births.

Nearctic Region. The faunal region that includes North America south to central Mexico. See Figure 1.4.

necropsy. Partial dissection and inspection of a dead animal to gather information on causes of death, to determine reproductive condition, or to collect samples of organs for study.

nectarivorous. Feeding on nectar.

neopallium. Nonolfactory portion of the cerebral cortex.

Neotropical Region. The faunal region that includes South America, the West Indies, Central America, and southern Mexico. See Figure 1.4.

nest. A structure (of grass, leaves, or some other material) built by a mammal for shelter.

nestling. A young individual, that is still confined to a nest.

new capture. In population studies, the first capture of an individual.

niche. See **ecological niche**.

nictitating membrane. Thin membrane at the inner angle of the eye in some species, which can be drawn over the surface of the eyeball. The "third eyelid."

nidic territory. A territory that includes only the immediate area of and around a home site and that is often possessed only by a female.

nipples. Protuberances of a mammary gland that have numerous ducts leading directly to the surface, to serve as outlets for the milk.

nocturnal. Pertaining to nighttime, the hours without daylight. In particular, pertaining to animals that are active only or primarily at night. Opposite of diurnal.

nomadism. The tendency for individuals to make erratic or wandering long-distance movements.

nomenclature. The system for giving distinctive names to different taxa.

North American Datum (1927). A geodetic control network for maps and used for most United States topographic maps prior to 1983. This datum has been updated to the North American Datum (1983).

North American Datum (1983). The latest geodetic control network for mapping in the United States. This datum supercedes the older North American Datum (1927).

nose leaf. A structure on the noses of some bats. Ranges from a small, simple flap to a highly complex structure of numerous projections and chambers. Apparently aids in echolocation. See Figure 14.4.

notochord. In the phylum Chordata, a long, slender skeletal rod composed of large, vacuolated cells and lying between the dorsal nerve cord and the digestive system. In mammals, it is present only in the embryonic stages and later is surrounded and supplanted by the vertebral column.

O

occipital bone. The bone surrounding the foramen magnum and bearing the occipital condyles. Formed from four embryonic elements: a ventral *basioccipital,* a dorsal *supraoccipital,* and two lateral *exoccipitals.* See Figure 2.1B, C.

occipital condyles. The knobs (condyles) on the occipital bone, flanking the skull's foramen magnum that articulate with the atlas, the first cervical vertebra.

occipital crest. Confluent ridges extending laterally from the posterior end of the sagittal crest and more or less along the dorsal edge of the occipital bone. See Figure 2.1A, C.

occiput. General term for the posterior end of the skull. See Figure 2.1.

occlusal. Pertaining to the surfaces of contact in the upper and lower teeth. The contact of these tooth surfaces is termed *occlusion.*

occlusion. See **occlusal.**

Oceanic. Region consisting of oceanic islands not included in one of the defined faunal regions (e.g., New Zealand and Hawaii).

-oidea. Suffix added to the stem of the name of a type genus, in order to form the name of a superfamily.

olecranon process. Process of the ulna, which projects proximally beyond the articulation surface with the humerus. See Figure 6.1.

omnivorous. Pertaining to those animals that eat both animal and vegetable food.

optic foramen. Opening (in the medial wall of each orbit) through which the optic nerve and ophthalmic artery pass.

orbit. The eyeball socket in the skull. See Figure 2.1A, B, C.

orbitosphenoid. That portion of the presphenoid that is visible in the wall of the orbit. See Figure 2.1C.

order. The major category, in scientific classification, that is just below class; or a taxon at that level and containing one or more families.

Oriental Region. The faunal region that includes, roughly, the tropical portions of Asia. Extends from Pakistan south of the Himalayas through Indochina, southeastern China, and Indonesia to Wallace's Line. See Figure 1.4.

os clitoris. A small bone present in the clitoris of some species of mammals. Homologous to the baculum or os penis of male mammals.

os penis. See **baculum.**

ossicle. Any small bone (e.g., the three middle ear ossicles of mammals).

ossify. To become bony.

otic capsule. The bony capsule enclosing the organs of hearing and balance of the inner ear.

ovary (pl., **ovaries**). The female gonad. The site of egg production and egg maturation.

overhairs. See **guard hairs.**

oviduct. One of the paired ducts that carry the eggs from the ovary to the uterus. In human anatomy, the term *fallopian tube* is used.

oviparous. Pertaining to species that lay eggs.

P

pad. A usually naked structure on the ventral (plantar) surface of the manus or pes, that comes in contact with the ground. May be several per manus or pes. Also see **tori** and Figures 4.2, 6.5, 6.6, and 6.7.

palatal branches. Portions or projections of the premaxillae and maxillae that contribute to the formation of the secondary palate. See Figure 2.1B.

palatal foramina. Incisive foramina. Paired perforations of the anterior end of the hard palate, at the junction of the premaxillae with the maxillae. See Figure 2.1B.

palate. The partition that separates the nasal passages from the oral cavity. Its bony component is formed by the palatine bones and palatal branches of the maxillae and premaxillae. See Figure 2.1B.

palatine. Either of the pair of bones that form the posterior portion of the bony palate. See Figure 2.1B.

Palearctic Region. The faunal region that includes Europe, North Africa, and Asia (except for the southern portion of the Arabian Peninsula and the tropical and subtropical regions included in the Oriental Region). See Figure 1.4.

palmate. 1. Pertaining to the presence of webbing between the digits. 2. Pertaining tto palmate antlers.

palmate antlers. The broadened tines found in the antlers of some deer, e.g., the moose, *Alces alces,* Cervidae. See Figure 5.9A.

palp (=**palpus**, pl. **palpi**). A feelerlike mouthpart structure located on a maxilla or on the labium.

palpal. Of or pertaining to a palp.

palpal apotele. The modified claw of the palp pretarsus of Mesostigmata (Acarina). See Figure 32.13C.

Pangaea. A supercontinent or aggregation of continents that existed some 180 million years ago.

papilla. Any blunt, rounded, or nipple-shaped projection of soft tissue.

paraconid. A cusp on the anterior, lingual side of the trigonid portion of the crown in lower cheek teeth.

paraconule. A cusp, in some upper cheek teeth, that is situated between the protocone and paracone. Sometimes termed the protoconule.

parapatric. Pertaining to the ranges of species that are contiguous but not overlapping.

parasite. Any organism that spends all or part of its life cycle living on or in the living body of another species of organism (the host species) and that derives its food from the dead and/or living tissues of its host, or from the contents of the host's digestive tract.

parastyle. The anterior cusp on the stylar shelf. See Fig. 3.3

paratypes. Certain specimens, other than the holotype, taken to be representative of a new species or subspecies, and that are explicitly designated as "paratypes" in the original scientific description in which the new taxon is named. Otherwise, all the specimens mentioned of the new taxon, other than the holotype, in such a description.

paraxonic foot. Type of foot in which the main axis of weight passes between a pair of similarly sized digits. See Figure 27.2.

parietal. Either of the pair of bones contributing to the roof of the braincase, posterior to the frontals and anterior to the occipital bone. See Figure 2.1A, C.

paroccipital process. A process on the exoccipital portion of the occipital bone. It extends ventrally from just posterior to the tympanic bulla and the mastoid process. See Figure 2.1C.

parous. Pertaining to a female mammal that is pregnant or shows evidence of previous pregnancies (e.g., with placental scars).

parturition. The process of giving birth in therian mammals.

patagium (pl., **patagia**). A web of skin for flight or gliding, such as the parachutelike skin extensions of a colugo, the gliding membranes of a flying squirrel, or the membranes of the wing of a bat. See Figures 7.17, 7.18.

patchy dispersion. See **clumped dispersion.**

patella. The kneecap. A sesamoid bone protecting the knee joint.

pectinate. Comb-shaped. With or consisting of a series of projections like the teeth of a comb.

pectoral girdle. The shoulder girdle. In most mammals, formed by the scapula and clavicle or by the scapula alone. See Figure 6.1.

pectoral limb. The forelimb. Its skeleton includes the humerus, radius and ulna, carpals, metacarpals, and manual phalanges. See Figure 6.1.

pedal. Pertaining to the pes or hindfoot.

pedicel. Also "pedicle." Any stalk or stem supporting an organ or other structure (e.g., the pedicel of an antler). See Figure 5.7.

pelage. Collectively, all the hairs on a mammal.

pellets. 1. The undigested and regurgitated remains of the prey of hawks and owls. 2. The fecal remains of mammals that are deposited in the habitat and help provide evidence of the presence of these mammals, of their density, or of their feeding habits.

pelvic girdle. The hip girdle. Composed of the paired ischia, ilia, and pubic bones. See Figure 6.1.

pelvic limb. The hindlimb. Its skeleton is composed of the femur, patella, fibula and tibia, tarsals, metatarsals, and pedal phalanges. See Figure 6.1.

pentadactyl. Five-digited. See Figures 6.3 and 6.4, for examples.

periotic bone. A particular one of the bones forming the otic capsule.

permanent teeth. Teeth in the replacement dentition. Teeth in the second of the two sets of dentition in diphyodont mammals. Succeed the milk or deciduous teeth. The molars of most eutherian mammals are also "permanent teeth" because they do not replace a deciduous set of teeth. Some mammals, such as manatees and elephants, have a groove in the jaw in which teeth slide forward as they are worn—thus complicating the definition of a set of teeth in these mammals.

pes. The hindfoot. Skeletally, includes the tarsals, metatarsals, and pedal phalanges of the pelvic limb. See Figure 6.4.

Petersen Index. A method for estimating population size, using the formula $N = Mn/m$, and in which there is only one marking occasion.

phalanx (pl., **phalanges**). Any one of the bones (usually two or three) forming the skeleton of each digit and distal to a metapodial. See Figures 6.3 and 6.4.

pheomelanin. The pigment in the mammalian integument that produces shades and tints of red, orange, and yellow.

pheromone. A substance that is secreted by an animal and that influences the physiological processes and/or behavior of other individuals of the same species.

piebald. A mammal that exhibits patches of fur or hair that lack pigment. These white patches are often the result of somatic mutations and are not passed on to offspring.

piercing-sucking. In arthropods, mouth parts adapted for piercing tissue and for sucking fluid(s).

pigment. The constituents of the minute granules that impart color to an organism and that are especially abundant in the integument. Pigments are usually metabolic wastes and in mammals may be black, brown, yellow, or red.

pinna (pl., **pinnae**). External ear. The flap located around the opening of the ear canal. It gathers sound vibrations and channels them toward the opening. Absent in many aquatic and fossorial mammals.

piscivorous. Pertaining to animals that eat fish.

PIT transponder. A special transmitter with a unique frequency, that can be injected into animals to provide unique identification numbers when read by a special receiver.

placenta. An apposition or fusion of the fetal membranes to the uterine wall, for exchange of nutrients and removal of waste products. See **yolk-sac placenta** and **chorioallantoic placenta.**

placental scar. The scar that remains on the uterine wall after a deciduate placenta has detached at parturition.

plagiaulacoid. Pertaining to an enlarged blade-like, grooved, shearing cheek tooth found in the potoroid marsupials, the mountain pygmy possum, and certain of the extinct multituberculates. See Figure 11.20A.

plagiopatagium. One of the patagia forming the wing of a bat. Between the fifth finger and the leg. See Figure 14.1.

plantigrade. Pertaining to feet in which the toes and the entire sole of the pes, including the heel, all touch the ground in walking. The basic structure for ambulatory (walking) locomotion.

plate tectonics. The movement and mechanics of movement of the plates making up the outer solid portions of the earth. Their margins are often located along volcanic ridges and deep trenches. The motions of these plates create and eliminate ocean basins and result in new continental configurations.

point-area count. Census of individuals that are observed to pass within a certain distance of a single given point.

pollex. The first (most medial) digit on the manus. In humans, the thumb. See Figure 6.3.

polyprotodont. Condition found in didelphimorphian, microbiotherian, dasyuromorphian, paramelemorphian, and notoryctemorphian marsupials, in which the lower jaw is not shortened and the anterior lower incisors are not greatly elongated. See Figure 11.2. Compare **diprotodont.**

polytypic. Pertaining to a taxon that contains two or more taxa at the immediately lower major level (e.g., a species with two or more named and recognized subspecies).

population. 1. In statistics, all possible values (observations) of a particular variable (e.g., character) in all individuals of a particular group within a specified space or time interval. 2. In biology, all the individuals, collectively, that form a single interbreeding group.

population density. A measure of the number of individuals per unit area.

postcanine teeth. See **cheek teeth.**

posterior. Of or pertaining to the rear portion.

postorbital bar. A bony rod separating the orbit from the temporal fossa. Formed by a union of the postorbital process of the frontal bone with that of the zygomatic arch.

postorbital process (of the frontal). A projection of the frontal bone, which marks the posterior margin of the orbit. See Figure 2.1C.

prebaiting. Placing bait at trapping stations prior to the initiation of trapping.

precoracoid. A bone in the pectoral girdle. Found in mammals only in Monotremata. Inherited from ancestral reptiles.

preference index. A measure of the food preferences of a sampled population. Determined by comparing the relative occurrence of dietary items in the individuals sampled, with the relative occurrence of the resource upon which they feed. The *electivity index* is another statistical measure of dietary preferences. See Chapter 34.

prehensile. Pertaining to structures adapted for grasping or seizing by curling or wrapping around, such as the tail of some Neotropical monkeys and of most opossums.

premaxilla (pl., **premaxillae**). One of the paired bones, at the anterior end of the rostrum, that frequently bear teeth. *Palatal branches* of the premaxillae form the anterior part of the secondary palate, and the *nasal branches* contribute to the sides of the rostrum. See Figure 2.1A, B, C.

premolar. One of the four kinds of teeth in mammals. Located anterior to the molars and posterior to the canine. Some are the only cheek teeth normally represented in both the milk and replacement dentitions. Usually do not exceed four per quadrant.

presphenoid. A bone of complex shape, that contributes to the ventral and lateral walls of the braincase. Ventrally, it is visible between the pterygoids, from where it then passes beneath other bones to reappear in the wall of the orbit, where it is termed the *orbitosphenoid.* See Figure 2.1B, C.

pretarsus. See Figure 32.7.

primary data. The initial data recorded by an investigator before conversion to other media. For example, handwritten field notes or tape recordings before these data are transferred to computer files. See Figures 29.1, 29.2, 29.5, 29.6, 29.7, and 29.8.

probability of capture. The likelihood or probability that an animal will be caught in a trap on a given occasion. This probability is affected by weather, behavior, interactions between individuals and other species, and other factors. See Chapter 36.

proboscis (pl., **proboscises, proboscises,** or **proboscides**). 1. A long, more or less flexible snout, as in tapirs and elephants. 2. The flexible beak of certain insects such as butterflies, moths, and flies, and others that allow the animal to secure food.

procumbent. Pertaining to teeth that slant forward, such as the incisor teeth of a horse.

prolegs. The tubular, fleshy, leglike appendages located ventrally on some abdominal segments of some immature insects. See Figure 32.6A.

pronate. To rotate the palm of the primate manus (hand) to face downward or posteriorly.

pronghorn. 1. A type of horn (in both sexes of Antilocapridae) that grows over a permanent bony core and is shed annually. Each horn is slightly curved and has one anterolateral prong. See Figure 5.4. 2. *Antilocapra americana,* the North American pronghorn "antelope." See Figure 27.13.

propatagium. In bats, a thin web of skin that extends from the shoulder to the wrist, anterior to the upper arm and forearm. See Figure 14.1.

prothorax. The anterior segment of the insect thorax, bearing the anterior pair of legs but no wings.

protocone. In upper cheek teeth, a cusp on the lingual side of the crown, at the apex of the trigon area of the tooth.

protoconid. In lower teeth, a cusp on the posterior lingual side of the trigonid area of the crown.

protoconule. See **paraconule.**

protogomorphous. Pertains to the primitive type of jaw muscle found today only in the sewellel (*Aplodontia rufa,* Aplodontidae, Rodentia).

Prototheria. A subclass of mammals. Among living mammals, contains only the monotremes (egg-laying mammals).

protoungulate. A representative of an extinct group of pre-ungulate mammals.

proximal. Situated nearer to or nearest to the main part of the body (e.g., the proximal end of a limb is the end closest to [and attached to] the rest of the body). At or near the base or area of attachment. Contrast with **distal.**

pterygoid. Either of the paired bones in the ventral wall of the braincase and that are posterior to the palatine bones. See Figure 2.1B, C.

pubic bone. See **pubis.**

pubic symphysis. Midventral plane of contact, or near contact, between the two halves of the pelvic girdle.

pubis. Either of the pair of bones forming the anterior ventral portion of the pelvic girdle. See Figure 6.1.

pulp. Collectively, the nerves, blood vessels, and connective tissue occupying the pulp chamber and root canals of a tooth. See Figure 3.1.

Q

quadrant. In mammalian dentition, pertaining to each of the four tooth-bearing sections of the skull: the right and left premaxillae/maxillae and the two dentaries.

quadrat. A plot used for ecological sampling; strictly speaking, always square in outline; in practice, may be of other shapes.

quadrate. A bone of the mandibular fossa of many vertebrates, which, in mammals, becomes the incus, one of the middle ear ossicles.

quadrate tooth. A euthemorphic cheek tooth that has a square outline.

quadritubercular tooth. Any upper cheek tooth with four major cusps: the paracone, metacone, protocone, and hypocone.

quadrupedal. Pertaining to an animal that uses all four limbs for terrestrial locomotion.

R

rack. The pair of antlers on a cervid. See Figure 5.9.

radioisotope. A radioactive isotope.

radiolocation telemetry. Technique for obtaining location and movement information on an animal by use of a transmitter (on the animal) and a receiver (used by observer).

radius (pl., **radii**). One of the two bones in the lower foreleg, between the humerus and the carpals. Generally the more medial of the two. See Figure 6.1.

ramus (pl., **rami**). See **dentary** and Figure 2.1D.

random dispersion. Distribution of individuals in a given area such that there is an equal probability that an individual will occupy any given point in space, and the presence of another individual nearby will not affect this probability. See Figure 35.1C.

range. 1. The spread of the smallest and largest values in a statistical distribution or set of data. 2. The geographic area inhabited by a particular taxon—especially a species.

range length. The distance between the most widely separated capture points. See Figure 35.3D.

rat trap. A commercial snap trap especially designed to kill mammals the size of the Norway rat, *Rattus norvegicus*. See Figure 30.1A.

recapture. In population studies, the capture of a previously marked or tagged individual.

re-entrant angle. Inward-pointing angle of enamel along the margin of a cheek tooth. See Figure 23.11.

re-entrant fold. Curved invagination of enamel along the margin of a cheek tooth. See Figure 23.16.

reingestion. The ingestion, by an individual, of caecotrophic fecal pellets that the animal has previously voided. Occurs in many lagomorphs. See **caecotrophic pellets.**

relative age. Age of an individual in terms relative to those of others in the population. Compare **absolute age.**

relative dominance. The dominance of one species as a decimal fraction or percentage of the combined density of all species. (Cox 1996).

relative estimate. In contrast to density estimates, a measure of (relative) numbers of individuals, that has not been standardized to number of individuals per unit area.

removal trapping. Trapping practice in which individuals are permanently (by kill trapping) or temporarily (by live trapping) removed from an area.

representative fraction. Expression of a map scale as a ratio or fraction (Campbell 1999). For example, with a representative fraction of 1:24,000, the first number represents the number of units on the map, and the 24,000 indicates how many units these represent on the surface of the earth.

reprint. A printed separate copy of a journal article or chapter of a book, that is given or sold to its author or authors for distribution to other scientists.

reprint-exchange system. The practice of sending reprints on request or through a preset distribution scheme to other scientists.

ricochetal. Pertaining to bipedal movement involving propulsion into the air by both pelvic limbs simultaneously, and landing on both pelvic limbs without involvement of pectoral limbs (e.g., rapid locomotion in a kangaroo). See Figure 7.4.

roadside count. Census of individuals seen along a roadway.

root. A portion of a tooth that lies below the gum line and fills an alveolus. See Figure 3.1.

rooted tooth. A tooth that has definitive growth (is not evergrowing). See Figure 3.1.

rootless tooth. A tooth that is evergrowing, having a permanently, and widely, open root canal.

rostrum. The facial region of the skull, anterior to a cross-sectional plane through the anterior margins of the orbits. See Figure 2.1A, C.

rough out. To remove the large muscles and viscera when preparing a skeleton for later cleaning by dermestids or by other means.

rover. Common name for a global positioning system (GPS) reveiver that is carried about by the user and is not fixed to a single location.

rudimentary. Relatively undeveloped and nonfunctional. Vestigial.

runway. A worn, cut, or otherwise detectable pathway produced by and repeatedly used by small mammals.

rutting season. Season when copulation occurs. Particularly applied to deer and other artiodactyls.

S

sacral vertebrae. The vertebrae articulating with, and often fused to, the pelvic girdle. See Figure 6.1.

sagittal crest. A medial dorsal ridge on the braincase, formed by coalescense of the temporal ridges. See Figure 2.1C.

Saharo-Sindian Region. The arid and semiarid areas extending across northern Africa and southwest Asia to the Sind area of northwestern India. Sometimes recognized as a distinct faunal region, but usually included in the Palearctic.

saltation. Form of locomotion characterized by leaping. Includes ricochetal and springing locomotor movements. See **spring.**

saltatorial. Progressing by means of saltation (bipedal leaping).

sample. In statistics, a subset of individuals, observations, or values selected from a population. See also **population** and **sample size.**

sample mean. See **mean.**

sample size (N). The number of individuals, observations, or values that comprise a sample. See **sample** and **population.**

sampled count. Count of individuals occupying several sample areas within a larger area. **Sampling unit.** An object, specimen, or other entity upon which measurements are taken.

sanguinivorous. Feeding on blood.

scale. 1. In mammals, one of many flattened, epidermal plates that may cover the tail, feet, and (in pangolins and armadillos only) most of the body. 2. In insects, a flat outgrowth of the body wall, that forms plates or sclerites.

scansorial. Pertaining to arboreal or semiarboreal animals that climb nimbly by means of sharp, curving claws (e.g., tree squirrels).

scapula. The shoulder blade. The dorsalmost bone in the pectoral girdle of mammals. See Figure 6.1.

scapular spine. A ridge of bone on each scapula that increases the area for muscle attachment. See Figure 6.1.

scat. Mammal feces or droppings. See Figure 28.3.

scavenger. An animal that feeds on portions of dead animals that it has not killed.

scent glands. Sweat or sebaceous glands, or a combination of these two gland types, modified for the production of odoriferous secretions.

Sciurognathi. A suborder of rodents. In these rodents, the masseter complex of muscles is either the protrogomorphous (Fig. 23.3A), sciuromorphous (fig. 23.3B), or myomorphous (Fig. 23.3C) type and the mandible of the sciurognathous type (e.g., Fig. 23.7). See Table 23.1.

sciurognath mandible. Type of mandible found in sciuromorph rodents in which the angular process arises medial to the outer border of the incisive alveolar sheath. See Figure 23.7.

sciuromorph. Formerly used to refer to some members of a suborder of rodents that are now included in the Suborder Sciurognathi. Pertains to those rodents that have a sciuromorphous type of jaw musculature (Fig. 23.B).

sclerite. Any section of the arthropod body wall, bounded by "sutures," which are seams, impressed lines, or areas of more flexible body wall.

scrotal. 1. Pertaining to the scrotum. 2. A description of the reproductive state of male mammals experiencing a seasonal descent of the testes into the scrotum during the period of active spermatogenesis.

scrotum. The pouch of skin in which the testes may be situated outside of the abdominal cavity proper. Permanently present in some mammal species, seasonally present in others, and never present in still others.

seasonal migration. Periodic to-and-fro movements of individuals, groups, or populations of animals, in response to seasonal patterns of resource availability or other seasonal factors.

seasonal molts. Molts in species that replace all guard hairs and underhair more than once a year. For example, the winter and summer molts of some long-tailed weasels or the Arctic hare.

sebaceous gland. One of a number of epidermal glands that secrete a fatty substance and usually open into a hair follicle. See Figure 4.1.

secodont dentition. With cheek teeth that have a cutting or shearing action adapted for a carnivorous diet. The carnassial teeth, found in the order Carnivora, are an example of this type of dentition. See Figure 19.2.

secondary data. Data that have been transformed from their primary state into another medium. For example, a computer file that was typed from handwritten data sheets or field notes.

selective availability (SA). The intentional error that is placed in the global positioning satellite (GPS) signals, by the U.S. Department of Defense, to alter, for national security purposes, the true locational position indicated on civilian GPS receivers. See also **differential correction.**

selenodont. Type of dentition characterized by molariform teeth with a crown pattern of longitudinally oriented, crescent-shaped ridges formed by elongated cusps. See Figure 3.8.

selenolophodont. Pertaining to cheek teeth that have features of both **selenodont** and **lophodont** teeth.

semiaquatic. Pertaining to mammals partially, but not fully, adapted to life in water (e.g., river otters, beavers, water shrews).

semifossorial. Pertaining to mammals that are partially, but not fully, adapted for life underground (e.g., ground squirrels, badgers, European rabbits).

series. In general, all of the specimens of a species that were collected from a single locality by a collector or collectors during a single visit or field season.

sesamoid bone. Any ossification in a tendon. For example, the patella.

Sherman live trap. A commercial brand of live trap, constructed of galvanized iron or of aluminum. See Figure 30.7.

sibling species. Species that are reproductively isolated but morphologically indistinguisable, or virtually so.

sign. Any indication of an animal's presence (e.g., footprints, scats, burrows, runways).

skull. The skeleton of the head. Includes the cranium and the mandible. See Figure 2.1.

slip. A skin is said to "slip" when decomposition results in large patches of hair and epidermis becoming detached.

sloth movement. A quadrupedal "walk" by an animal that is suspended upside down by all four limbs. See Figure 7.16.

small-scale map. A map in which a large geographic area is shown and which, in turn, does not provide a lot of detail. For exapmle, maps of the scales 1;250,000 or 1:2,000,000 are small-scale maps.

snap trap. Kill traps that usually consist of a wooden base, a wire bail, a spring, and a trigger mechanism. Designed primarily to catch small rodents and insectivores. See **rat trap, mousetrap,** and **Museum Special.** See Figure 30.1.

spatial territory. An area temporarily or permanently defended by an animal. This is the most common type of territory found in species that are territorial.

speciation. Pertaining to processes that form new species of organisms.

species. 1. Actually or potentially interbreeding populations that are intrinsically reproductively isolated from all other kinds of organisms. 2. The taxonomic category between the category subspecies and genus.

species accounts. A chronological listing of all field observations pertaining to a species; a portion of the field notes. See Figure 29.5.

specific name. The second word of a binomen or trinomen, which, in conjunction with the generic name, forms the name of a species.

sphincter. Any muscle having its fibers in a circular arrangement around an opening or passage so that, upon contraction, that opening or passage is closed.

spines. Thick, sharply pointed guard hairs that serve primarily for defense.

spring. A quadrupedal leap in which a mammal propels itself forward with both hindlimbs and lands first on both forefeet (e.g., the leap of a rabbit).

spur. In mammals, a pointed structure of bone covered with keratinized material and projecting from the ankle of (usually) male monotremes.

squamosal. The bone that forms the major portion of the lateral wall of the braincase and the posterior root of the zygomatic arch. See Figure 2.1B.

standard deviation. The square root of the variance(s^2), expressed in the same units as the original observations.

standard deviation of capture radii. Used by Calhoun and Casby (1958) and others to estimate the dimensions of the home range of an animal. See **standard deviation**.

standard error of the mean. The square root of the quotient obtained by dividing the variance by the sample size.

standard minimum grid. A trapping grid consisting of 16 rows and 16 columns, yielding 256 trap stations, each of which has two traps.

stapes. The innermost of the three middle ear ossicles. A small, stirrup-shaped bone (see Fig. 1.1) derived from the columella of reptiles.

state plane coordinates. Locational positions, based on the plane rectangular coordinate system, that are individually applied to zones in each state in the United States. See Chapter 29 and Campbell (1999).

steel leg trap. A trap primarily used by fur trappers and designed to catch the animal by the leg. Not considered humane for scientific collecting and not recommended.

sternum. The breast bone, which articulates with the distal ends of the costal ribs. See Figure 6.1.

stigmal. Of or pertaining to a stigma.

stigmata (sing., **stigma**). In arthropods, the spiracles or breathing pores.

stratum basale. See **stratum germinativum**.

stratum corneum. The relatively well-cornified surface layer of the epidermis. See Figure 4.1.

stratum germinativum. Stratum basale. The basal epidermal layer where mitosis actively produces the outer layers of cells. See Figure 4.1.

stratum granulosum. A layer in the epidermis, between the stratum lucidum (above) and the stratum spinosum (below). See Figure 4.1.

stratum lucidum. The layer of the epidermis immediately underneath the stratum corneum, although not always evident or present. See Figure 4.1.

stratum spinosum. The layer of the epidermis immediately above the stratum germinativum and below the stratum granulosum. See Figure 4.1.

study skin. A mammal (or bird) skin prepared for placement in a research collection. See Figures 31.19, 31.20, and 31.21.

stylar shelf. Expanded horizontal portion of the labial cingulum of a tooth. Frequently bears stylar cusps.

stylettiform. Pertaining, in arthropods, to bladelike mouth parts; long, thin piercing structures.

subadult. A young individual, generally not fully grown, that may be a young-of-the-year and may or may not be in reproductive condition.

subadult pelage. A pelage that is characteristic of subadults of a species. This term is used only when an obvious difference exists between the juvenile and adult pelages.

subcutaneous. Pertaining to a location immediately beneath the skin.

subunguis. The ventral, cornified portion of a claw, nail, or hoof. It is softer than the unguis. See Figures 6.5 through 6.7.

subungulates. Traditionally, the proboscideans, sirenians, and hyracoideans, collectively.

sudoriferous (= apocrine sweat) gland. See **sweat glands**.

supine. Pertaining to when the palm of the primate manus (hand) is upward or forward.

supraoccipital. See **occipital bone**.

suture. An immovable contact zone between two bones, particularly in the skull. See Figure 2.1.

sweat. Perspiration. A very dilute aqueous solution that contains small amounts of inorganic salts and certain nitrogenous excretory products.

sweat glands. Long, tubular epidermal glands that extend into the dermis and that secrete perspiration and/or various scents. *Apocrine sweat glands* or *sudoriferous glands* generally empty their secretions into a hair follicle and produce the odor component of perspiration. *Eccrine sweat glands* (see Figure 4.1) open directly onto the skin surface, produce most of the fluid component of perspiration, and function primarily to regulate body temperature by evaporation of this liquid.

sympatric. Pertaining to two or more species with overlapping geographic ranges. Compare with **allopatric**.

symphysis. A relatively immovable articulation between two bones (e.g., the pubic symphysis of the pelvic girdle).

syndactylous. Pertaining to two or more digits that are fused together. See Figures 11.3, 11.4.

synopsis. Summarized description of a taxon.

systematics. The field of science concerned with taxonomy and phylogeny.

systematist. Scientist who does research in systematics.

T

tagging. The process of uniquely marking animals so that they can be identified when recaptured or observed.

talon. An expansion of the posterior cingulum of an upper cheek tooth. Frequently identifiable only as a cusp, the hypocone.

talonid. An extension of the posterior cingulum of a lower cheek tooth. It squares the outline of the tooth. See Figure 3.3B.

talonid basin. A region of the talonid, frequently forming a basin, surrounded by the hypoconulid, hypoconid, and entoconid.

tarsal bones. Tarsals. Series of bones in the ankle. They are distal to the fibula and tibia and proximal to the metatarsals. See Figure 6.4.

tarsus (pl., tarsi). In arthropods, the jointed structure attached to the distal end of the tibia, consisting of 1 to 5 segments and bearing the pretarsus. See Fig. 32.6.

taxon (pl., taxa). The actual sort(s) of organism(s) that is/are designated by a scientific name. For example, the *taxon* to which the *name Homo sapiens* applies consists of the actual humans designated by the name. Similarly, all the actual bats in the world constitute the taxon designated by the name Chiroptera.

taxonomic character. Any attribute of a member of a taxon by which it differs or may differ from a member of a different taxon (Mayr 1969:121).

taxonomic category. A named *level* in a taxonomic scheme (e.g., "order," "genus," etc.) as opposed to an actual, particular *group* (or taxon) of actual organisms classified at that level (e.g., Chiroptera, *Homo,* etc.).

taxonomist. A scientist who classifies organisms.

taxonomy. The discipline devoted to classifying organisms.

teat. A protuberance of the mammary glands and in which numerous small ducts empty into a common collecting structure that in turn opens to the exterior through one pore. See Figure 4.10C.

temporal bone. A paired bone in the braincase of certain mammals. It results from the fusion of a squamosal and a tympanic bone.

temporal fossa. The portion of the space bounded laterally by the zygomatic arch and that is posterior to the orbit. See Figure 2.1A, B.

temporal ridges. A pair of ridges on the top of the braincase of many mammals. Usually originate on the frontal bones near the postorbital processes and converge posteriorly to form the middorsal sagittal crest. See Figure 2.1A. They mark the dorsal edges of the origins of the external portions of the masseter muscles.

tendon. Cord or band of dense connective tissue attaching a muscle to a skeletal element or to another muscle.

terete. "Cylindrical" but tapering toward the end(s).

territoriality. The pattern of behavior associated with the defense of a territory. The persistent attachment to a specific territory.

territory. An area defended by an individual or group. According to some ecologists (e.g., Pitelka 1959), an exclusive area possessed by an individual or group.

testis (pl., testes). Gonad of the male. The organ of sperm formation. Also produces testosterone.

tetrapod. A mammal, bird, reptile, or amphibian; vertebrates other than fishes and agnathans.

†Therapsida. The order of Permian and Triassic reptiles (of the subclass Synapsida) considered to be ancestral to mammals.

therapsid reptiles. See **†Therapsida.**

Theria. The subclass that includes marsupial (metatherian) and placental (eutherian) mammals.

three-dimensional home range. Home range in three dimensions, including the horizontal plane and the vertical. See Koeppl et al. (1977a) for the general model. See Figure 35.4A, B.

thoracic vertebrae. The set of vertebrae, that provide points of attachment for ribs, and which are located between the cervical and lumbar sets of vertebrae.

thorax. In insects, the middle body region bearing the jointed legs and, if present, the wings.

thumb. The pollex, or first medial digit, of a hand.

tibia. More medial of the two bones between the knee joint and ankle, in the lower hindlimb. The shin bone. See Figure 6.1.

tine. A spike or prong on an antler. See Figure 5.8.

tissues biopsy. Pertaining tto the removal of tissue from an anumal. See Chapter 31.

toe. A pedal digit.

torus (pl., tori). One of the epidermal pads on the bare surfaces of the manus and pes of mammals other than ungulates and primates. See **pad** and Figure 4.2A.

total count. A complete enumeration of all individuals, of a species, occupying a given area.

tragus. In most microchiropteran bats, the projection from the lower medial margin of the pinna. Also found in some other mammals. See Figure 14.3A–C.

trail. A pathway created by repeated use by animals.

transect. 1. In population studies, a line of traps spaced at regular intervals (see Figure 37.1B) or a route along which counts of animals are made. 2. In ecological studies, a line along which measurements are made such as the plant cover intercepted by the measuring tape (see Figure 28.10).

trap line. A line of traps placed at regular (= transect) or irregular intervals to secure specimens for identification, study specimens, or necropsy purposes.

trap night. A trap set for one night.

trap nights. The number of traps set multiplied by the number of nights they were set.

tribe. 1. A level of classification between family and genus. 2. An actual taxon classified at that level. Names of the tribe end in the suffix *-ini.*

tribosphenic tooth. An upper molar with three main cusps in a triangular pattern (these constitute the trigon) or a lower molar with a triangular arrangement of the trigonid and an adjacent talonid (see Fig. 3.3A). Modifications of this type of tooth have produced the great diversity seen in the cheek teeth of present-day mammals.

triconodont tooth. Elongate cheek tooth characterized by three to five main cusps arranged in a straight line, particularly in members of the extinct order †Triconodonta.

tricuspid. 1. A tooth with three major cusps. 2. Having three points or cusps.

trigon. A triangle formed by three main cusps of an upper molar, with the protocone oriented along the lingual edge of the tooth.

trigonid. A triangle formed by three main cusps of a lower molar with the protoconid side of the triangle oriented along the labial edge of the crown. See Fig. 3.3.

trinomen. A series of three words (the generic, specific, and subspecific names) that constitute the scientific name of a subspecies.

tritubercular tooth. An upper tribosphenic molar or premolar.

true horns. See **horn.**

turbinals. Convoluted or scroll-shaped bones in the nasal passages of the skull.

tympanic bone. The bone that forms the ring holding the eardrum or tympanic membrane and that usually forms the major portion of the auditory bulla. See Figure 2.1B, C.

tympanic bulla. See **auditory bulla.**

tympanic membrane. Eardrum. The thin membranous structure that receives external vibrations from the air and transmits them to the middle ear ossicles.

type. See **holotype.**

type genus. One particular genus, contained within a specific, named family, that will always be associated with that nominal family, even if all other genera that have ever been included within it have been removed in the course of taxonomic reclassification. The name of a new family is expected to designate a contained genus as the type genus.

type species. A species, contained within a specific, named genus, which, even if all other species that have been classified at one time or another in that nominal genus have been removed from it, will still always belong to that nominal genus. The namer of a new genus is expected to designate a contained species as the type species.

U

ulna. One of the two long bones in the lower portion of the pectoral limb, and between the humerus and the carpals. Usually, the more lateral of the two. See Figure 6.1.

underhairs. Shorter hairs that serve primarily for insulation: See **fur, wool,** and **velli.**

unguligrade. Hoofed. Having a foot structure in which only the unguis (or hoof) is in contact with the ground. See Figures 6.7, 26.1, and 27.1.

unguis. The hard, dorsal, keratinized portion of a claw, nail, or hoof. See Figures 6.5, 6.6, and 6.7.

unicuspid. A tooth having a single cusp. Often used to designate certain teeth of uncertain homology in some insectivores.

uniform dispersion. Dispersion in which the points in space occupied by individuals are approximately equidistant from one another. See Figure 35.1A.

upper maxillary process. See **maxillary process.**

uropatagium. A web of skin extending between the hindlegs and frequently enclosing the tail—especially in bats. See Figures 14.1 and 14.2.

uterus (pl., **uteri**). In female mammals, a muscular expansion of the reproductive tract, in which the embryo and then the fetus develop; opens externally by way of the vagina. Usually forked, but the uterine "horns" may be partially or completely fused.

UTM coordinates. Locational positions in meters, between the 80°S and 84°N, based on the Universal Transverse Mercator (UTM) system. In the civilian version of this system, the zones are designated by numbers.

V

valvular. Capable of being closed by flaps (e.g., a valvular nostril).

variance. In statistics, the sample variance is a measure of the dispersion of a set of data about the mean, always expressed in squared units. The sample variance (s^2) is the best estimate of the population variance (∂^2).

velli (sing., **vellus**). Very fine, short, dense underhairs.

velvet. The skin and pelage covering a growing antler. See Figure 5.5A.

venter. The ventral surface of an animal.

ventrad. Toward the venter. Ventral.

ventral. Pertaining to the under or lower surface.

vermiform. Wormlike, worm-shaped (e.g., the vermiform tongue of many ant-eating mammals). See Figure 18.1A.

vernier calipers. See **calipers.**

vertebrae. In vertebrate animals (agnathans, fishes, amphibians, reptiles, birds, and mammals), the cartilaginous or bony structures that collectively form the backbones of these animals.

Vertebrata. The subphylum of the phylum Chordata that includes all agnathans, fishes, amphibians, reptiles, birds, and mammals. Except for present-day agnathans, the brain is enclosed in a skeletal braincase, and a segmented vertebral column supports the body.

vestigial. Having become functionless and greatly reduced in size in the course of evolution.

vibrissae. Long, stiff hairs that serve primarily as tactile receptors. See Figure 4.5.

viviparous. Pertaining to animals, such as all eutherian and metatherian mammals, that give birth to their young.

volant. Pertaining to the ability to fly. Often used (incorrectly) to describe the locomotion of mammals that glide, but are not capable of sustained flight.

vomer. Unpaired bone that forms part of the septum between the nasal passages of the skull.

W

Wallace's Line. Imaginary line through Indonesia. It separates the Australian from the Oriental Faunal Regions and runs between Bali and Lombok, between Borneo and Sulawesi (= Celebes), and then continues east of the Philippines. Conceived of by Alfred Russell Wallace (1823–1913), who independently developed the same theory of evolution as Charles Darwin. See Figure 1.4.

warning coloration. Any striking coloration in animals with powerful mechanisms of defense, such as the musk of skunks and zorillas (see Fig. 4.8), that may give predators and other species advance warning of the potential danger.

webbed. Having a membrane or patagium between digits.

weight. A measure of gravitational force in Newtons (SI units) or pounds (English system). Compare with **mass.**

Wiley Mill. Machine for grinding food samples into uniform particle sizes for microscopic dietary analysis.

wing. 1. A forelimb modified for sustained flight. Among mammals, found only in bats. See Figure 14.1. 2. An insect wing.

wool. Certain hair or underhair with angora growth. Serves primarily for insulation.

wrist (joint). The joint between the manus and the rest of the forelimb.

X

xenarthrales. Extra articular surfaces found on the posterior trunk vertebrae of xenarthrans. See Figure 17.2.

Y

yolk-sac placenta. The type of placenta found in most marsupials, in which the yolk sac and chorion of the embryo make contact with the maternal uterine lining.

young of the year. General age description for an animal that was born in the most recent breeding season and is less than a year old. Frequently used when it is difficult to assign individuals to more precise age categories or when a collective term is needed for this group of young animals.

Z

zalambdodont tooth. A tooth characterized by a V-shaped ectoloph on the occlusal surface. See Figure 3.4.

Zippin method. A census method based on removal trapping on two occasions. See Chapter 36.

zygapophysis (pl., **zygapophyses**). Any one of the processes of the dorsal portion of a vertebra and with an articulation surface for contacting a zygapophysis of an adjacent vertebra.

zygomatic arch. An arch of bone that encloses the orbit and temporal fossa laterally and is formed by the jugal bone and processes of the maxilla and the squamosal. See Figure 2.1.

zygomatic plate. In rodents, the expanded and flattened lower maxillary process. See Figure 23.2C(dd).

zygomatic process of the maxilla. Maxillary process; that projection of the maxilla that forms the anterior root of the zygomatic arch. See Figure 2.1.

zygomatic process of the squamosal. Zygomatic process; that projection of the squamosal bone that forms the posterior root of the zygomatic arch. See Figure 2.1.

Literature Cited

ad hoc Committee for Acceptable Field Methods in Mammalogy. 1987. Acceptable field methods in mammalogy: preliminary guidelines approved by the American Society of Mammalogists. *J. Mammal.* 68(4) Supplement: 1–18.

ad hoc Committee for Animal Care Guidelines. 1985. Guidelines for the use of animals in research. *J. Mammal.* 66:834.

Adamczewska-Andrzejewska, K. 1973. Growth, variations and age criteria in *Apodemus agrarius* (Pallas, 1771). *Acta Theriologica* 18: 353–394.

Adams, L. 1957. A way to analyze herbivore food habits by fecal examination. *Trans. N. American Wildlife Conf.* 22:152–159.

Adams, L., and S. D. Davis. 1967. The internal anatomy of home range. *J. Mammal.* 48:529–536.

Afanasev, A. V., V. S. Barzhanov, et al. 1953. [*The animals of Kazakhstan.*] Akademia Nauk Kazkhskot SSR. Alma-Alta. [In Russian]

Alcoze, T. M., and E. G. Zimmerman. 1973. Food habits and dietary overlap of two heteromyid rodents from the mesquite plains of Texas. *J. Mammal.* 54:900–908.

Allen, J. A. 1924. Carnivora collected by the American Museum Congo Expedition. *Bull. Amer. Mus. Nat. Hist.* 47:1–283.

Altevogt, R. 1975. Elephants, pp. 478–480. *In* B. Grzimek (Ed.). *Animal life encyclopedia, Vol. 12, Mammals III.* Van Nostrand Reinhold, New York.

Altevogt, R, and F. Kurt. 1975. Asiatic elephants, pp. 484–500. *In* B. Grzimek (Ed.). *Animal life encyclopedia, Vol. 12, Mammals III.* Van Nostrand Reinhold, New York.

Altmann, J. 1974. Observational study of behavior: sampling methods. *Behaviour* 49:227–267.

American Society of Mammalogists Committee on Information Retrieval. 1996. *Documentation standards for automatic data processing in mammalogy. Version 2.0.*

Andersen, K. 1912. *Catalog of the Chiroptera in the collection of the British Museum, Vol. I, Megachiroptera.* British Museum (Nat. Hist.), London.

Anderson, P. K., G. E. Heinsohn, P. H. Whitney, and J-P Huang. 1977. *Mus musculus* and *Peromyscus maniculatus:* homing ability in relation to habitat utilzation. *Canad. J. Zool.* 55:169–182.

Anderson, R. M. 1965. Methods of collecting and preserving vertebrate animals, 4th ed., revised. *Bull. Nat. Mus. Canada* 69:1–199.

Anderson, S. 1961. A new method of preparing lagomorph skins. *J. Mammal.* 42:409–410.

Anderson, S. 1969. Taxonomic status of the woodrat, *Neotoma albigula,* in southern Chihuahua, Mexico. *Univ. Kansas Mus. Natur. Hist., Misc. Publ.* 51:25–50.

Anderson, S. 1972. Two semiautomatic systems for linear measurements. *Curator* 15:220–228.

Anderson, S., and J. K. Jones, Jr. 1984. *Orders and families of Recent mammals of the world.* John Wiley & Sons, New York.

Anderson, T. J. C., A. J. Berry, J. N. Amos, and J. M. Cook. 1988. Spool-and-line tracking of the New Guinea spiny bandicoot, *Echymipera kalubu* (Marsupialia, Permelidae). *J. Mammal.* 69:114–120.

Animal Care and Use Committee. 1998. *Guides for the capture, handling, and care of mammals as approved by the American Society of Mammalogists.* Amer. Society of Mammalogists.

Armitage, K. B. 1974. Male behavior and territoriality in the yellow-bellied marmot. *J. Zool., London* 172:233–265.

Arnett, R. H., Jr. (Ed.). 1985. *The naturalists' directory and almanac (international),* 44th ed. E. J. Brill, Gainesville, Florida.

Arnett, R. H., Jr. (Ed.). 1993. *The naturalists' directory and almanac (International),* 46th ed. E. J. Brill, Gainesville, Florida.

Association for the Study of Animal Behaviour. 1996. Guidelines for the treatment of animals in behavioural research and teaching. *Animal Behaviour* 51:241–256.

August, P. V. 1993. GIS in mammalogy: building a database, pp. 11–26. *In* S. B. McLaren and J.K. Braun (Eds.). *GIS applications in mammalogy.* Oklahoma Mus. Nat. History, Spec. Publ.

Bailey, N. T. J. 1952. Improvements in the interpretation of recapture data. *J. Animal Ecol.* 21:120–127.

Baker, E. W., J. H. Camin, F. Cunliffe, T. A. Woolley, and C. E. Yunker. 1958. Guide to the families of mites. *Institute of Acarology Contributions* 3:1–242.

Baker, R. J., and M. Haiduk. 1985. Collections of tissue cultured cell lines suspended by freezing. *Acta Zoologica Fennica* 170:91–92.

Baker, R. J., and S. L. Williams. 1972. A live trap for pocket gophers. *J. Wildl. Manag.* 36:1320–1322.

Bander, R. B., and W. W. Cochran. 1969. Radio-location telemetry, pp. 95–103. *In* R.H. Giles, Jr. (Ed.). *Wildlife management techniques,* 3rd ed. The Wildlife Society, Washington, D.C.

Banks, E. M., R. J. Brooks, and J. Schnell. 1975. A radiotracking study of home range and activity of the brown lemming (*Lemmus trimucronatus*). *J. Mammal.* 56:888–901.

Barbour, R. W., and W. H. Davis. 1969. *Bats of America.* Univ. Press of Kentucky, Lexington.

Barkley, L., and J. O. Whitaker, Jr. 1984. Confirmation of *Caenolestes* in Peru with information on diet. *J. Mammal.* 65:328–330.

Barrett-Hamilton, G. E. H. 1910. *A history of British mammals, Vol. I, Bats.* Gurney and Jackson, London.

Bartholomew, G. A., and P. G. Hoel. 1953. Reproductive behavior of the Alaska fur seal, *Callorhinus ursinus. J. Mammal.* 34:417–436.

Baumgartner, L. L., and A. C. Martin. 1939. Plant histology as an aid in squirrel food-habit studies. *J. Wildl. Manag.* 3:266–268.

Beddard, F. E. 1902. Mammalia, Vol. 10. *In* S. F. Harmer and A. E. Shipley (Ed.). *The Cambridge natural history.* Macmillian and Co., Ltd., New York.

Benedict, F. A. 1957. Hair structure as a generic character in bats. *Univ. Calif. Publ. Zool.* 59:285–548.

Bennett, A. F., and B. J. Baxter. 1989. Diet of the long-nosed potoroo, *Potorous tridactylus* (Marsupialia: Potoroidae), in south-western Victoria. *Australian Wildl. Res.* 16:263–271.

Berger, T. J., and A. M. Neuner. 1981. *Directory of state protected species: a reference to species controlled by non-game regulations.* Assoc. Systematics Collections, Lawrence, Kansas (looseleaf).

Berger, T. J., and J. D. Phillips (Compilers). 1977. *Index of U.S. federal wildlife regulations.* Assoc. Systematics Collections, Lawrence, Kansas (looseleaf).

Birney, E. C., R. Jenness, and D. D. Baird. 1975. Eye lens proteins as criteria of age in cotton rats. *J. Wildl. Manag.* 39:718–728.

Bland, R. G., and H. E. Jaques. 1978. *How to know the insects,* 3rd ed. Wm. C. Brown Company Publishers, Dubuque, Iowa.

Bobrinskii, N. A., B. A. Kuznekov, and A. P. Kuzyakin. 1965. [*Key to the mammals of the USSR,* 2nd ed.] Moscow. [In Russian]

Boellstorff, D. E., and D. H. Owings. 1995. Home range, population stucture, and spatial organization of California ground squirrels. *J. Mammal.* 76:551–561.

Bolkovic, M. L., S. M. Caziani, and J. J. Protomastro. 1995. Food habits of the three-banded armadillo (Xenarthra: Dasypodidae) in the dry Chaco, Argentina. *J. Mammal.* 76:1119–1204.

Bonaccorso, F. J., N. Smythe, and S. R. Humphrey. 1976. Improved techniques for marking bats. *J. Mammal.* 57:181–182.

Boonstra, R., and F. H. Rodd. 1982. Another potential bias in the use of the Longworth trap. *J. Mammal.* 63:672–675.

Booth, E. S. 1982. *How to know the mammals,* 4th ed. Wm. C. Brown Company Publishers, Dubuque, Iowa.

Boulanger, J. G., and G. C. White. 1990. A comparison of home-range estimators using Monte Carlo simulation. *J. Wildl. Manag.* 54:310–315.

Bram, R. A. (Compiler). 1978. Surveillance and collection of arthropods of veterinary importance. *USDA Agric. Handbook* 518:1–125.

Brander, R. B., and W. W. Cochran. 1969. Radio-location telemetry, pp. 95–108. *In* R. H. Giles, Jr. (Ed.). *Wildlife management techniques,* 3rd ed. The Wildlife Society, Washington, D.C.

Braun, S. E. 1985. Home range and activity patterns of the giant kangaroo rat, *Dipodomys ingens. J. Mammal.* 66:1–12.

Brazenor, C. W. 1950. *The mammals of Victoria.* National Museum of Victoria, Melbourne.

Brower, J., J. Zar. 1974. *Field and laboratory methods for general ecology.* Wm. C. Brown Company, Dubuque, Iowa.

Brower, J., J. Zar, and C. von Ende. 1998. *Field and laboratory methods for general ecology,* 4th ed. McGraw-Hill Publishers, Dubuque, Iowa.

Brown, J. C., and D. M. Stoddart. 1977. Killing mammals and general postmortem methods. *Mammal Review* 7:63–94.

Brown, J. H., G. A. Lieberman, and W. F. Dengler. 1972. Woodrats and cholla: dependence of a small mammal community on the density of cacti. *Ecology* 53:310–313.

Brown, J. L., and G. H. Orians. 1970. Spacing patterns in mobile animals. *Ann. Rev. Ecol. Syst.* 1:239–262.

Brown, L. E. 1966. Home range and movement of small mammals. *Symp. Zool. Soc. London* 18:111–142.

Bruner, H., and B. J. Coman. 1974. *The identification of mammalian hair.* Inkata Press, Melbourne, Australia.

Buchler, E. R. 1976. A chemiluminescent tag for tracking bats and other small nocturnal animals. *J. Mammal.* 57:173–176.

Buechner, H. K. 1961. Territorial behavior in Uganda kob. *Science* 133:698–699.

Burt, W. H. 1927. A simple live trap for small mammals. *J. Mammal.* 8:302–304.

Burt, W. H. 1943. Territoriality and home range concepts as applied to mammals. *J. Mammal.* 24:346–352.

Burt, W. H., and R. P. Grossenheider. 1964. *A field guide to the mammals,* 2nd ed. Houghton Mifflin Company, Boston.

Butler, P. M. 1941. A theory of the evolution of mammalian molar teeth. *Amer. J. Sci.* 239:421–450.

Butler, S. R., and E. A. Rowe. 1976. A data acquisition and retrieval system for studies of animal social behaviour. *Behaviour* 57:281–287.

Cabrera, A. 1914. *Fauna iberica. Mamíferos.* Museo Nacional de Ciencias Naturales, Madrid.

Cabrera, A. 1919. *Genera mammalium: Monotremata Marsupialia.* Museo Nacional de Ciencias Naturales, Madrid.

Cabrera, A. 1925. *Genera mammalium: Insectivora Galeopithecia.* Museo Nacional de Ciencias Naturales, Madrid.

Cabrera, A. 1932. Los mamiferos de Marruecos. *Trabajos del Museo Nacional de Ciencias Naturales (ser. Zool.)* 57:1–361.

Cahalane, V. H. 1932. Age variation in the teeth and skull of the white-tail deer. *Cranbrook Inst. Sci., Scient. Publ.* 2:1–14.

Calaprice, J. R., and J. S. Ford. 1969. Digital calipers—an inexpensive electronic measuring and recording device. *Fisheries Res. Board Canada, Tech. Rept.* 141:1–6.

Calhoun, J. B., and J. U. Casby. 1958. Calculation of home range and density of small mammals. USDHEW, *Pub. Health Monog.* 55:1–24.

Cameron, G. N., and D. G. Rainey. 1972. Habitat utilization by *Neotoma lepida* in the Mohave Desert. *J. Mammal.* 53:251–266.

Campbell, C. B. G. 1974. On the phyletic relationships of the tree shrews. *Mammal Review* 4:125–143.

Campbell, J. 1998. *Map use and analysis,* 3rd ed. WCB/McGraw-Hill, Dubuque, Iowa.

Carey, A. B., and J. W. Witt. 1991. Track counts as indices to abundance of arboreal rodents. *J. Mammal.* 72:192–194.

Carleton, M. D. 1984. Introduction to rodents, pp. 255–265. *In* S. Anderson and J. K. Jones, Jr. *Orders and families of Recent mammals of the world.* John Wiley & Sons, New York.

Carleton, M. D, and G. G. Musser. 1984. Muroid rodents, pp. 289–379. *In* S. Anderson and J. K. Jones, Jr. *Orders and families of recent mammals of the world.* John Wiley & Sons, New York.

Carroll, R. L. 1988. *Vertebrate paleontology and evolution.* W. H. Freeman, New York.

Case, L. D. 1959. Preparing mummified specimens for cleaning by dermestid beetles. *J. Mammal.* 40:620.

Caughley, G. 1965. Horn rings and tooth eruption as criteria of age in the Himalayan thar, *Hemitragus jemlahicus. New Zealand J. Sci.* 8:333–351.

Caughley, G. 1966. Mortality patterns in mammals. *Ecology* 47:906–918.

Chiasson, R. B. 1969. *Laboratory anatomy of the white rat,* 3rd ed. Wm. C. Brown Company Publishers, Dubuque, Iowa.

Choate, J. R. 1975. Review of: A. F. DeBlase and R. E. Martin. 1974. A manual or mammalogy: with keys to families of the world. *J. Mammal.* 56:281–283.

Christensen, I. 1973. Age determination, age distribution and growth of bottlenose whales, *Hyperoodon ampullatus* (Forster), in the Labrador Sea. *Norwegian J. Zool.* 21:331–340.

Chu, H. F. 1949. *How to know the immature insects.* Wm. C. Brown Company Publishers, Dubuque, Iowa.

Chudoba, S., and S. Huminski. 1980. Estimating numbers of rodents and edge effect using a modified version of the Standard Minimum Method. *Acta Theriologica* 25:365–376.

Coffey, D. J. 1977. *Dolphins, whales and porpoises: an encyclopedia of sea mammals.* Macmillian Publishing Co., Inc., New York.

Cole, L. C. 1954. The population consequences of life history phenomena. *Quart. Rev. Biol.* 29:103–137.

Colwell, R. K., and E. R. Fuentes. 1975. Experimental studies of the niche. *Ann. Rev. Ecol. Syst.* 6:281–310.

Constantine, D. G. 1958. An automatic bat-collecting device. *J. Wildl. Manag.* 22:17–22.

Constantine, D. G. 1969. Trampa portatil para vampiros usada en programas de campana antirrabica. *Bol. Of. Sanit. Panamer.* 67:39–42.

Cormack, R. M. 1968. The statistics of capture-recapture methods. *Oceanography Marine Bio.* 6:455–506.

Couch, L. K. 1942. Trapping and transplanting live beavers. *U.S.D.I. Fish and Wildlife Serv. Cons. Bull.* 30:1–20.

Council of Biology Editors, Committee on Form and Style. 1972. *CBE style manual,* 3rd ed. American Inst. Biol. Sciences, Washington, D.C.

Cox, G. W. 1996. *Laboratory manual of general ecology,* 7th ed. Wm. C. Brown Company Publishers, Dubuque, Iowa.

Cox, M. K., and W. L. Franklin. 1990. Premolar gap technique for aging black-tailed prairie dogs. *J. Wildl. Manag.* 54:143–146.

Crandall, L. S. 1964. *The management of wild mammals in captivity.* Univ. Chicago Press, Chicago.

Crawford, R. L. 1983. Grid systems for recording specimen collection localities in North America. *Syst. Zool.* 32:389–402.

Crouch, W. E. 1933. Pocket-gopher control. *U.S.D.A. Farmers' Bull.* 1709:1–20.

Dapson, R. W. 1973. Letter to the editor. *J. Mammal.* 54:804.

Dapson, R. W., and J. M. Irland. 1972. An accurate method of determining age in small mammals. *J. Mammal.* 53:100–106.

Davis, D. D. 1964. The giant panda; a morphological study of evolutionary mechanisms. *Fieldiana: Zool. Mem.* 3:1–339.

Davis, W. B. 1974. *The mammals of Texas,* 2nd ed. Texas Parks and Wildlife Dept., Austin.

Day, M. G. 1966. Identification of hair and feather remains in the gut and faeces of stoats and weasels. *J. Zool. London* 148:201–217.

Deevey, E. S. 1947. Life tables for natural populations of animals. *Quart. Rev. Biol.* 22;283–314.

Dekeyser, P. L. 1955. *Les mammiferes de l'Afrique Noire Francaise,* 2nd ed. Initiations Africaines, Institut Francais D'Afrique Noire.

Delany, M. J. 1974. The ecology of small mammals. *Inst. Biol. Studies Biology* 51:1–60.

de la Torre, L. 1951. A method for cleaning skulls of specimens preserved in alcohol. *J. Mammal.* 32:231–232.

Dickman, C. R., and C. Huang. 1988. The reliability of fecal analysis as a method for determining the diet of insectivorous mammals. *J. Mammal.* 69:108–113.

Dieterlen, F. 1993. Rodentia (Anomaluridae), pp. 757–758; Rodentia (Pedetidae & Ctenodactylidae), pp. 759–761. *In* D. E. Wilson and D. M. Reeder (Eds.). *Mammal species of the world: a taxonomic and geographic reference.* Smithsonian Institution Press, Washington, D.C.

Dimmick, R. W., and M. R. Pelton. 1996. Criteria of sex and age, pp. 169–214. *In* T.A. Bookhout (Ed.). *Research and management techniques for wildlife and habitats,* 5th ed., rev. The Wildlife Society, Bethesda, Maryland.

Dixon, K. R., and J. A. Chapman. 1980. Harmonic mean measure of animal activity areas. *Ecology* 61:1040–1044.

Dobson, G. E. 1876. On the peculiar structures in the feet of certain species of mammals which enable them to walk on smooth perpendicular surfaces. *Proc. Zool. Soc. London* 1876:526–535.

Dobson, G. E. 1878. *Catalogue of the Chiroptera in the collection of the British Museum.* British Museum (Nat. Hist.), London.

Dorst, J., and P. Dandelot. 1970. *A field guide to the larger mammals of Africa.* Houghton Mifflin Company, Boston.

Doty, R. L. (Ed.). 1976. *Mammalian olfaction, reproductive processes, and behavior.* Academic Press, New York.

Douglas-Hamilton, I., and O. Douglas-Hamilton. 1975. Among the elephants. Collins & Harvill Press, London.

Doutt, J. K. 1967. Polar bear dens on the Twin Islands, James Bay, Canada. *J. Mammal.* 48:468–471.

Dowler, R. C., and H. H. Genoways. 1976. Supplies and suppliers for vertebrate collections. *Museology,* Texas Tech University 4:1–83.

Duncan, P. M. 1877–83. *Cassell's natural history,* 3 vols. Casell & Co., London.

Dusi, J. L. 1949. Methods for the determination of food habits by plant microtechnics and histology and their application to cottontail rabbit food habits. *J. Wildl. Manag.* 13:295–298.

Eberhardt, L. L. 1969. Population analysis, pp. 457–495. *In* R. H. Giles, Jr. (Ed.). *Wildlife management techniques,* 3rd ed. The Wildlife Society, Washington, D.C.

Egoscue, H. J. 1962. The bushy-tailed wood rat: a laboratory colony. *J. Mammal.* 43:328–337.

Eisenberg, J. F. 1966. The social organizations of mammals. *Handbuch der Zoologie* 10(7):1–92.

Eisenberg, J. F., and E. Gould. 1970. The tenrecs: a study in mammalian behavior and evolution. *Smithsonian Contrib. Zool.* 27:1–137.

Eisenberg, J. F., and D. G. Kleiman. 1972. Olfactory communication in mammals. *Ann. Rev. Ecol. Syst.* 3:1–32.

Elder, W. H. 1951. The baculum as an age criterion in mink. *J. Mammal.* 32:43–50.

Elder, W. H., and C. E. Shanks. 1962. Age changes in tooth wear and morphology of the baculum in muskrats. *J. Mammal.* 43:144–150.

Ellis, B. A., J. N. Mills, G. E. Glass, K. T. McKee, Jr., D. A. Enria, and J. E. Childs. 1998. Dietary habits of the common rodents in an agrosystem in Argentina. *J. Mammal.* 79:1203–1220.

Emry, R. J. 1970. A North American Oligocene pangolin and other additions to the Pholidota. *Bull. Amer. Mus. Nat. Hist.* 142:459–510.

Erickson, J. A., and W. G. Seliger. 1969. Efficient section of incisors for estimating ages of mule deer. *J. Wildl. Manag.* 33:384–388.

Ewer, R. F. 1968. *Ethology of mammals.* Plenum Press, New York.

Ewer, R. F. 1973. *The carnivores.* Cornell Univ. Press, Ithaca, New York.

Flagerstone, K. A., and B. E. Johns. 1987. Transponders as permanent identification markers for domestic ferrets, black footed ferrets, and other wildlife. *J. Wildl. Manag.* 51:294–297.

F & W Publications.

Feldhamer, G. A., L. C. Drickamer, S. H. Vessey, and J. F. Merritt. 1999. *Mammalogy: adaptation, diversity, and ecology.* WCB/McGraw-Hill, Dubuque, Iowa.

Fielder, W. 1975. Guenons and their relatives, pp. 396–441. In B. Grzimek (Ed.). *Animal life encyclopedia, Vol. 10, Mammals I.* Van Nostrand Reinhold, New York.

Fiero, B. C., and B. J. Verts. 1986. Comparison of techniques for estimating age in raccoons. *J. Mammal.* 67:392–395.

Finn, F, 1929. *Sterndale's mammals of India.* Thacker, Spink and Co., Bombay.

Firchow, K. M., M. R. Vaughan, and W. R. Mytton. 1990. Comparison of aerial survey techniques for pronghorns. *Wildl. Soc. Bull.* 18:18–23.

Fisler, F. F. 1969. Mammalian organizational systems. *Los Angeles County Museum Contrib. Sci.* 167:1–31.

Fitch, H. S. 1950. A new style live-trap for small mammals. *J. Mammal.* 31:364–365.

Flower, W. H. 1885. *An introduction to the osteology of the Mammalia.* Macmillian and Co., London.

Flower, W. H., and R. Lydekker. 1891. *An introduction to the study of mammals living and extinct.* Adam and Charles Black, London.

Flowerdew, J. R. 1976. Ecological methods. *Mammal Review* 6:123–159.

Foreyt, W. J., and W. C. Glazner. 1979. A modified trap for capturing feral hogs and white-tailed deer. *Southwestern Natur.* 24:377–380.

Fracker, S. B., and H. A. Brischle. 1944. Measuring the local distribution of *Ribes. Ecology* 25:283–303.

Fradrich, H. 1972. Tapirs, pp. 17–33 In B. Grzimek (Ed.). *Animal life encyclopedia, Vol. 13, Mammals IV.* Van Nostrand Reinhold, New York.

Francis, C. M. 1989. A comparioson of mist nets and two designs of harp traps for capturing bats. *J. Mammal.* 70:865–870.

Franson, J. C., P. A. Dahm, and L. D. Wing. 1975. A method for preparing and sectioning mink (*Mustela vison*) mandibles for age determination. *Amer. Midl. Natur.* 93:507–508.

Franz, C. E., O. J. Reichman, and K. M. Van DeGraff. 1973. Diets, food preferences and reproductive cycles of some desert rodents. *US/IBP/Desert Biome Research Memorandum RM* 73-24:1–128.

Free, J. C., and R. M. Hansen, and P. L. Sims. 1970. Estimating dryweights of foodplants in feces of herbivores. *J. Range Manage.* 23:300–302.

French, N. R., Stoddart, D. M., and B. Bobek. 1975. Patterns of demography in small mammal populations, pp. 73–102. In F. B. Golley, K. Petrusewicz, and L. Ryszkowski. (Eds.). *Small mammals: their productivity and population dynamics.* I.B.P. Handbook No. 5, Cambridge Univ. Press, Cambridge.

Friend, M. 1968. The lens techniques. *Trans. 33rd. N. Amer. Wild. Res. Conf.* 33:279–298.

Friend, M., D. E. Toweill, R. L. Brownell, D. S. Davis, and W. J. Forey. 1996. Guidelines for proper care and use of wildlife in field research, pp. 96–105. In T. A. Bookhout (Ed.). *Research and management techniques for wildlife and habitats,* 5th ed., rev. The Wildlife Society, Bethesda, Maryland.

Friley, C. E., Jr. 1949. Use of the baculum in age determination of Michigan beavers. *J. Mammal.* 30:261–266.

Furman, D. P., and E. P. Catts. 1970. *Manual of medical entomology.* National Press Books, California.

Furrer, R. K. 1973. Homing of *Peromyscus maniculatus* in the channeled scablands of east-central Washington. *J. Mammal.* 54:466–482.

Gandal, C. P. 1954. Age determination in mammals. *New York Acad. Sci. Trans.* (Ser. 2) 16:312–314.

Gardner, A. L. 1993. Didelphimorphia, Paucituberculata, and Microbiotheria, pp. 15–72; Xenarthra, pp. 63–68. In D. E. Wilson and D. M. Reeder (Eds.). 1993. *Mammal species of the world: a taxonomic and geographic reference.* Smithsonian Institution Press, Washington, D.C.

Garlough, F. E., J. F. Welch, and H. J. Spencer. 1942. Rabbits in relation to crops. *U.S.D.I. Fish and Wildl Serv., Conserv. Bull.* 11.

Gebcynska, Z., and A. Myrcha. 1966. The method of quantitative determining of the food composition of rodents. *Acta Theriologica* 11:385–390.

Geist, V. 1966. Validity of horn segment counts in aging bighorn sheep. *J. Wildl. Manag.* 30:634–635.

Genoways, H. H., and J. R. Choate. 1976. Federal regulations pertaining to collection, import, export, and transport of scientific specimens of mammals. *J. Mammal.* 57(2) Supplement:1–9.

Gervais, M. P. 1855. *Histoire naturelle des mammiferes.* L. Curmer, Paris.

Getz, L. L., and G. O. Batzli. 1974. A device for preventing disturbance of small mammal live-traps. *J. Mammal.* 55:447–448.

Getz, L. L., and M. L. Prather. 1975. A method to prevent removal of trap bait by insects. *J. Mammal.* 56:955.

Giebel, C. G. 1859. *Die Naturgeschichte des Thierreichs. Book 1: Die Saugethiere.* Verlag von Otto Wigand, Leipzig.

Giebel, C. G., and W. Leche. 1874–1900. *Dr. H. G. Bronn's Klassen und Ordnungen des Thier-Reichs wissenchaftlich dargestellt in Wort und Bild. Sechster Band. V. Abtheilung.*

Säugethiere: Mammalia. C. F. Wintersche Verlagshandlung, Leipzig.

Gilmore, R. M. 1943. Mammalogy in an epidemological study of jungle yellow fever in Brazil. *J. Mammal.* 24:144–162.

Giuliani, B. 1993. *Animal illustrations.* Dover Publications, Inc., New York.

Giuliani, B. 1995. *Illustrations of Marine Mammals.* Dover Publications, Inc., New York.

Goodwin, G. G., and A. M. Greenhall. 1961. A review of the bats of Trinidad and Tobago: descriptions, rabies infection, and ecology. *Bull. Amer. Mus. Nat. Hist.* 122:189–301.

Goss, R. J. 1983. *Deer antlers—regeneration, function, and evolution.* Academic Press, New York.

Grant, P. R. 1970. A potential bias in the use of Longworth traps. *J. Mammal.* 51:831–835.

Grant, P. R. 1972. Interspecific competition among rodents. *Ann. Rev. Ecol. Syst.* 3:79–106.

Gray, J. E. 1865. Notices of some apparently undescribed species of sapajous (*Cebus*) in the collection of the British Museum. *Proc. Zool. Soc. London* 1865:824–828.

Gray, J. E. 1869. *Catalogue of carnivorous, pachydermatous, and edentate Mammalia in the British Museum.* British Museum (Nat. Hist.), London.

Gray. J. E. 1873. *Hand-list of the edentate, thick-skinned and ruminant mammals in the British Museum.* British Museum (Nat. Hist.), London.

Green, E. L., L. H. Blankenship, V. F. Cougar, and T. McMahon. 1985. *Wildlife food plants: a microscopic view.* Kleberg Studies in Natural Resources, Texas A&M University.

Green, G. A., G. W. Witmer, and D. S. deCalesta. 1986. NaOH preparation of mammalian predator scats for dietary analysis. *J. Mammal.* 67:742.

Green, H. L. H. H. 1934. A rapid method of preparing clean bone specimens from fresh or fixed material. *Anat. Rec.* 61:1–3.

Greenhall, A. M. 1976. Care in captivity, pp. 89–131. *In* R. J. Baker, R. J., J. K. Jones, and D. C. Carter, (Eds.). Biology of bats of the New World family Phyllostomatidae, Part 1. *Texas Tech Univ. Spec. Publ. Museum* 10:1–218.

Greenhall, A. M., and J. L. Paradiso. 1968. Bats and bat banding. *U. S. Dept. Interior, Bureau Sport Fisheries and Wildlife, Resource Publ.* 72:1–47.

Gregory, W. K. 1910. The orders of mammals. *Bull. Amer. Mus. Nat. Hist.* 27:1–524.

Griffin, D. R. 1970. Migrations and homing of bats, pp. 233–264. *In* W. A. Wimsatt (Ed.). *Biology of bats,* Vol. 1. Academic Press, New York.

Griffiths, M. 1978. *The biology of the monotremes.* Academic Press, New York.

Grodzinski, W., Z. Pucek, and L. Ryskowski. 1966. Estimation of rodent number by means of prebaiting and intensive removal. *Acta Theriologica* 11:297–314.

Gromov, I. M., A. A. Guryev, G. A. Novikov, I. I. Sokolov, P. P. Strelkov, and K. K. Chapskiy. 1963. [*Mammalian fauna of the USSR,* 2 vols.] USSR Acad. Sci., Moscow. [In Russian]

Gromova, V. I. (Ed.). 1962. *Fundamentals of paleontology: a manual for paleontologists and geologists of the USSR, Vol. XIII, Mammals.* Trans. from Russian, 1968, Israel Prog. Scient. Transl.

Groves, C. P. 1993. Monotremata, p. 13; Dasyuromorphia, Peramelemorphia, Notoryctemorphia, and Diprotodontia, pp. 29–62; Primates, pp. 243–277. *In* D. E. Wilson and D. M. Reeder (Eds.). *Mammal species of the world: a taxonomic and geographic reference.* Smithsonian Institution Press, Washington, D.C.

Grubb, P. 1993. Perissodactyla, pp. 369–372; Artiodactyla, pp. 377–414. *In* D. E. Wilson and D. M. Reeder (Eds.). *Mammal species of the world: a taxonomic and geographic reference.* Smithsonian Institution Press, Washington, D.C.

Grzimek, B. 1975. African elephants, pp. 500–512. *In* B. Grzimek (Ed.). *Animal life encyclopedia, Vol. 12, Mammals III.* Van Nostrand Reinhold, New York.

Guggisberg, C. A. W. 1960. *Simba.* Hallwag, Berlin. (English translation, 1963, Chilton Books).

Gurnell, J., and J. H. W. Gipps. 1989. Inter-trap movement and estimating rodent densities. *J. Zool., London* 217:241–254.

Guryev, A. A. 1964. [*Fauna USSR, Vol. 3, No. 10. Lagomorpha.*] USSR Acad. Sci., Moscow. [In Russian]

Guyer, M. F. 1953. *Animal micrology,* 5th ed. Univ. Chicago Press, Chicago.

Haffner, M., and V. Ziswiler. 1989. Tasthaare als diagnostisches Merkmal bei mitteleuropäischen Vespertilionidae (Mammalia, Chiroptera). *Revue Suisse Zoology* 96:663–672.

Hafner, D. J., J. C. Hafner, and M. S. Hafner. 1984. Skin-plus-skeleton preparation as the standard mammalian museum specimen. *Curator* 27:141–145.

Hall, A. V. 1970. A computer-based system for forming identification keys. *Taxon* 19:12–18.

Hall, E. R. 1962. Collecting and preparing study specimens of vertebrates. *Univ. Kansas Mus. Natur. Hist., Misc. Publ.* 30:1–46.

Hall, E. R., and W. C. Russell. 1933. Dermestid beetles as an aid in cleaning bones. *J. Mammal.* 14:372–374.

Hamaker, C., and J. W. Koeppl. 1984. Estimation of the latitude and longitude coordinates of points on maps. *Univ. Kansas Mus. Natur. Hist., Misc. Publ.* 108:1–9

Hansen, R. M., and J. T. Flinders. 1969. Food habits of North American hares. *Colorado State Univ. Sci. Series, Range Sci. Dept.* 1:1–18.

Hansen, R. M., and D. N. Ueckert. 1970. Dietary similarity of some primary consumers. *Ecology* 51:640–648.

Hanson, W. R. 1963. Calculation of productivity, survival, and abundance of selected vertebrates from sex and age ratios. *Wildlife Monog.* 9:1–60.

Hanson, W. R., and F. Graybill. 1956. Sample size in food-habits analysis. *J. Wildl. Manag.* 20:64–68.

Hansson, L. 1969. Home range, population structure and density estimates at removal catches with edge effect. *Acta Theriologica* 14:153–160.

Hansson, L. 1970. Methods of morphological diet micro-analysis in rodents. *Oikos* 21:255–266.

Haresign, T. 1960. A technique for increasing the time of dye retention in small mammals. *J. Mammal.* 41:528.

Harestad, A. S., and F. L. Bunnel. 1979. Home range and body weight—a reevaluation. *Ecology* 60:389–402.

Harper, F. 1955. The barren ground caribou of Keewatin. *Univ. Kansas Mus. Natur. Hist., Misc. Publ.* 6:1–164.

Harris, R. H. 1959. Small vertebrate skeletons. *Museums Journal* 58:223–224.

Harris, R. H. 1978. Age determination in the red fox (*Vulpes vulpes*)—an evalutaion of technique efficiency as applied to a sample of suburban foxes. *J. Zool., London* 184:91–117.

Harris, S., W. J. Cresswell, P. G. Forde, W. J. Trewhella, T. Woollard, and S. Wray. 1990. Home-range analysis using radio-tracking data—a review of problems and techniques particularly as applied to the study of mammals. *Mammal Review* 20:97–123.

Harris, V. T. 1952. An experimental study of habitat selection by prairie and forest races of the deermouse, *Peromyscus maniculatus. Univ. Mich. Contrib. Lab. Vert. Biol.* 56:1–53.

Harrison, D. L. 1964. *The mammals of Arabia, Vol. 1, Insectivora, Chiroptera, Primates.* Ernest Benn, London.

Harthoorn, A. M. 1965. Application of pharmacological and physiological principles in restraint of wild animals *Wildlife Monog.* 14:1–78.

Harthoorn, A. M. 1975. *The chemical capture of animals.* Bailliere-Tindall, London.

Hatt, R. T. 1934. The pangolins and aardvarks collected by the American Museum Congo Expedition. *Bull. Amer. Mus. Nat. Hist.* 66:643–672 + 8 pls.

Hatt, R. T. 1946. Guide to the hall of biology of mammals. *Amer. Mus. of Nat. Hist., Sci. Guide* 76:1–49.

Hayne, D. W. 1949a. Calculation of size of home range. *J. Mammal.* 30:1–18.

Hayne, D. W. 1949b. Two methods for estimating populations from trapping records. *J. Mammal.* 30:399–411.

Hayne, D. W. 1950. Apparent home range of *Microtus* in relation to distance between traps. *J. Mammal.* 31:26–39.

Hayne, D. W. 1978. Experimental design and statistical analyses, pp. 3–13. *In* D. P. Snyder (Ed.) *Populations of small mammals under natural conditions.* Univ. Pittsburgh Pymatuning Lab. Ecol. Spec. Publ. Ser., Vol. 5.

Heidt, G. A., R. H. Baker, and I. O. Ebert. 1967. Magnetic detection of small mammal activity. *J. Mammal.* 48:330–331.

Helfer, J. R. 1953. *How to know the grasshoppers, cockroaches and their allies.* Wm. C. Brown Company Publishers, Dubuque, Iowa.

Henche, J. E., G. L. Kirkland, Jr., H. W. Setzer, and L. W. Douglass. 1984. Age classification for the gray squirrel based on eruption, replacement, and wear of molariform teeth. *J. Wildl. Manag.* 48:1409–1414.

Henderson, F. R. 1960. Beaver in Kansas. *Univ. Kansas Mus. Nat. Hist., Misc. Publ.* 26:1–85.

Henshaw, R. E., and R. O. Stephenson. 1974. Homing in the gray wolf (*Canis lupus*). *J. Mammal.* 55:234–237.

Henson, O. W., Jr. 1970. The central nervous system, pp. 58–152. *In* W. A. Wimsatt (Ed.). *Biology of bats,* Vol. 2. Academic Press, New York.

Hershkovitz, P. 1954. *Collecting and preserving mammals for study: a provisional account for museum personnel and field associates.* Chicago Natural History Museum.

Hershkovitz, P. 1962. Evolution of Neotropical cricetine rodents (Muridae) with special reference to the phyllotine group. *Fieldiana: Zoology* 46:1–524.

Hershkovitz, P. 1971. Basic crown patterns and cusp homologies of mammalian teeth, pp. 95–150. *In* A. A. Dahlberg (Ed.). *Dental morphology and evolution.* Univ. Chicago Press, Chicago.

Hershkovitz, P. 1977. *Living New World monkeys (Platyrrhini) with an introduction to the Primates,* Vol. 1. Univ. Chicago Press, Chicago.

Hess, W. M., J. T. Flinders, C. L. Pritchett, and J. V. Allen. 1985. Characterization of hair morphology in families Tayassuidae and Suidae with scanning electron microscopy. *J. Mammal.* 66:75–84.

Hickey, M. B., L. Acharya, and S. Pennington. 1996. Resource partitioning by two species of vespertilionid bats (*Lasiurus cinerus* and *Lasiurus borealis*) feeding around street lights. *J. Mammal.* 77:325–334.

Hickman, C., Hickman, F., and L. Kats. 1997. *Laboratory Studies in Integraged Principles of Zoology.* McGraw-Hill Publishers, Dubuque, Iowa.

Hildebrand, M. 1952. The integument of Canidae. *J. Mammal.* 33:419–428.

Hildebrand, M. 1968. *Anatomical preparations.* Univ. Calif. Press, Berkeley.

Hill, J. E. 1974. A new family, genus and species of bat (Mammalia: Chiroptera) from Thailand. *Bull. British Museum (Nat. Hist.)* 27:301–336.

Hinton, H. E. 1945. A monograph of the beetles associated with stored products. *Bull. British Mus. (Nat. Hist.)*

Hoffman, R. S. 1993. Lagomorpha, pp. 807–827. *In* D. E. Wilson and D. M. Reeder (Eds.). *Mammal species of the world: a taxonomic and geographic reference.* Smithsonian Institution Press, Washington, D.C.

Hoffmann, R. S., C. G. Anderson, R. W. Thompson, Jr., and L. R. Heaney. 1993. Sciuridae, pp. 419–465. *In* D. E. Wilson and D. M. Reeder (Eds.). *Mammal species of the world: a taxonomic and geographic reference.* Smithsonian Institution Press, Washington, D.C.

Hoffmeister, D. F., and E. G. Zimmerman. 1967. Growth of the skull in the cottontail. (*Sylvilagus floridanus*) and its application to age-determination. *Amer. Midl. Natur.* 78:198–206.

Hoffmeister, D. F., and M. R. Lee. 1963. Cleaning mammalian skulls with ammonium hydroxide. *J. Mammal.* 44:283–284.

Holden, M. E. 1993. Rodentia (Dipodidae), pp. 487–499; Rodentia (Myoxidae), pp. 763–770. *In* D. E. Wilson and D. M. Reeder (Eds.). *Mammal species of the world: a taxonomic and geographic reference.* Smithsonian Institution Press, Washington, D.C.

Holechek, J. L., B. Gross, S. M. Dabo, and T. Stephenson. 1982. Effects of sample preparation, growth stage, and observer on microhistological analysis of herbivore diets. *J. Wildl. Manag.* 46:502–505.

Hoogland, J. L., and J. M. Hutter. 1987. Using molar attrition to age live prairie dogs. *J. Wildl. Manag.* 51:393–394.

Hooper, E. T. 1950. Use dermestid beetles instead of cooking pots. *J. Mammal.* 31:100.

Hooper, E. T. 1956. Selection of fats by dermestid beetles. *J. Mammal.* 37:125–126.

Howard, W. E. 1952. A live trap for pocket gophers. *J. Mammal.* 33:61–65.

Howard, W. E. 1953. A trigger mechanism for small mammal live traps. *J. Mammal.* 34:513–514.

Hrdy, S. B. 1977. *The langurs of Abu: female and male strategies of reproduction.* Harvard Univ. Press, Cambridge.

Hsia Wu-ping, et al. 1964. [*Illustrated description of animals in China—Mammals.*] Peking. [In Chinese]

Humason, G. L. 1967. Animal tissue techniques, 2nd ed. W. H. Freeman & Co., San Francisco.

Humphrey, S. R. 1971. Photographic estimation of population size in the Mexican free-tailed bat (*Tadarida brasiliensis*). *Amer. Midl. Natur.* 86:220–223.

Hutterer, R. 1993. Insectivora, pp. 69–130. *In* D. E. Wilson and D. M. Reeder (Eds.). *Mammal species of the world: a taxonomic and geographic reference.* Smithsonian Institution Press, Washington, D.C.

Inglis, J. M., and C. J. Barstow. 1960. A device for measuring the volume of seeds. *J. Wildl. Manag.* 24:221–222.

Jackson, H. H. T. 1915. A review of the American moles. *N. American Fauna* 38:1–100 + 6 pls.

Jaeger, E. C. 1955. *A source book of biological names and terms,* 3rd ed. Chas. C. Thomas, Springfield, Illinois.

Jennrich, R. I., and F. B. Turner. 1969. Measurement of non-circular home range. *J. Theoret. Biol.* 22:227–237.

Jewell, P. A. 1966. The concept of home range in mammals. *Symp. Zool. Soc. London* 18:85–109.

Jobling, B. 1939. On some American genera of the Streblidae and their species, with the description of a new species of *Trichobius. Parasitology* 31:486–497.

Johnson, D. H. 1996. Population analysis, pp. 419–444. *In* T. A. Bookhout (Ed.). *Research and management techniques for wildlife and habitats,* 5th ed., rev. The Wildlife Society, Bethesda, Maryland.

Johnson, M. K., H. Wofford, and H. A. Pearson. 1983. Microhistological techniques for food habits analysis. *USDA, Forest Service Research Paper* SO199:1–40.

Jolly, G. M. 1965. Explicit estimates from capture-recapture data with both death and immigration—stochastic model. *Biometrika* 52:225–247.

Jones, C., W. J. McShea, M. J. Conroy, and T. H. Kunz. 1996. Capturing mammals, pp. 115–122 + 146–155. *In* D. E. Wilson, F. R. Cole, J. D. Nichols, R. Rudran, and M. S. Foster (Eds.). *Measuring and monitoring biological diversity. Standard methods for mammals.* Smithsonian Institution Press, Washington, D.C.

Jones, J. K., Jr., and S. Anderson. 1970. Readings in mammalogy. *Univ. Kansas Monog. Mus. Natur. Hist.,* 2:1–586.

Jones, J. K., Jr., S. Anderson, and R. S. Hoffman. 1976. Selected readings in mammalogy. *Univ. Kansas Monog. Mus. Nat. Hist.* 5:1–640.

Jones, J. K., Jr., and R. R. Johnson. 1967. Sirenians, pp. 366–373. *In* S. Anderson and J. K. Jones, Jr. (Eds.). *Recent mammals of the world: a synopsis of families.* Ronald Press Co., New York.

Jonsgård, Å. 1969. Age determination of marine mammals, pp. 1–30. *In* H. T. Andersen (Ed.). *The biology of marine mammals.* Academic Press, London.

Jorgensen, C.D. 1968. Home range as a measure of probable interactions among populations of small mammals. *J. Mammal.* 49:104–112.

Justice, K.E. 1961. A new method for measuring home range of small mammals. *J.Mammal.* 55:309–318.

Kardong, K. V. 1998. *Vertebrates: comparative anatomy, function, evolution,* 2nd. ed. WCB/McGraw-Hill, Dubuque, Iowa.

Kaufman, D. W., G. C. Smith, R. M. Jones, J. B. Gentry, and M. H. Smith. 1971. Use of assessment lines to estimate density of small mammals. *Acta Theriologica* 16:127–147.

Kaufmann, J. H. 1962. The ecology and social behavior of the coati, *Nasua narica,* on Barro Colorado Island, Panama. *Univ. Calif. Publ. Zoo.* 60:95–222.

Kawamichi, T. 1976. Hay territory and dominance rank of pikas (*Ochotona princeps*). *J. Mammal.* 57:133–148.

Keith, J. O., R. M. Ansen, and A. L. Ward. 1959. Effect of 2,4D on abundance and foods of pocket gophers. *J.Wildl. Manag.* 23:137–145.

Kent, G. C., and L. Miller. 1997. *Comparative anatomy of the vertebrates* 8th ed. Wm. C. Brown Company Publishers, Dubuque, Iowa.

Kie, J. G., J. A. Baldwin, and C. J. Evans. 1996. CALHOME: a program for estimating animal home ranges. *Wildl. Soc. Bull.* 24:342–344.

King, J. A. 1955. Social behavior, social organization, and population dynamics in a black-tailed prairiedog town in the Black Hills of South Dakota. *Univ. Mich. Contrib. Lab.Vert. Biol.* 67:1–123 + 4 pls.

Kingsley, J. S. (Ed.) 1884. *The riverside natural history, Vol. V Mammals.* Houghton Mifflin, Boston.

Kirkpatrick, C. M., and R. A. Hoffman. 1960. Ages and reproductive cycles in a male gray squirrel population. *J. Wild. Manag.* 24:218–221.

Klevezal, G. A., and S. E. Kleinenberg. 1967. *Age determination of mammals from annual layers in teeth and bones.* USSR Acad. Sci. Moscow. Trans. From Russian, 1969, Israel Prog. Scient. Transl.

Klingener, D. 1984. Gliroid and dipodoid rodents, pp. 381–388. *In* S. Anderson and J. K. Jones, Jr. *Orders and families of Recent mammals of the world.* John Wiley & Sons, New York.

Knowlton, F .F., E. D. Michael, and W. C. Glazener. 1964. A marking technique for field recognition of individual turkeys and deer. *J. Wild.Manag.* 28:167–170.

Koeppl, J. W., N. A. Slade, and R. S. Hoffman. 1975. A bivariate home range model with possible application to ethological data analysis. *J. Mammal.* 56:81–90.

Koeppl, J. W., N. A. Slade, and R. S. Hoffman. 1977. Distance between observations as an index of average home range size. *Amer.Midl.Natur.* 98:476–482.

Koeppl, J. W., N. A. Slade, K. S. Harris, and R. S. Hoffman. 1977. A three-dimensional home range model. *J. Mammal.* 58:213–220.

Koopman, K. F. 1993. Chiroptera, pp. 137–241. *In* D. E. Wilson and D. M. Reeder (Eds.). *Mammal species of the world: a taxonomic and geographic reference.* Smithsonian Institution Press, Washington, D.C.

Korschgen, L. J. 1969. Procedures for food habits analysis, pp. 233–250. *In* R. H. Giles, Jr. (Ed.). *Wildlife management techniques,* 3rd ed. The Wildlife Society, Washington, D.C.

Korschgen, L. J. 1980. Procedures for food-habits analyses, pp. 113–127. *In* S.D. Schemnitz. (Ed.). *Wildlife management techniques manual,* 4th ed. The Wildlife Society, Bethesda, Maryland.

Kosin, I. L. 1972. The growing importance of Russian as a language of science. *BioScience* 22:723–724.

Krantz, G. W. 1970. *A mammal of acarology.* Oregon State Univ. Bookstores, Corvallis.

Krantz, G. W. 1978. *A manual of acarology,* 2nd ed. Oregon State Univ. Bookstores, Corvallis.

Krebs, C. J. 1966. Demographic changes in fluctuating populations of *Microtus californicus. Ecol. Monog.* 36:239–273.

Krebs, C. J. 1972. *Ecology: the experimental analysis of distribution and abundance.* Harper and Row, Publishers, New York.

Krebs, C. J., B. L. Keller, and R. H. Tamarin. 1969. *Microtus* population biology: demographic changes in fluctuating populations of *M. ochrogaster* and *M. pennsylvanicus* in southern Indiana. *Ecology* 50:577–607.

Kronfeld, N., and T. Dayan. 1998. A new method to determine diets of rodents. *J. Mammal.* 79:1198–1202.

Kunz, T. H. (Ed.). 1988. *Ecological and behavioral methods for the study of bats.* Smithsonian Institution Press, Washington, D.C.

Kunz, T. H. 1996. Methods for marking bats, pp. 304–310. In D. E. Wilson, F. R. Cole, J. D. Nichols, R. Rudran, and M. S. Foster (Eds.). *Measuring and monitoring biological diversity.* Standard methods for mammals. Smithsonian Institution Press, Washington, D. C.

Kunz, T. H., and G. Gurri-Glass. 1996. Human health concerns, pp. 255–264. *In* D. E. Wilson, F. R. Cole, J. D. Nichols, R. Rudran, and M. S. Foster (Eds.). *Measuring and monitoring biological diversity. Standard methods for mammals.* Smithsonian Institution Press, Washington, D.C.

Kunz, T. H., C. R. Tidemann, and G. C. Richards. 1996. Capturing mammals: small volant mammals, pp. 122–146. *In* D. E. Wilson, F. R. Cole, J. D. Nichols, R. Rudran, and M. S. Foster (Eds.). *Measuring and monitoring biological diversity. Standard Methods for mammals.* Smithsonian Institution Press, Washington, D.C.

Kurt, F., and H. Wendt. 1975. Sirenians, pp. 523–533. *In* B. Grzimek (Ed.) *Animal life encyclopedia, Vol. 12, Mammals III.* Van Nostrand Reinhold, New York.

Kuzyakin, A. P. 1950. [*Bats.*] Moscow, [In Russian]

Lancia, R. A., and J. W. Bishir. 1996. Removal methods, pp. 210–217. *In* D. E. Wilson, F. R. Cole, J. D. Nichols, R. Rudran, and M. S. Foster (Eds.). *Measuring and monitoring biological diversity. Standard methods for mammals.* Smithsonian Institution Press, Washington, D.C.

Lancia, R. A., J. D. Nichols, and K. H. Pollock. 1996. Estimating the numbers of animals in wildlife populations, pp. 215–253. *In* T. A. Bookhout (Ed.). *Research and management techniques for wildlife and habitats.* 5th ed., rev. The Wildlife Society, Bethesda, Maryland.

Landry, S. O., Jr. 1957. The interrelationships of the New and Old World hystricomorph rodents. *Univ. Calif. Publ. Zool.* 56:1–118.

Lane-Petter, W., et al., (Eds.). 1967. *The UFAW handbook on the care and management of laboratory animals,* 3rd ed. Williams & Wilkins, Baltimore.

Lang, E. M. 1972. Introduction [in part] and Asiatic rhinoceros, pp. 36–48. *In* B. Grzimek (Ed.). *Animal life encyclopedia, Vol. 13, Mammals IV.* Van Nostrand Reinhold, New York.

Lang, H., and J. P. Chapin. 1917. Notes on the distribution and ecology of central African Chiroptera. *In* American Museum Congo Expedition Collection of Bats. *Bull. Amer. Mus. Nat. Hist.* 37:479–563.

Larson, J. S., and R. D. Taber. 1980. Criteria of sex and age, pp. 143–202. *In* S. D. Schemnitz (Ed.). *Wildlife management techniques manual,* 4th ed. The Wildlife Society, Bethesda, Maryland.

Lawrence, M. J., and R. W. Brown. 1967. *Mammals of Britain—their tracks, trails and signs.* Blanford Press, London.

Laws, R. M. 1952. A new method of age determination for mammals. *Nature* 169:972–973.

Lazarus, A. B., and F. P. Rowe. 1975. Freeze-marking rodents with a pressurized refrigerant. *Mammal Review,* 5:31–34.

Lehner, P. N. 1996. *Handbook of ethological methods,* 2nd ed. Cambridge Univ. Press, New York.

Lemen, C. A., and P. W. Freeman. 1985. Tracking mammals with fluorescent pigments: a new technique. *J. Mammal* 66:134–136.

Leuthold, W. 1966. Territorial behavior of Uganda kob. *Behavior* 27:255–256.

Leuthold, W. 1977 African ungulates—comparative review of their ethology and behavioral ecology. *Zoophysiol. Ecology* 8:1–307.

Lidicker, W. Z., Jr. 1962. The nature of the subspecies boundaries in a desert rodent and its implications for subspecies taxonomy. *Syst. Zool.* 11:160–171.

Linnaeus, C. 1758. *Systema naturae per regna tri naturae, secundum classes, ordines, genera, species cum characteribus, differentiis, synonymis, locis.* Editio decima, reformata. Tom. I. Laurentii Salvii, Holmiae.

Litvaitis, J. A., K. Titus, and E. M. Anderson. 1996. Measuring vertebrate use of terrestrial habitats and foods, pp.254–274. *In* T. A. Bookhout (Ed.). *Research and management techniques for wildlife and habitats,* 5th ed. The Wildlife Society, Bethesda, Maryland.

Liu, J. K., G. O. Batzli, and L. L. Getz. 1988. Home ranges of prairie dogs as determined by radiotracking and by powder-tracking. *J. Mammal.* 69:183–186.

LoBue, J. P., and R. M. Darnell. 1958. An improved live trap for small mammals. *J. Mammal.* 39:286–290.

Lord, R. D., Jr. 1959. The lens as an indicator of age in cottontail rabbits. *J. Wildl. Manag.* 23:358–360.

Lord, R. D., Jr., A. M. Vilches, J. I. Maiztegui, and C. A. Soldini. 1970. The tracking board: a relative census technique for studying rodents. *J. Mammal.* 51:828–829.

Lutton, L. M. 1975. Notes on Territorial behavior and response to predators of the pika, *Ochotona princeps. J. Mammal.* 56:231–234.

Lydekker, R. 1909. *Guide to the whales, porpoises and dolphins (order Cetacea) exhibited in the Department of Zoology British Museum (Natural History).* British Museum (Nat. Hist.), London.

M'Closkey, R. T., and D. T. Lajoie. 1975. Determinants of local distribution and abundance in white-footed mice. *Ecology* 56:467–472.

Machado-Allison, C. E., and A. Barrera. 1964. Sobre *Megamblyopinus, Amblyopinus y Amblyopinodes* (ol., Staph.). *Revista de la Sociedad Mexicana de Historia Natural* 25:173–191.

Madson, R. M. 1967. *Age determination of wildlife—a bibliography.* U.S. Dept. Int., Bibliog. 2.

Mahoney, R. 1966. Techniques for the preparation of vertebrate skeletons, pp. 327–351. *In* R. Mahoney *Laboratory techniques in zoology.* Butterworth, Inc., Washington, D.C.

Mares, M. A., K. E., Streilein, and M. R. Willig. 1981. Experimental assessment of several population estimation techniques on an introduced population of eastern chipmunks. *J. Mammal.* 62:315–328.

Marshall, L. G. 1978. *Lutreolina crassicaudata. Mammalian Species* 91:1–4.

Marshall, L. G. 1984. Monotremes and marsupials, pp. 59–115. *In* S. Anderson and J. K. Jones, Jr. (Eds.). *Orders and families of recent mammals of the world.* John Wiley & Sons, New York.

Martin, A. C. 1946. The comparative internal morphology of seeds. *Amer. Midl. Natur.* 36:513–660.

Martin, A. C., and W. D. Barkley. 1961. *Seed identification manual.* Univ. California Press, Berkeley.

Martin, R. E. 1970. Cranial and bacular variation in populations of spiny rats of the genus *Proechumys* (Rodentia: Echimyidae) from South America. *Smithsonian Contrib. Zool.* 35:1–19.

Martinka, C. J. 1976. Population characteristics of grizzly bears in Glacier National Park, Montana. *J. Mammal.* 55:21–29.

Mathiak, H. A. 1938. A rapid method of cross-sectioning mammalian hairs. *J. Wildl. Manag.* 2:162–164.

Mawbey, R. B. 1989. A new trap design for the capture of sugar gliders, *Petaurus breviceps. Australian Wildl. Res.* 16:425–428.

Mayer, W. V. 1952. The hair of California mammals with keys to the dorsal guard hairs of California mammals. *Amer. Midl. Natur.* 48:480–512.

Mayr, E. 1969. *Principles of systematic zoology.* McGraw-Hill Book Co., New York.

Maza, A. E., B. G. French, and N. R. Aschwanden. 1973. Home range dynamics in a population of heteromyid rodents. *J. Mammal.* 54:405–424.

McDaniel, B. 1979. *How to know the mites and ticks.* Wm. C. Brown Company Publishers, Dubuque, Iowa.

McDiarmid, R. W., R. P. Reynolds, and R. I. Crombie. 1996. Permits, pp. 68–69. *In* D. E. Wilson, F. R. Cole, J. D. Nichols, R. Rudran, and M. S. Foster (Eds.). *Measuring and monitoring biological diversity. Standard methods for mammals.* Smithsonian Institution Press, Washington, D.C.

McGaugh, M. H., and H. H. Genoways. 1976. State laws as they pertain to scientific collecting permits. *Museology, Texas Tech Univ.* 2:1–81.

McKay, G. M. 1973. Behavior and ecology of the Asiatic elephant in southeastern Ceylon. *Smithsonian Contrib. Zool.* 125:1–113.

McLaren, S. B. (Chair). 1998. Guidelines for computer usage of computer-based collection data. *J. Mammal.* 69:217–218.

McLaren, S. B., and J. K. Braun (Eds.). 1993. GIS applications in mammalogy. *Oklahoma Mus. Natur. Hist., Spec. Publ.* Norman, Oklahoma.

McLaughlin, C. A. 1984. Protrogomorph, sciuromorph, castorimorph, myomorph (geomyoid, anomaluroid, pedetoid, and ctenodactyloid) rodents, pp. 267–288. *In* S. Anderson and J. K. Jones, Jr. 1984. *Orders and families of Recent mammals of the world.* John Wiley & Sons, New York.

McNab, B. K. 1963. Bioenergetics and the determination of home range size. *Amer. Natur.* 97:133–140.

Mead, J. G., and R. L. Brownell, Jr. 1993. Cetacea, pp. 349–364. *In* D. E. Wilson and D. M. Reeder (Eds.). *Mammal species of the world: a taxonomic and geographic reference.* Smithsonian Institution Press, Washington D.C.

Meester, J., and H. W. Setzer (Eds.). 1971. *The mammals of Africa: an identification manual.* Smithsonian Institution Press, Washington, D.C. [looseleaf]

Melton, D. A. 1976. The biology of aardvark (Tubulidentata—Orycteropodidae). *Mammal Review* 6:75–88.

Merriam, C. H. 1895. Monographic revision of the pocket gophers family Geomyidae *North Amer. Fauna* 8:1–262.

Meserve, P. L. 1976. Food relationships of a rodent fauna in a California costal sage scrub community. *J. Mammal.* 57:300–319.

Meserve, P. L. 1977. Three-dimensional home ranges of cricetid rodents. *J. Mammal.* 58:549–558.

Meserve, P. L., B. K. Lang, and B. D. Patterson. 1988. Trophic relationships of small mammals in a Chilean temperate rain forest. *J. Mammal.* 69:721–730.

Meserve, P. L., R. Murua, O. Lopetegui N., and J. R. Rau. 1982. Observations on the small mammal fauna of a primary temperate rain forest in southern Chile. *J. Mammal.* 63:315–317.

Metcalf, Z. P. 1954. The construction of keys. *Syst. Zool.* 3:38–45.

Metzgar, L. H. 1973. A comparison of trap- and track-revealed home ranges in *Peromyscus. J. Mammal.* 54:513–515.

Miles, W. B. 1965. Studies of the cuticular structure of the hairs of Kansas bats. *Search, Univ. Kansas Publ.* 5:48–50.

Millar, J. S., and F. C. Zwickel. 1972. Determination of age, age structure, and mortality of the pika, *Ochotona princeps* (Richardson). *Canad. J. Zool.* 50:229–232.

Miller, E. H. 1975. Walrus ethology. I. The social role of tusks and applications of multidimensional scaling. *Canad. J. Zool.* 53:590–613.

Miller, F. L. 1974. Age determination of caribou by annulations in dental cementum. *J. Wildl. Manag.* 38:47–53.

Miller, G. S., Jr. 1907. The families and genera of bats. *Bull. U.S. Nat. Mus.* 57:1–282, 14 pls.

Mills, J. N., T. L. Yates, J. E. Childs, R. R. Parmenter, T. G. Ksiazek, P. E. Rollin, and C. J. Peters. 1995. Guidelines for working with rodents potentially infected with hantavirus. *J. Mammal.* 76:716–722.

Mills, R. S., G. W. Barrett, and M. P. Farrell. 1975. Population dynamics of the big brown bat (*Eptesicus fuscus*) in southwestern Ohio. *J. Mammal.* 56:591–604.

Mineau, P., and D. Madison. 1977. Radiotracking of *Peromyscus leucopus. Canad. J. Zool.* 55:465–468.

Moeller, W. 1975. Edentates, pp. 149–181. *In* B. Grzimek (Ed.). *Animal life encyclopedia, Vol. 11, Mammals II.* Van Nostrand Reinhold, New York.

Moore, A. W. 1940. A live mole trap. *J. Mammal.* 21:223–225.

Morisita, M. 1962. Iδ index, a measure of dispersion of individuals. *Res. Pop. Ecol.* 4:1–7.

Morris, P. 1972. A review of mammalian age determination methods. *Mammal Review* 2:69–104.

Mores, L. G. 1971. Specimen identification and key construction with time-sharing computers. *Taxon* 20:269–282.

Morse, L. G. 1974. Computer programs for specimen identification, key construction and description printing using taxonomic data matrices. *Michigan Sate Univ. Mus. Publ.* 5:1–128.

Mörzer Bruyns, W. F. J. 1971. *Field guide of whales and dolphins.* C. A. Mees Uitgeuerij, v.h., Amsterdam.

Mosby, H. S. 1969a. Making observations and records, pp. 61–72. *In* R. H. Giles, Jr. (Ed.). *Wildlife management techniques,* 3rd ed. The Wildlife Society, Washington, D.C.

Mosby, H. S. 1969b. Reconnaissance mapping and map use, pp. 119–134. *In* R. H. Giles, Jr. (Ed.). *Wildlife management techniques,* 3rd ed. The Wildlife Society, Washington, D.C.

Mosby, H. S. 1980. Reconnaissance mapping and map use, pp. 277–290. *In* S. D. Schemnitz (Ed.). *Wildlife management techniques manual,* 4th ed. The Wildlife Society, Bethesda, Maryland.

Mosby, H. S., and I. McT. Cowan (revised by Lars Karstad). 1969. Collection and field preservation of biological materials, pp. 259–275. *In* R. H. Giles (Ed.). *Wildlife management techniques,* 3rd ed. The Wildlife Society, Washington, D.C.

Mullican, T. R. 1988. Radiotelemetry and fluorescent pigments: a comparison of techniques. *J. Wildl. Manag.* 52:627–631.

Murie, A. 1944. The wolves of Mt. McKinley. *U. S. Park Service, Fauna National Parks* 5:1–238.

Murie, O. J. 1954. *A field guide to animal tracks.* Houghton Mifflin Co., Boston.

Musil, A. F. 1963. Identification of crop and weed seeds. *USDA Handbook* 219:1–171 + 43 pls.

Musser, G. G., and M. D. Carleton. 1993. Rodentia (Muridae), pp. 501–755. *In* D. E. Wilson and D. M. Reeder (Eds.). *Mammal species of the world: a taxonomic and geographic reference.* Smithsonian Institution Press, Washington, D.C.

Myers, G. T., and T. A. Vaughan. 1964. Food habits of the plains pocket gopher in eastern Colorado. *J. Mammal.* 45:588–598.

Myers, K., J. Carstairs, and N. Gilbert. 1977. Determination of age of indigenous rats in Australia. *J. Wildl. Manag.* 41:322–326.

Myhre, R., and S. Myrberget. 1975. Diet of wolverines (*Gulo gulo*) in Norway. *J. Mammal.* 56:752–757.

Napier, J. R., and P. H. Napier. 1967. *A handbook of living primates.* Academic Press, London.

Nason, E. D. 1948. Morphology of hair of eastern North American bats. *Amer. Midl. Natur.* 39:345–361.

National Academy of Sciences. 1996. *Guide for the care and use of laboratory animals.* Institute of Laboratory Animal Resources, Commission of Life Sciences, National Research Council, Washington, D.C.

Neff, D. J. 1968. The pellet-group count technique for big game-trend, census, and distribution: a review. *J. Wildl. Manag.* 32:597–614.

New , J. G. 1958. Dyes for studying the movements of small mammals. *J. Mammal.* 39:416–429.

Nichols, J. D. 1986. On the use of enumeration estimators for inteterspecific comparisons, with comments on a 'trappability estimator.' *J. Mammal.* 67:590–593.

Nichols, J. D. 1992. Capture-recapture models—using marked animals to study population dynamics. *BioScience* 42:941–942.

Nichols, J. D., and C. R. Dickman. 1996. Capture-recapture methods, pp. 217–226. *In* D. E. Wilson, F. R. Cole, J. D. Nichols, R. Rudran, and M. S. Foster (Eds.). 1996. *Measuring and monitoring biological diversity. Standard methods for mammals.* Smithsonian Institution Press, Washington, D.C.

Nichols, J. D., and K. H. Pollock. 1983. Estimation methodology in contemporary small mammal capture-recapture studies. *J. Mammal.* 64:253–260.

Nietfeld, M. T., M. W. Barrett, and N. Silvy. 1996. Wildlife marking techniques, pp. 140–168. *In* S. D. Schemnitz (Ed.). *Wildlife management techniques manual.* 5th ed. The Wildlife Society, Bethesda, Maryland.

Norris, K. S. (Ed.). Committee on Marine Mammals, American Society of Mammalogists. 1961. Standardized methods for measuring and recording data on the smaller cetaceans. *J. Mammal.* 42:471–476.

Novick, A. 1977. Acoustic orientation, pp. 74–289. *In* W. A. Wimsatt (Ed.). *Biology of bats, Vol. III.* Academic Press, New York.

Novikov, G. A. 1956. [*Carnivorous animals of USSR*]. USSR Academy of Sciences, Moscow. [In Russian]

Nowak, R. M. 1999. *Walker's mammals of the world,* Vols. I, II, 6th ed. Johns Hopkins Univ. Press, Baltimore.

O'Gara, B. W., and G. Matson. 1975. Growth and casting of horns by pronghorns and exfoliation of horns by bovids. *J. Mammal.* 56:829–846.

Ognev, S. I. 1940. *Mammals of the U.S.S.R and adjacent countries, Vol. IV, Rodents.* English translation 1966, Israel Prog. Scient Transl.

Ognev, S. I. 1947. *Mammals of the U.S.S.R and adjacent countries, Vol. V, Rodents.* English translation 1966, Israel Prog. Scient Transl.

Ognev, S. I. 1951. [*Papers on the ecology of mammals.*] Moscow. [In Russian]

Oosting, H. J. 1956. *The study of plant communities,* 2nd ed. W. H. Freeman and Co., San Francisco.

Osborn, H. F. (W. K. Gregory, Ed.). 1907. *Evolution of mammalian molar teeth to and from the triangular type.* The Macmillan Co., New York.

Osgood, W. H. 1921. A monographic study of the American marsupial, *Caenolestes. Field Mus. Nat. Hist. Zool. Ser.* 14(1) 1–156, pls. I–XX.

Osgood, W. H. 1924. Review of living caenolestids with description of a new genus from Chile. *Field Mus. Nat. Hist. (Zool. Ser.)* 14:165–172, pl. XXIII.

Osgood, W. H. 1943. The mammals of Chile. *Field Mus. Nat. Hist. (Zool. Ser.)* 30:1–268.

Otis, D. L., K. P. Burnham, G. C. White, and D. R. Anderson. 1978. Statistical inference from capture data on closed animal populations. *Wildlife Monog.* 62:1–35.

Overton, W. S. 1969. Estimating the numbers of animals in wildlife populations, pp. 403–455. *In* R. H. Giles, Jr. (Ed.). *Wildlife management techniques, 3rd ed.* The Wildlife Society, Washington, D.C.

Owen, R. 1866. *On the anatomy of vertebrates, Vols. II. and III.* Longmans, Green, and Co., London.

Pankhurst, R. J. 1971. Botanical keys generated by computer. *Watsonia* 8:357–368.

Patterson, B. 1956. Early Cretaceous mammals and evolution of mammalian molar teeth. *Fieldiana: Geol.* 13:1–105.

Patterson, B. 1975. The fossil aardvarks (Mammalia: Tubulidentata). *Bull. Mus. Comp. Zool.* 147(5):185–237.

Patterson, B. D., and M. H. Gallardo. 1987. *Rhyncholestes raphanurus. Mammalian Species* 286:1–5.

Patton, J. L. 1993. Rodentia (Geomyidae & Heteromyidae), pp. 469–486. *In* D. E. Wilson and D. M. Reeder (Eds.). *Mammal species of the world: a taxonomic and geographic reference.* Smithsonian Institution Press, Washington D.C.

Paulick, G. J., and D. S. Robson. 1969. Statistical calculations for change-in-ratio estimators of population parameters. *J. Wildl. Manag.* 33:1–27.

Pearson, O. P. 1960. Habits of *Microtus californicus* revealed by automatic photographic records. *Ecol. Monog.* 30:231–249.

Pelikan, J. 1970. Testing and elimination of the edge effect in trapping small mammals, pp. 57–61. *In* K. Petrusewicz and L. Ruszkowski (Eds.). *Energy flow through small mammal populations.* Polish Scientific Publishers, Warsaw.

Perkins, H. H. 1954. *Animal tracks: the standard guide for identification and characteristics.* Stackpole, Harrisburg, Pennsylvania.

Peterle, T. J. 1969. Radioisotopes and their use in wildlife research, pp. 109–118. *In* R. H. Giles, Jr. (Ed.). *Wildlife management techniques,* 3rd ed. The Wildlife Society, Washington, D.C.

Peterson, R. S., and G. A. Bartholomew. 1967. The natural history and behavior of the California sea lion. *Spec. Publ., Amer. Soc. Mammalogist.* 1:1–79.

Petrusewicz, K., and R. Andrzejewski. 1962. Natural history of a free-living population of house mice (*Mus musculus* L.) with particular reference to groupings within the population. *Ekologia Polska*, Ser. A. 10:85–122.

Pettingill, O. S., Jr. 1956. *A laboratory and field manual of ornithology*, 3rd ed. Burgess Publishing Co., Minneapolis.

Phillips, C. J., B. Steinberg, and T. H. Kunz. 1982. Denin, cementum, and age determination in bats: a critical evaluation. *J. Mammal.* 63:197–207.

Pine, R. H. 1993. A new species of *Thyroptera* Spix (Mammalia: Chiroptera: Thyropteridae) from the Amazon Basin of northeastern Perú. *Mammalia* 57:213–225.

Pitelka, F. A. 1959. Numbers, breeding schedule, and territoriality in pectoral sandpipers of northern Alaska. *Condor* 61:233–264.

Pocock, R. I. 1924a. Some external characters of *Orycteropus afer*. *Proc. Zool. Soc. London* 1924:697–706.

Pocock, R. I. 1924b. The external characters of the South American edentates. *Proc. Zool. Soc. London* 1924:983–1031.

Pocock, R. I. 1926. The external characters of the flying lemur (*Galeopterus temminckii*). *Proc. Zool. Soc. London* 1926:429–444.

Pollock, K. H. 1982. A capture-recapture design robust to unequal probability of capture. *J. Wildl. Manag.* 46:752–757.

Pollock, K. H., J. D. Nicols, C. Brownie, and J. E. Hines. 1990. Statistical inference for capture-recapture experiments. *Wildlife Monog.* 107:1–97.

Popov, B. M. 1956. [*Fauna of the Ukraine. Vol. I.*] Ukrainian Acad. Sci., Kiev. [In Russian]

Pucek, Z., and V. P. W. Lowe. 1975. Age criteria in small mammals, pp. 55–72. *In* F. B. Golley, K. Petrusewicz, and L. Ryszkowski (Eds.). *Small mammals: their productivity and population dynamics.* Cambridge I. B. Handbook No. 5., Cambridge Univ. Press, Cambridge.

Rahm, V. 1975a. Pangolins, pp. 182–188. *In* B. Grzimek (Ed.). *Animal life encyclopedia, Vol. 11, Mammals II.* Van Norstrand Reinhold, New York.

Rahm, V. 1975b. Aardvarks, pp. 473–477. *In* B. Grzimek (Ed.). *Animal life encyclopedia, Vol 12, Mammals III.* Van Norstrand Reinhold, New York.

Rahm, V. 1975c. Hyraxes. pp. 513–522. *In* B. Grzimek (Ed.). *Animal life encyclopedia, Vol. 12, Mammals III.* Van Norstrand Reinhold, New York.

Ralls, K. 1971. Mammalian scent marking. *Science* 171:443–449.

Ramsey, C.W. 1968. A drop-net deer trap. *J. Wildl. Manag.* 32:187–190.

Ray, G. C. 1973. Underwater observations increases understanding of marine mammals. *Mar. Technol. Soc. J.* 7:16–20.

Reichman, O. J. 1975. Relation of desert rodent diets to available resources. *J. Mammal.* 56:731–751.

Reig, O. A. 1970. Ecological notes on the fossorial octodont rodent *Spalacopus cyanus* (Molina). *J. Mammal.* 51:592–601.

Reig, O. A. 1977. A proposed unified nomenclature for the enameled components of the molar teeth of the Cricetidae (Rodentia). *J. Zool., London.* 181:227–241.

Ribble, D. O., and S. Stanley. 1998. Home ranges and social organization of syntopic *Peromyscus boylii* and *P. truei*. *J. Mammal.* 79:932–941.

Rice, C. G., and P. Kalk. 1991. Evaluation of liquid nitrogen and dry ice—alcohol refrigerants for freeze marking three mammal species. *Zoo Biology* 10:261–272.

Rice, D. W. 1984. Cetaceans, pp. 447–490. *In* S. Anderson and J. Jones, Jr. (Eds.). *Orders and families of Recent mammals of the world.* John Wiley & Sons, New York.

Rice, D. W., and A. A. Wolman. 1971. The life history and ecology of the gray whale (*Eschrichtius robustus*). *Spec. Pub. Amer. Soc. Mammalogist* 3:1–142.

Roe, H. S. J. 1967. Seasonal formation of laminae in the ear plug of the fin whale. *Discover Rept.* 35:1–30.

Roff, D. A. 1973a. On the accuracy of some mark-recapture estimators. *Oecologia* 12:15–34.

Roff, D. A. 1973b. An examination of some statistical tests used in the analysis of mark-recapture data.. *Oecologia* 12:35–54.

Rohlf, F. J., and R. R. Sokal. 1969. *Statistical tables.* W. H. Freeman and Co., San Francisco.

Romer, A. S. 1966. *Vertebrate paleontology,* 3rd ed. Univ. Chicago Press, Chicago.

Root, D. A., and N. F. Payne. 1984. Evaluation of techniques for aging gray fox. *J. Wildl. Manag.* 48:926–933.

Rose, R. K. 1973. A small mammal live trap. *Trans. Kansas Acad. Sci.* 76:14–17.

Rose, R. W. 1989. Age estimation of the Tasmanian bettong (*Bettongia gaimardi*) (Marsupialia: Potoridae). *Australian Wildl. Res.* 16:251–261.

Rosenzweig, M. L. 1973. Habitat selection experiments with a pair of co-existing heteromyid rodent species. *Ecology* 54:111–117.

Rosenzweig, M. L., and J. Winkaur. 1969. Population ecology of desert rodent communities: habitats and environmental complexity. *Ecology* 50:558–572.

Rudran, R. 1996. Methods for marking mammals: general marking techniques, pp. 299–304. *In* D.E. Wilson, F. R. Cole, J. D. Nicols, R. Rudran, and M. S. Foster (Eds.). *Measuring and monitoring biological diversity. Standard methods for mammals.* Smithsonian Institution Press, Washington, D.C.

Rudran, R., and T. H. Kunz. 1996. Ethics in research, pp. 251–254. *In* In D. E. Wilson, F. R. Cole, J. D. Nicols, R. Rudran, and M. S. Foster (Eds.). *Measuring and monitoring biological diversity. Standard methods for mammals.* Smithsonian Institution Press, Washington, D.C.

Rupp, R. S. 1966. Generalized equation for the ratio method of estimating population abundance. *J. Wildl. Manag.* 30:523–526.

Ruschi, A. 1953. *Algumas observacoes realizadas sobre os Quiropteros do E. Esprito Santo.* Palestra realizada no Faculdade de Ciencias e Filosofia, Rio de Janeiro.

Russell, W. C. 1947. Biology of the dermestid beetle with reference to skull cleaning. *J. Mammal.* 28:284–287.

Ryder, M. L. 1962. Rhinoceros horn. *Turtox News* 40:274–277.

Samuel, M. D., and M. R. Fuller. 1996. Wildlife radiotelemetry, pp. 370–418. *In* S.D. Schemnitz (Ed.) *Wildlife management techniques manual,* 5th ed., rev. The Wildlife Society, Bethesda, Maryland.

Sanderson, G. C. 1966. The study of mammal movements—a review. *J. Wildl. Manag.* 30:215–235.

Saur, J. R., and N. A. Slade. 1987. Size-based demography of vertebrates. *Ann. Rev. Ecol. Syst.* 18:71–90.

Schaffer, W. M., and C. A. Reed. 1972. The co-evolution of social behavior, and cranial morphology in sheep and goats (Bovidae, Caprini). *Fieldiana: Zoology* 61:1–88.

Schaller, G. B. 1963. *The mountain gorilla: ecology and behavior.* Univ. Chicago Press, Chicago.

Schlitter, D. A. 1993, Hyracoidea, pp. 373–374; Tubulidentata, p. 375; Pholidota, p.415; Macroscelidea, pp. 829–830, *In* D. E. Wilson and D. M.Reeder (Eds.) *Mammal species of the world: a taxonomic and geographic reference.* Smithsonian Institution Press, Washington, D.C.

Schnabel, Z. E. 1938. The estimation of total fish in a lake. *Amer. Math. Mon.* 45:348–352.

Schroder, G. D. 1981. Using edge effect to estimate animal densities. *J.Mammal.* 62:568–573.

Schultze-Westrum. T. 1975. Colugos or flying lemurs, pp. 64–66. *In* B. Grzimek (Ed.). *Animal life encyclopedia, Vol. 11, Mammals II.* Van Nostrand Reinhold Co., New York.

Schumacher, F. X., and R. W. Eschmeyer. 1943. The estimation of fish populations in lakes and ponds. *J. Tenn. Acad. Sci.* 18:228–249.

Sclater, W. L. 1900. *The mammals of South Africa, Vol. I: Primates, Carnivora and Ungulata.* R. H. Porter, London.

Sclater, W. L. 1901. *The mammals of South Africa, Vol. II: Rodentia, Chiroptera, Insectivora, Cetacea and Edentata.* R.H. Porter, London.

Sclater, W. L., and P. L. Sclater. 1899. *The Geography of mammals.* Kegan Paul, Trench, Trubner & Co., London.

Scott, T. G. 1943. Some food coactions of the northern plains red fox. *Ecol. Monog.* 13:427–479.

Seber, G. A. F. 1965. A note on the multiple-recapture census. *Biometrika* 52:249–259.

Serber, G. A. F. 1973. *The estimation of animal abundance and related parameters.* Griffin, London.

Seton, E. T. 1958. *Animal tracks and hunter signs.* Doubleday, New York

Setzer, H. W. 1963. Directions for preserving mammals for museum study. *U.S. Nat. Mus. Info. Leaf.* 380:1–19.

Shadle, A. R., and D. Po-Chedley. 1949. Rate of penetration of porcupine spine. *J.Mammal.* 30:172–173.

Sharp, W. M. 1958. Aging gray squirrels by use of tail-pelage characteristics. *J. Wildl. Manag.* 22:2934.

Shemanchuk, J. A., and H. J. Bergen. 1968. The Gen trap, a simple, humane trap for Richardson's ground squirrels, *Citellus richarsonii* (Sabine). *J. Wildl. Manag.* 49:553–555.

Sherman, P. W., M. L. Morton, L. M. Hoopes, J. Bochantin, and J. M. Watts. 1985. The use of tail collagen strength to estimate age in Belding's ground squirrels. *J. Wildl. Manag.* 49:874–879.

Sherwin, W. B. 1991. Collecting mammalian tissue and data for genetic studies. *Mammal Review* 21:21–30.

Short, H. L. 1978. Analysis of cuticular scales on hairs using the scanning electron microscope. *J. Mammal.* 59: 261–268.

Silver, J., and A. W. Moore. 1941. Mole control. *U.S. Fish and Wildlife Serv., Cons. Bull.* 16:1–17.

Slijper, E. J., and O. Heinermann. 1975. Whales, pp. 457–476; Baleen whales, pp. 477–492; Toothed whales, pp. 493–524. *In* B. Grzimek (Ed.) *Animal life encyclopedia, Vol. 11, Mammals II.* Van Nostrand Reinhold, New York.

Slaughter, B. H., R. H. Pine, and N. E. Pine. 1974. Eruption of cheek teeth in Insectivora and carnivora. *J. Mammal.* 55:115–125.

Smith, J. D. 1972. Systematics of the chiropteran family Mormoopidae. *Univ. Kansas Mus. Natur. Hist., Misc. Publ.* 56:1–132.

Smith, M. H., B. J. Boize, and J. B. Gentry. 1973. Validity of the center of activity concept. *J.Mammals.* 54:747–749.

Smith, M. H., R. H. Gardner, J. B. Gentry, D. W. Kaufman, and M. H. O'Farrel. 1975. Density estimation of small mammal populations, pp. 25–53. *In* F. B. Golley, K. Petrusewicz, and L. Ryszkowski (Eds.). *Small mammals: their productivity and population dynamics.* I. B. P. Handbook No. 5, Cambridge Univ. Press, London.

Smith, M. H., J. B. Gentry, and F. B. Golley. 1969. A preliminary report on the examination of small mammal census methods, pp. 25–29. *In* K. Petrusewicz and L. Ryszkowski (Eds.). *Energy flow through small mammal populations.* Warsaw, Poland.

Smith, R. C., and W. M. Reid. 1972. *Guide to the literature of the life sciences,* 8th ed. Burgess Publishing Co., Minneapolis.

Sneath, P. H. A., and R. R. Sokal. 1973. *Numerical taxonomy,* 2nd ed. W. H. Freeman and Co., San Francisco.

Sokal, R. R., and F. J. Rohlf. 1969. *Biometry: the principles and practice of statistics in biological research.* W. H. Freeman and Co., San Francisco.

Sokolov, I. I. 1959. [*Fauna of the USSR. Vol. 1(3): Perssodactyla and Artiodactyla.*] USSR Acad. Sci., Moscow. [In Russian]

Sokolov, V. E. 1982. *Mammal skin.* Univ. California Press, Berkeley.

Sommer, H. G., and S. Anderson. 1974. Cleaning skeletons with dermestid beetles—two refinements in the method. *Curator* 17:290–298.

Southwood, T. R. E. 1966. *Ecological methods with particular reference to the study of insect populations.* Meuthen and Co., London.

Sparks, D. R., and J. C. Malechek. 1967. Estimating percentage dry weights in diets. *J. Range Manage.* 21:203–208.

Spinage, C. A. 1973. A review of age determination of mammals by means of teeth, with especial reference to Africa. *E. African Wildl. J.* 11:165–187.

Spinge, C. A., and G. M. Jolly. 1974. Age estimation of warthog. *J Wildl. Manag.* 38:229–233.

Stains, H. J. 1962. *Game biology and game management: a laboratory manual.* Burgess Publishing Co., Minneapolis.

Stehlin, H. G., and S. Schaub. 1951. Die Trigonodontie der simplicidenten Nager. *Schweiz. Palaeont. Abhand.* 67:1–385.

Stenseth, N. C., A. Hagen, E. Ostbye, and H. J. Skar. 1974. A method for calculating the size of the trapping area in capture-recapture studies on small rodents. *Norwegian J. Zool.* 22:253–271.

Stephenson, G. R., D. P. B. Smith, and T. W. Roberts. 1975. The SSR system: an open format event recording system with computerized transcription. *Behav. Res. Meth. Instrumentation* 7:497–515.

Stickel, L. F. 1954. A comparison of certain methods of measuring ranges of small mammals. *J.Mammal.* 35:1–15.

Stokes, A. W. (Ed.). 1974. *Territory.* Dowden Hutchinson & Ross, Stroudsburg, Pennsylvania.

Stoneberg, R. P., and C. J. Jonkel. 1966. Age determination of black bears by cementum layers. *J. Wildl. Manag.* 30:411–414.

Stonehouse, B. (Ed.) 1978. *Animal marking: recognition marking of animals in research.* Univ. Park Press, Baltimore.

Storr, G. M. 1961. Microscopic analysis of faeces, a technique for ascertaining the diet of herbivorous mammals. *Australian J. Biol.* 14:157–164.

Stroganov, S. 1962. [*Carnivora of Siberia.*] USSR Acad. Sci., Moscow. [In Russian]

Stroganov, S.U. 1957. [*Insectivora of Siberia.*] USSR Acad. Sci., Moscow. [In Russian]

Sullivan, E. G., and A. O. Haugen. 1956. Age determination of foxes by X-ray of forefeet. *J. Wildl. Manag.* 20:210–212.

Suttie, J. M. 1980. The effect of antler removal on dominance and fighting behaviour on farmed red deer stags. *J. Zool.* 190:217–224.

Szalay, F.S. 1969. Mixodectidae, Microsyopidae, and the insectivore-primate transition. *Bull. Amer. Mus. Nat. Hist.* 140:193–330, pls. 17–57.

Taber, R. D. 1969. Criteria of sex and age, pp. 325–401 *In* R. H. Giles, J, (Ed.). *Wildlife management techniques.* 3rd ed. The Wildlife Society, Washington, D.C.

Taber, R. D., and McT. Cowan. 1969. Capturing and marking wild animals, pp. 277–317. *In* R. H. Giles (Ed.). *Wildlife management techniques,* 3rd ed. The Wildlife Society, Washington, D.C.

Talbot, L. M., and D. R. M. Stewart. 1964. First wildlife census of the entire Serengeti-Mara region. *East Africa J. Wildl. Manag.* 28: 815–827.

Taylor, K. D., and R. J. Quy. 1973. Marking systems for the study of rat movements. *Mammal Review* 3:30–34.

Thomas, O. 1904. On the osteology and systematic position of the rare Malagasy bat *Myzopoda aurita. Proc. Zool. Soc. London.* 1904:2–6.

Timm, R. M. 1982. Dermestids. *Field Mus. Nat. Hist. Bull.* 53(2):1418.

Timm, R. M., and L. H. Kermott. 1982. Subcutaneous and cutaneous melanins in *Rhabdomys:* complementary ultraviolet radiation shields. *J. Mammal.* 63:16–22.

Tinkle, D. W. 1977. The dynamics of populations of squamates, crocodilians and rhynchocephalians, pp. 157–264. *In* C. Gans (Ed.). *Biology of the Reptilia,* Vol.7. Academic Press, London.

Tomilin, A. G. 1962. *Mammals of the U.S.S.R. and adjacent countries, Vol.9, Cetacea.* English translation 1967, Israel Prog. Scient. Transl.

Tullberg, T. 1899. *Uber das System der Nagethiere: eine phylogenetische Studie.* Akademischen Buchdrükerei, Upsala.

Tumlison, R., and V. R. McDaniel. 1983. A reliable celloidin technique for dental cemetum analysis. *J. Wildl. Manag.* 47:274–278.

Tumlison, R., and V. R. McDaniel. 1984. Gray fox age classifications by canine tooth pulp cavity readiograhs. *J. Wildl. Manag.* 48:228–230.

Turnbull, W. D. 1971. The Trinity therians: their bearing on evolution in marsupials and other therians, pp. 151–179. *In* A. A. Dahlberg (Ed.). *Dental morphology and evolution.* Univ. Chicago Press, Chicago.

Tuttle, M. D. 1974. An improved trap for bats. *J. Mammal.* 55:475–477.

Twigg, G. I. 1975a. Finding mammals—their signs and remains. *Mammal Review* 5:71–82.

Twigg, G. I. 1975b. Catching mammals. *Mammal Review* 5:83–100.

Twigg, G. I. 1975c Marking mammals. *Mammal Review* 5:101–116.

USDA. 1948. Woody-plant seed manual. *Forest Service, USDA Misc. Publ.* 654:1–416.

van den Brink, F. H. 1967. *A field guide to the mammals of Britain and Europe.* Colins, London.

Van Valen, 1996. Deltatherida, a new order of mammals. *Bull. American Mus. Natur. Hist.* 132:1–126, 8 pls.

Vinogradov, B. S. 1937. [*Fauna of the USSR. Vol.3(4): Dipodidae.*] USSR Acad. Sci., Moscow. [In Russian]

Vinogradov, B. S., and A. I. Argiropulo. 1941. Fauna of the USSR: Mammals: Key to rodents. USSR Acad. Sci., Moscow. English translation, 1968, Israel Prog. Scient. Transl.

Volf, J. 1975. Horses, pp. 539–580. *In* B. Grzimek (Ed.) *Animal life encyclopedia, Vol. 12, Mammals III.* Van Nostrand Reinhold, New York.

Voorhies, C.T. 1948. A chest for dermestid cleaning of skulls. *J. Mammal.* 29:188–189.

Voss, E. G. 1952. The history of keys and phylogenetic trees in systematic biology. *J. Sci. Labs., Denison Univ.* 43:1–25.

Walker, E. P., F. Warnick, K. I. Lance, H. E. Uible, S. E. Hamlet, M. A. Davis, and P. F. Wright (revised by J. L. Paradiso). 1975. *Mammals of the world,* 3rd. ed. Johns Hopkins Press, Baltimore.

Wauters, L. A., and A. A. Dhondt. 1988. The use of red squirrel (*Sciurus vulgaris*) dreys to estimate population density. *J. Zool., London* 214:179–187.

Weber, M. 1928. *Die Säugetiere. Einführung in die Anatomie und Systematik der recenten und fossilen Mammalia. Band 2, systematischer Teil.,* 2nd. ed. Gustav Fischer, Jena.

Wenzel, R. L., and V. J. Tipton (Eds.). 1966 Ectoparasites of Panama. *Publication, Field Museum of Natural History,* Chicago 1010:1–861.

West, S. D. 1985. Differential capture between old and new models of the Museum Special snap trap. *J. Mammal.* 66:798–800.

Western, D., C. Moss, and N. Georgiadis. 1983. Age estimation and population age structure of elephants by footprint dimensions. *J. Wildl. Manag.* 47:1192–1197.

Westoby, M., G. H. R. Rost, and J. A. Weis. 1976. Problems with estimating herbivore diets by microscopically identifying plant fragments from stomachs. *J. Mammal.* 57:167–172.

White, E. G. 1971. A versatile Fortran computer program for the capture-recapture stochastic model of G. M. Jolly. *J. Fish. Res. Board Canada* 28:443–445.

White, G. C., D. R. Anderson, K. P. Burnham, and D. L. Otis. 1982. Capture-recapture and removal methods for sampling closed populations. *Los Alamos National Lab.* LA-8787-NERP.

White, R. E. 1983. A field guide to the beetles of North America. Houghton Mifflin Company, Boston.

Whitehead, P. J. P. 1971. Storage and retrieval of information in systematic zoology. *Biol. J. Linn. Soc.* 3:211–220.

Wight, H. M., and C. H. Conaway. 1962. A comparison of methods for determining age of cottontails. *J. Wildl. Manag.* 26:160–163.

Wiley, R. W. 1971. Activity periods and movements of the eastern woodrat. *Southwest. Natur.* 16:43–54.

Willcox, W. R., S. P. Lapage, S. Bascomb, and M. A. Curtis, 1973. Identification of bacteria by computer: theory and programming. *J. Gen. Microbiol.* 77:317–330.

Williams, D. F., and S. E. Braun. 1983. Comparison of pitfall and conventional traps for sampling small mammal populations. *J. Wildl. Manag.* 47:841–845.

Williams, O. 1962. A technique for studying microtine food habits. *J. Mammal.* 43:365–368.

Williams, S. L., and C. A. Hawks. 1986. Inks for documentation in vertebrate research collection. *Curator* 29:93–108.

Williams, S. L., R. Laubach, and H. H. Genoways. 1977. A guide to the management of Recent mammal collections. *Carnegie Mus. Natur. Hist., Spec. Publ.* 4:1–105.

Williams, S. L., M. J. Smolen, and A. A. Brigida. 1979. *Documentation standards for automatic data processing in mammalogy.* The Museum of Texas Tech Univ., Lubbock.

Williamson, V. H. H. 1951. Determination of hairs by impressions. *J. Mammal.* 32:80–85.

Wilson, D. E. 1993. Scandentia, pp. 131–133; Dermoptera, p. 135; Sirenia, pp. 365–366; Proboscidea, p. 367; Rodentia (Aplodontidae), p. 417; Rodentia (Castoridae), p. 467. *In* D. E. Wilson and D. M. Reeder (Eds.). *Mammal species of the world: a taxonomic and geographic reference.* Smithsonian Institution Press, Washington, D.C.

Wilson, D. E., F. R. Cole, J. D. Nichols, R. Rudran, and M. S. Foster (Eds.). 1996. *Measuring and monitoring biological diversity. Standard methods for mammals.* Smithsonian Institution Press, Washington, D.C.

Wilson, D. E., and D. M. Reeder (Eds.). 1993. *Mammal species of the world: a taxonomic and geographic reference.* Smithsonian Institution Press, Washington, D.C.

Wilson, E. O. 1975. *Sociobiology: the new synthesis.* Harvard Univ. Press, Cambridge.

Wobeser, G. A., T. R. Spraker, and V. L. Harms, 1980. Collection and field preservation of biological materials, pp. 537–551. *In* S. D. Schemnitz (Ed.). *Wildlife management techniques manual.* 4th ed. The Wildlife Society, Washington, D.C.

Wood, D. M. 1965. Studies on the beetles *Leptinillus validus* (Horn) and *Platypsyllus castoris* Ritsema (Coleoltera: Leptinidae) from beaver. *Proc. Entomol. Soc.* Ontario 95:33–63.

Woods, C. A. 1984. Hystricognath rodents, pp. 389–446. *In* S. Anderson and J. K. Jones, Jr. *Orders and families of Recent mammals of the world.* John Wiley & Sons, New York.

Woods, C. A. 1993. Rodentia (Hystricognathi), pp. 771–806. *In* D. E. Wilson and D. M. Reeder (Eds.). *Mammal species of the world: a taxonomic and geographic reference.* Smithsonian Institution Press, Washington, D.C.

Wozencraft, W. C. 1993. Carnivora, pp. 279–348. *In* D. E. Wilson and D. M. Reeder (Eds.). *Mammal species of the world: a taxonomic and geographic reference.* Smithsonian Institution Press, Washington, D.C.

Wydeven, P., and R. B. Dahlgren. 1982. A comparison of prairie dog stomach contents and feces using a microhistological technique. *J. Wildl. Manag.* 46:1104–1108.

Zerov, S. A., and E. N. Pavlovskii. 1953. [*Atlas of economic and game birds and animals of the U.S.S.R. Vol. 2. Beasts.*] USSR Acad. Sci., Moscow. [In Russian]

Zippin, C. 1956. An evaluation of the removal method of estimating animal populations. *Biometrics* 12:163–189.

CREDITS

Chapter 4
Figure 4.1: From K. Kardong, Vertebrates: Comparative Anatomy, Function, Evolution, 2nd ed. Copyright 1998 by The McGraw-Hill Companies, Inc., Dubuque, Iowa. All rights reserved. Reprinted by permission; **Figure 4.2:** From G.C. Kent and L. Miller, Comparative Anatomy of the Vertebrates, 8th ed. Copyright 1997 by The McGraw-Hill Companies, Inc., Dubuque, Iowa. All rights reserved. Reprinted by permission; **Figure 4.3:** From G.A. Feldhamer, et al., Mammalogy. Copyright 1999 by The McGraw-Hill Companies, Inc., Dubuque, Iowa. All rights reserved. Reprinted by permission; **Figure 4.8:** From G.A. Feldhamer, et al., Mammalogy. Copyright 1999 by The McGraw-Hill Companies, Inc., Dubuque, Iowa. All rights reserved. Reprinted by permission; **Figure 4.9a-d:** From G.A. Feldhamer, et al., Mammalogy. Copyright 1999 by The McGraw-Hill Companies, Inc., Dubuque, Iowa. All rights reserved. Reprinted by permission.

Chapter 5
Figure 5.9a-e: From G.A. Feldhamer, et al., Mammalogy. Copyright 1999 by The McGraw-Hill Companies, Inc., Dubuque, Iowa. All rights reserved. Reprinted by permission; **Figure 5.10:** original rendering by Michael Gilliland; **Figure 5.12b:** From G.C. Kent and L. Miller, Comparative Anatomy of the Vertebrates, 8th ed. Copyright 1997 by The McGraw-Hill Companies, Inc., Dubuque, Iowa. All rights reserved. Reprinted by permission.

Chapter 6
Figure 6.1: From C.P. Hickman, Jr., et al., Laboratory Studies in Integrated Principles of Zoology, 9th ed. Copyright 1996 by The McGraw-Hill Companies, Inc., Dubuque, Iowa. All rights reserved. Reprinted by permission.

Chapter 7
Figure 7.7: From G.A. Feldhamer, et al., Mammalogy. Copyright 1999 by The McGraw-Hill Companies, Inc., Dubuque, Iowa. All rights reserved. Reprinted by permission.

Chapter 9
Figure 9.2: From G.A. Feldhamer, et al., Mammalogy. Copyright 1999 by The McGraw-Hill Companies, Inc., Dubuque, Iowa. All rights reserved. Reprinted by permission.

Chapter 12
Figure 12.10: From G.A. Feldhamer, et al., Mammalogy. Copyright 1999 by The McGraw-Hill Companies, Inc., Dubuque, Iowa. All rights reserved. Reprinted by permission.

Chapter 14
Figure 14.1: From G.A. Feldhamer, et al., Mammalogy. Copyright 1999 by The McGraw-Hill Companies, Inc., Dubuque, Iowa. All rights reserved. Reprinted by permission.

Chapter 17
Figure 17.5: Original rendering by Michael Gilliland.

Chapter 20
Figures 20.1, 20.5, 20.25: From G.A. Feldhamer, et al., Mammalogy. Copyright 1999 by The McGraw-Hill Companies, Inc., Dubuque, Iowa. All rights reserved. Reprinted by permission.

Chapter 23
Figure 23.25a-c: From E.S. Booth, How to Know the Mammals, 4th ed. Copyright 1982 by Wm. C. Brown Company Publishers, Dubuque, Iowa. All rights reserved. Reprinted by permission; **Figure 23.48:** Original rendering by Michael Gilliland.

Chapter 25
Figure 25.1: Original rendering by Michael Gilliland.

Chapter 26
Figure 26.2: From G.A. Feldhamer, et al., Mammalogy. Copyright 1999 by The McGraw-Hill Companies, Inc., Dubuque, Iowa. All rights reserved. Reprinted by permission.

Chapter 28
Figure 28.9: From Brower, et al., Field and Laboratory Methods for General Ecology, 4th ed. Copyright 1998 by The McGraw-Hill Companies, Inc., Dubuque, Iowa. All rights reserved. Reprinted by permission.

Chapter 31
Figure 31.9: Original rendering by Terry Maxwell.

Chapter 32
Figure 32.2: From R.E. White, A Field Guide to the Beetles of North America. Copyright 1983 by Houghton Mifflin Company, Boston. Used with permission; **Figure 32.3a:** From R.E. White, A Field Guide to the Beetles of North America. Copyright 1983 by Houghton Mifflin Company, Boston. Used with permission; **Figures 32.13a-b, 32.14a-c, e-f, 32.15a-c, 32.16c, f, 32.17, 32.19, 32.20, 32.21, 32.22, 32.23:** From B. McDaniel, How to Know the Mites and Ticks. Copyright 1979 by Wm. C. Brown Company Publishers, Dubuque, Iowa. All rights reserved. Reprinted by permission.

Chapter 33

Figure 33.1: From R.W. Rose, 1989. Age estimation of the Tasmanian bettong (*Bettongia gaimardi*) (Marsupialia: Potoroidae). In *Australian Wildlife Research* 16:251–261. Used with permission of CSIRO.

Chapter 34

Figure 34.1: From A.F. Bennett and B.J. Baxter, 1989. Diet of the long-nosed potoroo, *Potorous tridactylus* (Marsupialia: Potoroidae), in south-western Victoria. *Australian Wildlife Research* 16:263–271. Used with permission of CSIRO.

Chapter 35

Figure 35.5: From K.B. Armitage, 1974. Male behaviour and territoriality in the yellow-bellied marmot. *Journal of Zoology, London* 172:233–265. Used with permission of the Zoological Society of London.

Chapter 37

Figure 37.1: Original rendering by R.E. Martin.

Index

Note: The coverage in this index is supplemental to entries that appear in the Glossary. **Bold-faced** type indicates pages that provide substantial coverage of a particular subject; *italic* type refers to an illustration that occurs outside of related text citations.

A

Aardvark. *See* Tubulidentata
Aardwolf. *See* Hyaenidae
Abrocomidae, 142, 154, *155, 156*
Abundance. *See* Population estimates
Acari, 232–233, 236
Acouchi. *See* Dasyproctidae
Acrobatidae, 64, 72, 74–75, 76
Aerial locomotion, **48**
　Chiroptera, 85
　Dermoptera, 83
Age determination, 9, **244–252**
　growth indicators, 245–248
　growth lines, 248–50
　limitations, 251–252
　in living mammals, 250–251
　statistics and known-age samples, 245
　terminology, 244
Agouti. *See* Dasyproctidae
Agoutidae, 142, 156, 160
Albinism, 26
Allactaginae, 150
Analysis
　age. *See* Age determination
　diet. *See* Diet
　distribution. *See* Spatial distribution, analysis
　habitat, **180, 184–188,** 267
　literature. *See* Literature research
　population. *See* Population estimates
　recordkeeping. *See* Data, recording
　sign, **180–184**
Angwantibo. *See* Loridae
Anomaluridae, 47, 139, 142, 152
Anoplura, 231, 235
Anteater. *See* Myrmechophagidae; Xenarthra
Antelope. *See* Bovidae
Antilocapridae, 174
　horns, 30, *31,* 34–35
　key, 176, 177, 179

Antlers. *See also* Horns
　age determination, 250
　anatomy, **30–33**
Aplodontidae, 139, 141, 152, *153,* 161
　trapping, 206
Appendages. *See also* Claws; Digits; Hoofs
　locomotion and. *See* Locomotion; *specific locomotive mode*
　measurement, 194–195, **212–214**
　skeletal structure, **37–40**
Aquatic locomotion, **45–46**
　Carnivora, 110
　Cetacea, 121
　Sirenia, 166
Arachnida, 232–233, 236, *237–243*
Arboreal locomotion, **46–48**
　Carnivora, 110
　Hyracoidea, 165
　Marsupialia, 64
　Pholidota, 108
　Primates, 96
　Tupaiidae, 94
　Xenarthra, 103
Armadillo. *See* Dasypodidae; Xenarthra
Armor. *See* Scales
Arthropoda
　as ectoparasites. *See* Ectoparasites
　as food resource, 256, 259
Artiodactyla, **172–179**
　age determination, *247,* 249, 250
　anatomy, 172–174
　families, 174
　function, antlers, horns, etc., 34–35, 184
　identification, 60, 179
　key, 56, 57, 174–179
　scats, *181*
Arvicolinae, 141, 143–144, *146,* 160–161
Ass. *See* Equidae
Astigmata, 233, 236, *240, 241*
Axial skeleton, **36–37**
Aye-aye. *See* Daubentoniidae

B

Baculum, 37, 246–247
Badger. *See* Mustelidae
Baits and scents, 207
Balaenidae, *123,* 125, 126

323

Balaenopteridae, 125, *126*
Baleen plates. *See* Mysticeti
Bat. *See* Chiroptera
Bat bug. *See* Heteroptera
Bathyergidae, 142, *143,* 153, *154*
Beaked whale. *See* Odontoceti, Ziphiidae
Bear. *See* Ursidae
Beaver. *See* Castoridae
Bed bug. *See* Heteroptera
Beetle. *See* Coleoptera
Behavior. *See also specific behavior*
    Cetacea, 123
    data, recording, 198
    investigative methods, **209–210, 273, 275–276**
Beluga. *See* Monodontidae; Odontoceti
Bleaching specimens, 228
Blesmol. *See* Bathyergidae
Body size, age determination, 245, 250–251
Boiling specimens, 228
Bone, age determination, **246–247, 248–250**
Bovidae, 174
    horns, 30, *31,* 34–35, 250
    key, 176, 177, *178,* 179
Brachiation, 46, *47*
Bradypodidae, *47,* 103, 105, 106
Burramyidae, 64, 72, 74, 76
Burrows, **182**
    Geomyidae, *183*
    Lagomorpha, 136
Bushbaby. *See* Galagonidae

C

Caenolestidae, **67–68**
CALHOME, 264, 266
Callitrichidae, 96, 97, 99–101
Camel. *See* Camelidae
Camelidae, 174
    anatomy, 172, 173
    key, 175, *176,* 179
Canidae, 112
    age determination, 247, 248, 249
    anatomy, *7, 24*
    key, 115, 116–117, 120
Canines, 14–15. *See also* Dentition
Capromyidae, 142, 158, 160
Capybara. *See* Hydrochaeridae
Cardiocraniinae, 150
Carnivora, **109–120**
    age determination, 247, 248, 249
    anatomy, 36–37, 110–111
    distribution, 109–110, 111, 112
    economic significance, 110
    families, 111, 112
    identification, 59, 120
    key, 56, 57, 112–120
Carnivorous diet
    Carnivora, 109
    Cetacea, 126
    Chiroptera, 85
    dentition and, 14, 15, 17
    Insectivora, 78
    Monotremata, 61
    Xenarthra, 103

Cartilage, age determination, 246
Cased skins, 222
Castoridae, 141
    anatomy, 139
    key, 143, *144,* 160
    sign, *181, 184*
    trapping, 206
Cat. *See* Felidae
Catalog entry, 190, *191,* 194–196, 212–214
Catarrhini, 96, 97, *98*
Cattle. *See* Bovidae
Caviidae, 142, 153, *154, 155*
Cavy. *See* Caviidae
Cebidae, 96, 97, 101, 102
Cementum, 249–250
Census, population estimates, 272–273, 274
Cercopithecidae, 97, 101, 102
Cervidae, 174
    antlers, **30–35,** *173,* 184
    dentition, 17
    key, 175–176, *177, 178,* 179
    scent glands, 28
Cetacea, **121–132.** *See also* Mysticeti; Odontoceti
    age determination, 249, 250
    anatomy, 121, *122*
    distribution, 124, 125, 127–128
    economic significance, 121, 123
    identification, 59, 132
    key, 55, 56, 124–126, 128–132
    measurement, 214, *215*
Change-in-ratio methods, 277
Cheek pouch, contents, 254
Cheek teeth, **15–18.** *See also* Dentition
Cheirogaleidae, 97, 102
Chemicals
    degreasing agents, 227–228
    fixatives. *See* Fixatives
Chevrotain. *See* Tragulidae
Chigger. *See* Acari
Chimpanzee. *See* Hominidae
Chinchilla. *See* Chinchillidae
Chinchillidae, 142, 154, *155*
Chiroptera, **85–93**
    anatomy, 36–37, 85–86
    collecting, 208–209
    diet, 85, *254*
    distribution, 86, 87
    echolocation, 85
    economic significance, 85
    families, 86, 87
    identification, 59, 93
    key, 56, 57, 87–93
    locomotion, 85
    specimen measurement, 194–195, 213–214
Chordata, 1
Chrysochloridae, *44,* 79
    key, 81
Civet. *See* Viverridae
Classification
    hair, **23–24**
    taxonomic hierarchies, 2–3
Claws, **39**
    Primate nails, 97
    Xenarthra, 103
Climate, trapping and, 207

Coleoptera, *231*, 232
   key, 233, 236
   specimen preparation with, **225–227**
Collagen fibers, 251
Collecting, **200–210**
   alternative methods, 208, 209, 210
   conservation, 201–202
   ectoparasites. *See* Ectoparasites
   ethics and laws, 201–202
   health and safety, 200–201
   hunting, 207–208
   recordkeeping. *See* Data, recording
   sites, 202
   trapping. *See* Trapping
Coloration, **25–27**
Colugo. *See* Dermoptera
Communication, Cetacea, 123, 126
Computers
   databases, 198, **281–283**
   population estimation, 270–271
Conibear trap, 203
Conservation, **201–202**
Corral traps and nets, 206, 208
Coruro. *See* Octodontidae
Coypu. *See* Myocastoridae
Cranium. *See* Skull
Craseonycteridae, 87, 89, 93
Cricetinae, 141, *147*, 148
Cricetomyinae, 141, 148
Crossbreeding, Equidae, 169
Ctenodactylidae, 139, 142, *146*, 151, *152*
Ctenomyidae, 142, 157, *158*
Cynocephalidae, **83–84**

D

Dasypodidae, 103, *104*, 105, 106
Dasyproctidae, 142, 154, *155*, 160
Dasyuridae, 64, *65*, **68–70**
Dasyuromorphia, 58, **68–70**
Data, recording, **189–199**, 202
   computerized, 198
   data forms, *196*, **198**, 264, 265
   equipment, 189–190, 198
   field notes, 190–191, 212–214
   labels, 190, 195–197, 229
   locality, 191–194, 202
   movement patterns, 268
   museum documentation standards, 198, 199
   population estimates, 272
   specimen characteristics, 194–195, 196, 198, **212–215**
Databases
   data entry, 198
   literature research, **281–282**
Date, recording, 194
Datum, 192
Daubentoniidae, 97, 98, *99*
Dead on the road, 195
Deer. *See* Cervidae
Degreasing, 227–228
Delphinidae, 29, 127, *130*, 131, 132
Dendromurinae, 141, *147*, 148, 151
Dens, 182–183, *184*, 208
Density estimation. *See* Population estimates
Dental formula, general, **18–20**

Dentaries. *See* Mandible
Dentition, **13–20**
   age determination, **247–248, 249–250, 251**
   anatomy, 13, *14*
   Artiodactyla, 172, 179
   canines, 14–15
   Carnivora, 110–111
   cheek teeth, 15–18
   dental formulas, 18–20
   Dermoptera, 83
   diet and, 13–18
   Hyracoidea, 166
   incisors, 14
   Insectivora, 78
   Lagomorpha, 136
   Macroscelidea, 133
   Marsupialia, 64, 76
   measurement, 10, 11
   molars and premolars, 15–18
   Monotremata, 61
   Odontoceti, 126
   Perissodactyla, 169–170
   Primates, 97
   Proboscidea, 165
   replacement, 13, *14*
   Rodentia, 138, 139
   Sirenia, 167
   Tupaiidae, 94
   Xenarthra, 103
Dermaptera, 231, 233
Dermestidae, **225–227**
Dermoptera, 47, 57, 59, **83–84**
Desman. *See* Talpidae
Didelphidae, 64, 65, **66–67**
Didelphimorphia, 58, 65, **66–67**
Diet, **253–260**. *See also specific diet, i.e.*
   Insectivorous diet
   Carnivora, 109
   Cetacea, 123, 126
   Chiroptera, 85
   dentition and, **13–18**
   feeding residues, 183–184
   Insectivora, 78
   Marsupialia, 64
   preference indices, 259–260
   Primates, 96
   resource levels, 259
   Xenarthra, 103
Diet samples
   analysis, 258–259
   collection, 254–255
   identification, 255–258
Digits, **38–40**
   Artiodactyla, 172
   Cetacea, 121
   Hyracoidea, 166
   locomotion and, 41, 43, 46, 48
   Perissodactyla, 169
   Primates, 96
   Proboscidea, 165
   Rodentia, 139
Dinomyidae, 142, 158, *159*
Dipodidae, 141
   anatomy, *43*, 139
   key, *146*, 150, 151, 161
Dipodinae, 150

Diprotodontia, **71–76**
   age determination, *245*
   anatomy, 47, 64
   diet, *253*
   distribution, 71–72
   families, 71–72
   key, 55, 58, 72–76
Diptera, 231, *232,* 233, 234, 235
Disease, collecting and, **200–201,** 227
Dispersion, index of, 261
Distribution, geographic
   Artiodactyla, 174
   Carnivora, 109–110, 111, 112
   Cetacea, 124, 125, 127–128
   Chiroptera, 86, 87
   Dasyuromorphia, 68
   Dermoptera, 84
   Didelphimorphia, 65
   Diprotodontia, 71–72
   global, 4
   Hyracoidea, 166
   Insectivora, 78
   Macroscelididae, 133
   Marsupialia, 64
   Microbiotheria, 68
   Monotremata, 62
   Notoryctemorphia, 77
   Paucituberculata, 67
   Peramelemorphia, 71
   Perissodactyla, 171
   Pholidota, 108
   Primates, 97
   Proboscidea, 16
   Rodentia, 139, 142
   Sirenia, 167
   Tubulidentata, 163
   Tupaiidae, 94
   Xenarthra, 105
Distribution, spatial. *See* Spatial distribution, analysis
Dog. *See* Canidae
Dolphin. *See* Cetacea; Delphinidae
Domestication
   Artiodactyla, 172, 174
   Carnivora, 110
   Lagomorpha, 135
   Perissodactyla, 171
   Proboscidea, 164
DOR (dead on the road), 195
Dormouse. *See* Myoxidae
Drug immobilization, 208
Dugongidae, **166–168**
Dye marking, 209
Dzhalman. *See* Myoxidae

E

Ear
   echolocation, 85
   inner and middle, 1, *2*
   Lagomorpha, 135
   measurement, 194–195, 212, *213*
Earplug, age determination, 250
Earwig. *See* Dermaptera
Echidna. *See* Monotremata; Tachyglossidae
Echimyidae, 142, 157, *158,* 160, *247*

Echolocation, 85
Ectoparasites, **230–243**
   collection supplies, 236–237, 239
   detection and removal, 230, 239–240, 242–243
   identification/key, 233–236, 243
   types, 230–233
Edentata. *See* Xenarthra
Egg-laying mammals. *See* Monotremata
Elephant. *See* Proboscidea
Elephant shrew. *See* Macroscelidea
Elephantidae, **164–165**
Emballonuridae, 87, 90, 93
Endangered species
   Carnivora, 110
   Primates, 97
   Proboscidea, 164
   Rhinocerotidae, 169
   Rodentia, 138
Enumeration method, 276–277
Enzymes, specimen preparation, 228–229
Epiphyseal cartilage, 246
Epithelial earplug, 250
Equidae, *42,* **169–171**
Erethizontidae, 142, *158,* 159, 160, 221
   specimen preparation, 221
Erinaceidae, 79, *80, 81*
Eschrichtiidae, 124–125, *126*
Euchoreutinae, 150
Euthanasia, 210
Eutheria, 53–54
Extinct species, 110, 164
Eye
   lens, age determination, 248
   primate vision, 96–97

F

Faunal regions, global, 4
Feces
   dietary analysis, 254, 259
   reingestion, 135
Feeding. *See* Diet
Feeding residues, 183–184
Feet. *See* Foot
Felidae, 112
   anatomy, *37, 42, 110*
   key, 115–116, *117,* 120
Field notes, **190–191, 212–214.** *See also* Catalog entry
Fin, 46, 121
Fitch trap, 205
Fixatives, 237, 239
   diet sample, 255
   labels and, 197
   reference sample, **255–258**
   specimen sample, 220, 222, 229
Flea. *See* Siphonaptera
Flight. *See* Aerial locomotion
Flipper, 45
Fluids. *See* Fixatives
Fluorescent pigments, 269
Fly. *See* Diptera
Food caches, 254. *See also* Diet
Foot. *See also* Locomotion
   Hyracoidea, 165, 166
   Insectivora, 78

Foot—*Cont.*
   Lagomorpha, 135, 136
   Marsupialia, *65, 245*
   measurement, 194–195, 212, *213*
   prints/tracks, **180–181,** 251
   Proboscidea, 165
   skeletal structure, **38–39**
Fossa. *See* Viverridae
Fossorial locomotion, **43–44**
   Carnivora, 110
   Lagomorpha, 135
   Marsupialia, 64
   Xenarthra, 103
Fox. *See* Canidae
Freeze marking, 209
Friction ridges, 21, *22*
Fumigation, 227
Fur. *See* Hair
Fur seal. *See* Otariidae
Furipteridae, 87, 92, 93

G

Galago. *See* Galagonidae
Galagonidae, 97, 99, *100, 101*
Gen trap, 205
Genet. *See* Viverridae
Geographic information systems, **191–194**
Geomyidae, 141, 142, *143*
   anatomy, 139, 160
   burrow system, *183*
   traps, 203–204, 205
Gerbil. *See* Gerbillinae
Gerbillinae, 141, 148
Gibbon. *See* Hylobatidae
Giraffe. *See* Giraffidae
Giraffidae, 33–34, 174, 175, *176,* 179
GIS (geographic information systems), **191–194**
Glands
   integumentary, **27–28**
   mammary, 1, **28–29,** 53, *54*
Glider. *See* Acrobatidae; Petauridae; Pseudocheiridae
Global positioning system, 191, 194
Gloger's Rule, 26
Glycyphagidae, *242*
Goat. *See* Bovidae
Golden mole. *See* Chrysochloridae
Gopher. *See* Geomyidae
Gorilla. *See* Hominidae
Government regulations, **201–202**
GPS (global positioning system), 191, 194
Gray whale. *See* Eschrichtiidae; Mysticeti
Grid method, 206
Growth analysis. *See* Age determination
Growth lines, 248–250
Guanaco. *See* Camelidae
Guard hairs, 24, 255
Gymnure. *See* Erinaceidae

H

Habitat analysis, **180, 184–188,** 267
Hair, 1, **22–27**
   anatomy, 22–23
   classification, 23–24
   color, 25–27, 251
   echidna, 61
   Insectivora, 78
   Perissodactyla, 170
   reference samples, 255
   replacement, 24–25
   Xenarthra, 103
Hamster. *See* Cricetinae
Hand, 38. *See also* Arboreal locomotion
Hand capture, 208
Hantavirus, 201
Haplorhini, **96–97**
Hare. *See* Lagomorpha
Harmonic mean method, 264
Havahart trap, 204
Health issues, in collecting, **200–201,** 227, 228
Hearing, Cetacea, 126
Heart, *2*
Hedgehog. *See* Erinaceidae
Hemiptera. *See* Heteroptera
Herbivorous diet
   Artiodactyla, 172
   Carnivora, 109
   dentition and, 13, 14, 15, 17
   Dermoptera, 83
   Lagomorpha, 135
   Perissodactyla, 165
   Primates, 96
   Sirenia, 166
   Xenarthra, 103
Herpestidae, 112, 119–120
Heteromyidae, 139, 141, 142, *143,* 160
Heteroptera, 231, *235*
Hindfoot. *See* Foot
Hippopotamidae, 174
   anatomy, 172, 173
   key, 175, *176,* 179
Histoplasmosis, 201
Hog. *See* Suidae
Home range
   analysis, **262–266**
   characteristics, **261–262**
Hominidae, *17,* 97, 102
Hoofs, **39–40**
   Artiodactyla, 172
   Hyracoidea and, 165
   Perissodactyla, 165
   Proboscidea and, 165
Horns, **30–35,** 172–173. *See also* Antlers
   age determination, 250
   function, 34–35
   giraffe, 33–34
   pronghorn, 30, *31*
   rhinoceros, 34
   true horns, 30, *31*
Horse. *See* Equidae
Hoyer's medium, 256
Human. *See* Hominidae
Humpback whale. *See* Balaenopteridae; Mysticeti
Hunting, 207–208
Hutia. *See* Capromyidae
Hyaenidae, 112, 113, *114,* 115
Hydrochaeridae, 142, 153, *154,* 159–160
Hydromyinae, 141, 145, *146*
Hyena. *See* Hyaenidae

Hylobatidae, 46, *47*, 97, 101, *102*
Hyracoidea, 55, 59, **165–166**
Hyrax. *See* Hyracoidea
Hystricidae, 139, 142, *143*
  key, *155*, 157–158, *159*
  specimen preparation, 221
Hystricognathi, 142. *See also specific family*

I

Identification
  diet samples, **255–258**
  gestalt characteristics, **58–60**
  keys. *See* Keys, biological identification
Incisors, 14. *See also* Dentition
Index of dispersion, 261
Indridae, 97, 98–99
Inks, 189–190
Insecta, as ectoparasites, **231–236**
Insectivora, **78–82**
  anatomy, *22,* 36–37, 78
  distribution, 78
  families, 78, 79
  identification, 59, 82
  key, 58, 79–81
Insectivorous diet, **78**
  Carnivora, 109
  Chiroptera, 85
  Pholidota, 108
  Primates, 96
  Xenarthra, 103
Integument, **21–29**
  Dermoptera, 84
  gland types, 27–28
  hair. *See* Hair
  Marsupialia, 76
  Perissodactyla, 170
  scales. *See* Scales
Intelligence, Cetacea, 123
Invertebrate prey samples, 255–256

J

Jackal. *See* Canidae
Jerboa. *See* Dipodidae
Jolly-Seber method, 276
Journal, 190. *See also* Catalog entry; Field notes

K

Kangaroo. *See* Macropodidae
Keys, biological identification, **49–52**. *See also specific order or family*
  character selection, 49–51
  construction, 51–52
  ectoparasites, **233–236**
  living mammals, **55–58**
  use, 49, 52, 58
Killer whale. *See* Delphinidae; Odontoceti
Kinkajou. *See* Procyonidae
Koala. *See* Phascolarctidae

L

Labeling
  diet samples, 255
  specimens, 190, **195–197**

Lactophenol, 256
Lagomorpha, **135–137**
  age determination, 248
  identification/key, 55, 59
  specimen preparation, 221
Legislation, 201–202
Lemming. *See* Arvicolinae
Lemur. *See* Cheirogaleidae; Indridae; Lemuridae; Megaladapidae
Lemuridae, 97, *100,* 102
Lens protein, 248, *249*
Lepidoptera
  as ectoparasites, 232, 233, 235–236
  as food source, *254*
Leporidae, **135–137**
Lice. *See* Anoplura; Mallophaga
Light transmitters, 268
Limbs. *See* Appendages
Lincoln Index, 275–276
Line transect sampling, 185–186, 187
Linsang. *See* Viverridae
Literature research, **278–283**
  bibliographies and abstracts, 279–280
  computer databases, 281–282
  search procedures, 280–281, 282, 283
Locality, **190–194**
Locomotion, **41–48**. *See also specific mode*
  aerial, 48
  aquatic, 45–46
  arboreal, 46–48
  brachiating, 46
  Carnivora, 110
  Dermoptera, 83
  fossorial, 43–44
  gliding, 47
  Insectivora, 78
  Macroscelidea, 133
  Marsupialia, 64, 76
  Rodentia, 138
  saltatorial, 42
  scansorial, 46
  semiaquatic, 45
  semifossorial, 44
  sloth movement, 46
  terrestrial, 41–43, *44*
  Tupaiidae, 94
Longworth trap, 205
Lophiomyinae, 141, 147–148
Loridae, 96, 97, 99, *100*
Loris. *See* Loridae
Lyme disease, 201

M

Maceration, 228
Macropodidae, 72
  anatomy, *65,* 71
  key, *73,* 76
  locomotion, *43,* 64
Macroscelidea, 58, 59, **133–134**
Macroscelididae, 58, 59, **133–134**
Mallophaga, 231, 235
Mammalia, **1–5, 53–60**
  age determination, 250–251
  anatomy, 1, 53
  distribution. *See* Distribution, geographic; Spatial distribution, analysis
  identification, 58–60
  key, 55–58. *See also* Keys

parasites on. *See* Ectoparasites
  specimen care, 5, 210
  subclasses and orders, 53–54
  systematics and nomenclature, 2–4
Mammary glands, 1, **28–29**, 53, *54*
Mandible, 1, **8–9**, 11
Manidae, **107–108**
Manus, 38
  Cetacea, 121
Mapping
  applications, **191–194**
  movement patterns, 268
Mara. *See* Caviidae
Mark and recapture method, 273, 275
Marking
  by mammals, 28, 184
  of mammals, 209–210, 273, 275–276
Marmoset. *See* Callitrichidae
Marmot. *See* Sciuridae
Marsupialia, **64–77**
  age determination, *245*
  anatomy, 64–65
  Dasyuromorphia, 68–70
  Didelphimorphia, 66–67
  Diprotodontia, 71–76
  distribution, 64
  economic significance, 65–66
  as food source, 66
  identification, 58–59
  locomotion, 64
  Microbiotheria, 68
  Notoryctemorphia, 77
  Paucituberculata, 67–68
  Peramelemorphia, 70–71
  reproduction, 64–65
Marsupium
  Diprotodontia, 71
  Marsupialia, 64, *65*
  Monotremata, 61
Maturity. *See* Age determination
Measurements
  cranial, **9–11**
  data, recording, 194–195, 196, 198, 214
  taking, **212–215**
Megachiroptera, 87
Megadermatidae, 87, 88–89, 93
Megaladapidae, 97, 102
Megalonychidae, 103, *104*, 105, 106
Melanism, 26–27
Mesostigmata, 232–233, 236, *239*
Metastigmata, 232, 236, *237, 238*
Metatheria, 53–54
Microbiotheria, 58, **68**
Microbiotheriidae, 64, **68**
Microchiroptera, **87–93**
Migration, analysis, **267–269**
Mite. *See* Acari
Molars, **15–18**. *See also* Dentition
Mole. *See* Talpidae
Molossidae, 85, 87, 91–92, 93
Molting, 24–25
Mongoose. *See* Herpestidae
Monkey. *See* Callitrichidae; Cebidae; Cercopithecidae
Monodontidae, 127, 128, 131, 132
Monotremata, 58, **61–63**
Mormoopidae, 87, 88
Moschidae, 173, 174, *178*, 179

Moth. *See* Lepidoptera
Mouse, families, 141–142. *See also specific family*
Movements, analysis, **267–269**
Multiple marking method, 276
Muridae, 141
  age determination, *246, 248, 249*
  anatomy, 139, *140*
  dens, *183*
  key, 143–151, 160–161
Murinae, 141, 146, *147*, 148, 150, 160
Museum Special, 203
Museum specimens, 198, 199, 210
Musk deer. *See* Moschidae
Muskrat. *See* Arvicolinae
Mustelidae, 112
  anatomy, *37, 111*
  key, 118, *119*, 120
  locomotion, 110
  warning coloration, *26*
Myocastoridae, 142, 157, 160
Myospalacinae, 141, 144
Myoxidae, 139, 142
  key, 145–146, *147*, 152, *153*
  nest, *180*
Myrmecobiidae, 64, **68–70**
Myrmecophagidae, 103, *104*, 105, 106
Mystacinidae, 87, 88, *89*, 93
Mysticeti, **123–126**
  age determination, 250
  anatomy, 121, *122*, 123
  baleen plates, *122*, 123, *124*, 250
  families, 124, 125
  key, 55, 124–126
Myzopodidae, 87, 92–93

N

Nails. *See* Claws
Narwhal. *See* Monodontidae; Odontoceti
Nasal measurements, 11
Natalidae, 87, 91, 93
Neobalaenidae, 125
Nesomyinae, 141, 149
Nests, *180, 182*–183
New World monkey. *See* Callitrichidae; Cebidae
Noctilionidae, 87–88, 93
Nomenclature, 3–4
Notoryctemorphia, **77**
  anatomy, 64, 65
  key, 56, 58
Notoryctidae, **77**
Nutria. *See* Myocastoridae
Nycteridae, 87, 89–90, 93

O

Ochotonidae, **135–137**
Octodont. *See* Octodontidae
Octodontidae, 142, 156–157, *158*
Odobenidae, *111*, 112
Odontoceti, **126–132**
  anatomy, *46*, 121, *122*, 126, 127
  families, 127–128
  key, 55, 128–131

Okapi. *See* Giraffidae
Old World monkey. *See* Cercopithecidae
Olfaction, Primate, 96
Omnivorous diet
 Artiodactyla, 172
 Carnivora, 109
 dentition and, 17
 Insectivora, 78
 Primates, 96
 Rodentia, 138
 Tupaiidae, 94
Opossum. *See* Didelphidae
Orangutan. *See* Hominidae
Ornithorynchidae, 45, **61–63**
Orycteropodidae, **162–163**
Otariidae, 109, 110, 112, *113,* 120
Otomyinae, 141, 147
Otter. *See* Mustelidae

P

Paca. *See* Agoutidae
Pacarana. *See* Dinomyidae
Paceline method, 206
Pads (tori), 21, *22*
Palate measurement, 10
Panda. *See* Ursidae
Pangolin. *See* Pholidota
Parasites, 195, *196,* 212
Patagium, 47
 Chiroptera, 85, *86*
 Dermoptera, 83
Paucituberculata, 58, 64–66, **67–68**
Peccary. *See* Tayassuidae
Pedetidae, 139, 142, 151, *152*
Pelage. *See* Hair
Pellets, raptor, 255, 259
Peramelemorphia, 64–65, **70–71**
Peramelidae, 64, **70–71**
Peramelina, 58
Perissodactyla, *17,* 56–57, 59–60, **169–171**
Peroryctidae, **70–71**
Pes, 39
Petauridae, 64, 72, 74, *75,* 76
Petersen Index, 275–276
Petromuridae, 142, 156, *157*
Petromyscinae, 141, 148
Phalangeridae, 64, 71–72, 75, 76
Phascolarctidae, 72
 anatomy, 64, *65,* 71
 key, 73–74, 76
Pheromone, 28
Phocidae, 109, 112
 anatomy, *18, 46,* 110
 key, 112, *114,* 120
Phocoenidae, 127, 130–131, 132
Pholidota, **107–108**
 identification/key, 55, 59
 scales, *22*
 specimen preparation, 221
Photography, 269
Phyllostomidae, 87, 88, 93
Physeteridae, 127, 128, *129,* 130, 132
Pig. *See* Suidae
Pigmentation, **25–27**
Pigments, fluorescent, 269

Pika. *See* Lagomorpha
PIT transponders, 209
Pitfalls, 206
Plants. *See* Vegetation
Platacanthomyinae, 141, 149, *150*
Platanistidae, 127, 130, *131,* 132
Platypus. *See* Monotremata;
 Ornithorynchidae
Platyrrhini, 97, *98*
Polygon and range length methods, 262, *263*
Population estimates, **270–277**
 accuracy, 272
 computers and, 270–271
 data, recording, 272
 marking techniques, **209–210,** 273, 275–276
 methodologies, 272–277
 relative, 273
 sampling, 271–272
 terminology, 270
Porcupine. *See* Erethizontidae; Hystricidae
Porpoise. *See* Cetacea; Odontoceti; Phocoenidae
Potoroidae, 72, *73,* 76
 age determination, *245*
 diet, *253*
Potto. *See* Loridae
Premolars, 15–18. *See also* Dentition
Preparation and preservation, **211–229**
 diet samples, **255–258**
 fluids, 197, 220, 222, 229
 initial processing, 212–215
 special techniques, 221–229
  skeletal cleaning, 225–229
  skinning and tanning, 223–225
 standard study skin, 215–221, 222
 supplies and equipment, 211
 types of preservation, 195
Preservatives. *See* Fixatives
Primary data, 189
Primates, **96–102**
 anatomy, *22,* 36–37, 96–97
 diet, 96
 distribution, 97
 economic significance, 97
 families, 97
 as food source, 97
 identification, 59, 102
 key, 55, 56, 98–102
 locomotion, 96, 97
 senses, 96–97
Proboscidea, **164–165**
 age determination, 251
 anatomy, *14, 43, 44*
 identification/key, 55, 57, 59
Procaviidae, **165–166**
Procyonidae, 112, 117, 118–119, 120
Pronghorn. *See* Antilocapridae
Prostigmata, 232, 236, *242*
Prototheria, 53, **61–63**
 reproduction, 53, *54*
Pseudocheiridae, 64, 72, 75, 76
Pteropodidae, *48,* **86,** *87,* 88
Pygmy right whale. *See* Mysticeti; Neobalaenidae

Q

Quadrat sampling, 186, 188

R

Rabbit. *See* Lagomorpha
Rabies, 200–201
Raccoon. *See* Procyonidae
Radioisotopes, 268
Radio-location telemetry, 210
Radiotelemetry, 268
Range length method, 262
Raptor pellets, 255, 259
Rat
 families, 141–142. *See also specific family*
Recapture radii method, **263–266**
Recordkeeping. *See* Data, recording
Reference samples, **255–258**
Removal trapping and catch per effort methods, 276–277
Reprints, **282–283**
Reproductive anatomy and modes, 53
 Artiodactyla, 172–173
 Carnivora, 111
 Cetacea, 121
 data, recording, 195, *196*, 214–215
 female organs, *54*
 Hyracoidea, 166
 Lagomorpha, 136
 mammary glands, 1, **28–29**, 53, *54*
 Marsupialia, 64–65
 Monotremata, 62
 Perissodactyla, 171
 Pholidota, 108
 Primates, 97
 Proboscidea, 165
 Rodentia, 139
 Sirenia, 166, 167
Resource levels, 259
Rhinoceros. *See* Rhinocerotidae
Rhinocerotidae, 34, **169–171**
Rhinolophidae, 85, 87, 90, 93
Rhinopomatidae, 87, 90, *91*, 93
Rhizomyinae, 141, 144–145
Right whale. *See* Mysticeti
Rodentia, 52, **138–161**
 age determination, *246, 247, 248, 249,* 251
 anatomy, 138, 139, *140–141*
  baculum, 36–37, *246*
  foot, *38, 39, 42*
  molt pattern, *25*
 diet, 254
 economic significance, 138
 families, 139, 141–142
 as food resource, 138
 identification/key, 55, 59, 171
  global families, 142–159
  North American families, 159–161
 taxonomic characteristics, 139
Rorqual. *See* Balaenopteridae; Mysticeti
Ruminantia, **172–174**
Runways, 182

S

Salvage, 209
Sampling. *See* Analysis
Scales, *22*
 epidermal, **21–22**
 hair, *23*
 Pholidota, 107, 108
 specimen preparation and, 221
 Xenarthra, 103
Scandentia, 56, **94,** *95*
Scats, 181–182
Scent baits, 207
Scent glands, 27–28
Sciuridae, 141
 age determination, *247,* 251
 anatomy, *47,* 139
 key, 143, *144,* 160
Sciurognathic, 139
 families, 141–142. *See also specific family*
Sea lion. *See* Otariidae
Seal. *See* Otariidae; Phocidae
Sebaceous glands, 27
Secondary data, 189
*See*ds, 256, 259
Semifossorial locomotion, 44
Sewellel. *See* Aplodontidae
Sex organs. *See* Reproductive anatomy
 and modes
Sheep. *See* Bovidae
Sherman trap, 205
Shrew. *See* Soricidae
Sicistinae, 151
Sight, primate, 96–97
Sigmodontinae, 142
 age determination, *248, 249*
 dens, *183*
 key, *146,* 148, 149, 150, 161
Sign analysis, **180–184**
Sign method (trap placement), 206
Siphonaptera, 231, 233–234
Sirenia, **166–168**
 identification/key, 56, 59
 measurement, 214, *215*
Skeleton, 1, *2,* **36–40**
 age determination, 245–250
 appendicular, 37–39
 axial, 36–37
 cleaning techniques, **225–229**
 labeling, 197
 locomotor adaptations. *See* Locomotion
 reference samples, 255
 skull. *See* Skull
 specimen care, 5, 210
Skin(s). *See* Integument; Preparation and
 preservation
Skull, **6–11.** *See also* Dentition
 age determination, 9, 245
 cranium, 6–8, *7*
 labeling, 197
 mandible, 8–9
 measurement, 9–11
 reference samples, 255
 specimen care, 5
 variation, 9
Skunk. *See* Mustelidae
Sloth. *See* Bradypodidae; Megalonychidae; Xenarthra
Smell, primate, 96
Snap traps, 203
Social structure
 Lagomorpha, 135–136
 Rodentia, 138–139
Soil sampling, 185
Solenodon. *See* Solenodontidae

Solenodontidae, 79, *80*
Soricidae, *16, 45,* 78, 79
Spalacinae, 142, 144
Spatial distribution, analysis, **261–269**
   dispersion, 261
   habitat utilization and preference, 267
   home range, 261–266
   movements, 267–269
   territory, 266–267
Species, 3
Species accounts, 190–191
Specimen care, 5, 210. *See also* Collecting; Data, recording; Diet; Preparation and preservation
Sperm whale. *See* Odontoceti; Physeteridae
Spines, echidna, 61
Spool-and-line tracking, 269
Springhaas. *See* Pedetidae
Squirrel. *See* Sciuridae
Statistics. *See also* Data, recording
   age determination, 245
   dietary analysis, 253, 258–260
   spatial distribution, **261–269**
Steel traps, **203**
Stomach contents, 254, 259
Strepsirhini, **96–97**. *See also* Primates
Study skins, 5, **215–225**
   labeling, **196–197**
   removal and tanning, 223–225
   special preparation, 221–225
   standard preparation, 215–221, 222
Subspecies, 3
Subungulates, **164–168**. *See also* Hyracoidea; Proboscidea; Sirenia
Suidae, 172–173, 174, *175,* 247
Suiformes, 57, **172–175**
Sweat glands, 27
Swine. *See* Suidae
Systematics, 2–4

T

Tachyglossidae, **61–63**
Tagging
   live mammals, 209, 268
   specimen tags. *See* Labeling
Tail
   Cetacea, 121
   Lagomorpha, 135
   locomotion and, 43–47, 180, *181*
   Marsupialia, 76, *245*
   measurement, 194–195, 212, *213*
   Pholidota, 108
   Primates, 96
   Rodentia, 139
   specimen preparation, 221
   Xenarthra, 103
Talpidae, 79
   distribution, 78, 79
   key, 50, 81, *82*
   traps, 204, 205
   tunnels, *182*
Tamarin. *See* Callitrichidae
Tapir. *See* Tapiridae
Tapiridae, **169–171**
Tarsier. *See* Tarsiidae
Tarsii, 97
Tarsiidae, 96, 97, 99
Tarsipedidae, 64, 71, 72, *73,* 76

Taxidermy. *See* Preparation and preservation
Taxonomy
   classification hierarchies, 2–3
   living mammalia, subclasses and orders, **53–60**
Tayassuidae, 172, 173, 174, *175,* 179
Teeth. *See* Dentition
Telemetry, 210
Tenrec. *See* Tenrecidae
Tenrecidae, *16,* **78–79**, *80,* 82
Terrestrial locomotion, **41–43**, *44*
   Carnivora, 110
   Hyracoidea, 165
   Lagomorpha, 135
   Pholidota, 108
   Primates, 96
   Tupaiidae, 94
   Xenarthra, 103
Territory, 266–267
Theria, 53–54
Thryonomyidae, 142, 155, *156*
Thumb, 96
Thylacinidae, 65, **68–70**
Thyropteridae, 87, 90–91, 93
Tick. *See* Acari
Tissue samples, 222
Tongue
   Pholidota, 108
   Xenarthra, 103
Tooth. *See* Dentition
Tori (pads), 21, *22*
Tracking movements, **267–269**
Tracks, **180–181**, 251
Tragulidae, 171, 174, 177, 179
Trails, 182, *184*
Transmitters, 268
Trapping, **203–207**, 208
   baits and scents, 207
   climatic adjustment, 207
   fecal sampling via, 254
   movement analysis via, 268
   population sampling via, 273
   success and specimen damage, 207
   timing, 206
   trap placement, 206
   trap types, 203–206
Tree shrew. *See* Tupaiidae
Trichechidae, **166–168**
True horns, 30, 34–35
Tubulidentata, 57, 59, **162–163**
Tuco tuco. *See* Ctenomyidae
Tunnels. *See* Burrows
Tupaiidae, 59, **94**, *95*
Tylopoda, 172, 174, 175, 179
Tyrosine, 248, *249*

U

Underhairs, 24
Ungulate hoofs. *See* Hoofs
Ungulates. *See* Artiodactyla; Perissodactyla
Universal Transverse Mercator, **192–193**
Ursidae, 109, 112, *115*
   anatomy, *42*
   identification/key, 113, *114,* 116, *117,* 120
   sign, *184*
UTM (Universal Transverse Mercator), **192–193**

V

Variation
  cranial, 9
  dental, 20
Vegetation
  dietary analysis, *253*
  line transect sampling, 185–186, *187*
  quadrat sampling, 186, *188*
  reference samples, **256–258**
  resource levels, 259
Vertebrae, Sirenia, 167
Vertebrata, 1
Vespertilionidae, 87
  diet, 85, *254*
  key, 91, *92*, 93
Vibrissae, 24
Viscacha. *See* Chinchillidae
Vision, primate, 96–97
Viverridae, 112, *114*, 120
Vole. *See* Arvicolinae
Vombatidae, 64, *66*, 72, 76

W

Wallaby. *See* Macropodidae
Walrus. *See* Odobenidae
Warrens. *See* Burrows
Weasel. *See* Mustelidae
Whale
  baleen. *See* Cetacea; Mysticeti
  toothed. *See* Cetacea; Odontoceti
Whaling, 121, 123, 126
Wing, Chiroptera, 85–86
Wolf. *See* Canidae
Wombat. *See* Vombatidae

X

Xenarthra, **103–106**
  anatomy, 103, 105
  distribution, 105
  identification/key, 55, 57, 59, 105, 106
  specimen preparation, 221

Z

Zapodinae, 151, 161
Zebra. *See* Equidae
Ziphiidae, 127, 128–129, *130*, 132, 249
Zippin method, 276
Zokor. *See* Myospalacinae